Biological Basis of Detoxication

BIOCHEMICAL PHARMACOLOGY AND TOXICOLOGY

A Series of Monographs

WILLIAM B. JAKOBY, Editor

National Institutes of Health
Bethesda, Maryland

William B. Jakoby (editor). ENZYMATIC BASIS OF DETOXICATION,
Volumes I and II, 1980

William B. Jakoby, John R. Bend, and John Caldwell (editors). METABOLIC BASIS
OF DETOXICATION: METABOLISM OF FUNCTIONAL GROUPS, 1982

John Caldwell and William B. Jakoby (editors). BIOLOGICAL BASIS
OF DETOXICATION, 1983

Biological Basis of Detoxication

EDITED BY

John Caldwell
Department of Pharmacology
St. Mary's Hospital Medical School
University of London
London, England

William B. Jakoby
National Institutes of Health
Bethesda, Maryland

1983

ACADEMIC PRESS
A Subsidiary of Harcourt Brace Jovanovich, Publishers
New York   London
Paris   San Diego   San Francisco   São Paulo   Sydney   Tokyo   Toronto

COPYRIGHT © 1983, BY ACADEMIC PRESS, INC.
ALL RIGHTS RESERVED.
NO PART OF THIS PUBLICATION MAY BE REPRODUCED OR
TRANSMITTED IN ANY FORM OR BY ANY MEANS, ELECTRONIC
OR MECHANICAL, INCLUDING PHOTOCOPY, RECORDING, OR ANY
INFORMATION STORAGE AND RETRIEVAL SYSTEM, WITHOUT
PERMISSION IN WRITING FROM THE PUBLISHER.

ACADEMIC PRESS, INC.
111 Fifth Avenue, New York, New York 10003

United Kingdom Edition published by
ACADEMIC PRESS, INC. (LONDON) LTD.
24/28 Oval Road, London NW1 7DX

Library of Congress Cataloging in Publication Data

Main entry under title:

Biological basis of detoxication.

   (Biochemical pharmacology and toxicology)
   Includes index.
   1. Metabolic detoxication. I. Caldwell, John,
Date   . II. Jakoby, William B., Date
III. Series.
QP171.B693 1982   599'.0133   82-18933
ISBN 0-12-155060-5

PRINTED IN THE UNITED STATES OF AMERICA

83 84 85 86   9 8 7 6 5 4 3 2 1

Contents

Contributors xi
Preface xiii

1. Metabolic Formation of Toxic Metabolites
Scott W. Cummings and Russell A. Prough

    I. Introduction   2
    II. General Examples of Metabolic Activation   7
    III. Specific Examples of Metabolic Activation   12
    IV. Comments   22
         References   23

2. Integration of Xenobiotic Metabolism in Carcinogen Activation and Detoxication
Colin R. Jefcoate

    I. Introduction   32
    II. Identification of Reactive Electrophiles Derived from Xenobiotics   33

III. Cell Specificity of Xenobiotic Effects   36
IV. Classification of Xenobiotic Activation Pathways   37
V. Partition of Metabolic Pathways   39
VI. Regioselectivity and Stereoselectivity of Cytochrome P-450   44
VII. Epoxide Hydrolase   52
VIII. Absolute Monooxygenase Activity in the Activation Pathway   54
IX. Cellular Cosubstrates of Xenobiotic Metabolism   54
X. UDPglucuronyltransferases and Sulfotransferases   60
XI. Glutathione S-Transferases   62
XII. Inhibitors of Cellular Transferase Activity   63
XIII. Xenobiotic Metabolism in the Nuclear Envelope   65
XIV. Transport of Reactive Intermediates and Precursors   65
XV. Comments   66
References   67

3. Ontogenesis
Julian E. A. Leakey

I. Introduction   77
II. Developmental Profiles of Drug-Metabolizing Enzymes   83
III. Factors Influencing Drug-Metabolizing Enzyme Development   91
IV. Comments: The Biological and Clinical Consequences of Drug-Metabolizing Enzyme Ontogenesis   95
References   97

4. Intratissue Distribution of Activating and Detoxicating Enzymes
Jeffrey Baron and Thomas T. Kawabata

I. Introduction   105
II. Distribution within the Liver   107
III. Distribution within the Lung   121
IV. Distribution within the Skin   124
V. Comments   127
References   128

5. Nonenzymatic Biotransformation
Bernard Testa

I. Introduction   137
II. Reactions with Macromolecules as Borderline Cases   139
III. Reactions with Endogenous Nucleophiles   141

IV. Reactions with Endogenous Electrophiles   144
  V. Breakdown and Rearrangement Reactions of Xenobiotics and Prodrugs in Acidic and Neutral Aqueous Media   145
  VI. Reactions between Two Xenobiotics   146
  VII. Comments   148
       References   149

6. Unmetabolized Compounds
A. G. Renwick

  I. Introduction   151
  II. Criteria for Classification of a Foreign Compound as Being Unmetabolized   152
  III. Properties of Unmetabolized Compounds   158
  IV. Toxicological Implications   173
  V. Comments   175
       References   175

7. Biological Basis of Detoxication of Oxygen Free Radicals
Helmut Sies and Enrique Cadenas

  I. Introduction   182
  II. Oxygen Free Radicals   182
  III. Defense Mechanisms: Enzymatic and Nonenzymatic   187
  IV. Factors Influencing Defense Mechanisms and the Production of Oxygen Free Radicals   190
  V. Biological Systems Associated with Increased Oxygen Free Radical Production   195
  VI. Comments   201
       References   201

8. Fate of Xenobiotics: Physiologic and Kinetic Considerations
K. Sandy Pang

  I. Introduction   214
  II. The Xenobiotic   215
  III. Hemodynamics   223
  IV. Xenobiotic Metabolism   225
  V. Xenobiotic Excretion   237
  VI. Eliminating Organs   239
  VII. Comments   246
       References   246

9. Excretion Mechanisms
Walter G. Levine

    I. Introduction   251
   II. Renal Handling of Organic Anions and Cations   252
  III. Biliary Excretion of Xenobiotics   254
  IV. Salivary Excretion   270
   V. Excretion into Milk   271
  VI. Excretion into Expired Air   272
 VII. Comments   273
      References   274

10. Impact of Nutrition on Detoxication
Juanell N. Boyd and T. Colin Campbell

    I. Introduction   287
   II. Fasting/Starvation   289
  III. Protein   290
  IV. Carbohydrate   293
   V. Lipid   294
  VI. Trace Nutrients—Vitamins and Minerals   297
 VII. Comments   302
      References   302

11. Relationships between the Enzymes of Detoxication and Host Defense Mechanisms
Kenneth W. Renton

    I. Introduction   307
   II. Nonspecific Immunostimulants   308
  III. Adjuvant-Induced Arthritis   311
  IV. Reticuloendothelial System   312
   V. Interferon   313
  VI. Infection in Animals   316
 VII. Viral Infection in Humans   318
      References   321

12. Metabolic Basis of Target Organ Toxicity
Gerald M. Cohen

    I. Introduction   325
   II. Distribution of the Toxin as a Factor in Target Organ Toxicity   329

III. Role of Metabolism in Determining Target Organ Toxicity 333
IV. Role of Specific Function of the Tissue 340
V. Comments 344
References 345

13. Enzymes in Selective Toxicity
C. H. Walker and F. Oesch

I. Introduction 349
II. The Toxicological Significance of Enzymatic Conversion 350
III. The Enzymatic Factor in Selective Toxicity 352
IV. The Enzymatic Factor in Resistance 361
V. Comments 363
References 365

14. Intraindividual and Interindividual Variations
Elliot S. Vesell and M. B. Penno

I. Introduction 369
II. Sources of Intraindividual Variability 384
III. Sources of Interindividual Variability 391
References 406

Index 411

Contributors

Numbers in parentheses indicate the pages on which the authors' contributions begin.

Jeffrey Baron (105), The Toxicology Center, Department of Pharmacology, University of Iowa College of Medicine, Iowa City, Iowa 52242

Juanell N. Boyd (287), Institute for Comparative and Environmental Toxicology, Cornell University, Ithaca, New York 14853

Enrique Cadenas (181), Institut für Physiologische Chemie I, Universität Düsseldorf, D-4000 Düsseldorf, F.R.G.

T. Colin Campbell (287), Division of Nutritional Sciences, Cornell University, Ithaca, New York 14853

Gerald M. Cohen (325), Toxicology Unit, Department of Pharmacology, School of Pharmacy, University of London, London, England

Scott W. Cummings (1), Department of Biochemistry, The University of Texas Health Sciences Center, Dallas, Texas 75235

Colin R. Jefcoate (31), Department of Pharmacology, University of Wisconsin Medical School, Madison, Wisconsin 53706

Thomas T. Kawabata (105), The Toxicology Center, Department of Pharmacology, University of Iowa College of Medicine, Iowa City, Iowa 52242

Julian E. A. Leakey (77), Department of Biochemistry, Medical Sciences Institute, University of Dundee, Dundee DD1 4HN, Scotland

Walter G. Levine (251), Department of Molecular Pharmacology, Albert Einstein College of Medicine, The Bronx, New York 10461

F. Oesch (349), Department of Toxicology and Pharmacology, Institute of Pharmacology, University of Mainz, 6500 Mainz, F.R.G.

K. Sandy Pang* (213), Department of Pharmaceutics, University of Houston, Houston, Texas 77030

M. B. Penno (369), Department of Pharmacology, The Pennsylvania State University College of Medicine, Hershey, Pennsylvania 17033

Russell A. Prough (1), Department of Biochemistry, The University of Texas Health Sciences Center, Dallas, Texas 75235

Kenneth W. Renton (307), Department of Pharmacology, Dalhousie University, Halifax, Nova Scotia B3H 4H7, Canada

A. G. Renwick (151), Clinical Pharmacology Group, University of Southampton, Medical and Biological Sciences Building, Southampton SO9 3TU, England

Helmut Sies (181), Institut für Physiologische Chemie I, Universität Düsseldorf, D-4000 Düsseldorf, F.R.G.

Bernard Testa (137), School of Pharmacy, University of Lausanne, CH-1005 Lausanne, Switzerland

Elliot S. Vesell (369), Department of Pharmacology, The Pennsylvania State University College of Medicine, Hershey, Pennsylvania 17033

C. H. Walker (349), Department of Physiology and Biochemistry, University of Reading, Whiteknights, Reading RGG 2AJ, England

---

* Present address: Faculty of Pharmacy, University of Toronto, Toronto, Ontario M5S 1A1, Canada.

Preface

The intent of this group of monographs on the ". . . Basis of Detoxication" is to offer pharmacologists, toxicologists, and biochemists a detailed summary of our knowledge of those processes that lead to the removal of xenobiotics, i.e., foreign compounds, from the organism. The disposition of xenobiotics is viewed throughout from a biochemical point of view, with the earlier volumes in the series covering the individual enzymes that participate in detoxication and the pathways along which the functional groups are altered.

This volume is oriented more toward the biological processes participating in detoxication. The range of articles covers topics as diverse as the formation of toxic metabolites and compounds that are not metabolized at all, tissue distribution and nutritional considerations, and the kinetics and mechanisms of the metabolic and excretory processes. Given this range, uniformity of presentation is not appropriate. Nevertheless, in terms of the purpose of each of the volumes in this series, the authors were asked to provide the pharmacologist and toxicologist with the biochemical aspects of the field and the biochemist with the pharmacological insight that is necessary for the study of detoxication.

John Caldwell
William B. Jakoby

CHAPTER 1

Metabolic Formation of Toxic Metabolites*

Scott W. Cummings and
Russell A. Prough

I. Introduction	2
A. Historical Development	2
B. Discovery of Activating Enzymes	2
C. Objective	7
II. General Examples of Metabolic Activation	7
A. Epoxidation	7
B. C-Hydroxylation	8
C. N-Oxidation	8
D. Free Radical Formation	10
E. Redox Cycling and Active Oxygen	10
F. S-Oxidation	11
III. Specific Examples of Metabolic Activation	12
A. Polycyclic Aromatic Hydrocarbons	12
B. Nitrogenous Compounds	14
C. Halogenated Compounds	20
D. Sulfur-Containing Compounds	21
IV. Comments	22
References	23

* Portions of the work presented were supported by the American Cancer Society Grant BC-336 and the Robert A. Welch Foundation Grant I-616.

I. INTRODUCTION

A. Historical Development

Since the observations of P. Pott in 1775 that linked scrotal cancer in English chimney sweeps with exposure to soot[1] and of L. Rehn in 1895 that linked bladder cancer in German dye workers with exposure to aromatic amines,[2] considerable effort has been expended in characterizing the biological and chemical processes that lead to human cancer. Although it was realized that exposure to certain chemicals could result in cancer, the concept that the parent chemicals themselves were not biologically active, but rather required a chemical transformation to form reactive intermediates capable of altering cell growth and expression, was not established. The early work of individuals such as E. Boyland,[3-5] Miller and Miller,[6,7] and R. T. Williams[8] demonstrated the biotransformation of foreign compounds and stimulated interest in the enzymatic processes involved. For example, Boyland's group described the enzymatic formation of *trans*-dihydrodiols that were assumed to be formed through epoxide intermediates. Concomitantly, Miller and Miller were interested in arylamine and arylamide metabolism resulting in the covalent binding of these compounds to proteins. Williams and colleagues focused their attention on metabolic conjugation steps that altered the disposition of foreign compounds.

At the same time, interest arose as to the chemical nature of the active principle of the aniline compounds that leads to methemoglobinemia.[9,10] It had been suggested by Heubner in 1913[11] that N-oxidation might account for the biological conversion of anilines to phenylhydroxylamines, derivatives known to oxidize oxyhemoglobin. On the basis of this suggestion, many investigators focused on nitrogen metabolism as an area that could lead to an understanding of the molecular mechanism of aromatic amine carcinogenesis and toxicity.

These areas of interest focused on the chemical aspects of carcinogenesis and toxicity that for further progress required the elucidation of the enzymatic basis of metabolic activation. The following section will describe the development of the enzymatic basis for the metabolic activation of foreign compounds to form reactive chemical intermediates.

B. Discovery of Activating Enzymes

1. Cytochrome P-450

During the 1950s, a number of investigators became involved in the study of the oxidative metabolism of certain endogenous and exogenous compounds.[12,13] As seen in Table I, various compounds ranging from steroids, aromatic amines, and polycyclic aromatic hydrocarbons to numerous drugs were found to be metabolized by NADPH- and O_2-dependent enzyme systems localized largely in the endoplasmic reticulum fraction of most animal tissues. For the steroids, the enzyme systems were found both in the endoplasmic reticulum and in the mitochondrial fraction.[14,15]

The availability of stable isotopes allowed investigators to inquire whether molecular oxygen could be directly incorporated into organic molecules as the result of enzyme-catalyzed oxidation. The classic experiments performed independently by Hayaishi's[16] and Mason's[17] research groups utilized $^{18}O_2$ to establish the concept of mono- and dioxygenase reactions. Hayano et al.[18] also applied this method to mammalian steroid metabolism. During the conversion of 11-deoxycortisol to cortisol, one atom of ^{18}O is incorporated into the organic molecule from molecular oxygen ($^{18}O_2$) and not from $H_2^{18}O$. Similar experiments on C-hydroxylation of drugs were performed by McMahon's group,[19,20] who established that some N-dealkylation reactions catalyzed by liver microsomes incorporated an atom of oxygen from molecular oxygen to form an unstable carbinolamine. These studies pointed to the existence of NADPH- and O_2-dependent enzyme systems that catalyze a number of C-, N-, and S-oxidation reactions.

TABLE I

Reactions Catalyzed by Microsomal Monooxygenases

Reaction type	Substrate	Product
N-Dealkylation	3'-Methyl-4-methylaminoazobenzene	3'-Methyl-4-aminoazobenzene
O-Dealkylation	Phenacetin	Acetaminophen
S-Oxidation	Thioacetamide	Thioacetamide S-oxide
N-Hydroxylation	2-Acetylaminofluorene; 4-aminoazobenzene	N-Hydroxy-2-acetylaminofluorene; 4-hydroxylaminoazobenzene
C-Hydroxylation	Acetanilide; lauric acid	4-Hydroxyacetanilide; 11- and 12-hydroxylauric acid
Epoxidation	Benzo[a]pyrene	Benzo[a]pyrene 4,5-oxide
Steroid hydroxylation	11-Deoxycortisol	Cortisol
Azo reduction	4-Dimethylaminoazobenzene	4-Dimethylaminoaniline and aniline
Nitro reduction	4-Nitroquinoline N-oxide	4-Hydroxylaminoquinoline N-oxide

Early studies by Klingenberg[21] and Garfinkel[22] demonstrated that a unique cytochrome exists in mammalian liver capable of binding carbon monoxide upon reduction. Omura and Sato[23] characterized this microsomal cytochrome by optical spectroscopy and noted that upon addition of detergents or other chemicals, it could be converted from a cytochrome forming an Fe^{2+}–CO species and absorbed light maximally at approximately 450 nm to one absorbing light at 420 nm. Mason et al.[24] detected a unique electron spin resonance signal associated with liver microsomes; this species was designated as microsomal FeX. Finally, Estabrook, Cooper, and Rosenthal,[25,26] in a classic study, utilized Warburg's technique of the reversal of inhibition of CO-bound cytochromes by monochromatic light[27] to link this biochemical entity, cytochrome P-450 (now classified as xenobiotic monooxygenase, EC 1.14.19.1), to the metabolic activity of liver microsomes and FeX. Subsequent reports have described many of the enzymatic and physical properties of this unique cytochrome.[12,13,28,29]

Our current state of knowledge, based on the contributions of many research groups, is that the mammalian cytochrome found principally in the endoplasmic reticulum of most cells consists of a flavoprotein reductase, NADPH–cytochrome c (P-450) reductase (EC 1.6.2.4), of 76,000–78,000 molecular weight,[30] and the terminal oxidase, cytochrome P-450, of 48,000–58,000 molecular weight.[31-33] When purified and reconstituted *in vitro*, the binary enzyme complex normally requires added lipid (e.g., dilauroylphosphatidyl choline) to function at maximal catalytic turnover.[34] A number of isoenzymes of the cytochrome have been demonstrated to exist in rat and rabbit liver; some are constitutive and others are induced by animal treatment with either barbiturates, polycyclic aromatic hydrocarbons, steroid derivatives,[35] or isosafrole.[36] The multiple forms of the cytochrome probably account for the broad substrate specificity of the monooxygenase (see Chapter 2, this volume). A plethora of studies have shown that these enzymes catalyze the oxidation of numerous compounds, as seen in Table I.

2. FAD-containing Monooxygenase

Initially, the oxidation of amines and sulfur-containing compounds was thought to be catalyzed exclusively by the cytochrome P-450–dependent monooxygenase. In 1962, Ziegler and Pettit[37] demonstrated the existence of a different enzyme in liver that catalyzed N-oxidation reactions. The participation of this enzyme in carcinogen and drug metabolism has, until recently, been underestimated because of the heat lability of the enzyme in the absence of NADP(H) and the difficulties in assaying its activity.

1. Metabolic Formation of Toxic Metabolites

Ziegler's group[38,39] has purified the microsomal FAD-containing monooxygenase (EC 1.14.13.8, dimethylaniline monooxygenase) to homogeneity and thoroughly characterized it biochemically.

The enzyme catalyzes the N-oxidation of tertiary, secondary, and some primary amines (arylamines), hydroxylamines, and hydrazines, as well as the S-oxidation of sulfides, thiols, thioamides, and thiocarbamates (Table II). Several reviews describe the substrate specificity, product chemistry, and those aspects related to developmental and species differences.[40–42] It is of interest that the carcinogenic aromatic amines such as 2-aminofluorene and 2-naphthylamine can be metabolized to reactive N-hydroxylamines by this enzyme.[43,44] Because this result establishes that both cytochrome P-450 and the FAD-containing monooxygenase can N-hydroxylate aromatic amines, studies on the metabolic activation of

TABLE II

Reactions Catalyzed by the Microsomal FAD–containing Monooxygenase[a]

Substrate type	Example	Product
N-Oxidation		
1°-Amines	2-Naphthylamine	2-Naphthylhydroxylamine
2°-Amines	4-Chloro-N-methylaniline	4-Chloro-N-methylphenylhydroxylamine
1°-Hydroxylamines	Benzylhydroxylamine	Benzaldehyde oxime
2°-Hydroxylamines	N-Methyl-N-benzylhydroxylamine	α-Phenyl-N-methylnitrone or α-methyl-N-phenylnitrone
3°-Amines	N,N-Dimethylaniline	N,N-Dimethylaniline N-oxide
Monosubstituted hydrazines	Methylhydrazine	(Methyldiazene) methane and formaldehyde
1,1-Disubstituted hydrazines	1,1-Dimethylhydrazine	(1,1-Dimethyldiazenium ion) formaldehyde and methylhydrazine
1,2-Disubstituted hydrazines	1,2-Dimethylhydrazine	Azomethane
S-Oxidation		
Thiols	Cysteamine	Cysteamine disulfide
Sulfides	Dimethyl sulfide	Dimethyl sulfoxide
Disulfides	Dibenzyl sulfide	Dibenzylsulfinic acid
Thioamides	Thioacetamide	Thioacetamide S-oxide and S-dioxide
Thioureas	Phenylthiourea	Phenylformamidine sulfinic and sulfenic acids

[a] Data obtained from a review by Ziegler.[40]

aromatic amines in various species and organs should consider the existence of both enzymes. Prough and Ziegler[45] have noted that both enzymes function simultaneously in the metabolism of N-methylamines in a number of species. In the case of $tert$-N-methylamines, two reactions proceed during metabolism: N-dealkylation by cytochrome P-450 and N-oxide synthesis by the FAD-containing monooxygenase. sec-N-Methylamines are metabolized by the FAD-containing monooxygenase to sec-N-methylhydroxylamine products that form formaldehyde when treated with acid. Because of the similarities of the products of the enzymes, one must use specific inhibitors and inducing agents to distinguish between the reactions of these two enzymes.

3. Other Redox Enzymes

Several other enzyme processes have been described that may lead to reactive intermediates. Some of these processes are dependent upon oxidation of certain organic compounds in peroxide-dependent reactions.[46] The possibility of H_2O_2-dependent peroxidase activity in extrahepatic tissues[47] or cooxidation reactions with prostaglandin synthetase[48] may account for some target–tissue reactions in organs with low concentrations of the monooxygenases. In addition, reduction reactions of nitro compounds have been demonstrated that may lead to such ultimate carcinogens as 4-hydroxylaminoquinoline N-oxide.[49,50] Other nitro compounds may form the nitro anion radical, which can be rapidly reoxidized by molecular oxygen to yield superoxide anion.[51] This process has been demonstrated with nitrofurantoin, which in certain avian species is a lung and liver toxin. This toxic phenomenon may be due to the active oxygen species generated as a result of formation of the nitro anion radical. It is assumed that a number of flavin-dependent systems can reduce nitro compounds to the nitro anion radical or hydroxylamine derivatives.[51]

4. Conjugating Enzymes

As shown by the pioneering work of R. T. Williams,[8] many oxidized products formed in the enzyme-catalyzed reactions are further metabolized by enzymes that conjugate the oxidized compound with water-soluble agents. Many metabolites of xenobiotics are conjugated with either sulfate, glucuronic acid, or glutathione, to name three of the more common reactions, producing the more readily excreted sulfate esters, glucuronides, or mercapturic acid derivatives.[52] However, the products of other conjugation reactions, for example, the sulfate derivatives of aromatic amines[53,54] and perhaps the glucuronic acid derivative of the reduced quinones of polycyclic aromatic hydrocarbons[55] have been reported to be responsible for the formation of reactive intermediates.

1. Metabolic Formation of Toxic Metabolites

C. Objective

The purpose of this chapter is to examine the formation of reactive chemical intermediates by enzymatic processes and to provide a definition of which enzyme systems may be functional with the various reaction types. The authors are aware of the abundant literature relevant to these topics and ask the reader's indulgence for choosing a relatively small number of examples to document these concepts.

II. GENERAL EXAMPLES OF METABOLIC ACTIVATION

Given the structural heterogeneity of chemical carginogens or toxins requiring metabolic transformation to elicit their toxic responses, it is not surprising that there are a variety of mechanisms by which these compounds are metabolically activated. Among such processes are epoxidation of double bonds, C-hydroxylation of certain nitroso and hydrazine derivatives, N-hydroxylation of organic molecules containing the nitrogen heteroatom, the generation of free radicals, and S-oxidation of organic molecules containing the sulfur heteroatom (Table III). The importance of each mechanism in terms of carcinogenicity and toxicity has been the subject of detailed study. It is the purpose of this chapter to examine both generally and specifically each area of metabolic activation and its relationship to deleterious biological reactions. This exposition is complemented by that of Chapter 2, this volume.

A. Epoxidation

The production of phenols from aromatic compounds in mammals has long been considered a mode of detoxication; these reactions are catalyzed by NADPH- and O_2-dependent monooxygenases predominantly located in the liver of mammals. In the early 1950s, certain aromatic hydrocarbons, for example, naphthalene, were found to be excreted in the form of dihydrodiol or mercapturate derivatives.[3,4] On the basis of earlier chemical work of Pullman and Pullman,[56] it was assumed that reactive arene oxides (epoxides) were formed, thereby accounting for the several metabolic products (see Table III). Final proof for the existence of arene oxides came from the observation that specific hydrogen atoms on an aromatic ring, experimentally labeled with tritium for this purpose, would migrate and be retained in a different position on the ring during hydroxylation; this phenomenon was termed the NIH shift.[57] In addition,

TABLE III

Mechanisms of Metabolic Activation of Carcinogens

Type of metabolic activation	Theoretical active metabolite	Specific examples
Epoxidation	Arene oxide	Benzo[a]pyrene
C-Hydroxylation	Carbonium ion	Dimethylnitrosamine and 1,2-Dimethylhydrazine
N-Hydroxylation	Nitrenium ion	Naphthylamine and N-Acetylaminofluorene
Free radical formation and redox cycles	Carbon-centered radical	Carbon tetrachloride
	Oxygen-centered radical	Quinones and nitro compounds
S-Oxidation	Sulfinic acid	Thioacetamide

the stable K-region epoxide metabolites for phenanthrene, benz[a]anthracene, benzo[a]pyrene, and 7,12-dimethylbenz[a]anthracene have been isolated and characterized.[58-61] These observations set the chemical framework for defining the reactions of arene oxides: isomerization to form phenols, hydration to form dihydrodiols, and nucleophilic addition to form adducts with glutathione, protein, RNA, and DNA.

B. C-Hydroxylation

The importance of C-hydroxylation in metabolic activation is not fully understood. N-Nitrosamines, a group of compounds that exhibit a wide range of carcinogenic and toxic effects, are believed to be activated by C-hydroxylation at the α-carbon of the nitrosamine. Direct evidence for this reaction has been obtained using cyclic nitrosamines.[62,63] In these studies, the unstable products of hydroxylation (decomposition products of α-hydroxynitrosamines) were isolated, identified, and shown to be reflective of the activation of these compounds to this ultimate carcinogenic form. The importance of C-hydroxylation in metabolic activation of 1,2-dimethylhydrazine in forming the putative diazomethane intermediate is discussed in Section III.

C. N-Oxidation

The concept of N-oxidation was discovered after trimethylamine administered to humans was observed to be excreted as the trimethylamine

1. Metabolic Formation of Toxic Metabolites

N-oxide.[64] With naturally occurring alkaloids and some synthetic medicinal amines, N-oxidation results in less reactive derivatives. However, several studies have shown that N-oxidation of arylamines yields products that are more carcinogenic than the parent compounds.[65]

The N-oxidation of primary, secondary, and tertiary amines yields a diverse group of products.[40] The primary amines can be oxidized to hydroxylamines, and if a free α-hydrogen exists, they can be further oxidized to yield oximes. Secondary amines are N-oxidized in one or two steps to give either hydroxylamines or nitrones, respectively. The initial oxidation of tertiary amines yields an amine oxide that can be protonated under appropriate reaction conditions to the N-hydroxyammonium ion. In general, N-oxidation of tertiary amines results in formation of a more polar, less toxic compound that is readily excreted. However, aromatic primary amines, for example, arylhydroxylamines, decompose under acidic conditions to yield a very reactive amino cation (nitrenium ion, $ArNH^+$) that may be responsible for their toxicity.[5]

Aromatic hydroxylamines have been implicated in a number of toxicological responses. Their ability to interact with cellular nucleophiles such as proteins, thiols, and nucleic acids is believed to be the cause of a number of toxic reactions. The basic reactivity of aromatic hydroxylamines is proposed to lie in the ability of the OH (or OH_2^+) to act as a good leaving group yielding the nitrenium cation, which is subject to nucleophilic attack.[5,65]

The reactivity of N-hydroxyamides is similar to that of hydroxylamines. Several N-acetylarylamines have been shown to exhibit both carcinogenic and toxic properties. The biological oxidation of these compounds to form N-hydroxy derivatives, which can subsequently lose OH (or OH_2^+) and generate the reactive nitrenium ion, has been suggested.[5,65] The positive charge generated can be delocalized to the adjacent aromatic ring, thereby allowing nucleophilic attack by cellular components.

In addition to amine and amide derivatives, dinitrogen compounds, the hydrazines, also undergo N-oxidation. In general, the initial oxidation product of 1,2-disubstituted hydrazines is the chemically stable azo derivative.[66] The apparent toxicity of certain hydrazines does not appear to be due to the azo derivative itself but rather to the further oxidative products of the azo which can form either reactive free radicals[67] or reactive carbonium ions by C-hydroxylation of the azoxy intermediates.[68] The mono- and 1,1-disubstituted hydrazines are oxidized to diazene intermediates.[69] A major route of decomposition of the diazenes proceeds through free radical intermediates.[70]

D. Free Radical Formation

The role of radical formation in carcinogenic and toxic reactions is much more speculative, although specific examples can be cited. For example, the initiation of lipid peroxidation by carbon tetrachloride is thought to involve formation of the trichloromethyl radical.[71] This chemical species apparently propagates the radical oxidation chain reactions involving lipid peroxidation *in vitro*.[72] The concept of cooxidation of substrates by a heme-dependent peroxidative mechanism[46,48] has provided a means for forming carbon-centered radicals by one-electron oxidation. Griffin has demonstrated that aminopyrine is oxidized by H_2O_2 in such a mechanism with a transient radical species as an intermediate.[73] Oxygen-centered radicals have been demonstrated for reactions involving air oxidation of hydroquinones, for example, 3,6-dihydroxybenzo[a]pyrene[74] and the uncoupling of monooxygenases to yield the superoxide anion radical.[75] Each of these radical species would lead to such deleterious processes as lipid peroxidation; a few specific examples are discussed in Section III.

E. Redox Cycling and Active Oxygen

Several reactions allow redox cycling of the oxidative enzymes. One of the first examples of this process described was the interaction of menadione (vitamin K_3) with NADPH–cytochrome c (P-450) reductase.[76,77] The quinone appears to be reduced in a one-electron process to form a semiquinone intermediate. The semiquinone form rapidly reduces molecular oxygen to superoxide anion, ultimately forming 0.5 mole of hydrogen peroxide and 0.5 mole of molecular oxygen. The reoxidized quinone can subsequently be reduced by the flavoprotein. During the course of these reactions, the otherwise tightly coupled flavoprotein, one possessing low reactivity with molecular oxygen, is uncoupled via the quinone, and large amounts of O_2^- (H_2O_2) are formed at the expense of NADPH and oxygen.

Other agents that appear to cause similar "oxidant" stress include paraquat,[78] nitrofurantoin,[51] benzo[a]pyrene quinones,[74,79] and hydroquinone.[80] The high rate of production of superoxide anion and hydrogen peroxide can have several deleterious effects. First, one of the major protective mechanisms of the cell against hydrogen peroxide, that is, cytosolic glutathione (GSH), may be depleted or its steady-state concentration reduced to a level that allows the initiation of such processes as lipid peroxidation. The necessity of maintaining reduced GSH levels by use of glutathione reductase, an NADPH-dependent cytosolic enzyme,

1. Metabolic Formation of Toxic Metabolites

may perturb cytosolic and mitochondrial levels of reducing equivalents such as NADH, NADPH, and GSH. This oxidant stress may place the cell in jeopardy by preventing its normal physiological role.

F. S-Oxidation

Organic sulfur compounds containing nucleophilic divalent sulfur atoms exist in four states: sulfides, thiols, disulfides, and thiones. Ziegler[40] has reviewed the reactions of sulfur compounds in this regard (Table IV). All of them are capable of undergoing oxidation to a wide variety of compounds. In the case of sulfides, oxidation to sulfoxides or further oxidation to sulfones results in relatively stable products that are more polar than the parent sulfide. As was suggested for N-oxides, the oxidative products of sulfides are believed to be detoxication products because once formed they are readily excreted. However, oxidation of certain thiols, disulfides, and thiones can yield highly reactive electrophiles capable of eliciting toxic responses.

Both aryl and alkyl thiols are easily oxidized to yield very reactive sulfenic acid derivatives that rapidly react with free thiols to yield disulfides and, in turn, can be further oxidized to sulfonic acids through

TABLE IV

S-Oxidation of Organic Sulfur Compounds[a]

Compound	Oxidation reaction
Sulfides	$R\text{—}S\text{—}R \xrightarrow{[O]} R\text{—}S(\text{=O})\text{—}R \xrightarrow{[O]} R\text{—}S(\text{=O})_2\text{—}R$
Thiols	$R\text{—}SH \xrightarrow{[O]} R\text{—}S(\text{=O})H \xrightarrow{RSH} RSSR + H_2O$
Disulfides	$RSSR \xrightarrow{[O]} RS(\text{=O})SR \xrightarrow{[O]} 2RSO_2H \xrightarrow{[O]} 2RSO_3H$
Thioamides	$RC(\text{=S})NH_2 \xrightarrow{[O]} RC(\text{=S})NH_2 \xrightarrow{[O]} RC(SO_2H)\text{=}NH$
Thiocarbamates	$RN\text{=}C(SH)\text{—}NH_2 \xrightarrow{[O]} RN\text{=}C(SOH)\text{—}NH_2 \xrightarrow{[O]} RN\text{=}C(SO_2H)\text{—}NH_2 \xrightarrow{[O]} RN\text{=}C(SO_3H)\text{—}NH_2$

[a] Data taken from Ziegler.[40]

thiosulfenic and sulfinic acid intermediates. The toxicity of many thiols has been described, but the mechanism(s) of action is unclear. Whereas binary complex formation with metalloenzymes may explain their toxicity in part, the ability of sulfenic acid to react with free intracellular thiols, particularly GSH, may be more critical in toxicity.

Thiones, either the thioamides or thiocarbamides, undergo S-oxidation but have quite different properties relative to the other S-oxidized derivatives. Thioamides undergo oxidation to sulfoxides (sulfines), which under vigorous oxidizing conditions form iminosulfenic acids (sulfenes). The dioxygenated sulfene product exists transiently because of its extreme reactivity with nucleophiles. Sulfenes are strong electrophiles and are thus easily hydrolyzed to amides and thiosulfate. In contrast, thiocarbamides (thioureas) are similar in structure to thioamides but exhibit uniquely different properties. Thiocarbamides are readily oxidized to form formamidine sulfenic acids. Like alkylsulfenic acids, formamidine sulfenic acids are reduced by thiols, including GSH, to form the free thiol and glutathione disulfide (GSSG). If low molecular weight thiols are not present, thiol groups on proteins may be oxidized to form internal disulfide bonds.

III. SPECIFIC EXAMPLES OF METABOLIC ACTIVATION

The following examples of specific xenobiotics that are metabolically activated were chosen as representative of the various classes of chemical carcinogens and toxins.

A. Polycyclic Aromatic Hydrocarbons

As described in the introduction, the metabolism of aromatic hydrocarbons has been extensively investigated. The observations that dihydrodiols and mercapturic acid derivatives are formed metabolically, as well as the existence of the NIH shift, provide evidence that arene oxides generally represent important reactive intermediates in various carcinogenic and toxic processes. During the last decade, considerable effort has been expended by the scientific community in the attempt to understand the events contributing to the initiation of cancer by aromatic hydrocarbons. Because this subject is reviewed in Chapter 2 of this volume, we will deal only with certain salient features of the metabolic activation of polycyclic aromatic hydrocarbons in this chapter.

A critical issue in understanding the reactivity of polycyclic aromatic

1. Metabolic Formation of Toxic Metabolites

hydrocarbons is related to the means by which a hydrocarbon such as benzo[a]pyrene is metabolized to reactive arene oxides and the routes of disposition of these epoxides. Two basic reactions can occur with arene oxides: rearrangement to regain aromaticity of the ring (Eq. 1a) or nucleophilic attack at a carbon center of the epoxide possessing a partial positive charge (Eq. 1b).

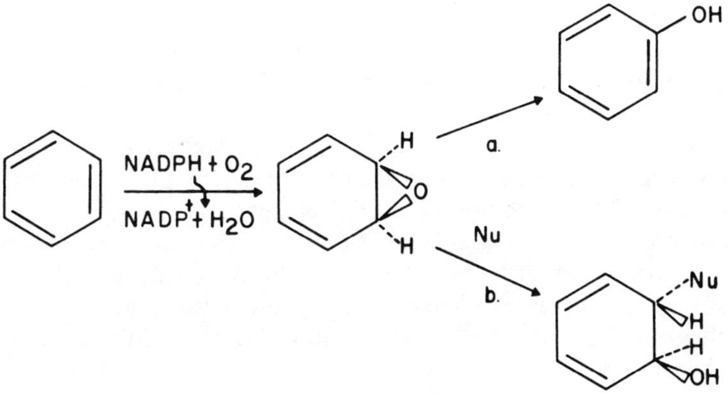

(1a)

(1b)

The balance between these two reaction types and the stability of the arene oxide dictate the products formed. For example, the arene oxides of ring systems of lesser complexity tend to rearrange to phenols because of the thermodynamic factors, which add to the stability of aromatic rings compared to rings containing only a *cis*-diene structure. The best examples of this are benzene and biphenyl[81]; both yield only phenols as products. Some evidence suggests that arene oxides are intermediates of the hydroxylation reactions of these compounds, but hydration to *trans*-dihydrodiols or nucleophilic addition of cellular nucleophiles to these rings are reasonably uncommon.

Aromatic hydrocarbons that possess more stable arene oxides obviously can undergo the second reaction with cellular nucleophiles such as water, thiols such as GSH, proteins, and nucleic acids. Each of the nucleophiles reacts to a small extent without need for enzyme catalysis if the arene oxide is sufficiently stable to form an appreciable steady-state concentration. Normally, nucleophilic attack on an epoxide leads predominantly to the trans addition products as follows: water forms *trans*-dihydrodiols, thiols form the *trans*-thiol adducts, and proteins or nucleic acids form trans adducts. Depending on the thermodynamics of the ring system, the phenolic group of these adducts can be eliminated to regain the aromatic ring system; this is a very common reaction.

Whereas the formation of dihydrodiols or GSH adducts can occur chemically, enzyme systems[52] exist that facilely catalyze the addition of water (epoxide hydrolase, EC 3.3.2.3) and glutathione (glutathione S-transferase, EC 2.5.1.18). As might be expected, metabolism of benzo[a]pyrene in the presence of a purified, reconstituted cytochrome P-450–dependent monooxygenase system leads to a large amount of phenolic products.[82] The product ratio for the total phenols : total *trans*-dihydrodiols formed shifts from a value of 13 in the absence of epoxide hydrolase to a minimum value of 0.4 in the presence of epoxide hydratase. Certain arene oxides, such as benzo[a]pyrene 4,5-oxide, are sufficiently stable so that inhibition of epoxide hydrolase allows recovery of the intact epoxide.[83]

A major role for epoxide hydrolase[84] and glutathione S-transferase[85] is thought to be detoxication. However, Sims and Grover have reported that addition of the 7,8-*trans*-dihydro-7,8-diol of benzo[a]pyrene, that is, the hydrolysis product of the epoxide, led to more DNA alkylation in Chinese hamster embryo cells than did addition of the parent hydrocarbon.[86] A series of elegant studies by Conney, Levin, and Jerina's group[87] and by Gelboin's group[88] documented that three steps are required in the conversion of benzo[a]pyrene, a nonreactive molecule, to its "ultimate" carcinogenic form, the 7,8-*trans*-dihydro-7,8-diol 9,10-oxide: epoxidation to form the 7,8-oxide catalyzed by the cytochrome P-450 system, hydration to form the *trans*-7,8-dihydrodiol catalyzed by epoxide hydrolase, and a subsequent epoxidation also catalyzed by cytochrome P-450 to form the 7,8-*trans*-dihydrodriol 9,10-oxide. For most polycyclic hydrocarbons that have been studied, a testable hypothesis exists that implicates the unique chemical reactivity of arene oxides containing certain benzylic carbons in a benzo ring that has previously undergone an epoxidation–hydration reaction.[89] This proposition is termed the *Bay-region* theory. The benzo rings of most polycyclic aromatic hydrocarbons contain an area defined by the unique nuclear magnetic resonance characteristic of the benzylic carbon (Fig. 1). It is ultimately this carbon that becomes a potent electrophilic center allowing attack by nucleophiles such as nucleic acids or proteins. A number of excellent reviews describe the details of the metabolic activation of the polycyclic hydrocarbons.[89–91]

B. Nitrogenous Compounds

1. Nitrosamines

Nitrosamines are a ubiquitously occurring group of nitrogenous compounds; dimethylnitrosamine (DMN) has been the most thoroughly studied member of this group. Initially, DMN was considered to have industrial potential, but the early work of Magee and Barnes demonstrated it to

1. Metabolic Formation of Toxic Metabolites

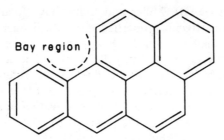

Fig. 1. Bay region of benzo[a]pyrene.

be a potent hepatocarcinogen.[92] DMN administration *in vivo* led to methylation of liver DNA, suggesting that a metabolic process converted the compound to an active methylating agent.[93] Specific DNA–carcinogen adducts have been observed; the N^7-methylguanine and O^6-methylguanine adducts have been considered important adducts possibly leading to carcinogenesis.[94] Because formaldehyde was noted as one of the products, C-hydroxylation was proposed to lead to formation of diazomethane (Eq. 2). Other nitrosamines with longer aliphatic side chains, such as butyl groups, can be C-hydroxylated at positions other than the α-carbon. All of these reactions appear to be catalyzed exclusively by the cytochrome *P*-450 monooxygenases, and the alkyl groups can be subsequently oxidized *in vivo* to CO_2.

$$(CH_3)_2NNO \rightarrow \begin{bmatrix} CH_2OH \\ | \\ CH_3N - NO \end{bmatrix} \rightarrow HCHO + [CH_3N_2^+ \; OH^-] \qquad (2)$$

On the basis of two types of evidence, it would appear that α-hydroxylation is a significant reaction leading to nitrosamine carcinogenesis. First, it was noted that dideuteration at the α-positions led to marked reduction in the carcinogenicity of dimethylnitrosamine, nitrosoazetidine, and 4-nitrosomorpholine.[95–97] Second, α-hydroxynitrosamines were found to be very unstable *in vitro* and to decompose, forming an aldehyde and alkyldiazohydroxide.[62,63] The α-hydroxylated derivative was synthesized with a blocked hydroxyl group and its degradation followed after enzymatic deacetylation; the products were nearly identical to those obtained following microsomal metabolism. Such alkyldiazonium compounds appear to be the most likely reactive intermediates of nitrosamine metabolism.

2. Arylamines and Arylamides

The metabolism of arylamines to form reactive intermediates can best be exemplified by the activation of the aminoazo dye, dimethyl-4-

aminoazobenzene (DAB). Miller and Miller have focused considerable attention on the tumorigenic properties of this carcinogen.[65] The formation of a reactive intermediate from DAB was suggested because administration of the dye to rats resulted in yellow coloration of the isolated liver proteins. The color appears to be due to the retention of the azo chromophore covalently bound to the protein.[6] These studies prompted considerable interest in the mechanism of activation of aromatic amines. Several groups have described the products formed upon metabolism of arylamines such as 2-naphthylamine and DAB.[6,98] Ring hydroxylation appeared to be a major reaction *in vivo* and *in vitro*. For the arylamide, 2-acetylaminofluorene (2-AAF), N- and C-hydroxylated products can be formed, and many of the C-hydroxylated products are conjugates of glucuronic acid.[99] Conjugated intermediates are rapidly excreted; on the basis of this observation, ring hydroxylation products were suggested as the products of detoxication reactions for arylamines and arylamides.

The arylamines can also be N-acetylated to form amides, and the amide products, such as *N*-acetyl-4-aminoazobenzene, can be N-hydroxylated *in vivo*. For example, administration of either dimethyl-, monomethyl-, or 4-aminoazobenzene to rats led to excretion of *N*-hydroxy-*N*-acetyl-4-aminoazobenzene.[65] Thus, arylamines can be converted to arylamides that are subsequently N-hydroxylated to yield hydroxamic acids (*N*-hydroxyarylamides). In 1960, Cramer, Miller, and Miller noted that the hydroxamic acid, 2-AAF, could be N-hydroxylated[100] and that *N*-hydroxy-2-AAF was more carcinogenic than the parent compound.[101] Other acetylated aromatic amines were also shown to be metabolically activated to *N*-hydroxy derivatives that were also more carcinogenic than the parent compound.[65]

During metabolism of aniline *in vitro*, a considerable amount of *o*-hydroxyaniline was formed.[5] Boyland and others had also noted that mercapturic acid derivatives of arylamines were formed *in vivo*.[102] On the basis of these observations, Booth and Boyland[5] proposed that the N-hydroxylation of aromatic amines to give hydroxylamines could lead to a nitrenium intermediate that subsequently can rearrange to *o*-hydroxy products (Eq. 3).

$$\text{(3)}$$

Collectively, these studies suggested that many aromatic amines can be converted to the *N*-acetyl derivative that may serve as "proximal" carcinogens and, upon N-hyroxylation, may form the nitrenium ion capable of serving as a reactive electrophile. However, the role of hydroxamic acids as ultimate carcinogens is still not clear. For example, Kriek[103] demonstrated that at an acid pH, the nonenzymatic alkylation of guanine residues of DNA and RNA by N-hydroxy-2-aminofluorene occurred, but not by N-hydroxy-2-AAF. In addition, Miller and Miller[65,104] demonstrated the acid-catalyzed alkylation of methionine and free guanosine by N-hydroxyaminofluorene, but not by N-hydroxy-2-AAF.

That hydroxylamines but not hydroxamic acids are reactive *in vitro* is paradoxical and leads to the conclusion that N-hydroxylation may be required to form an N-hydroxy intermediate, but depending upon the compound's structure, the N-oxidized intermediate may not readily form a nitrenium ion. N-Oxidation reactions can be catalyzed by one of two enzyme systems depending upon the structure of the carcinogenic substrate: arylamides appear to be uniquely N-hydroxylated by cytochrome P-450 monooxygenase, and arylamines are N-hydroxylated by both the FAD and cytochrome P-450 monooxygenases. Although one-electron oxidation (peroxidase) reactions may play a role in metabolism in extrahepatic target organs, insufficient data exist to support this contention.

There is an apparent dilemma regarding the ability of N-hydroxyarylamines to alkylate nucleic acids at slightly acid pH and the inability of N-hydroxyarylamides to do the same. A complicating factor is that certain aromatic hydroxylamines, for example, phenylhydroxylamines, easily form o-hydroxy derivatives but are not inherently carcinogenic. It must be presumed that the chemistry, that is, stability, of a given N-hydroxy derivative dictates the facility with which the compounds can form a putative nitrenium ion.

The concept of stability may be related to the possibility that some N-hydroxy derivatives can be converted rapidly in a concerted reaction to the o-phenol at neutral pH and other aromatic amines, for example, N-hydroxy-4-methylaminoazobenzene, may require acid pH in order for the hydroxyl group to become a good leaving group. It would seem that at or near physiological pH, a small amount of reactive nitrenium ion may be formed to account for DNA alkylation by aromatic amines and that in the bladder, the urinary pH may be sufficient to enhance aromatic hydroxylamine decomposition.[105] However, the pK_a of the hydroxamic acids may be sufficiently low so that the OH is not a good leaving group. Miller and Miller[54] and Irving[53] have demonstrated that the N-benzoyl or sulfate esters of the hydroxamic acids are much better leaving groups than OH. The reactivity of the esters is thought to allow the arylamides to form electrophilic centers. The role of conjugating enzymes in activation of the

N-hydroxy compounds to reactive esters is an intriguing mechanism but remains controversial.

An additional toxic response to the formation of aromatic hydroxylamines is methemoglobinemia. While searching for the mechanism by which aniline derivatives cause methemoglobinemia, phenylhydroxylamines were described as the active agents in this persistent problem seen in mammals exposed to aniline and aniline dyes.[10] Kiese and co-workers have demonstrated that aniline and N-alkylanilines are oxidized to N-hydroxy derivatives in the presence of liver microsomes, NADPH, and oxygen.[9,10] The resultant phenylhydroxylamines can interact with oxyhemoglobin to form methemoglobin, H_2O_2, and the corresponding nitrosobenzene derivative. However, this deleterious process involves a reduced heme(Fe^{2+})-oxygen-dependent process, not a concerted rearrangement of the hydroxylamine to a nitrenium ion. Rather, a redox reaction described by Wallace and Caughey[106] as a reductive displacement reaction seems to be involved.

3. Nitro Compounds

The mechanisms of formation of toxic metabolites of nitro compounds appear to be limited to two phenomena: reduction to the aromatic hydroxylamine or reduction to the nitro anion radical followed by redox cycling. Nitro compounds can be reduced in a four-electron step to form the hydroxylamine derivatives.[50,51] If the OH is a sufficient leaving group at the pH of the tissue or if specific mechanisms exist to aid in formation of the putative nitrenium ion, this process would account for one mechanism of formation of reactive intermediates from nitro compounds. A number of flavoproteins or flavoprotein systems supplemented with free flavin can, indeed, catalyze the reduction of nitro compounds.[50] Because the carcinogen 4-nitroquinoline N-oxide is reduced to 4-hydroxylaminoquinoline N-oxide, the hydroxylamine derivative most likely is the ultimate carcinogenic form of the compound.[49] It is of interest that an intact N-oxide functional group is required for biological reactivity but probably does not participate directly in the alkylation reaction.

The possible toxicity of nitro compounds by redox cycling is not well established. Boyd's group demonstrated that the lung toxin, nitrofurantoin, has increased covalent binding potential with liver microsomal protein under anaerobic conditions *in vitro,* suggesting a mechanism related to nitrenium ion formation.[107] This observation rules out a furan epoxide as a potential reactive intermediate, because covalent binding by the epoxide derivative of the furan ipomeanol was inhibited under these conditions. However, nitrofurantoin forms a nitro anion radical *in vitro* under anaerobic conditions but not aerobically,[51] suggesting that nitrofurantoin may rapidly redox cycle between its one-election reduced form and the parent compound concomitantly generating appreciable amounts of superoxide anion radical and hydrogen peroxide. In the presence of nitrofurantoin and oxygen, NADPH is rapidly oxidized and hydrogen peroxide produced *in vitro*.[51] It is assumed that the radical species reacts rapidly with oxygen and precludes formation of the four-electron reduction product, the hydroxylamine. The possible role of redox cycling in the oxidant stress of nitro compounds is supported by the observation that selenium-deficient chicks are more susceptible to nitrofurantoin-induced lung toxicity than are selenium-sufficient chicks.[108] In addition, the concentration of GSH in the liver was decreased relative to controls. Because of the preliminary nature of these results, more direct evidence will be required to establish whether redox cycling of nitro compounds can occur *in vivo*. One precedent exists: the most likely effect of paraquat in lung[78] is the redox cycle established between several flavin-dependent reductases and oxygen by the catalytic function of the dye acting as a one-electron shuttle.

4. Hydrazines

The principal reactions of hydrazines are in serving either as strong nucleophiles or as reducing agents. The existence of two nitrogen heteroatoms bonded together makes hydrazines reactive toward the carbonyl groups of intermediary metabolites such as α-keto acids and pyridoxal phosphate. Certain hydrazines undoubtedly exert part of their biological effect by interacting with pyridoxal phosphate and altering amine metabolism.

The main redox reaction of hydrazines is the loss of two electrons to form azo or diazene derivatives. 1,2-Disubstituted hydrazines form stable azo compounds and are chemically unreactive.[69] Mono- or 1,1-disubstituted hydrazines are readily oxidized to form diazenes; diazenes normally are unstable and either eliminate molecular nitrogen to form free radicals (Eq. 4a), or are further oxidized to form unstable diazoalkanes (Eq. 4b), or rearrange to form hydrazones (Eq. 4c).

$$RCH_2NHNH_2 \xrightarrow{[O]} [RCH_2N=NH] \begin{array}{c} \xrightarrow{a.} [R + N_2 + H] \quad (4a) \\ \xrightarrow[b.]{[O]} [RCH_2N_2^+ \ ^-OH] \quad (4b) \\ \xrightarrow{c.} RCH=N\ NH_2 \quad (4c) \end{array}$$

The biological oxidation reactions that result in azo and diazene compounds involve a number of oxidative enzymes: the cytochrome *P*-450[66] and FAD-dependent monooxygenases,[109] mitochondrial[110] and plasma monoamine oxidases,[111] and certain peroxidative reactions.[112]

Free radicals formed from diazenes can react with tissue macromolecules to form covalent adducts. For example, the diazene intermediate of phenylhydrazine has been shown to react with flavin and heme prosthetic groups of several enzymes and heme proteins including oxyhemoglobin.[113-115] Alkylation may occur because of the formation of the phenyl radical and its reactivity with the prosthetic groups. During metabolic oxidation of 2-methylbenzylazo compounds, methane is produced from the N-methyl group.[67,116] As another consequence of this reaction, microsomal protein is methylated by the N-methyl group of the azo, but not azoxy derivatives, indicating that the azo compound is the proximal substrate leading to methane formation and protein alkylation. The study[116] implicated the cytochrome P-450 monooxygenase in metabolic activation of the azo derivative because inhibitors of the function of the heme protein decreased the products of oxidative metabolism of the azo compound. In addition, methane formation was stimulated at the expense of protein methylation by addition of GSH. The results are suggestive of participation of a diazene intermediate or a methyl free radical in both methane formation and protein methylation. The effect of thiol may be to terminate the radical by allowing hydrogen abstraction, thereby favoring methane formation. Although the implications of these reactions regarding the deleterious effects of 2-methylbenzylhydrazines remain unclear, the putative methyl radical may represent a reactive intermediate formed during the N-oxidation of hydrazines and their azo(diazene) intermediates.

The cytochrome P-450–dependent monooxygenase also converts azo compounds to stable azoxy derivatives.[65,117] In 1970, Druckrey[68] provided a rationale for the metabolic activation of 1,2-disubstituted hydrazines, particularly 1,2-dimethylhydrazine. The C-hydroxylation of methylazoxymethane leads to the unstable methylazoxymethanol, which decomposes to diazomethane and formaldehyde. Subsequent work confirmed this hypothesis,[118] and others have suggested that procarbazine, a 2-methylbenzylhydrazine derivative, may also be metabolically activated by C-hydroxylation of the azoxy metabolites,[119] thereby accounting for its carcinogenic and toxic effects. However, clear differences exist between the chemistry of the two 1,2-disubstituted hydrazines because 1,2-dimethylhydrazine does not yield CH_4 upon metabolism.

C. Halogenated Compounds

At least two mechanisms of metabolic activation have been described for halogenated compounds. Some evidence suggests that halogenated

1. Metabolic Formation of Toxic Metabolites

aromatic hydrocarbons are metabolized to reactive intermediates, perhaps epoxides.[120] Although the mechanism of activation of the hepatotoxin, carbon tetrachloride (CCl_4) has been disputed, Butler[71] suggested that CCl_4 may be converted to the trichloromethyl radical ($\cdot CCl_3$). The most likely mechanism of activation was suggested to require the cytochrome P-450–dependent monooxygenase,[71] and its metabolism to the trichloromethyl free radical has been substantiated by two major observations. First, oral administration of CCl_4 leads to a unique metabolite of $\cdot CCl_3$, hexachloroethane, that is, the dimerization product expected from the radical.[121] Second, radical spin-trapping reactions, performed *in vitro* using CCl_4, NADPH, and oxygen in the presence of liver microsomal protein,[122] yielded an ESR spectrum similar to the one that would result from $\cdot CCl_3$.

The role of $\cdot CCl_3$ in CCl_4 toxicity is less clear. When treated with CCl_4, one major finding in laboratory animals is the formation of fatty livers.[123] It has been assumed that the processes of lipid synthesis and of repackaging into lipoproteins are altered in such a manner as to increase the concentration of free and esterified lipids in that organ. It is not clear whether this process is directly related to the formation of the trichloromethyl radical. However, it has been demonstrated that $^{14}CCl_4$ is metabolically activated to form an intermediate that can be bound covalently to protein and lipid.[124] The formation of a reactive intermediate capable of alkylating certain key enzymes could account for the alteration of hepatic lipid metabolism and disposition.

It is also clear that *in vitro* formation of the $\cdot CCl_3$ radical initiates a radical chain reaction in which lipid peroxidation of unsaturated fatty acids follows. As part of the process, small amounts of the $\cdot CCl_3$ radical abstract a methylene hydrogen from fatty acids, for example, arachidonic acid, to form a carbon-centered fatty acid radical.[72,125] The intermediate can react with molecular oxygen to form the lipid peroxy radical capable of sustaining and propagating the radical chain reaction. Under *in vitro* conditions, a small amount of CCl_4 and NADPH leads to massive utilization of molecular oxygen and unsaturated fatty acids.[72] The products of lipid peroxidation can also be measured *in vitro* and *in vivo* as respired hydrocarbons such as ethane and pentane.[126]

D. Sulfur-Containing Compounds

The hepatotoxic effects of the thiono–sulfur compound, thioacetamide, are expressed only after metabolic conversion of the parent compound to the thioacetamide S-oxide, thioacetamide sulfine.[127] This S-oxide is a more potent hepatotoxin than thioacetamide at equal doses, but the toxicity of either compound can be enhanced by pretreatment of animals with phenobarbital or be decreased by treatment with inhibitors of the cytochrome P-450 monooxygenase. Other studies using [^{14}C]- or [^{35}S]thioacetamide *in vivo*[128] indicated that an appreciable amount of the carbonyl moiety of the molecule was covalently bound to protein and other cellular macromolecules, but that negligible radioactive sulfur was bound. The results suggest that thioacetamide is S-oxidized repetitively by cytochrome P-450 to form thioacetamide S-oxide and thioacetamide S-dioxide. It is presumed that the S-dioxide is the ultimate toxic metabolite.

The toxicity of thioureas was once believed to be due to the release of hydrogen sulfide by reductive desulfuration. However, Boyd and Neal[129] demonstrated an oxidative attack on sulfur catalyzed by microsomal protein fractions. Poulsen *et al.*[130] have shown that unlike the situation seen with thioamides, the FAD-containing monooxygenase can generate the toxic sulfenic and sulfinic acid intermediates from the thioureas. The ability of formamidine sulfenic acids to be rapidly reduced by GSH may explain the glycogen-depleting effects demonstrated by thioureas.[131] The formamidine sulfinic acid esters have also been predicted to react with cellular nucleophiles to form covalent bonds and release the oxidized sulfur atom.[127]

IV. COMMENTS

An attempt has been made to summarize the evidence that characterizes the chemical nature of the reactive intermediates formed from carcinogens and toxins. A number of enzyme systems have been described as catalyzing the biological oxidation reactions that are available for metabolic activation of xenobiotics. As will be pointed out in Chapter 2 of this volume, a number of oxidative and conjugative reactions can alter the equilibrium between reactions, leading to reactive intermediates and reactions that detoxify foreign compounds. In fact, a number of compounds can be shown to be metabolically activated to chemical species that alkylate DNA *in vitro* but not *in vivo;* 9-hydroxybenzo[*a*]pyrene is an example.[132,133]

It is clear that metabolic activation is a prerequisite for the initiation of chemically induced tumors. The second stage required for chemical carcinogenesis involves the process of promotion of tumor formation, a new and exciting area of research.[134] A challenging problem exists in the relationship of compounds that are *complete carcinogens* (initiate and promote tumor formation) and others that are *incomplete carcinogens* (only initiate tumor formation). Because many nontumorigenic metabolites of carcinogens are effective in initiating hyperplasia (9-hydroxybenzo[a]pyrene, for example[135]), it will be of interest to correlate the two phases of tumor formation (initiation and promotion) to the biological activity of the various metabolites of a given carcinogen. Such studies may have to delineate the role of some metabolites, for example, diol epoxides, in the initiation process and others, for example, 9-hydroxybenzo[a]pyrene, in the promotion process in order to understand clearly why some carcinogens are complete and others are incomplete.

ACKNOWLEDGMENTS

The authors wish to thank Ms. Sabina Dougherty for her aid in preparation of the manuscript.

REFERENCES

1. Pott, P. (1775). Cancer scroti. In "Chirurgical Observation," pp. 63–68. Hawes, Clarke, & Collins, London.
2. Rehn, L. (1895). Blasengeschwulste bei Fuchsin-Arbeitern. Arch. Klin. Chir. **50**, 588–600.
3. Booth, J., and Boyland, E. (1949). Metabolism of polycyclic compounds. 5. Formation of 1,2-dihydroxy-1,2-dihydronaphthalenes. Biochem. J. **44**, 361–365.
4. Boyland, E. (1950). The biological significance of metabolism of polycyclic compounds. Biochem. Soc. Symp. **5**, 40–54.
5. Booth, J., and Boyland, E. (1964). The biochemistry of aromatic amines 10. Enzymic N-hydroxylation of arylamines and conversion of arylhydroxylamines into o-amino phenols. Biochem. J. **91**, 362–369.
6. Miller, E. C., and Miller, J. A. (1947). The presence and significance of bound aminoazo dyes in the livers of rats fed p-dimethylaminoazobenzene. Cancer Res. **7**, 468–480.
7. Miller, J. A., Cramer, J. W., and Miller, E. C. (1960). The N- and ring-hydroxylation of 2-acetylaminofluorene during carcinogenesis in the rat. Cancer Res. **20**, 950–962.
8. Williams, R. T. (1959). "Detoxification Mechanisms," 2nd ed. Chapman & Hall, London.
9. Kiese, M., and Uehleke, H. (1961). Der ort dei N-oxydation des anilins in hoheren tier. Naunyn-Schmiedebergs Arch. Exp. Pathol. Pharmakol. **242**, 117–129.
10. Kiese, M. (1966). The biochemical production of ferrihemoglobin-forming derivatives from aromatic amines and mechanisms of ferrihemoglobin formation. Pharmacol. Rev. **18**, 1091–1161.
11. Heubner, W. (1913). Arbeiten aus dem pharmkologischen Institut zu Göttingen. 2. Reihe. 14. Studien über methämoglobinidung. Arch. Exp. Pathol. Pharmakol. **72**, 241–281.
12. Gillette, J. R., Conney, A. H., Cosmides, G. J., Estabrook, R. W., Fouts, J. R., and Mannering, G. J., eds. (1969). "Microsomes and Drug Oxidations." Academic Press, New York.

13. Estabrook, R. W., Gillette, J. R., and Leibman, K. C., eds. (1973). "Microsomes and Drug Oxidations." Williams & Wilkins, Baltimore.
14. Ryan, K. J., and Engel, L. L. (1956). Steroid 21-hydroxylation by adrenal microsomes and reduced triphosphopyridine nucleotide. *J. Am. Chem. Soc.* **78**, 2654–2655.
15. Sweat, M. L., and Lipscomb, M. D. (1955). A transhydrogenase and reduced triphosphopyridine nucleotide involved in the oxidation of desoxycorticosterone to corticosterone by adrenal tissue. *J. Am. Chem. Soc.* **77**, 5185–5187.
16. Hayaishi, O., Katagiri, M., and Rothberg, S. (1955). Mechanism of the pyrocatechase reactions. *J. Am. Chem. Soc.* **77**, 5450–5451.
17. Mason, H. S., Fowlkes, W. L., and Peterson, E. (1955). Oxygen transfer and electron transport by the phenolase complex. *J. Am. Chem. Soc.* **77**, 2914–2915.
18. Hayano, M., Lindberg, M. C., Dorfman, R. I., Hancock, J. E. H., and von E. Doering, W. (1955). On the mechanism of the C-11β-hydroxylation of steroids; a study with $H_2^{18}O$ and $^{18}O_2$. *Arch. Biochem. Biophys.* **59**, 529–532.
19. McMahon, R. E., Culp, H. W., and Occulowitz, J. C. (1969). Studies on hepatic microsomal N-dealkylation reaction. Molecular oxygen as the source of the oxygen atom. *J. Am. Chem. Soc.* **91**, 3389–3390.
20. McMahon, R. E., Sullivan, H. R., Craig, J. C., and Pereirea, W. E. (1969). The microsomal oxygenation of ethyl benzene: Isotopic, stereochemical, and induction studies. *Arch. Biochem. Biophys.* **132**, 575–577.
21. Klingenberg, M. (1958). Pigments of rat liver microsomes. *Arch. Biochem. Biophys.* **75**, 376–386.
22. Garfinkel, D. (1958). Studies on pig liver microsomes. I. Enzymic and pigment composition of different microsomal fractions. *Arch. Biochem. Biophys.* **77**, 493–509.
23. Omura, T., and Sato, R. (1962). The carbon monoxide-binding pigment of liver microsomes. I. Evidence for its hemoprotein nature. *J. Biol. Chem.* **239**, 2370–2385.
24. Mason, H. S., North, J. C., and Vanneste, M. (1965). Microsomal mixed-function oxidations: The metabolism of xenobiotics. *Fed. Proc., Fed. Am. Soc. Exp. Biol.* **24**, 1172–1180.
25. Estabrook, R. W., Cooper, D. Y., and Rosenthal, O. (1963). The light reversible carbon monoxide inhibition of the steroid C-21-hydroxylase system of the adrenal cortex. *Biochem. Z.* **338**, 741–755.
26. Cooper, D. Y., Levin, S., Narasimhula, S., Rosenthal, O., and Estabrook, R. W. (1965). Photochemical action spectrum of the terminal oxidase of mixed function oxidase systems. *Science* **147**, 400–402.
27. Warburg, O. (1949). "Heavy Metal Prosthetic Groups and Enzyme Action," p. 73. Oxford Univ. Press, London/New York.
28. Ullrich, V., Hildebrandt, A., Roots, I., Estabrook, R. W., and Conney, A. H., eds. (1976). "Microsomes and Drug Oxidation." Pergamon, Oxford.
29. Sato, R., and Omura, T., eds. (1978). "Cytochrome *P*-450." Academic Press, New York.
30. Masters, B. S. S., and Okita, R. T. (1979). The history, properties, and function of NADPH-cytochrome *P*-450 reductase. *In* "Enzymatic Basis of Detoxication" (W. B. Jakoby, ed.), Vol. 1, pp. 183–199. Academic Press, New York.
31. Lu, A. Y. H., and Coon, M. J. (1968). Role of hemoprotein *P*-450 in fatty acid omega-hydroxylation in a soluble enzyme system from liver microsomes. *J. Biol. Chem.* **243**, 1331–1332.
32. Kawalek, J. C., Lewis, W., Ryan, D., Thomas, P. E., and Lu, A. Y. H. (1975). Purification of liver microsomal cytochrome *P*-448 from 3-methylcholanthrene-treated rabbits. *Mol. Pharmacol.* **11**, 874–878.

33. Imai, Y., and Sato, R. (1974). A gel-electrophoretically homogeneous preparation of cytochrome P-450 from liver microsomes of phenobarbital pretreated rabbits. *Biochem. Biophys. Res. Commun.* **60,** 8–14.
34. Strobel, H. W., Lu, A. Y. H., Heidema, J., and Coon, M. J. (1970). Phosphatidylcholine requirements in the enzymatic reduction of hemoprotein P-450 and in fatty acid, hydrocarbon, and drug hydroxylation. *J. Biol. Chem.* **245,** 4851–4854.
35. Solymoss, B., Werringloer, J., and Toth, S. (1971). The influence of pregnenolone-16-α-carbonitrile on hepatic mixed-function oxygenase. *Steroids* **17,** 427–433.
36. Dickins, M., Bridges, J. W., Elcombe, C. R., and Netter, K. J. (1978). A novel haemoprotein induced by isosafrole pretreatment in the rat. *Biochem. Biophys. Res. Commun.* **80,** 89–96.
37. Ziegler, D. M., and Pettit, F. H. (1966). Microsomal oxidases. I. The isolation and dialkylarylamine oxygenase activity of pork liver microsomes. *Biochemistry* **5,** 2932–2938.
38. Ziegler, D. M., and Mitchell, C. H. (1972). Microsomal oxidase. IV. Properties of a mixed function amine oxidase isolated from pig liver microsomes. *Arch. Biochem. Biophys.* **150,** 116–125.
39. Poulsen, L. L., and Ziegler, D. M. (1979). The liver microsomal FAD-containing monooxygenase. Spectral characterization and kinetic studies. *J. Biol. Chem.* **254,** 6449–6455.
40. Ziegler, D. M. (1980). Microsomal flavin-containing monooxygenase: Oxygenation of nucleophilic nitrogen and sulfur compounds. *In* "Enzymatic Basis for Detoxication" (W. B. Jakoby, ed.), Vol. 1, pp. 201–227. Academic Press, New York.
41. Ziegler, D. M. (1982). Metabolic oxidation of organic nitrogen and sulfur compounds. *In* "Drug Metabolism and Drug Toxicity" (J. Mitchell and M. Horning, eds.). Waverly Press, Baltimore, Maryland (in press).
42. Prough, R. A. (1982). Nitrogen oxidation by flavin enzymes, *In* "N-Oxidation of Drugs and Related Compounds" (A. K. Cho, ed.). SP Medical and Scientific Press, Jamaica, New York (in press).
43. Pelroy, R. A., and Gandolfi, A. J. (1980). Use of a mixed function amine oxidase for metabolic activation in the Ames/salmonella assay system. *Mutat. Res.* **72,** 329–334.
44. Frederick, C. B., Mays, J. B., Baetchke, K. P., and Kadlubar, F. F. (1981). Microsomal cytochrome P-450 and flavoprotein amine oxidase catalyzed N-oxidation of 2-aminofluorene. *Proc. Am. Assoc. Cancer Res.* **22,** 382.
45. Prough, R. A., and Ziegler, D. M. (1977). The relative participation of liver microsomal amine oxidase and cytochrome P-450 in N-demethylation reactions. *Arch. Biochem. Biophys.* **180,** 363–373.
46. Rahimtula, A. D., and O'Brien, P. J. (1977). The role of cytochrome P-450 in the hydroperoxide-catalyzed oxidation of alcohols by rat-liver microsomes. *Eur. J. Biochem.* **77,** 201–208.
47. Reigh, D. L., Stuart, M., and Floyd, R. A. (1978). Activation of the carcinogen N-hydroxy-2-acetylaminofluorene by rat mammary peroxidase. *Experientia* **34,** 107–108.
48. Marnett, L. J., Wlodawer, P., and Samuelson, B. (1975). Co-oxygenation of organic substrates by the prostaglandin synthetase of sheep vesicular gland. *J. Biol. Chem.* **250,** 8510–8517.
49. Tada, M., and Tada, M. (1976). Metabolic activation of 4-nitroquinoline N-oxide and its binding to nucleic acid. *In* "Fundamentals of Cancer Prevention" (P. N. Magee, S. Takayama, T., Sugimura, and T. Matsushima, eds.), pp. 217–227. Univ. Park Press, Baltimore, Maryland.

50. Kato, R., Takahashi, M., and Oshina, T. (1970). Characteristics of nitro reduction of the carcinogenic agent, 4-nitroquinolone-N-oxide. Biochem. Pharmacol. **19**, 45–55.
51. Mason, R. P., and Holtzman, J. L. (1975). The role of catalytic superoxide formation in the O_2 inhibition of nitroreductase. Biochem. Biophys. Res. Commun. **67**, 1267–1274.
52. Jakoby, W. B., ed. (1981). "Methods in Enzymology," Vol. 77. Academic Press, New York.
53. Irving, C. C. (1971). Metabolic activation of N-hydroxy compounds by conjugation. Xenobiotica **1**, 387–398.
54. Kadlubar, F. F., Miller, J. A., and Miller, E. C. (1976). Hepatic metabolism of N-hydroxy-N-methyl-4-aminoazobenzene and other N-hydroxy arylamines to reactive sulfuric acid esters. Cancer Res. **36**, 2350–2359.
55. Lind, C., Vadi, H., and Ernster, L. (1978). Metabolism of benzo[a]pyrene-3,6-quinone and 3-hydroxybenzo[a]pyrene in liver microsomes from 3-methylcholanthrene-treated rats. Arch. Biochem. Biophys. **190**, 97–108.
56. Pullman, A., and Pullman, B. (1955). Electronic structure and carcinogenic activity of aromatic molecules. Adv. Cancer Res. **3**, 117–169.
57. Daly, J. W., Jerina, D. M., and Witkop, B. (1972). Arene oxides and the NIH shift: The metabolism, toxicity, and carcinogenicity of aromatic compounds. Experientia **28**, 1129–1149.
58. Jerina, D. M., and Daly, J. W. (1974). Arene oxides: A new aspect of drug metabolism. Science **185**, 573–582.
59. Grover, P. L., Hewer, A., and Sims, P. (1971). Epoxides as microsomal metabolites of polycyclic hydrocarbons. FEBS Lett. **18**, 76–80.
60. Grover, P. L., Hewer, A., and Sims, P. (1972). Formation of K-region epoxides as microsomal metabolites of pyrene and benzo[a]pyrene. Biochem. Pharmacol. **21**, 2713–2726.
61. Wang, I. Y., Rasmussen, R. E., and Crocker, T. T. (1972). Isolation and characterization of an active DNA-binding metabolite of benzo[a]pyrene from hamster living microsomal incubation systems. Biochem. Biophys. Res. Commun. **49**, 1142–1149.
62. Hecht, S. S., Chen, C. B., and Hoffmann, D. (1978). Evidence for metabolic α-hydroxylation of N-nitrosopyrrolidine. Cancer Res. **38**, 215–218.
63. Chen, C. B., Hecht, S. S., and Hoffmann, D. (1978). Metabolic α-hydroxylation of the tobacco-specific carcinogen N'-nitrosonornicotine. Cancer Res. **38**, 3639–3645.
64. Lintzel, W. (1934). Untersuchungen uber trimethylammoniumbase. III. Trimethylammoniumbasen im menschlichen harn. Biochem. Z. **273**, 243–261.
65. Miller, J. A., and Miller, E. C. (1969). The metabolic activation of carcinogenic aromatic amines and amides. Prog. Exp. Tumor Res. **11**, 273–301.
66. Dunn, D. L., Lubet, R. A., and Prough, R. A. (1979). The oxidative metabolism of N-isopropyl-α-(2-methylhydrazo)-p-toluamide hydrochloride (procarbazine) by rat liver microsomes. Cancer Res. **39**, 4555–4563.
67. Prough, R. A., Coomes, M. W., Cummings, S. W., and Wiebkin, P. (1982). Metabolism of procarbazine. In "Biological Reactive Intermediates 2: Chemical Mechanisms and Biological Effects" (R. Snyder, D. V. Parke, J. Kocsis, D. J. Jollow, G. G. Gibson, and C. M. Witmer, eds.), pp. 1067–1076. Plenum, New York.
68. Druckrey, H. (1970). Production of colonic carcinoma by 1,2-dialkylhydrazines and azoxyalkanes. In "Carcinoma of the Colon and Antecedent Epithelium" (W. J. Burdette, ed.), pp. 267–279. Thomas, Springfield, Illinois.
69. Smith, P. A. S. (1966). "The Chemistry of Open-Chain Organic Nitrogen Compounds," Vol. 2, pp. 119–209. Benjamin, New York.

70. Kosower, E. M. (1971). Monosubstituted diazenes (diimides). Surprising intermediates. *Acc. Chem. Res.* **4**, 193–198.
71. Butler, T. C. (1961). Reduction of carbon tetrachloride *in vivo* and reduction of carbon tetrachloride and chloroform *in vitro* by tissues and tissue homogenates. *J. Pharmacol. Exp. Ther.* **134**, 311–319.
72. Hochstein, P., and Ernster, L. (1963). ADP-Activated lipid peroxidation coupled to the TPNH oxidase system of microsomes. *Biochem. Biophys. Res. Commun.* **12**, 388–394.
73. Griffin, B. W., and Ting, P. L. (1978). Mechanism of N-demethylation of aminopyrine catalyzed by horseradish, metmyoglobin, and protohemin. *Biochemistry* **17**, 2206–2211.
74. Capdevila, J., Estabrook, R. W., and Prough, R. A. (1978). The existence of a benzo[a]pyrene-3,6-quinone reductase in rat liver microsomal fractions. *Biochem. Biophys. Res. Commun.* **83**, 1291–1298.
75. Ullrich, V., and Diehl, V. (1971). Uncoupling of monooxygenation and electron transport by fluorocarbons in liver microsomes. *Eur. J. Biochem.* **20**, 509–512.
76. Nishibayashi, H., Omura, T., and Sato, R. (1963). A flavoprotein oxidizing NADPH isolated from liver microsomes. *Biochim. Biophys. Acta* **67**, 520–522.
77. Nishibayashi-Yamashita, H., and Sato, R. (1970). Vitamin K_3-dependent NADPH oxidase of liver microsomes. *J. Biochem. (Tokyo)* **67**, 199–210.
78. Bus, J. S., Aust, S. D., and Gibson, J. E. (1974). Superoxide- and singlet oxygen-catalyzed lipid peroxidation as a possible mechanism for paraquat (methyl viologen) toxicity. *Biochem. Biophys. Res. Commun.* **58**, 749–755.
79. Shen, A. L., Fahl, W. E., Wrightson, S. A., and Jefcoate, C. R. (1979). Inhibition of benzo[a]pyrene and benzo[a]pyrene 7,8-dihydrodiol metabolism by benzo[a]pyrene quinones. *Cancer Res.* **39**, 4123–4129.
80. Iyanagi, T., and Yamazaki, I. (1969). One-electron-transfer reactions in biochemical systems III. One-electron reduction of quinones by microsomal flavin enzymes. *Biochim. Biophys. Acta* **172**, 370–381.
81. Tomaszewski, J. E., Jerina, D. M., and Daly, J. W. (1975). Deuterium isotope effects during formation of phenols by hepatic monooxygenases. Evidence for an alternative to the arene oxide pathway. *Biochemistry* **14**, 2024–2031.
82. Holder, G., Yagi, H., Dansette, P., Jerina, D. M., Levin, W., Lu, A. Y. H., and Conney, A. H. (1974). Effects of inducers and epoxide hydrase on the metabolism of benzo[a]pyrene by liver microsomes and a reconstituted system: Analysis by high pressure liquid chromatography. *Proc. Natl. Acad. Sci. U.S.A.* **71**, 4356–4360.
83. Selkirk, J. K. (1977). Benzo[a]pyrene carcinogenesis: A biochemical selection mechanism. *J. Toxicol. Environ. Health* **2**, 1245–1258.
84. Oesch, F. (1973). Mammalian epoxide hydrases: Inducible enzymes catalysing the inactivation of carcinogenic and cytotoxic metabolites derived from aromatic and olefinic compounds. *Xenobiotica* **3**, 305–340.
85. Jakoby, W. B. (1978). The glutathione S-transferases: A group of multifunctional detoxification proteins. *Adv. Enzymol.* **46**, 383–414.
86. Sims, P., Grover, P. L., Swaisland, A., Pal, K., and Hewer, A. (1974). Metabolic activation of benzo[a]pyrene proceeds by a diol-epoxide. *Nature (London)* **252**, 326–328.
87. Thakker, D. R., Yagi, H., Lu, A. Y. H., Levin, W., Conney, A. H., and Jerina, D. M. (1976). Metabolism of benzo[a]pyrene: Conversion of (±)-*trans*-7,8-dihydro-7-8-dihydroxybenzo[a]pyrene to a highly mutagenic 7,8-diol-9,10-epoxide. *Proc. Natl. Acad. Sci. U.S.A.* **73**, 3381–3385.
88. Huberman, E., Sachs, L., Yang, S. K., and Gelboin, H. V. (1976). Identification of

mutagenic metabolites of benzo[a]pyrene in mammalian cells. *Proc. Natl. Acad. Sci. U.S.A.* **73,** 607–611.
89. Kapitulnik, J., Wislocki, P. G., Levin, W., Yagi, H., Jerina, D. M., and Conney, A. H. (1978). Tumorigenicity studies with diol-epoxides of benzo[a]pyrene indicate that ($\pm$)-*trans*-7β,8α-dihydroxy-9α,10α-epoxy-7,8,9,10-tetrahydrobenzo[a]**pyrene** is an ultimate carcinogen in newborn mice. *Cancer Res.* **38,** 354–358.
90. Conney, A. H., Levin, W., Wood, A. W., Yagi, H., Lehr, R. E., and Jerina, D. M. (1978). Biological activity of polycyclic hydrocarbon metabolites and the bay region theory. *Adv. Pharmacol. Ther.* **9,** 41–52.
91. Wood, A. W., Levin, W., Chang, R. L., Yagi, H., Thakker, D. R., Lehr, R. E., Jerina, D. M., and Conney, A. H. (1979). Bay region activation of carcinogenic polycyclic hydrocarbons. *In* "Polynuclear Aromatic Hydrocarbons" (P. W. Jones and P. Leber, eds.), pp. 531–551. Ann Arbor Sci. Publ., Ann Arbor, Michigan.
92. Magee, P. N., and Barnes, J. M. (1962). Carcinogenic nitroso compounds. *Adv. Cancer Res.* **10,** 163–246.
93. Magee, P. N., and Farber, E. (1962). Toxic liver injury and carcinogenesis of rat-liver nucleic acids by dimethylnitrosamine *in vivo. Biochem. J.* **83,** 114–124.
94. Pegg, A. E. (1977). Formation and metabolism of alkylated nucleosides: Possible role in carcinogenesis by nitroso compounds and alkylating agents. *Adv. Cancer Res.* **25,** 195–269.
95. Keefer, L. K., Lijinsky, W., and Garcia, H. (1973). Deuterium isotope effect on the carcinogenicity of dimethylnitrosamine in rat liver. *JNCI, J. Natl. Cancer Inst.* **51,** 299–302.
96. Lijinsky, W., and Taylor, H. W. (1977). Carcinogenicity of nitrosoazetidine and tetradeuteronitrosoazetidine in Sprague-Dawley rats. *Z. Krebsforsch.* **89,** 215–219.
97. Lijinsky, W., Taylor, H. W., and Keefer, L. K. (1976). Reduction of rat liver carcinogenicity of 4-nitrosomorpholine by α-deuterium substitution. *JNCI, J. Natl. Cancer Inst.* **57,** 1311–1313.
98. Boyland, E., Dukes, C. E., and Grover, P. L. (1963). Carcinogenicity of 2-naphthylhydroxylamine and 2-naphthylamine. *Br. J. Cancer* **17,** 79–84.
99. Weisburger, J. H., and Weisburger, E. K. (1973). Biochemical formation and pharmacological, toxicological, and pathological properties of hydroxylamines and hydroxamic acids. *Pharmacol. Rev.* **25,** 1–65.
100. Cramer, J. W., Miller, J. A., and Miller, E. C. (1960). N-hydroxylation: A new metabolic reaction observed in the rat with the carcinogen 2-acetylaminofluorene. *J. Biol. Chem.* **235,** 885–888.
101. Irving, C. C., Wiseman, R., Jr., and Young, J. M. (1967). Carcinogenicity of 2-acetylaminofluorene and *N*-hydroxy-2-acetylaminofluorene in the rabbit. *Cancer Res.* **27,** 838–848.
102. Boyland, E., and Chasseauld, L. F. (1969). The role of glutathione and glutathione *S*-transferases in mercapturic acid biosynthesis. *Adv. Enzymol.* **32,** 173–219.
103. Kriek, E. (1965). On the interaction of *N*-2-fluorenylhydroxylamine with nucleic acids *in vitro. Biochem. Biophys. Res. Commun.* **20,** 793–799.
104. Miller, E. C., Juhl, U., and Miller, J. A. (1966). Nucleic acid guanine: Reaction with the carcinogen *N*-acetoxy-2-acetylaminofluorene. *Science* **153,** 1125–1127.
105. Miller, E. C. (1978). Some current perspectives on chemical carcinogenesis in humans and experimental animals: Presidential Address. *Cancer Res.* **38,** 1479–1496.
106. Wallace, W. J., and Caughey, W. S. (1975). Mechanism for the autoxidation of hemoglobin by phenols, nitrite and "oxidant" drugs. Peroxide formation by one electron donation to bound dioxygen. *Biochem. Biophys. Res. Commun.* **62,** 561–567.

107. Boyd, M. R., Stiko, A., and Sasame, H. A. (1979). Metabolic activation of nitrofurantoin-possible implications for carcinogenesis. *Biochem. Pharmacol.* **28**, 601–606.
108. Peterson, F. J., Mason, R. P., and Holtzman, J. L. (1980). The effect of selenium and vitamin E deficiency on the toxicity of nitrofurantoin in the chick. *In* "Microsomes, Drug Oxidation and Chemical Carcinogenesis" (M. J. Coon, A. H. Conney, R. W. Estabrook, H. V. Gelboin, J. R. Gillette, and P. O'Brien, eds.), Vol. 2, pp. 873–876. Academic Press, New York.
109. Prough, R. A., Freeman, P. C., and Hines, R. N. (1981). The oxidation of hydrazine derivatives catalyzed by the purified liver microsomal FAD-containing monooxygenase. *J. Biol. Chem.* **256**, 4178–4184.
110. Coomes, M. W., Wiebkin, P., and Prough, R. A. (1980). The metabolism of 1,2-disubstituted hydrazines in isolated hepatocytes. *In* "Microsomes, Drug Oxidation and Chemical Carcinogenesis" (M. J. Coon, A. H. Cooney, R. W. Estabrook, H. V. Gelboin, J. R. Gillette, and P. J. O'Brien, eds.), Vol. 2, pp. 1133–1136. Academic Press, New York.
111. Hucko-Haas, J. E., and Reed, D. J. (1970). Hydrazines as substrates for bovine plasma amine oxidase (PAO). *Biochem. Biophys. Res. Commun.* **39**, 396–400.
112. Nelson, S. F. (1981). One electron oxidation of tetralkylhydrazines. *Acc. Chem. Res.* **14**, 131–138.
113. Nagy, J., Kenney, W. C., and Singer, T. P. (1979). The reaction of phenylhydrazine with trimethylamine dehydrogenase with free flavins. *J. Biol. Chem.* **254**, 2684–2688.
114. Saito, S., and Itano, H. A. (1981). β-*meso*-Phenylbiliverdin IXα and *N*-phenylprotoporphyrin IX, products of the reaction of phenylhydrazine with oxyhemoproteins. *Proc. Natl. Acad. Sci. U.S.A.* **78**, 5508–5512.
115. Ortiz de Montellano, P. R., and Kunze, K. L. (1981). Formation of *N*-phenylheme in the hemolytic reaction of phenylhydrazine with hemoglobin. *J. Am. Chem. Soc.* **103**, 6534–6536.
116. Prough, R. A., Cummings, S. W., Coomes, M. W., Wiebkin, P., and Moloney, S. J. (1981). Evaluation of possible reactive intermediates formed from the antitumor agent, procarbazine. *Proc. Am. Assoc. Cancer Res.* **22**, 0395.
117. Wiebkin, P., and Prough, R. A. (1980). Oxidative metabolism of *N*-isopropyl-α-(2-methylazo)-*p*-toluamide (azo-procarbazine) by rodent liver microsomes. *Cancer Res.* **40**, 3524–3529.
118. Fiala, E. S., Kulakis, C., Christiansen, G., and Weisburger, J. H. (1978). Inhibition of the metabolism of the colon carcinogen, azomethane. *Cancer Res.* **38**, 4515–4521.
119. Weinkam, R. J., and Shiba, D. A. (1978). Metabolic activation of procarbazine. *Life Sci.* **22**, 937–946.
120. Reid, W. D. (1973). Mechanism of renal necrosis induced by bromobenzene or chlorobenzene. *Exp. Mol. Pathol.* **19**, 197–214.
121. Fowler, J. S. L. (1969). Carbon tetrachloride metabolism in the rabbit. *Br. J. Pharmacol.* **37**, 733–737.
122. Keller, F., Snyder, A. B., Petracek, F. J., and Sancier, K. M. (1971). Hepatic free radical levels in ethanol-treated and carbon tetrachloride-treated rats. *Biochem. Pharmacol.* **20**, 2507–2511.
123. Lombardi, B. (1966). Considerations on the pathogenesis of fatty liver. *Lab. Invest.* **15**, 1–20.
124. Reynolds, E. S. (1967). Liver parenchymal cell injury. IV. Pattern of incorporation of carbon and chlorine from carbon tetrachloride into chemical constituents of liver *in vivo*. *J. Pharmacol. Exp. Ther.* **155**, 117–126.

125. Horning, M. G., Earle, M. J., and Maling, H. M. (1962). Changes in fatty acid composition of liver lipids induced by carbon tetrachloride and ethionine. *Biochim. Biophys. Acta* **56,** 175–177.
126. Riely, C. A., and Cohen, G. (1974). Ethane evolution: A new index of lipid peroxidation. *Science* **183,** 208–210.
127. Hunter, A. L., Holscher, M. A., and Neal, R. A. (1977). Thioacetamide-induced hepatic necrosis. I. Involvement of the mixed-function oxidase enzyme system. *J. Pharmacol. Exp. Ther.* **200,** 439–448.
128. Porter, W. R., Gudzinowicz, M. J., and Neal, R. A. (1979). Thioacetamide-induced hepatic necrosis. II. Pharmacokinetics of thioacetamide and thioacetamide S-oxide in the rat. *J. Pharmacol. Exp. Ther.* **208,** 386–391.
129. Boyd, M. R., and Neal, R. A. (1976). Studies on the mechanism of toxicity and of development of tolerance to the pulmonary toxin, α-naphthylthioureau (ANTU). *Drug. Metab. Disp.* **4,** 314–322.
130. Poulsen, L. L., Hyslop, R. M., and Ziegler, D. M. (1974). S-Oxidation of thioureylenes catalyzed by a microsomal flavoprotein mixed-function oxidase. *Biochem. Pharmacol.* **23,** 3431–3440.
131. Giri, S. N., and Combs, A. (1970). Sulfur-containing compounds and tolerance in the prevention of certain metabolic effects of phenylthiourea. *Toxicol. App. Pharmacol.* **16,** 709–717.
132. Lubet, R. A., Capdevila, J., and Prough, R. A. (1979). The metabolic activation of benzo[a]pyrene and 9-hydroxybenzo[a]pyrene by liver microsomal fractions. *Int. J. Cancer* **23,** 353–357.
133. Ashurst, S. W., and Cohen, G. M. (1980). A benzo[a]pyrene-7,8-dihydrodiol-9,10-epoxide is the major metabolite involved in the binding of benzo[a]pyrene to DNA in isolated viable rat hepatocytes. *Chem.-Biol. Interact.* **29,** 117–127.
134. Slaga, T. J., Sivak, A., and Boutwell, R. K., eds. (1978). "Mechanisms of Tumor Promotion and Carcinogenesis," Vol. 2. Raven Press, New York.
135. Bresnick, E., McDonald, T. F., Yagi, H., Jerina, D. M., Levin, W., Wood, A. W., and Conney, A. H. (1977). Epidermal hyperplasia after topical application of benzo[a]pyrene, benzo[a]pyrene diol epoxides, and other metabolites. *Cancer Res.* **37,** 984–989.

CHAPTER 2

Integration of Xenobiotic Metabolism in Carcinogen Activation and Detoxication

Colin R. Jefcoate

I.	Introduction	32
II.	Identification of Reactive Electrophiles Derived from Xenobiotics	33
III.	Cell Specificity of Xenobiotic Effects	36
IV.	Classification of Xenobiotic Activation Pathways	37
V.	Partition of Metabolic Pathways	39
VI.	Regioselectivity and Stereoselectivity of Cytochrome P-450	44
	A. Multiplicity of Cytochrome P-450	44
	B. Multiplicity and Carcinogen Metabolism	46
	C. Alternative Modes of Activation	48
	D. Selective Inhibition	50
	E. Cross-Inhibition between Pathways	51
VII.	Epoxide Hydrolase	52
VIII.	Absolute Monooxygenase Activity in the Activation Pathway	54
IX.	Cellular Cosubstrates of Xenobiotic Metabolism	54
	A. Regulation of NADPH	55
	B. UDPGA	57
	C. 3′-Phosphoadenosine 5′-Phosphosulfate	58
	D. Glutathione	59
X.	UDPglucuronyltransferases and Sulfotransferases	60

XI. Glutathione S-Transferases 62
XII. Inhibitors of Cellular Transferase Activity 63
XIII. Xenobiotic Metabolism in the Nuclear Envelope 65
XIV. Transport of Reactive Intermediates and Precursors 65
XV. Comments 66
References 67

I. INTRODUCTION

A substantial proportion of xenobiotics are lipid-soluble molecules that are not excreted without metabolism. The integrated process of xenobiotic metabolism frequently consists of a functionalization step catalyzed by the microsomal polysubstrate monooxygenase system of the liver, which consists of several forms of cytochrome P-450 and the associated P-450 reductase. This step is typically followed by conjugation of the new functional group through reaction at one or more of three classes of enzyme: UDPglucuronyltransferase, sulfotransferase, and glutathione S-transferase.

It has become increasingly evident that xenobiotic-metabolizing enzymes function in an ambivalent manner. On the one hand, lipid-soluble chemicals are converted to water-soluble forms that are readily excreted via the kidney or bile duct; on the other hand, the same enzymes may generate reactive electrophilic molecules that modify essential cellular macromolecules, with resulting toxic or carcinogenic effects. The latter are frequently derived from metabolic pathways that are minor with respect to the gross metabolic clearance of the chemical. Although most chemicals are preferentially metabolized in the liver, chemical tumorigenesis exhibits remarkable tissue specificity. For example, polycyclic hydrocarbons such as benzo[a]pyrene (BP) are carcinogenic in the lung or breast, 2-naphthylamine induces bladder tumors, N-acetyl-2-aminofluorene (AAF) is tumorigenic in the breast and liver, and aflatoxin B_1 is relatively specific to the liver. In each case, DNA in the target tissue is extensively modified, consistent with the hypothesis that electrophilic modification of specific genomic sites may initiate cell transformation through one or more mutational events.[1]

Much research on xenobiotic metabolism has focused on the detection and modulation of these side reactions that generate reactive electrophiles. Because DNA-adduct formation provides a quantifiable endpoint that correlates with a biological process, namely, initiation of tumorigenesis, this discussion of integrated xenobiotic metabolism will focus on the metabolism of selected carcinogens to illustrate approaches

2. Integration of Xenobiotic Metabolism

that can be generally applied to problems in xenobiotic metabolism and toxicity.

Carcinogenesis requires, in addition to initiation by DNA modification, a complex, multistep process. Tissue and species differences in the repair and replication of chemically modified DNA, in the promotion of transformed cells and in tumor progression, may all contribute in major ways to the diversity of chemical carcinogenesis. Nevertheless, although differences in DNA modification provide only one of many explanations for tissue or species differences in chemical susceptibility, increases in the formation of certain identified adducts in specific tissues may correlate well with increased incidence of tumor formation.

In attempting to understand the contribution of xenobiotic metabolism to the chemical initiation of cancer, four goals are of major importance. These goals are (1) identification of the ultimate carcinogenic intermediate(s); (2) identification of DNA modifications that correlate closely with the extent of tumor formation; (3) determination of the extent to which differences in DNA modifications account for selectivity in tumorigenesis between different cells, tissues, or species; and (4) evaluation of the enzymatic characteristics of the formation and removal of carcinogenic electrophiles (or modified DNA). The same strategy could equally be applied to chemical toxicity, although with less emphasis on DNA as the primary target. This chapter will focus on the integrated cellular metabolism of certain carcinogens that illustrate distinct features of metabolic activation and detoxication processes.

II. IDENTIFICATION OF REACTIVE ELECTROPHILES DERIVED FROM XENOBIOTICS

Chemical identification of the products of metabolite attack on polynucleotides provides essential information on the nature of the reactive electrophile and on the selectivity of attack upon different bases. Most laboratories now use a three-step enzymatic hydrolysis to yield nucleosides and chemical adducts that are fractionated either by chromatography on Sephadex LH-20[2] or by reverse-phase, high-pressure liquid chromatography (HPLC).[3] Deoxyguanine is the base that is generally most extensively modified, and Fig. 1 indicates some of the diversity of base modifications.

Reaction selectivity at DNA is determined by molecular orbital characteristics of the attacking electrophile such as polarizability and charge delocalization, along with the attacking steric factors. In addition, reac-

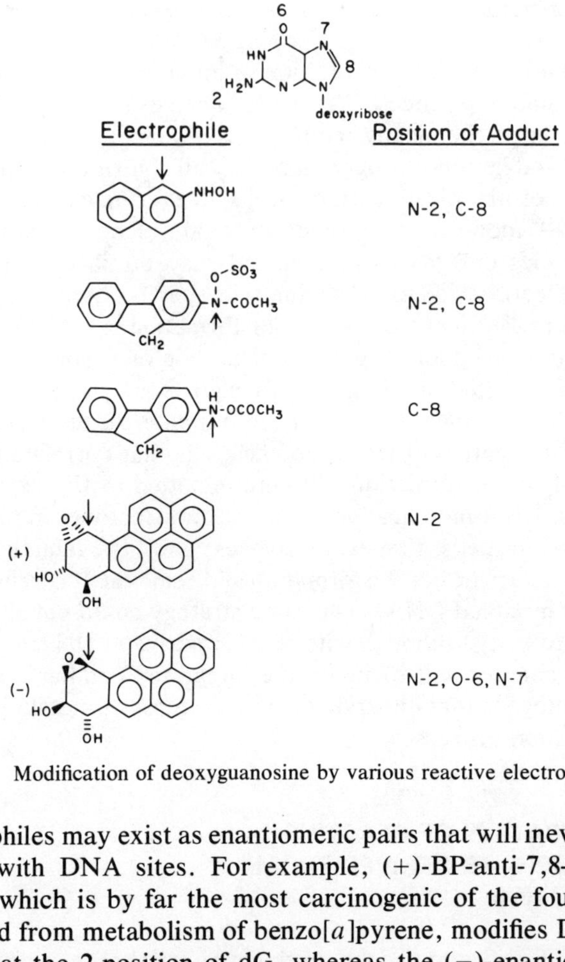

Fig. 1. Modification of deoxyguanosine by various reactive electrophiles.

tive electrophiles may exist as enantiomeric pairs that will inevitably react selectively with DNA sites. For example, (+)-BP-anti-7,8-dihydrodiol 9,10-oxide, which is by far the most carcinogenic of the four stereoisomers derived from metabolism of benzo[a]pyrene, modifies DNA almost exclusively at the 2-position of dG, whereas the (−)-enantiomer reacts with different selectivity between dG, dA, and dC.[4] Simple alkylating agents with less steric demands modify bases in a manner determined more by the electrophilicity of the alkylating species. Thus, the O-6/N-7 ratio for alkylation of dG in DNA by a series of direct alkylating agents increases with the $S_N 1$ character of the alkylating agent.[5] Some DNA adducts may be overlooked because of instability. For example, (−)-BP-anti-7,8-dihydrodiol 9,10-oxide modifies dG in DNA at N-7, but this adduct undergoes depurination with strand breakage and liberation of an N-7–guanine adduct without prior hydrolysis of the DNA.[4]

The fluorescence spectrum of modified DNA, particularly at low temperature, is extremely sensitive and diagnostic of the chemical structure

2. Integration of Xenobiotic Metabolism

of the hydrocarbon moiety in the adduct. The fluorescence spectrum of DNA after incubation with benzo[a]pyrene and methylcholanthrene-induced rat liver microsomes has established that the major modification is derived from 9-phenol 4,5-epoxide.[6] Similarly, fluorescence analysis of DNA isolated from tissue after dimethylbenzanthracene metabolism clearly indicates a bay-region dihydrodiol oxide adduct rather than addition through direct substitution on a hydroxymethyl metabolite of dimethylbenzanthracene.[7] Caution must be exercised in the use of fluorescence spectra because major differences in the quenching of DNA adducts by the bases may lead to erroneous quantitation. However, when the extent of modification is known from other methods, the fluorescence characteristics (such as spectra, quantum yield, and polarization) can provide major insight into the environment of the adducts.

DNA adducts have also been quantitated by the use of antibodies to specific metabolite–DNA adducts.[8] Using this technique, fmole levels of BP-7,8-dihydrodiol 9,10-oxide adducts have been quantitated,[8] and antibodies can discriminate between complete and deacetylated adducts of 2-AAF.[9] The method should be adaptable to many immunological techniques to provide both great sensitivity and direct tissue quantitation.

Isotope labeling techniques have proved invaluable in the elucidation of activation pathways. Specific ^{14}C labeling of the acyl, aryl, or ethyl groups of phenacetin has been used to examine the modification of protein resulting from metabolism of phenacetin by hamster liver microsomes.[10] This approach indicates that covalent binding of phenacetin metabolites to microsomal protein is accompanied by complete loss of the acetyl group to acetamide, in addition to 50% deethylation. Very different results have been obtained *in vivo*, where phenacetin seems to be substantially metabolized to acetaminophen prior to binding.

The electrophile modifying DNA may be tissue dependent. HPLC and antibody analysis of the nucleoside adducts formed as a consequence of cellular metabolism of 2-AAF indicates that the acyl group is predominantly retained in the modification of DNA in hepatocytes, whereas almost complete deacetylation is found in fibroblasts and epidermal cells.[9] Acyltransferase or deacetylation pathways, although possibly of less importance in the liver, may be crucial in extrahepatic tissues such as the mammary gland, where sulfotransferase activity is low.

The extent of modification of DNA in living tissue may reflect the balance between reaction of DNA with electrophiles and the repair of this damage. Chemically modified bases are removed from DNA by a family of excision repair enzymes.[11] This process is generally carried out with perfect fidelity so that mutations are not generated in the excision process. Excision repair rates differ from one cell type to another and are highly dependent on the chemical modification. For example, although metabolism of 2-AAF in rat hepatocytes leads to a predominance of adducts retaining the N-acetyl group, the dominant C-8 AAF adduct is rapidly repaired, resulting in a large increase in the proportion of deacetylated adducts. Because various types of mutation can be introduced into cellular DNA during the replication of chemically modified sequences, some correlation is likely between resistance to excision repair and mutation or transformation frequency. Significantly, O-6 alkyl–dG and N-2 dihydrodiol oxide–dG adducts are both relatively resistant to repair, and formation of these adducts correlates with both mutation and morphological transformation.[12,13] The rapid excision of adducts forms part of the liver cells' protection against reactive drug metabolites. Some comparisons show more rapid excision of DNA adducts in the liver than in the lung or bladder.[14,15] Clearly, the net effect of chemical activation and modification on the one hand and excision on the other, particularly at the time of DNA replication, is central for any quantitative understanding of the initiation of carcinogenesis.

III. CELL SPECIFICITY OF XENOBIOTIC EFFECTS

A major goal in investigating the relationship of xenobiotic metabolism to either toxicity or initiation of cancer is the tissue specificity of the effect. Certainly, lower protection against reactive intermediates contributes to the prevalent polycyclic hydrocarbon carcinogenesis in the lung rather than the liver, even though metabolism is far higher in the liver. For example, the level of nucleoside adducts bound per milligram of DNA in mice injected with methylcholanthrene is two- to eightfold higher in the lung than in the liver and is much more slowly excised in the lung. Indeed, in the strain of mouse that is most susceptible to lung neoplasia, 40% of the adducts remained 28 days after exposure to methylcholanthrene.[14] Similarly, major differences in the chemical activation by liver and lung cells have been observed in culture. For example, aflatoxin B_1 and dimethylnitrosamine are both converted to more mutagens by the liver cells, whereas polycyclic hydrocarbons generate more mutagens during metabolism by lung cells.[16]

A second example of tissue specificity in both DNA binding and tumorigenesis is provided by the metabolism and binding of naphthylamines.[17] 2-Naphthylamine is metabolized in the liver but preferentially binds to DNA and is tumorigenic in the bladder. In contrast, 1-naphthylamine produces only a low level of liver DNA adducts and is not tumorigenic in the bladder. Reasons for the dissociation of metabolism and DNA modification will be developed in this chapter.

Although cytotoxic effects of xenobiotics can be observed soon after metabolism, the initial sites of modification leading to cell damage have rarely been identified. The gross modifications of DNA, RNA, protein, or lipid provide an indicator of the extent of formation, reactivity, and selectivity of the electrophilic species.[17] The toxicity of ipomeanol, a furan isolated from moldy sweet potatoes, illustrates the close correlation between the xenobiotic toxic response and the extent of covalent binding of the compound to protein. In the uninduced rat, covalent binding and toxicity are far higher in the lung than in the liver.[18] This difference in covalent binding is derived in part from differences in the selectivity of product formation resulting from metabolism at cytochrome P-450. This presumably is derived from differences in the proportions of various forms of cytochrome P-450 in the rat lung and liver.[19] Changes in cytochrome P-450 caused by induction with methylcholanthrene also alter the tissue specificity for toxicity and for metabolism of ipomeanol. Binding of metabolites to protein in the lung decreases by about 2.5 times but nearly doubles in the liver, and toxicity changes accordingly.

Cytotoxic processes may contribute significantly to postinitiation tumorigenesis through selective cytotoxic effects against normal cells as compared to initiated cells with accompanying proliferation of the initiated cells and replication of the chemically modified DNA.[20] Indeed, it has been suggested that the major role of reactive sulfates in 2-AAF carcinogenesis is through this type of cytotoxic stimulation of initiated cells.[21]

Most organs consist of several different cell types that exhibit characteristic metabolism of xenobiotics. For example, separation of the Clara and alveolar II cells from rabbit lung has indicated that xenobiotic metabolism is usually more active in the Clara cells,[22] although substrate specificity differs considerably between the two cell types. Because the cells responsible for generation of reactive intermediates may be the most susceptible targets, specific cellular characterization of the metabolism of the suspected toxic compound may be crucial.

IV. CLASSIFICATION OF XENOBIOTIC ACTIVATION PATHWAYS

The analysis of the contributions of different enzymes to carcinogen activation can be simplified by classification of carcinogens according to

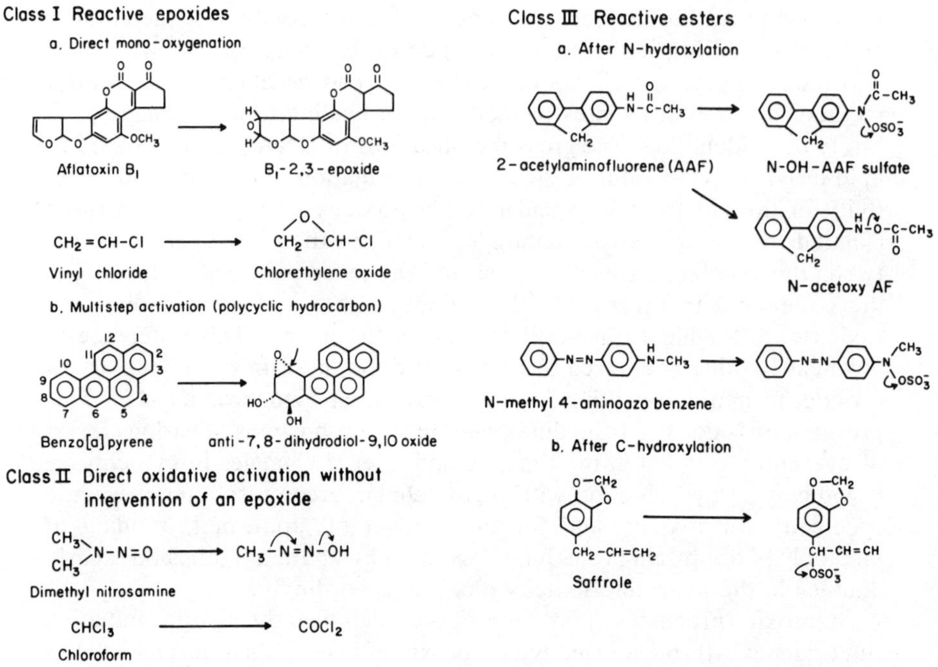

Fig. 2. Classification of reactive electrophiles according to the activation pathway.

the mode of generation of an ultimate carcinogen (Fig. 2). Information on specific enzymes has been developed in Chapter 1 of this volume.

Class Ia: Epoxides requiring direct formation. Aflatoxin B_1[23] and vinyl chloride[24] typify this class when activated to aflatoxin B_1 2,3-epoxide and chloroethylene oxide, respectively.

Class Ib: Epoxide formation requiring multiple metabolic steps. Polycyclic hydrocarbons form a separate group, differentiated from those of Class Ia by the required formation of a *trans*-dihydrodiol adjacent to the position of epoxidation.[25]

Class II: Direct activation without intervention of an epoxide. Dimethylnitrosamine[26] and chloroform[27] provide examples of activation to an electrophilic intermediate without formation of an epoxide. In both cases, initial hydroxylation leads to a very unstable intermediate that reacts further to form a reactive electrophile.

Class IIIa: Activation through ester formation following N-hydroxylation. Aryl hydroxylamine derivatives may be formed by microsomal N-hydroxylation mediated either by cytochrome *P*-450 with 2-AAF as an

2. Integration of Xenobiotic Metabolism

example[28] or by the flavoprotein monooxygenase with N-methyl-4-aminoazobenzene as an example.[29] Primary or acetyl arylamines react preferentially through the former pathway, whereas N-methylation increases the probability of reaction at the flavoprotein monooxygenase.[30] N-Hydroxy aryl acetamides are converted to electrophilic esters either by acyltransferase (deacetylated DNA adducts)[31] or by sulfation.[32] A deacylation pathway may also contribute to activation of N-hydroxy aryl acetamides and can be assessed from the effect of parathion, an inhibitor of this reaction.[33] N-Methyl-4-aminoazobenzene is also converted to an ultimate carcinogen by sulfation.[34] Aryl sulfotransferases therefore contribute to the activation of Class III carcinogens but can facilitate inactivation of Class Ib carcinogens by conjugating and inactivating precursor *trans*-dihydrodiols. Whereas O-glucuronidation of N-hydroxy aryl acetamides provides a detoxication pathway,[35] N-glucuronidation plays a major role in the carcinogenesis of arylamines and aryl hydroxylamines.[15] Aryl hydroxylamine N-glucuronides formed in the liver are stable at neutrality, leading to renal and biliary excretion. However, at the low pH of the urine, these glucuronides hydrolyze, liberating free aryl hydroxylamines in the bladder, where they react with DNA either directly or after activation by sulfation (see also Chapter 1, this volume).

This sequence provides an explanation of the specificity of 2-naphthylamine carcinogenesis for the bladder. HPLC analysis of DNA adducts in the liver and the bladder urothelium indicates a distribution of DNA adducts similar to that obtained by direct addition of N-hydroxy-2-naphthylamine to isolated DNA *in vitro* (N-2, C-8 of dG and N-6 of dA).[17]

Class IIIb: Activation through ester formation following C-hydroxylation. The best-documented example of this type of activation is safrole, which is a hepatocarcinogen in rats and mice.[36] 1'-Hydroxylation has been demonstrated in these animals, and 1'-hydroxysafrole is also a potent hepatocarcinogen. Carcinogenicity appears to correlate approximately with changes in aryl sulfotransferase activity, suggesting that the ultimate electrophile is safrole 1'-sulfate.[37]

V. PARTITION OF METABOLIC PATHWAYS

The contribution of xenobiotic metabolizing enzymes to biochemical activation as compared to detoxication depends on the type of activation pathway. For Class Ia compounds (see Fig. 2) epoxide hydrolase may be protective, but this depends on the aqueous hydrolysis rate of the labile epoxide and the specificity of epoxide hydrolases. Thus, there is probably little effect of the enzyme on hydrolysis of aflatoxin B_1 2,3-oxide but direct involvement in the hydration of chloroethylene oxide.[24] For Class Ib polycyclic hydrocarbons, epoxide hydrolase is required for generation of the active species but also contributes to hydrolysis of dihydrodiol oxides.[38] UDPglucuronyltransferases and sulfotransferases function in a protective role by decreasing the levels of *trans*-dihydrodiols that are available for secondary monooxygenation. Activation of Class II compounds depends only on the characteristics of the polysubstrate monooxygenase system and should not be affected by epoxide hydrolase, UDPglucuronyltransferases, or aryl sulfotransferases. For Class III compounds, sulfotransferase activity is generally activating, whereas UDPglucuronyltransferase has a protective role because of the greater stability of aryl hydroxylamine O-glucuronides. The relative importance of aryl sulfotransferase and acyltransferase in activation will again depend on the interplay of several competing reactions: ester hydrolysis, transport from the tissue, and the reaction of active esters with GSH as compared to the rate of modification of DNA. Glucuronides, sulfates, and acetates clearly differ in aqueous stability, rate of excretion, electrophilicity, and steric properties. For example, $t_{1/2}$ for hydrolysis of 2-N-acetoxy-AF is 7 min but is only 1 min for hydrolysis of 2-N-sulfonoxy-AAF.[35]

Most of the reactive electrophiles generated from compounds shown in Fig. 2 that modify DNA may be expected to react with GSH either with or without catalysis by glutathione S-transferases. This process will be referred to here as *glutathione detoxication*. The extent of DNA modification depends on the ratio of the rate of modification of DNA (k_{DNA}[DNA]) to the sum of the rates of reaction with water and with GSH. The effectiveness of glutathione S-transferase depends in part on the relative rates of the reactive electrophile with water and with GSH. The steady-state concentration of the intermediate will inevitably be low for reactive electrophiles, implying that enzymatic involvement will be significant only if [enzyme] $\times$ V_{max}/K_m for the enzymatic reaction (assuming GSH is saturating) is comparable to the rate of nonenzymatic breakdown (k_{GSH}[GSH] $+ k_{H_2O}$). Enzymatic catalysis of the hydration and GSH conjugation of BP-anti-7,8-dihydrodiol 9,10-oxide has been established with high levels of purified enzymes,[38,39] but the significance of the glutathione S-transferases in cellular detoxication of dihydrodiol oxides remains to be established.

Initiation of cell transformation *in vivo* usually involves treatment of cultured cells with a small dose of carcinogen that is fully metabolized. In most cases, carcinogen metabolism occurs in a time interval that does not allow significant DNA repair. DNA modification then depends primarily

2. Integration of Xenobiotic Metabolism

upon the fraction of carcinogen metabolism passing into the carcinogen activation pathway as compared to pathways leading to detoxication. This activation pathway may consist of the conversion of the carcinogen to the ultimate electrophile (U) through a sequence of one or more precursors. Several types of detoxication may occur prior to glutathione detoxication. These are represented in Fig. 3 for benzo[a]pyrene activation. First, partial detoxication occurs at the initial monooxygenation step. *Primary selectivity detoxication* will be used as the term to describe reactions of the initial carcinogen (C) at alternative positions (regioselectivity) or with different stereochemistry (stereoselectivity) that do not lead to carcinogenic metabolites. The term *diversionary detoxication* will be used to describe reactions of the precursors (P) that lead to noncarcinogenic products and that therefore compete with the activation pathway. This is represented at only a single step, but this type of partitioning of reaction intermediates between activation and detoxication may occur at several steps in the activation sequence. Strictly speaking, partitioning of benzo[a]pyrene metabolism also occurs at BP-7,8-oxide because GSH conjugation weakly competes with the hydration step.

Figure 4 represents the more complex activation pathway that has been worked out for 2-AAF. Here the precursor 2-N-(OH)AAF can be converted to two ultimate electrophiles, 2-acetoxy-AAF or 2-N-sulfonoxy-AAF, depending on the relative activities of the appropriate sulfotransferase and acyltransferase. O-Glucuronidation provides diversionary detoxication. There is strong correlative evidence for the involvement

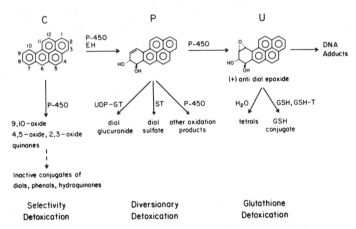

Fig. 3. Carcinogen activation pathway for benzo[a]pyrene and the three types of detoxication. UDP-GT = UDPglucuronyltransferase; ST = sulfotransferase; GSH-T = glutathione S-transferase; EH = epoxide hydrolase.

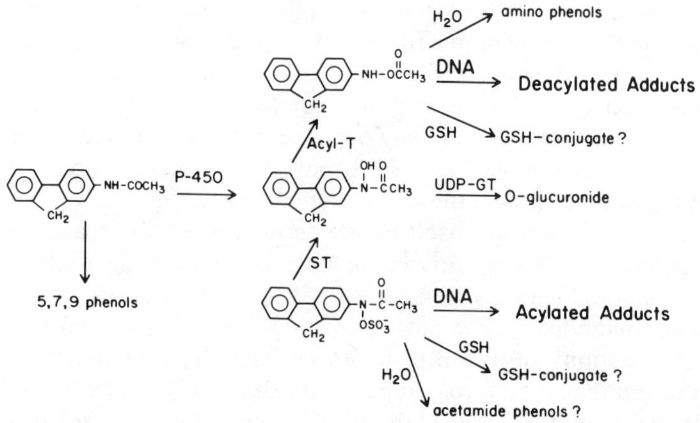

Fig. 4. Carcinogen activation and detoxication pathways for 2-acetylaminofluorene. Acyl-T = acyl transferase; ST = sulfotransferase.

of aryl sulfotransferase in the activation of 2-AAF as a hepatocarcinogen. Sulfotransferase activity, covalent binding to protein, and hepatocarcinogenicity are all higher in the male rat than in the female rat or in other tested animals. Thyroidectomy, hypophysectomy, and castration all lower sulfotransferase, protein binding, and hepatocarcinogenesis in the male rat.[35]

Both the carcinogenic activity and the metabolic reactions that generate an ultimate electrophile are typically stereoselective. In addition, steric factors are major determinants of the carcinogenicity of reactive electrophiles. For example, the stereoselectivity of DNA in the reaction with dihydrodiol oxides also appears to be highly conserved for many different hydrocarbons.

Primary selectivity detoxication is determined by the regioselectivity and stereoselectivity of multiple forms of cytochrome P-450 (or flavoprotein monooxygenase) that metabolize the carcinogen. The partition of metabolism between activation and detoxication pathways at a microsomal polysubstrate monooxygenase system is determined by the characteristics of the multiple forms of cytochrome P-450 present in the microsomes (forms 1, 2, 3, . . . present at concentrations a_1, a_2, a_3, . . .). The primary selectivity denoted by S is determined, in fact, by the sum of selectivities of each cytochrome-dependent reaction.

Primary selectivity $(S) = a_1 r_1 P_1 + a_2 r_2 P_2 + \ldots / a_1 r_1 + a_2 r_2 + \ldots$ (1)

In Eq. (1), r_1, r_2, and so on are the total rates of carcinogen metabolism at the separate forms, and P_1, P_2, and so on are the respective fractions metabolized to a carcinogenic precursor.

2. Integration of Xenobiotic Metabolism

The partition of the metabolism of precursor P between activation to an electrophile U and detoxication can be represented by the diversionary detoxication factor D.

$$D = k_{act} \bigg/ \left(\sum k_{inact} + k_{act} \right) \quad (2)$$

In Eq. (2), the rate of activation of the precursor by secondary monooxygenation k_{act} includes the respective regioselectivity terms $a_1 r_1' P_1' + a_2 r_2' P_2' + \ldots$ for metabolism of the precursor to the electrophile U by cytochrome P-450 forms 1, 2 ..., whereas $\sum k_{inact}$ also includes terms of form $a_1 r_1' (1 - P_1')$ for oxidative inactivation. Additional D-factors can be similarly constructed for each step in the activation pathway where diversionary reactions are significant.

Carcinogen activation is further partitioned in the competition between DNA modification and the combined effects of glutathione S-transferase–mediated detoxication and hydrolysis. These are quantitatively determined by G, the glutathione detoxication factor.

$$G = k_{DNA}[DNA]/(k_{H_2O} + k_{GSH}) \quad (3)$$

Glutathione conjugation is assumed to be the only reaction other than hydrolysis that is quantitatively significant in removal of ultimate electrophiles. Enzymatic and nonenzymatic reactions of GSH are included in k_{GSH}. The formation of DNA adducts may require initial rapid complex formation of the electrophile with DNA, followed by either DNA-mediated hydrolysis or adduct formation.[40] In this case, the numerator of G may require careful evaluation.

Combining the various detoxication factors, the fraction of initial carcinogen that ultimately modifies DNA (FM_{DNA}) is given by Eq. (4).

$$FM_{DNA} = S \times D \times G \quad (4)$$

Increased specificity can be included in this expression by adding a further factor to denote modification of DNA at particular sites. Current evidence suggests that initiation of carcinogenesis may be derived from slowly excised adducts,[41] which could readily be specified in Eq. (4).

The selectivity factor S can be directly determined from the distribution of products from microsomal metabolism of the carcinogen or from cellular metabolism in which conjugation reactions have been inhibited. Likewise, D and its components can be directly determined by measuring the short-term conversion of the precursor to hydrolysis products and GSH conjugates derived from the ultimate carcinogen in relation to other metabolites of the precursor. Glutathione detoxication G can be determined in cells from the ratio of DNA adducts to this sum of hydrolysis and GSH conjugation products derived from the ultimate carcinogen.

This model inevitably overlooks many complexities of cellular activation of carcinogens. However, the value of the analysis is that the contribution of various cellular processes to detoxication can be evaluated. Knowledge of the quantitative contribution to S, D, and G under one set of conditions should allow prediction of the impact of metabolic perturbations that affect individual components of the detoxication scheme. A simple example of this analysis of DNA modification is provided by conversion of benzo[a]pyrene to BP-anti-7,8-dihydrodiol 9,10-oxide-dG DNA adducts with methylcholanthrene-induced microsomes and 1 mg/ml DNA.[42] The fractional modification of DNA by benzo[a]pyrene in the absence of conjugation enzymes is 2.5×10^{-3}. Diversion of metabolism at each stage of the activation is shown by the three factors: $S = 0.1$, $D = 0.25$, and $G = 0.1$.

When a carcinogen that redistributes slowly is applied directly to a tissue (e.g., skin, lung, or breast), activation competes with the rate of redistribution and general clearance (see Chapters 8 and 9, this volume). This concept can be included in this simple model by use of a rate constant for redistribution of the carcinogen out of the tissue in the denominator of the primary selectivity factor. The preceding analysis reveals initiation of chemical carcinogenesis as a cascading sequence of chemical reactions in which an ever-diminishing amount of carcinogen seeks reaction with a DNA target. Enzymatic determinants of this cascade will now be reviewed in more detail.

VI. REGIOSELECTIVITY AND STEREOSELECTIVITY OF CYTOCHROME *P*-450

A. Multiplicity of Cytochrome *P*-450

Many distinct forms of cytochrome P-450, now designated as xenobiotic monooxygenase (EC 1.14.19.1), have been purified from both rat and rabbit liver microsomes.[43,44] Indeed, the purification of eight distinct forms of cytochrome P-450 from uninduced rabbit liver microsomes has been reported.[45] The relative concentrations of these forms can be dramatically altered by exposure *in vivo* to inducers such as phenobarbital, stilbene oxide, pregnenolone 16-carbonitrile, ethanol, isosafrole, and polycyclic aromatic compounds that also increase total cytochrome P-450 levels. For many of these inducers, one or two distinct forms predominate that may be difficult even to detect in uninduced microsomes. The topic of multiplicity of P-450 has been extensively reviewed.[46]

Immunological comparison of microsomal cytochrome P-450 from a single species indicates that although some forms are very similar, the

2. Integration of Xenobiotic Metabolism

majority share very few antigenic determinants.[47] This picture has been confirmed by sequence studies and peptide mapping.[44] However, it is important to note that a single amino acid substitution is sufficient to change substrate specificity, as has been observed in one form of phenobarbital-induced rabbit P-450.[47a]

Cytochrome P-450 is present in nearly all cells, although extrahepatic tissues have at least a 10-fold lower level.[46] The forms present in a tissue are characteristic of that tissue, as is the response to each type of inducer. For example, the major uninduced form in rabbit lung is indistinguishable from the major phenobarbital-induced hepatic form, P-450$_{LM2}$, which is undetectable in uninduced liver.[47] Phenobarbital induces cytochrome P-450 only in liver and intestines, and this tissue specificity has also been shown in the mouse for the far more potent inducer of the same class, 1,4-bis[2-(3,5-dichloropyridyloxy)]benzene.[48] Clearly, the failure to induce in a particular tissue is a property of gene regulation rather than access of the inducer to the tissue. 2,3,7,8-Tetrachlorodibenzo-p-dioxin (TCDD) and polycyclic hydrocarbons induce cytochrome P-450 in perhaps the widest variety of tissues. This induction is associated with binding to a cytosolic receptor whose synthesis is determined by the Ah locus[49] but whose concentration varies widely among tissues.[50]

When the concentration of the enzyme is very low, regioselectivity provides a useful indicator of the form of cytochrome P-450. For example, the specific activity and regioselectivity of benzo[a]pyrene metabolism in mouse embryo 10 $T_{1/2}$ fibroblasts before and after induction by benzanthracene strongly suggest a single form of the enzyme that is elevated by the inducer and that has an active site very different from those of the dominant liver forms that are inducible by benzo[a]anthracene.[51]

Despite the broad specificity of the various forms of cytochrome P-450, substrates may be metabolized at widely different rates with regioselectivities and stereoselectivities that are characteristic of the specific form of cytochrome P-450. Regioselectivity differences are exemplified by the metabolism of benzo[a]pyrene by cytochrome P-450 purified from rabbit.[52] The distinct regio/stereoselectivity of different forms of cytochrome P-450 has also been elegantly demonstrated by the metabolism of R- and S-enantiomers of warfarin to 4′-,6-,8-,7-, and benzylic hydroxywarfarins and dehydrowarfarin.[53] The stereoselectivity of purified rat liver cytochrome P-450$_C$ toward epoxidation of polycyclic hydrocarbons has also been determined by formation, separation, and quantitation of diastereomeric trans-addition products of the oxide products with glutathione. Over 97% of BP-4,5-oxide formed enzymatically at P-450$_C$ indicated oxygen transfer from one side of the ring to form the (+)-(4S,5R) enantiomer.[54] Presumably, both stereoselectivity and regioselectivity are determined by the constraints on the orientation of the substrate in the active site.

B. Multiplicity and Carcinogen Metabolism

The importance of cytochrome P-450 regioselectivity in carcinogen metabolism is indicated by the many positions of oxidative attack on benzo[*a*]pyrene, aflatoxin B_1, and 2-AAF (Fig. 5). The role of individual forms of cytochrome P-450 in the activation of polycyclic hydrocarbons is emphasized by the effect of induction on the regioselectivity for the carcinogen activation pathway (Table I). Phenobarbital-inducible forms of rat liver microsomal P-450 combine to metabolize benzo[*a*]pyrene at lower specific activity than methylcholanthrene-inducible forms and with lower regioselectivity for the precarcinogenic 7,8-bond.[55] Although 7,12-dimethylbenzanthracene is more rapidly metabolized by forms induced by

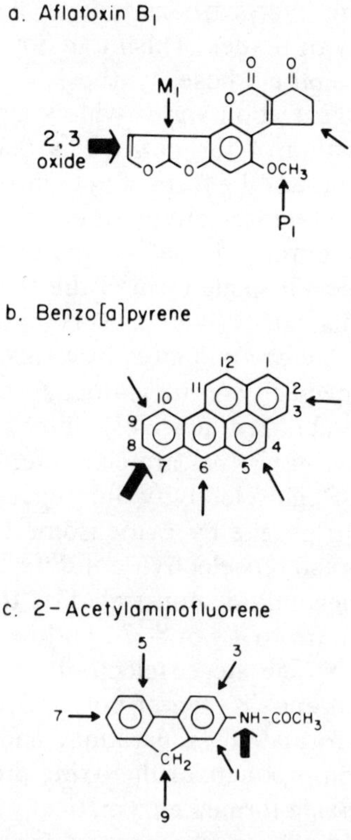

a. Aflatoxin B_1

b. Benzo[*a*]pyrene

c. 2-Acetylaminofluorene

Fig. 5. Positions of oxidation of three carcinogens at cytochrome P-450. The heavy arrow indicates the position for formation of the carcinogen precursor.

2. Integration of Xenobiotic Metabolism

TABLE I

Activation of Polycyclic Hydrocarbons in the Carcinogenic Pathway by Rat Liver Microsomes by Various Inducers[a]

Carcinogen	Inducer	Percentage of metabolism to carcinogen precursor	Total metabolism (nmol/ min/mg)	Rate of carcinogen activation (nmol/min $\times 10^2$)	Reference
Benzo[a]pyrene	Control	10	0.6	6	55
	PB	4.5	1.0	4.5	
	MC	12	3.6	43	
7,12-Dimethyl-	Control	3	1.0	3	56
benz[a]anthracene	PB	5	2.4	12	
	MC	0.2	3.2	0.6	
Benz[a]anthracene	MC	1.5	12	18	58
Dibenz[a,h]anthracene	MC	24	2.5	60	62
Chrysene	MC	26	2.4	63	61

[a] Carcinogen activation occurs at the following bonds: benzo[a]pyrene (7,8); 7,12-dimethylbenz[a]-anthracene; benzo[a]anthracene and dibenz[a,h]anthracene (3,4); and chrysene (1,2).

methylcholanthrene, 7-hydroxymethyl 12-methylbenzanthracene is preferentially metabolized by phenobarbital-induced forms.[56] Different forms of cytochrome P-450 may preferentially catalyze the secondary activation of the precursor dihydrodiol. Thus, rabbit P-450$_{LM4}$, which is induced in rabbit liver by polycyclic hydrocarbons, is far more effective in the metabolism of BP-7,8-dihydrodiol compared to benzo[a]pyrene.[57] Benzo[a]anthracene (BA) is only weakly carcinogenic, even though BA-3,4-dihydrodiol exhibits the typical high carcinogenic activity of a bay-region dihydrodiol precursor.[58] The regioselectivity of the forms of cytochrome P-450 that metabolize polycyclic hydrocarbons is such that very little 3,4-dihydrodiol is formed.[59] However, although 7,12-dimethyl BA is far more tumorigenic, this is apparently not due to the effect of the methyl groups on the regioselectivity of cytochrome P-450, which remains low for the 3,4-position.[56] Introduction of a bismethylene bridge and a 3-methyl group into BA to form methylcholanthrene again provides potent carcinogenic activity. However, in this case, prior 1-hydroxylation (a major position for methylcholanthrene metabolism) is necessary for providing the appropriate active site orientation on cytochrome P-450 for significant generation of the bay-region precursor epoxide.[60] Chrysene and dibenzanthracene, in contrast, exhibit high proportions of epoxidation to the carcinogen precursor.[61,62]

Regioselectivity of aflatoxin B_1 metabolism is also affected by induction of different forms of cytochrome P-450. Methylcholanthrene-induced mouse liver preferentially 4-hydroxylates aflatoxin B_1 (~90%) to a noncarcinogenic metabolite, whereas for phenobarbital-induced forms, regioselectivity changes with a predominance of carcinogenic B_1-2,3-oxygenase activity (60%).[63] The overall rate of aflatoxin B_1 metabolism is similar for both phenobarbital- and methylcholanthrene-induced microsomes.

Major differences have been reported in the regioselectivity of metabolism of 2-AAF by purified forms of cytochrome P-450 from rabbit liver microsomes.[64] P-450_{LM4} predominantly produces the carcinogenic precursor N-hydroxyAAF, whereas forms 3 and 6 effect exclusively ring hydroxylation. Form 2 does not metabolize 2-AAF. The prevalence of the various forms depends on the mode of induction. Forms 3 and 4 are major forms in uninduced rabbit liver, form 2 is the major form after phenobarbital induction, and form 4 is the major form in the adult rabbit liver after induction by TCDD or polycyclic hydrocarbons. Form 6 is the major form after TCDD induction in the neonate. The effect of induction by polycyclic hydrocarbons on 2-AAF carcinogenesis is species specific; fewer liver tumors are found in the rat compared to the hamster even though N-hydroxylation is elevated in both.[65] However, regioselectivity is shifted toward N-hydroxylation (activation) in the hamster and toward ring hydroxylation (detoxication) in the rat.

A different aspect of cytochrome P-450 heterogeneity is shown by dimethylnitrosamine demethylation. The reaction is characterized in rat liver microsomes by both high and low K_m values. The rate is elevated severalfold by induction with pyrazole concomitant with an increase in one form of cytochrome P-450 and a decrease in constitutive cytochrome P-450 isoenzymes.[66] This contrasts to the decrease in activity by inducers such as phenobarbital and methylcholanthrene. A single low K_m value is then observed in the microsomal dimethylnitrosamine metabolism, consistent with the involvement of this single form.

C. Alternative Modes of Activation

Regioselectivity may also be affected by the mechanism of activation of cytochrome P-450. Evidence suggests that the active species during NADPH-supported monooxygenation is $[FeO^{3+}]$.[67] However, cytochrome P-450 can also support monooxygenation by direct reaction of peroxides with ferric–cytochrome P-450[68] and by generation of hydroxyl radicals.[69]

Electron transport to cytochrome P-450 can be accelerated and partially uncoupled by many xenobiotics, resulting in reduction of O_2 to O_2^- and formation of hydrogen peroxide and hydroxyl radicals.[70] Clearly, regioselectivity and substrate selectivity are significantly dependent on the nature of the oxidizing species.

Peroxidative monooxygenation may also be catalyzed by hemoproteins other than cytochrome P-450. For example, prostaglandin synthetase, with addition of archidonic acid, mediates the oxidation of benzo[a]pyrene to quinones[71] and the activation of BP-7,8-dihydrodiol to BP-7,8-dihydrodiol 9,10-oxides.[72] This reaction occurs through a hydroperoxide-mediated autoxidation of the hydrocarbon and indeed can be catalyzed by other peroxidases. For example, the binding of benzidine to nucleic acids is activated by endoperoxide-mediated metabolism at prostaglandin synthetase.[73] The binding of N-OH-2-AAF to DNA is alternatively stimulated by a peroxidase-mediated disproportionation mechanism.[74] This peroxidase activation of xenobiotics may therefore be important in tissues that have low levels of cytochrome P-450, particularly when prostaglandin synthetase or other peroxidases are present at high levels. Addition of arachidonic acid to microsomes from several target organs for carcinogenesis stimulates BP-7,8-dihydrodiol oxidation, and in guinea pig and human lung microsomes, this rate is comparable to the rate of NADPH-supported monooxygenation.[72] This type of reaction now seems to mediate the oxidation of a wide range of xenobiotics.

Cytochrome b_5 can transfer the second electron of the monooxygenase cycle, leading to a synergistic stimulation of activity by NADH.[75] Investigation of the metabolism of many substrates by different purified forms of microsomal cytochrome P-450 indicates that involvement of cytochrome b_5 depends on both the substrate and the form of cytochrome P-450.[76] In particular, it requires that second-electron transfer is rate limiting in the monooxygenase cycle. Metabolism of nitroanisole by rabbit cytochrome P-450$_{LM3}$ exhibits a near absolute requirement for cytochrome b_5,[77] whereas many P-450–mediated reactions appear independent of this pathway. Even though it is possible for NADH/cytochrome b_5 synergism to discriminate between forms of cytochrome P-450, there have been no reports of NADH-induced changes in regioselectivity.

Cytochrome P-450 may act both as an oxygenase and as a reductant for some substrates, resulting in product ratios that depend on O_2 concentrations. This is particularly true of polyhalogen compounds such as halothane[78] and chloroform.[27] Oxidative metabolism of chloroform probably occurs via oxygen insertion and subsequent elimination of HCl to yield phosgene, which is a potent electrophile and potential source of toxicity or carcinogenicity.[27] Reductive metabolism generates either CCl_3, a radical derived by one-electron reduction, or CCl_2, a carbene derived by two-electron reduction. These intermediates exhibit different chemical reactivity. The metabolism of halothane probably proceeds similarly.[78] Significantly, there is evidence to suggest that for halothane, toxicity is more closely associated with reductive metabolism. The products from reductive metabolism of halothane are greatly perturbed by the additional involvement of cytochrome b_5 in second-electron transfer.[78] The ratio of 2-chloro-1,1,1-trifluoroethane to 2-chloro-1,1-difluoroethylene is decreased more than twofold by the addition of cytochrome b_5 to microsomes, whereas antibody against cytochrome b_5 increases the ratio.

D. Selective Inhibition

Microsomal regioselectivity can be redirected by inhibitors that bind preferentially to one or more forms of cytochrome P-450. Conversely, selective inhibitors have been used in cellular or microsomal metabolism of xenobiotics to assess the contribution of different forms of cytochrome P-450. The types of inhibitors have been comprehensively reviewed.[79] Two different types have commonly been used to probe function: (1) ligands that form reversible complexes with oxidized cytochrome P-450, for example, bases such as metyrapone or octylamine, and polycyclic compounds such as α- and β-naphthoflavone, and (2) compounds that require $NADPH/O_2$-dependent metabolic activation to form essentially irreversible complexes with cytochrome P-450. The two most common types of molecules within this category are amines, including compounds related to SKF 525A, amphetamines, and arylamines, and methylene dioxybenzene derivatives such as isosafrole and piperonyl butoxide.

The potential impact of such complex formation on regioselectivity and xenobiotic metabolism is demonstrated by the effect of piperonyl butoxide on 2-AAF metabolism in mice. Ring hydroxylation is inhibited, and 10 times more metabolism is diverted to N-hydroxylation (carcinogen activation pathway).[80]

Spectrophotometric studies of reversible complexes with oxidized microsomal cytochrome P-450 clearly indicate overlapping interactions with the multiple forms.[81] Consequently, the effects of inhibitors on metabolic rates and regioselectivity are difficult to interpret. For example, in uninduced rat microsomes, 1 mM metyrapone selectivity inhibits aflatoxin B_1 metabolism at the different positions of metabolism: 9-hydroxylation (92%) > 2,3-oxygenation (72%) >> 4-hydroxylation (no effect).[82] After both phenobarbital and methylcholanthrene induction, inhibition of 4-hydroxylation increases drastically to over 50%, strongly suggesting that different forms of cytochrome P-450 are involved in the metabolism at this position after induction. The effects of flavones are also complex.

2. Integration of Xenobiotic Metabolism

Broadly speaking, reactions in control and phenobarbital-induced rat liver microsomes are weakly inhibited and sometimes activated by flavones, whereas reactions in microsomes induced by polycyclic hydrocarbons are generally potently inhibited.[83] The effects are species specific and depend critically on the form of cytochrome P-450. Metabolism of benzo[a]pyrene by rabbit P-450$_{LM6}$ is potently inhibited by flavones and metabolism by P-450$_{LM4}$ and P-450$_{LM3c}$ is greatly enhanced, whereas the net effect on P-450$_{LM2}$ and P-450$_{LM3b}$ is dose dependent.

Several mechanisms can alter rates of metabolism and regioselectivity by decreasing the concentration of specific forms of P-450. Indeed, induction, although increasing some forms, may decrease others. Activation of heme oxygenase decreases cytochrome P-450 by decreasing the cellular level of heme. Enzyme activity is greatly elevated by cobalt salts[84] and by treatment with morphine.[85] Certain forms are likely to be more sensitive than others to such decreases in the heme pool.

Metabolism of three classes of xenobiotics generates reactive intermediates that covalently modify and inactivate the enzyme. Polyhalogenated compounds generate free radicals, carbenes, or acid chlorides,[27,78] sulfur compounds containing C=S or P=S release atomic sulfur during monooxygenation,[86] and olefins or acetylenes are converted to reactive intermediates that alkylate heme nitrogen.[87] In each case, the most susceptible target is frequently the form of cytochrome P-450 that catalyzes production of the active metabolite.

E. Cross-Inhibition between Pathways

Microsomal oxidation of xenobiotics may be limited by product inhibition.[88] Investigations of the metabolism of benzo[a]pyrene by methylcholanthrene-induced rat liver microsomes indicate that cross-inhibitory interaction may be dominant in determining reaction rates.[42] Stimulation of glucuronidation by addition of UDPGA increases by fourfold the rate of formation of BP-anti-7,8-dihydrodiol 9,10-oxide and DNA adducts derived from this oxide. Investigation of the individual steps in activation has shown that the secondary monooxygenation of 7,8-dihydrodiol to the dihydrodiol oxide is 90–95% inhibited by the combined effects of the primary substrate benzo[a]pyrene and benzo[a]pyrene quinones. The quinones are removed by reductive glucuronidation, thereby reducing their contribution to the inhibition. However, agents that increase benzo[a]pyrenequinone levels in hepatocytes do not decrease benzo[a]pyrene metabolism in these cells,[89] probably because the quinones are tightly sequestered by polynucleotides and by cytosolic proteins.[90] An examination of the metabolism of BA-3,4-dihydrodiol has shown that polycyclic quinones may also be formed by secondary cytochrome P-450–dependent oxidation of dihydrodiols.[91]

This effect of UDPGA on benzo[a]pyrene metabolism may be enhanced by an additional cross-inhibitory effect. Benzo[a]pyrene phenols are far better substrates of UDPglucuronyltransferase and of aryl sulfotransferase[92] than are the dihydrodiols[93] and inhibit the diversionary detoxication of 7,8-dihydrodiol. These effects apply generally to activation of polycyclic hydrocarbons, in which the quantitatively minor carcinogen activation pathway may be highly susceptible to cross-inhibitory effects.

VII. EPOXIDE HYDROLASE

Epoxide hydrolase activity is found in the cytosolic, microsomal, and nuclear fractions of cells.[94] The cytosolic enzyme exhibits different substrate specificity and presumably is a distinct protein.[95] Most notably, styrene oxide, which is used to assay the microsomal enzyme, is not a substrate for the soluble enzyme. However, if the epoxide is β to the aromatic ring, as in allylbenzene epoxide, rapid hydrolysis is effected by the cytosolic hydrolase.[95] Among other preferred substrates are polyunsaturated fatty acid and terpene epoxides. The ratio of microsomal and cytosolic enzymes is species specific, and indeed the very low cytosolic activity in the rat has proved unrepresentative.[95]

The properties of the microsomal enzyme have been reviewed previously in this series.[94] The microsomal activity is highest in liver; 2- to 3-fold lower in intestines (for rabbit and guinea pig; very low in the rat); 5- to 15-fold lower in the kidney; and 25- to 35-fold lower in the lung.[96] Liver activity is increased by stilbene oxide, phenobarbital, and pregnenolone carbonitrile[96] but not by such agonists of the Ah receptor as polycyclic hydrocarbons and TCDD.[94] The microsomal activity can be directly activated by both isoquinolines (norharman and ellipticine) and small aromatic ketones (metyrapone and benzil), although activation is found only with smaller substrates.[97,98]

The relationship of epoxide hydrolase to cytochrome P-450 and the competition with reaction at glutathione S-transferase and with nonenzymatic hydrolysis are particularly critical to xenobiotic metabolism. Competition from nonenzymatic epoxide reactions is shown in two different ways by the reactions of the primary and secondary oxide metabolites in the benzo[a]pyrene-activation pathway. The primary oxide metabolites of benzo[a]pyrene (2,3-, 4,5-, 7,8-, and 9,10-) all rearrange to phenols (3-, 5-, 7-, and 9-) but at very different rates (2,3- $>>$ 9,10- $>$ 7,8- $>>$ 4,5-). The demands on the epoxide hydrolase increase with the instability of the oxide. Thus, no dihydrodiol can be detected at the 2,3-position, and the level of hydrolase required to compete with the nonenzymatic rearrangement also increases with instability.[55] Two practical implications of the reaction kinetics can be noted. First, the ratio of dihydrodiol to phenol produced in microsomal metabolism of benzo[a]pyrene increases in the order, 9,10- > 7,8- > 4,5-. Second, when the activity of epoxide hydrolase is lowered by the addition of epoxide hydrolase inhibitors such as trichloropropylene epoxide, the sensitivity of dihydrodiol formation decreases in the same order: 9,10- > 7,8- > 4,5-.[99] Although each of these oxides is a good substrate for epoxide hydrolase, stereospecificity is dependent on the substrate. For example, racemic 7,8-oxide is hydrolyzed without stereospecificity, whereas 4,5-oxide almost exclusively forms (−)-4,5-dihydrodiol.[100]

The ratio of monooxygenase and epoxide hydrolase activities may also have a critical bearing on a metabolic pathway. For example, in mouse liver, microsomal epoxide hydrolase activity is relatively low, and the ratio of dihydrodiols to phenols generated from benzo[a]pyrene is also low. In lung microsomes, the absolute hydrolase activity is lower than in the liver, but the hydrolase/monooxygenase activity ratio is 70-fold higher.[101] As a consequence, benzo[a]pyrene is converted to a much higher proportion of dihydrodiols in the mouse lung.

A direct interaction between epoxide hydrolase and two forms of cytochrome P-450 has been revealed by a small type I spectral perturbation.[102] The ratio of dihydrodiol to phenol formed by monooxygenation of 2,3-dichlorobiphenyl at fixed concentrations of epoxide hydrolase and cytochrome P-450 depends on the forms of the cytochrome used in the reconstitution. Less effective formation of dihydrodiol is associated with a weaker hydrolase–cytochrome P-450 complex. Dissociation of a lipophilic aryl oxide from cytochrome P-450 into an aqueous environment is likely to be relatively slow. Consequently, direct transfer from one hydrophobic site to another may provide a facilitated or coupled pathway. Metabolism of benzo[a]pyrene in mouse embryo fibroblast microsomes exhibits a higher turnover number at cytochrome P-450 than in mouse liver and produces a high ratio of BP-9,10-dihydrodiol/BP-9-phenol even though the levels of the monooxygenase, the reductase, and epoxide hydrolase are 10- to 20-fold lower than in liver.[103] These enzymes may be located within a specific domain of the endoplasmic reticulum that facilitates their interaction.

The stereochemistry of the hydration reaction may also be indicative of epoxide hydrolase involvement. Hydration of (+)-BP-anti-7,8-dihydrodiol 9,10-oxide formed by monooxygenation of (−)-BP-7,8-dihydrodiol results in a different ratio of tetrol stereoisomers depending on the presence or absence of epoxide hydrolase.[38] An acceleration of oxide breakdown by epoxide hydrolase has been demonstrated by a decreased modification of added DNA that in effect acts as a trap for the intermediate oxide.

VIII. ABSOLUTE MONOOXYGENASE ACTIVITY IN THE ACTIVATION PATHWAY

For cells *in vitro* where metabolism occurs in a closed system, the primary selectivity between activation and detoxication at an initial monooxygenase step is a more critical determinant of DNA modification than the absolute rate of monooxygenation. For example, modification of DNA in cultured mouse embryo 10 $T_{1/2}$ cells by a transforming dose of benzo[*a*]pyrene is not increased when cytochrome *P*-450 is induced by two- to threefold (A. L. Shen, G. Gehly, C. R. Jefcoate, and C. Heidelberger, unpublished work). However, in most cases the total monooxygenase activity may be important for one of four reasons. (1) Monooxygenase activation and flux through the activation pathway must compete with redistributive transfer of either the starting compound or reaction intermediates away from the primary site of exposure. (2) The rate of modification of DNA may be comparable to the rate of excision repair of adducts. (3) At higher rates of primary monooxygenation, subsequent enzymes in the pathway, particularly the sulfotransferases, may become relatively less effective either through enzyme saturation or through depletion of the levels of limiting coreactants. (4) The xenobiotic monooxygenase system may be involved in secondary activation in competition with conjugation reactions, as has been described previously for metabolism of polycyclic hydrocarbons.

IX. CELLULAR COSUBSTRATES OF XENOBIOTIC METABOLISM

The balance between activation and detoxication depends critically on the balance between the activities of the various enzymes of metabolism. In cells, this involves not only total enzyme levels but also the concentrations of the coreactants. The xenobiotic monooxygenase system, that is, the cytochrome *P*-450 system, depends primarily on NADPH and to a lesser extent on NADH, whereas the three principal transferases require UDPGA, 3'-phosphoadenosine 5'-phosphosulfate (PAPS), and GSH, respectively. The generation of these coreactants depends to a varying extent on the cellular level of ATP and on the redox status of the cell. These

2. Integration of Xenobiotic Metabolism

pathways do not function exclusively or even principally in xenobiotic metabolism, so that depletion of these coreactants or other molecules in these pathways may critically affect other cell functions. For example, NADPH is not only required for monooxygenation but also determines the GSH/GSSG ratio. ATP is a determinant of the formation of UTP (the UDPGA precursor), is the immediate precursor of PAPS, and is required for the biosynthesis of GSH. A high ATP/ADP ratio may also limit the flow of reducing equivalents to NADPH through the glycolytic pathway by inhibition of phosphofructokinase.[104] High NADH may benefit certain monooxygenase reactions through reduction of cytochrome b_5 but inhibits production of UDPGA, which requires NAD for the rate-limiting oxidation of UDPglucose.

A. Regulation of NADPH

NADPH is generated in cells from the dehydrogenases of the pentose–phosphate pathway, from isocitrate dehydrogenase, and from the malate enzyme (Fig. 6). In hepatocytes prepared under most conditions, there is sufficient NADPH to fulfill the requirements of xenobiotic metabolism. Thus, in hepatocytes from uninduced starved rats, glucose and lactate, which increase the availability of substrates for NADPH generation, do not stimulate microsomal monooxygenase activity. However, in phenobarbital-induced starved animals, the cellular monooxygenase activity is increased twofold by administration of glucose or lactate.[105] Although these findings may be explained in part by increased utilization of

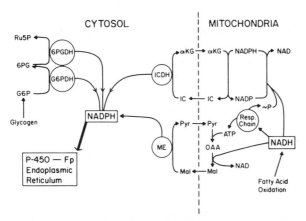

Fig. 6. Pathways participating in the generation of NADPH. G6PDH = glucose-6-phosphate dehydrogenase; 6PGDH = 6-phosphogluconate dehydrogenase; ICDH = isocitrate dehydrogenase; ME = malic enzyme.

NADPH through the xenobiotic monooxygenase system, phenobarbital further increases the levels of glucose-6-phosphate dehydrogenase and the malic enzyme.[104] Stilbene oxide also induces synthesis of glucose-6-phosphate dehydrogenase and 6-phosphogluconate dehydrogenase, whereas the effects of many other inducers remain to be studied.[106] At the other end of the activity scale, metabolism of polycyclic hydrocarbons in mouse embryo fibroblast cells occurs at the same rate as metabolism by a homogenate of these cells with added NADPH.[103] Again, NADPH generation in cells is not limiting in xenobiotic metabolism.

During metabolism of 4-nitroanisole, very different changes are produced in the concentrations of intermediary metabolites of control and phenobarbital-induced perfused rat liver. In control livers, nitroanisole metabolism halves the fructose 1,6-bisphosphate level, whereas other metabolites remain unchanged. In phenobarbital-induced livers, nitroanisole metabolism increases 6-phosphogluconate fivefold and also increases other glycolytic and pentose intermediates. These increases in phenobarbital-induced livers are consistent with the increased activity particularly of glucose-6-phosphate dehydrogenase resulting from the increased supply of NADPH to cytochrome P-450 required by xenobiotic metabolism.[104] Under conditions of lower demand for NADPH in uninduced liver, it has been reported that either the malic enzyme[105] or isocitrate dehydrogenase[107] is the preferred source of NADPH.[107] The preferred source of NADPH may also depend on the substrate. For example, in phenobarbital-induced rat livers, aminopyrine decreases the NADPH/NADP ratio at the malic enzyme, whereas p-nitroanisole produces no change.[104]

Respiratory chain inhibitors such as rotenone, antimycin A, or KCN each decrease xenobiotic oxidation in hepatocytes from starved animals (three- to fivefold), with relatively little effect on hepatocytes from fed animals. This reaction is probably derived from a decrease in the activity of mitochondrial pyruvate decarboxylase through diminished availability of ATP.[105] As a consequence, the generation of NADPH through the malic enzyme is impaired. However, this pathway is less significant in the supply of NADPH for fed animals. Indeed, the reverse effect of respiratory uncouplers may be observed in hepatocytes from phenobarbital-treated fed rats in which uncoupling of respiration increases NADPH, probably through a decrease in the inhibition of phosphofructokinase by ATP.[105]

The ratio of NADPH/NADP is an important determinant of the capability of the cell to turn over cytochrome P-450. The redox potential for the microsomal enzyme is -300 mV in the unbound form and increases to -220 mV when complexed by a substrate.[104a] As a consequence, reduction is unfavorable in the absence of substrate and even then requires a

2. Integration of Xenobiotic Metabolism

high ratio of NADPH/NADP. This ratio can be determined from the substrate/product ratio and the corresponding equilibrium constant for each of the enzymes that contribute to the formation of NADPH. The NADPH/NADP couple in equilibrium with each enzyme varies but is generally in the range of 10 to 100. For example, the NADPH/NADP ratio in the control rat liver is 30 for the malic enzyme and 80–90 for 6-phosphogluconate dehydrogenase.[104] 4-Nitroanisole metabolism does not perturb these values. Phenobarbital induction results in a large decrease in the NADPH/NADP ratio for 6-phosphogluconate dehydrogenase to 25. However, substrate/product ratios give only an approximate indication of NADPH/NADP ratios during xenobiotic metabolism. One problem is that this metabolism may decrease the cellular ATP, which directly modulates the activities of several of these enzymes.[104]

The ratio of NADH/NAD that determines the reduction of cytochrome b_5 has been calculated from the substrate/product ratio for lactate dehydrogenase to be 10^{-3}. The redox potential of cytochrome b_5 and the potential for donation of the second electron to cytochrome P-450 during monooxygenation are both far more positive than for first-electron reduction of cytochrome P-450. As a consequence, the low level of NADH can reduce cytochrome b_5. Low concentrations of ethanol increase aminopyrine and nitroanisole demethylation in phenobarbital-induced fasted rats comcomitant with increased reduction of NAD. This stimulation has been attributed to enhanced second-electron donation to cytochrome P-450.[108]

Substrate binding to cytochrome P-450 substantially increases the redox potential and accelerates the rate of reduction of the cytochrome. For many substrates (ethylmorphine and aminopyrine), the monooxygenation reaction is partially uncoupled, with the result that about half of the electron flux is diverted to the reduction of O_2, with production of O_2^- and H_2O_2.[109] This generation of reactive oxidants may lead to lipid peroxidation, even though the cell is protected against this by GSH (see the following discussion). This pool of peroxides may, however, also contribute to monooxygenation by a peroxidase mechanism (see the preceding discussion).

B. UDPGA

The synthesis of UDPGA is more sensitive to the nutritional and redox status of the cell than is the level of NADPH. This is revealed by the sensitivity of hepatic UDPGA levels and xenobiotic glucuronidation to nutritional deprivation and to the effects of ethanol or uncoupling agents. Nitrophenol conjugation is not sensitive to the nutritional status in control rat livers, but refeeding in phenobarbital-induced animals increases glucuronidation relative to fasted animals.[110] The nutritional sensitivity effected by phenobarbital induction reflects the elevated levels of UDPglucuronyltransferase and also of UDPglucose dehydrogenase, the rate-limiting enzyme in the production of UDPGA. UDPglucose dehydrogenase is also induced about twofold by stilbene oxide.[111] Interestingly, the utilization of NADPH for metabolism of p-nitroanisole directly perturbs glycogenolysis in fasted-refed, phenobarbital-induced livers, resulting in double the level of UDPglucose.[110] In fasted animals, a 30% depletion is observed. Ethanol greatly decreases glucuronidation by increasing the NADH/NAD ratio, with consequent inhibition of UDPglucose oxidation.[112] The ratio of NADH to NAD increases when O_2 declines below 30 μM, suggesting that glucuronidation may be sensitive to hypoxia.[113] At lower O_2 concentrations, a decline in ATP formation is also likely to decrease levels of UTP and consequently UDPglucose. However, UDPglucose levels do not necessarily directly determine formation of UDPGA. For example, D-glucosamine lowers liver UDPglucose but raises UDPGA.[114] Differential requirements for ATP in hepatocytes are indicated by the insensitivity of monooxygenation to respiratory inhibitors such as rotenone and 2,4-DNP, as compared to near elimination of sulfation and glucuronidation.

C. 3'-Phosphoadenosine 5'-Phosphosulfate

The hepatocyte level of 3'-phosphoadenosine 5'-phosphosulfate (PAPS) is very sensitive to depletion of both ATP and inorganic sulfate (see also Chapter 11, this volume). The apparent K_m for ATP in the synthesis in PAPS (1.6 mM) at ATP-sulfurylase (sulfate adenyltransferase) is about 10-fold higher than the ATP-requiring steps in the synthesis of UDPGA and GSH.[113] When ATP levels are decreased by inhibition of oxidative phosphorylation, sulfation of 4-hydroxybiphenyl[115] decreases far more than glucuronidation. Similar differences in sensitivity may also occur when ATP levels are perturbed by hypoxia or even by xenobiotic monooxygenation.

Sulfation of acetaminophen and harmol in rat hepatocytes[116] is negligible in the absence of a source of sulfate in the medium. Na_2SO_4 at 10 mM achieves near saturation of the sulfation reaction without significant effect on glucuronidation. Cysteine and methionine provide only 15 and 5%, respectively, of the direct sulfate-stimulated sulfation at equivalent concentrations (5 mM). In the isolated perfused liver, serum $^{35}SO_4$ rapidly equilibrates with the hepatic PAPS pool.[117] Again, very little sulfation of harmol occurs in the absence of sulfate in the perfusion medium. Interestingly, 13 mM sulfate increases sulfation from 7.5 to 19.5% while decreasing glucuronide formation from 68 to 40%. These experiments suggest an apparent K_m for sulfate at the sulfurylase enzyme of approximately 0.4 mM, comparable to the sulfate concentration in rat serum of 0.8 mM. This is consistent with the sensitivity of sulfate to changes in serum inorganic sulfate.

D. Glutathione

Glutathione not only provides the last line of defense against the actions of reactive electrophiles but also protects cells against the action of lipid peroxides (see Chapter 7, this volume). In addition, GSH may play a crucial role in cellular processes. For example, GSH is a cofactor for ribonucleotide reductase, which is required for synthesis of deoxyribonucleotides, and may facilitate amino acid transport through the γ-glutamyl cycle.[118] Indeed, when animals are given moderate doses of amino acids, turnover of the γ-glutamyl cycle depletes GSH from renal cells. GSH is synthesized by the successive actions of γ-glutamyl cysteine synthetase and glutathione synthetases, each of which utilizes ATP.[118] Because the K_m for ATP for both enzymes is low, GSH synthesis is relatively insensitive to changes in ATP.[113]

The ratio of GSH to GSSG is maintained at a normal level of >20 both by equilibration with NADPH/NADP at glutathione reductase and by efflux of GSSG from the cell during oxidation of GSH.[119] This enzyme is induced by methylcholanthrene (24%), phenobarbital (80%), and stilbene oxide (260%).[120] A Se-dependent glutathione peroxidase plays an important role in preventing the accumulation of H_2O_2 and lipid peroxides (Chapter 7). This enzyme is decreased 25-fold in Se-deficient animals, although with very little change in the hepatic GSH level.[121] Oxidative metabolism of acetaminophen, aminopyrine, and ethylmorphine depletes GSH both through reaction with reactive metabolites and through substrate-stimulated peroxide production at cytochrome P-450 (see the preceding discussion).[109] There is a much larger decline in GSH in Se-dependent animals concomitant with dramatically increased lipid peroxidation.[121]

Cells may adapt to chemical or nutritional depletion of GSH by producing a rebound elevation of GSH levels. For example, dimethylazoaminobenzene and other azo dyes cause a two- to threefold increase in liver GSH within 2 to 3 days.[122]

It seems likely that GSH depletion with ensuing lipid peroxidation is a significant contributor to drug-induced toxicity. Oxidation of GSH also results in the oxidation of mitochondrial NADPH with a concomitant release of Ca^{2+} from the mitochondria.[123] The ensuing increase in cytosolic Ca^{2+} may play an important role in drug toxicity.

X. UDP-GLUCURONYLTRANSFERASES AND SULFOTRANSFERASES

The characteristics of enzymes belonging to these classes have been reviewed in detail.[124,125] Multiple forms of both enzymes differ in substrate specificity, tissue distribution, and induction. UDPglucuronyltransferases are located in the endoplasmic reticulum, whereas the alcohol and aryl sulfotransferases are cytosolic enzymes. The regulation of the two groups is also very different. UDPglucuronyltransferases are induced by methylcholanthrene (or other Ah-receptor agonists), phenobarbital, and stilbene oxide. The aryl sulfotransferases have not been induced chemically but are very sensitive to hormonal modulation. For example, hepatic sulfation of N-OH-AAF is diminished by administration of estradiol to gonadectomized rats but is increased by testosterone.[35] This reaction is also stimulated by thyroid hormone in parallel with an increase in N-OH-AAF–initiated hepatocarcinogenesis.[35] The hormonal sensitivity of sulfation is not surprising because both steroids and thyroid hormones are substrates for the enzyme.[124]

Two classes of microsomal UDPglucuronyltransferase have been distinguished. One type, A, increases in the latefetal period and is induced threefold by methylcholanthrene in the adult rat; the second type, B, increases in the neonatal period and is induced twofold by phenobarbital.[126] The Class A enzyme(s) conjugates mostly phenols, and the Class B enzyme(s) conjugates steroids and bulky alcohols such as morphine. The relative activities for different substrates may also be very sensitive to the activating effect of membrane fluidization,[127] to changes in lipid environment,[128] and to the regulatory influences of UDP-N-acetylglucosamine.[129] The differences in regulation indicate that the ratio of UDPglucuronyltransferase to sulfotransferase activities may be readily varied within a specific cell type and will also change substantially between different types of cells.

Four different classes of sulfotransferase utilize PAPS as the sulfate donor: aryl, hydroxysteroid (primary and secondary alcohols), estrone, and bile acid. Xenobiotic metabolism primarily concerns the first two classes. Four types of aryl sulfotransferase are known, but only one (type IV) has been shown to activate hydroxamic acids such as N-OH-AAF.[125]

The differences in substrate specificity between positional isomers are very apparent for polycyclic hydrocarbon metabolites (Table II). It should

TABLE II

Sulfation and Glucuronidation of Benzo[a]pyrene Metabolism[a]

BP derivatives	Sulfate formation (nmol/mg protein/min)	Glucuronide formation (nmol/mg protein/min)
1-OH	0.48	0.99
2	0.29	0.32
3	0.59	1.06
4	0.45	0.32
6	0.85	0.40
7	0.46	1.18
8	0.66	0.23
9	0.47	1.21
10	0.08	1.19
12	0.86	0.67
trans-4,5-Diol	0.25	0.33
trans-7,8-Diol	0.03	0.08
trans-9,10-Diol	0.07	0

[a] Sulfate formation was measured on rat liver cytosol with a PAPS generating system,[92] and glucuronide formation was measured with rat liver microsomes with added UDPGA (1.5 mM).[93] Compiled from Nemoto et al.[92,93]

be noted that dihydrodiols, including the carcinogen precursor BP-7,8-dihydrodiol, are poorer substrates than phenols for both UDPglucuronyltransferase and aryl sulfotransferases.[92,93]

Phenol sulfation typically is associated with a lower K_m than is found for phenol glucuronidation. As the dose of drug increases, sulfation tends to saturate earlier, with a subsequent rise in the glucuronide/sulfate ratio.[130] This effect may be compounded by depletion of the cellular reserve of PAPS, which is more likely to have limited availability than UDPGA. When sulfation functions as a detoxication step, premature saturation at higher doses of carcinogen relative to the carcinogen activation pathway may increase the proportion of initial carcinogen ultimately bound to DNA. The variability of the ratio of glucuronyltransferase to sulfotransferase has been shown for BP-7,8-dihydrodiol. In cultured liver cells, BP-7,8-dihydrodiol is converted to a sulfate at fivefold the rate of glucuronidation, whereas glucuronidation is threefold faster in the lung.[16]

Quinones, which inhibit xenobiotic monooxygenation and stimulate oxygen radical generation and lipid peroxidation, are in part detoxified by both glucuronidation and sulfation.[92,93] The actual substrate is either the semiquinone or hydroquinone, thus requiring an intermediate reduction step that can be catalyzed either by a cytosolic menadione-sensitive quinone reductase[131] or by electron transfer from the cytochrome P-450 reductase.[132] The cytosolic enzyme is induced by both Ah-receptor agonists and 2(3)-*tert*-butyl-4-hydroxyanisole (BHA).[131] The great effectiveness of the integrated reduction–conjugation process in hepatocytes is demonstrated by the nearly complete removal of quinones during hepatocyte metabolism of benzo[*a*]pyrene. Administration of salicylamide, an inhibitor of both glucuronidation and sulfate, increases quinone levels in hepatocytes by three- to fourfold.[90]

XI. GLUTATHIONE S-TRANSFERASES

The characteristics of the several species of cytosolic glutathione S-transferase have been reviewed in detail.[133] Transferases B and C are present in liver cytosol at high concentration. The K_m for GSH is about two orders of magnitude below typical hepatic levels (10 mM), implying that even with very extensive depletion these enzymes should remain saturated with GSH. However, surprisingly, even a 50% depletion of GSH greatly limits cellular transferase activity.[90] As with other enzymes of xenobiotic metabolism, there are generally lower levels of glutathione S-transferases in extrahepatic tissues. Many of the arene oxides that are substrates for glutathione S-transferases are both lipophilic and rapidly hydrolyzed by water. The characteristic lipophilic binding sites and the high concentrations of the transferases combine to facilitate their effectiveness with arene oxide intermediates. Again, we find large differences in the reactivity of structural isomers. For benzo[*a*]pyrene oxides, the rates of GSH conjugation (4,5- > 7,8- > 9,10-) are in the reverse order from the competing hydrolysis rates.[134]

Many labile and lipophilic oxide intermediates are generated by cytochromes P-450 within the membranes of the endoplasmic reticulum. These intermediates are probably preferentially retained within the membrane, where solvolysis may be slower. An as yet uninducible form of glutathione S-transferase is a major component of liver microsomes.[135] This protein is much smaller than the cytosolic forms (MW 14,000) and comprises 2.5–3.0% of the microsomal protein.[136] This enzyme is also remarkable in that SH modification, including formation of oxidized —S—S— derivatives, increases the activity up to eightfold.

Glutathione S-transferases are induced in rat liver by methylcholanthrene, phenobarbital, and stilbene oxide. The most effective of these is stilbene oxide, which is also the least effective inducer for total cytochrome P-450.[137] The principal cytosolic forms, A, B, and C, are induced equally and by more than 3-fold. Phenobarbital induces form B by 2-fold. None of these inducers will increase microsomal glutathione S-

transferases. BHA is the most effective inducer of cytosolic transferase activity in the mouse but is less effective in rat.[138] The tissue specificity of the induction effect is demonstrated by two isomers of BHA. 2- and 3-BHA induce hepatic transferase by 2-fold and 5.5-fold, respectively, whereas the reverse inductive effect is observed in the forestomach (4-fold and 30% increases).[139] The importance of glutathione S-transferase in protection against carcinogenesis is suggested by the protective effect of BHA against many chemical tumor initiators.[140]

The effectiveness of glutathione S-transferase in preventing DNA modification by polycyclic hydrocarbon metabolites is determined in part by the competition with epoxide hydrolase for precursor polycyclic oxides. Our earlier considerations indicate that the ratio of these activities may be an important determinant of the diversionary detoxication. The ratio of epoxide hydrolase to cytosolic glutathione S-transferase activities has been determined for BP-4,5-oxide to be 0.3 and 0.5 in lung and liver, respectively.[141,142] However, this ratio is very dependent on the specific substrate and, indeed, is much larger for 7,8-oxide, a comparable substrate for epoxide hydrolase but a poor substrate for glutathione S-transferase; microsomal activities were not measured. Induction may greatly affect this ratio. For example, in the rat liver *trans*-stilbene oxide, 2(3)-BHA and 2-AAF induce epoxide hydrolase to a larger extent than glutathione S-transferases.[143] This of course in part reflects the initial high level of total cell protein in the glutathione S-transferase class (4% after induction in liver).

XII. INHIBITORS OF CELLULAR TRANSFERASE ACTIVITY

The availability of selective perturbants of each of the three transferase reactions provides useful probes for determining the contribution of each class of enzyme to cellular xenobiotic metabolism.

D-Galactosamine interferes with the synthesis of UDPglucuronic acid by removing uridine nucleotides as UDPgalactosamine, which then inhibits UDPglucose dehydrogenase.[144] A 30-min preincubation of hepatocytes with 2 mM D-galactosamine decreases glucuronidation of both β-naphthol[116] and acetaminophen[116] to about 20% of the initial value without affecting other transferase reactions.

Pentachlorophenol is a potent inhibitor of aryl sulfotransferases but not of UDPglucuronyltransferases.[145] Inhibition by pentachlorophenol is nontoxic and is effective at low doses. The conjugation of N-OH-AAF in perfused rat liver is switched from unidentified water-soluble products to glucuronides by the same extent whether sulfate depletion or administration of pentachlorophenol is used. These studies indicate that 20% of administered N-OH-AAF is conjugated as sulfate. Pentachlorophenol and sulfate depletion each decreases the total covalent binding of N-OH-AAF to DNA in perfused rat liver by about 1.7- to 3-fold. It is likely that much of the residual binding is due to the acetoxy N-OH-AAF.

The key steps in arylamine-induced bladder carcinogenesis have been previously discussed. Pentachlorophenol administration prior to 2-naphthylamine *in vivo* causes a two- to threefold increase in the urinary levels of both N-OH-2-naphthylamine N-glucuronide and N-OH-2-naphthylamine.[146] This consequence of decreased sulfation establishes that sulfation and glucuronidation compete for the intermediate N-OH-2-naphthylamine in the liver.

Salicylamide is an inhibitor of both sulfation and glucuronidation but is a far more potent inhibitor of sulfation than of glucuronidation. Sulfation of harmol generated from harmine in hepatocytes is 90% inhibited by 0.5 mM salicylamide, with a fourfold rise in glucuronide formation. At 2 mM salicylamide, glucuronidation is only 50% inhibited but remains double the initial level.[116]

After essentially complete metabolism of benzo[a]pyrene by hepatocytes from methylcholanthrene-induced rat liver, over half of the hydrocarbon is converted equally to glucuronides or sulfates. However, 70% of the glucuronides are formed from either 3-phenol or 3,6-quinone, whereas 70% of the sulfates are formed from 3,6-quinone, 1,6-quinone, or 9,10-dihydrodiol. Salicylamide elevates both dihydrodiol and phenol levels in addition to the quinones.[90] The fourfold stimulation of dihydrodiols is surprising in view of their slow conjugation in subcellular fractions.

Reactions with GSH have been probed by pretreatment of cells with either sulfoximes or diethyl maleate. The former inhibits synthesis of GSH[147]; the latter reacts rapidly with GSH by means of the glutathione S-transferases.[148] By either method, about 80% of the glutathione in the hepatocyte can be depleted. Diethyl maleate is less effective in depleting GSH in cells such as kidney, where glutathione S-transferase levels are lower.

The relative modification of protein (cysteine, methionine, lysine, histidine, and tyrosine), polynucleotide bases, and GSH depends on the reactivity and characteristics of the reactive electrophile. These considerations, together with the extent of transferase involvement, also determine the protective effectiveness of GSH. More reactive S_N1-type electrophiles are less selective and may react rapidly with GSH without enzymatic involvement. For less reactive species, specific target binding, as can be provided by DNA intercalation, may facilitate macromolecule modification and thus diminish the relative effectiveness of GSH detoxica-

2. Integration of Xenobiotic Metabolism

tion. The effectiveness of GSH detoxication is indicated by the effect of GSH depletion on DNA modification. For example, depletion with diethyl maleate enhances modification of DNA by both BP-7,8-dihydrodiol 9,10-oxide (2-fold) and by BP-9-phenol 4,5-oxide (3.5-fold).[90] This difference in GSH intervention is consistent with the apparently greater reactivity of the phenol–oxide and the probable involvement of intercalation in facilitating diol–epoxide attack on DNA.

GSH may also be depleted by administration of diethyl maleate to rats *in vivo* prior to isolation of hepatocytes.[148] Bromobenzene metabolism, which produces a substantial proportion of 3,4-oxide in hepatocytes, results in the diversion of 12% of total metabolism to protein adducts after depletion of GSH. This is decreased to 2.8% when N-acetylcysteine is added to stimulate additional GSH synthesis.[148] In contrast, diethyl maleate treatment of hepatocytes increased covalent binding of 2-AAF metabolites to protein by only 15–25%.[149] This suggests that N-OH-AAF sulfate reacts slowly with GSH and is a poor substrate for glutathione S-transferases. A different picture is seen for metabolism of 1,2-dibromo ethane in hepatocytes, in which diethyl maleate decreases covalent binding of the drug to DNA. This and other evidence indicates that a GSH conjugate is a reactive electrophile.[150]

XIII. XENOBIOTIC METABOLISM IN THE NUCLEAR ENVELOPE

The nuclear envelope also contains significant levels of the xenobiotic-metabolizing enzymes of the endoplasmic reticulum. Although the levels of these enzymes are substantially lower,[151] the notable difference between the enzymes of the nuclear envelope and those of the endoplasmic reticulum is that phenobarbital fails to produce more than marginal increases in nuclear envelope that is free of contamination from endoplasmic reticulum. Methylcholanthrene, in contrast, induces in both membranes. Xenobiotic metabolism in the nuclear envelope, even after induction, contributes little to total cell metabolism but may contribute relatively more to the modification of nuclear DNA. However, DNA modification resulting from metabolism in isolated nuclei provides little evidence that this contribution is an important means of cellular activation.[152]

XIV. TRANSPORT OF REACTIVE INTERMEDIATES AND PRECURSORS

Little is known about the extent to which reactive intermediates in one organ may be transported to a second organ or cell type (see Chapter 8, this volume). The most clearly documented example is provided by bromobenzene toxicity. The sites of major toxicity and covalent binding, in addition to the liver, are the kidney and lung. In spite of this, toxicity and covalent binding are both enhanced by phenobarbital, which stimulates metabolism in the liver but not in the kidney and lung.[153] Apparently, the reactive intermediate, considered to be bromobenzene 3,4-epoxide, is generated in the liver and transferred to the target tissues.

Activation of polycyclic hydrocarbons to dihydrodiol epoxides requires at least two monooxygenation steps and, in the case of methylcholanthrene, three steps. It seems possible that precursor dihydrodiols generated in the liver may be transported to the lung for further monooxygenation. Certainly, polycyclic dihydrodiols should escape the liver as readily as bromobenzene metabolites. Benzo[a]pyrene and metabolites are readily bound by serum lipoproteins.[154] This may provide an important mechanism for removing unconjugated metabolites from the liver, particularly for dihydrodiols, which are poor substrates for UDPglucuronyltransferases and aryl sulfotransferases.

XV. COMMENTS

An attempt has been made to present a logical examination of the contributions of specific enzymes to detoxication and carcinogen (or toxicity) activation. The approach can be presented as a sequence of steps:

1. Identify the major DNA adducts and the ultimate electrophile(s) leading to formation of these adducts
2. Establish the activation pathway for formation of the electrophilic intermediate(s)
3. Determine the regio/stereospecificity of monooxygenase reactions with respect to the activation and detoxication pathways
4. Determine the relative activities of enzymes involved in either activation or diversionary detoxication of precursors, and calculate the metabolic partition factor between activation and detoxication for each precursor of an ultimate electrophile
5. Measure the effect of changes in the concentration of GSH on the formation of DNA adducts, hydrolysis products, and GSH conjugates from the ultimate electrophile, and calculate the glutathione detoxication factor. Compare the contributions of each detoxication factor (S, D, or G) to the overall fractional modification of DNA when particular cells are exposed to a level of xenobiotic that is fully metabolized

6. Evaluate factors that will modulate the enzyme activities by enzyme induction or inhibition or through depletion of key cosubstrates, particularly glutathione UDPGA and PAPS

The partition model that has been discussed in this chapter should allow prediction of the effect of a change in a specific enzyme activity on the balance between xenobiotic detoxication and activation, as quantitated by the extent of DNA modification. The same model can be used to predict changes from one cell type to another on the basis of changes in monooxygenase selectivity and in the levels of various enzymes and cofactors. Conversely, this approach can be used as a starting point in understanding why DNA modification after exposure to a given xenobiotic varies substantially with cell type.

REFERENCES

1. Ts'o, P. O. P. (1979). Basic mechanism in chemical carcinogenesis. In "Chromatin Structure and Function" (C. A. Nicolini, ed.), Part B, pp. 751–769. Plenum, New York.
2. King, H. W. S., Thompson, M. H., and Brookes, P. (1975). The benzo[a]pyrene deoxyribonucleoside products isolated from DNA after metabolism of benzo[a]pyrene by rat liver microsomes in the presence of DNA. Cancer Res. **34**, 1263–1269.
3. Meehan, T., Straub, K., and Calvin, M. (1977). Benzo[a]pyrene diol epoxide covalently binds to deoxyguanosine and deoxyadenosine in DNA. Nature (London) **269**, 725–727.
4. Osborne, M. R., Jacobs, S., Harvey, R. G., and Brookes, P. (1981). Minor products from the reaction of (+) and (−)-benzo[a]pyrene-antidiolepoxide with DNA. Carcinogenesis **2**, 553–558.
5. Lawley, P. D. (1976). Carcinogenesis by alkylating agents. ACS Monogr. **173**, 83–244.
6. Dock, L., Undeman, O., Gräslund, A., and Jernström, B. (1978). Fluorescence study of DNA-complexes formed after metabolic activation of benzo[a]pyrene derivatives. Biochem. Biophys. Res. Commun. **85**, 1275–1282.
7. Dipple, A., Tomaszewski, J. E., Moschel, R. C., Bigger, C. A. H., Nebzydoski, J. A., and Egan, M. (1979). Comparison of metabolism-mediated binding to DNA of 7-hydroxymethyl-12-methylbenzo[a]anthracene and 7,12-dimethylbenz[a]anthracene. Cancer Res. **39**, 1154–1158.
8. Poirier, M. C., Santella, R., Weinstein, I. B., Grunberger, D., and Yuspa, S. H. (1980). Quantitation of benzo[a]pyrene-deoxyguanosine adducts by radioimmunoassay. Cancer Res. **40**, 412–416.
9. Poirier, M. C., Williams, G. M., and Yuspa, S. H. (1980). Effect of culture conditions, cell type, and species of origin on the distribution of acetylated and deacetylated deoxyguanosine C-8 adducts of N-acetoxy-2-acetylaminofluorene. Mol. Pharmacol. **18**, 581–587.
10. Nelson, S. D., Forte, A. J., Vaishnav, Y., Mitchell, J. R., Gillette, J. R., and Hinson, J. A. (1981). The formation of arylating and alkylating metabolites of phenacetin in hamsters and hamster liver microsomes. Mol. Pharmacol. **19**, 140–145.

11. Grossman, L. (1981). Enzymes involved in the repair of damaged DNA. *Arch. Biochem. Biophys.* **211**, 511–522.
12. Singer, B. (1975). The chemical effects of nucleic acid alkylation and their relation to mutagenesis and carcinogenesis. *Prog. Nucleic Acid Res. Mol. Biol.* **15**, 219–284.
13. Cohen, G. M., Bracken, W. M., Iyer, R. P., Berry, D. L., Selkirk, J. K., and Slaga, T. J. (1979). Anticarcinogenic effects of 2,3,7,8-tetrachlorodibenzo-p-dioxin on benzo[a]pyrene and 7,12-dimethylbenz[a]anthracene tumor initiation and its relationship to DNA binding. *Cancer Res.* **39**, 4027–4033.
14. Eastman, A., and Bresnick, E. (1979). Persistent binding of 3-methylcholanthrene to mouse lung DNA and its correlation with susceptibility to pulmonary neoplasia. *Cancer Res.* **39**, 2400–2405.
15. Kadlubar, F. F., Anson, J. F., Dooley, K. L., and Beland, F. A. (1981). Formation of urothelial and hepatic DNA adducts from the carcinogen 2-naphthylamine. *Carcinogenesis* **2**, 467–470.
16. Langenbach, R., Nesnow, S., Tompa, A., Gingell, R., and Kuszynski, C. (1981). Lung and liver cell-mediated metagenesis systems: Specificities in the activation of chemical carcinogens. *Carcinogenesis* **2**, 851–858.
17. Kadlubar, F. F., Miller, J. A., and Miller, E. C. (1978). Guanyl O^6-arylamination and O^6-arylation of DNA by the carcinogen N-hydroxy-1-naphthylamine. *Cancer Res.* **38**, 3628–3638.
18. Boyd, M. R., and Burka, L. T. (1978). *In vivo* studies on the relationship between target organ alkylation and the pulmonary toxicity of a chemically reactive metabolite of 4-ipomeanol. *J. Pharmacol. Exp. Ther.* **207**, 687–697.
19. Robertson, I. G. C., Philpot, R. M., Zeiger, E., and Wolf, C. R. (1981). Specificity of rabbit pulmonary cytochrome P-450 isozymes in the activation of several aromatic amines and aflatoxin B. *Mol. Pharmacol.* **20**, 662–668.
20. Farber, E. (1980). The sequential analysis of liver cancer induction. *Biochim. Biophys. Acta* **605**, 149–166.
21. Shirai, T., Lee, M. S., Wang, C. Y., and King, C. M. (1981). Effects of partial hepatectomy and dietary phenobarbital on liver and mammary tumorigenesis by two N-hydroxy-N-acylaminobiphenyls in female CD rats. *Cancer Res.* **41**, 2450–2456.
22. Devereux, T. R., and Fouts, J. R. (1981). Isolation of pulmonary cells and use in studies of xenobiotic metabolism. *In* "Methods in Enzymology" (W. B. Jakoby, ed.), Vol. 77, pp. 147–154. Academic Press, New York.
23. Swenson, D. H., Miller, E. C., and Miller, J. A. (1974). Aflatoxin B_1-2,3-oxide: Evidence for its formation in rat liver *in vivo* and by human liver microsomes *in vivo*. *Biochem. Biophys. Res. Commun.* **60**, 1036–1043.
24. Guengerich, F. P., Crawford, W. M., Jr., Mason, P. S., Mitchell, M. B., and Watanabe, P. G. (1980). Covalent binding of vinyl chloride metabolites *in vitro*: Roles of cytochrome P-450, NADPH-cytochrome P-450 reductase, epoxide hydratase, and alcohol dehydrogenase. *In* "Microsomes, Drug Oxidations and Chemical Carcinogenesis" (M. J. Coon and A. H. Conney, R. W. Estabrook, H. V. Gelboin, J. R. Gillette, and P. J. O'Brien, eds.), pp. 1211–1214. Academic Press, New York.
25. Jerina, D. M., Lehr, R. E., Schaefer-Ridder, M., Yagi, H., Karle, J. M., Thakker, D. R., Wood, A. W., Lu, A. Y. H., Ryan, D., West, S., Levin, W., and Conney, A. H. (1977). Bay region epoxides of dihydrodiols: A concept which explains the mutagenic and carcinogenic activity of benzo[a]pyrene and benzo[a]anthracene. *Cold Spring Harbor Conf. Cell Proliferation* **4**, 639–658.
26. Pegg, A. E. (1977). Formation and metabolism of alkylated nucleosides. Possible role in carcinogenesis by nitroso compounds and alkylating agents. *Adv. Cancer Res.* **25**, 195–269.

27. Pohl, L. R., George, J. W., Martin, J. L., and Krishna, G. (1979). Deuterium isotope effect in in vivo bioactivation of chloroform to phosgene. Biochem. Pharmacol. **28**, 561–563.
28. DeBaun, J. R., Miller, E. C., and Miller, J. A. (1970). N-Hydroxy-2-acetylaminofluorene sulfotransferase: Its probable role in carcinogenesis and in protein-(methion-S-yl) binding in rat liver. Cancer Res. **30**, 577–595.
29. Kadlubar, F. F., Miller, J. A., and Miller, E. C. (1976). Microsomal N-oxidation of the hepatocarcinogen N-methyl-4-aminoazobenzene and the reactivity of N-hydroxy-N-methyl-4-aminoazobenzene. Cancer Res. **36**, 1196–1206.
30. Ziegler, D. M. (1980). Microsomal flavin-containing monooxygenase oxygenation of nucleophilic nitrogen and sulfur compounds. In "Enzymatic Basis of Detoxication" (W. B. Jakoby, ed.), Vol. 2, pp. 201–277. Academic Press, New York.
31. King, C. M., and Olive, C. W. (1975). Comparative effects of strain, species and sex on the acyltransferase- and sulfotransferase-catalyzed activations of N-hydroxy-N-2-fluorenylacetamide. Cancer Res. **35**, 906–912.
32. DeBaun, J. R., Smith, J. Y. R., Miller, E. C., and Miller, J. A. (1970). Reactivity in vivo of the carcinogen N-hydroxy-2-acetylaminofluorene: Increase by sulfate ion. Science **167**, 184–186.
33. Vaught, J. B., McGarvey, P. B., Lee, M.-S., Garner, C. D., Wang, C. Y., Linsmaier-Bednar, E. M., and King, C. M. (1981). Activation of N-hydroxyphenacetin to mutagenic and nucleic acid-binding metabolites by acyltransfer, deacylation, and sulfate conjugation. Cancer Res. **41**, 3424–3429.
34. Kadlubar, F. F., Miller, J. A., and Miller, E. C. (1976). Hepatic metabolism of N-hydroxy-N-methyl-4-aminoazobenzene and other N-hydroxyarylamines to reactive sulfuric acid esters. Cancer Res. **36**, 2350–2359.
35. Miller, J. A. (1970). Carcinogenesis by chemicals: An overview. Cancer Res. **30**, 559–576.
36. Borchert, P., Miller, J. A., Miller, E. C., and Shires, T. K. (1973). 1'-Hydroxysafrole, a proximate carcinogenic metabolite of safrole in the rat and mouse. Cancer Res. **33**, 590–600.
37. Wislocki, P. G., Borchert, P., Miller, J. A., and Miller, E. C. (1976). The metabolic activation of the carcinogen 1'-hydroxysafrole in vivo and in vitro and the electrophilic reactivities of possible ultimate carcinogens. Cancer Res. **36**, 1686–1695.
38. Gozukara, E. M., Belvedere, G., Robinson, R. C., Deutsch, J., Coon, M. J., Guengerich, F. P., and Gelboin, H. V. (1981). The effect of epoxide hydratase on benzo[a]pyrene diol epoxide hydrolysis and binding to DNA and mixed-function oxidase proteins. Mol. Pharmacol. **19**, 153–161.
39. Cooper, C. S., Hewer, A., Ribeiro, O., Grover, P. L., and Sims, P. (1980). The enzyme-catalyzed conversion of anti-benzo[a]pyrene-7,8-diol-9,10-oxide into a glutathione conjugate. Carcinogenesis **1**, 1075–1080.
40. Geacintov, N. E., Ibanez, V., Gabliano, A. G., Yoshida, H., and Harvey, R. G. (1980). Kinetics of hydrolysis of tetraols and binding of benzo[a]pyrene 7,8-dihydrodiol-9,10-oxide and its tetraol derivatives to DNA. Biochem. Biophys. Res. Commun. **92**, 1335–1342.
41. Yang, L. L., Maher, V. M., and McCormick, J. J. (1980). Error-free excision of the cytotoxic, mutagenic N^2-deoxyguanosine DNA adduct formed in human fibroblasts by ($\pm$)-7β,8α-dihydroxy-9α,10α-epoxy-7,8,9,10-tetrahydrobenzo[a]pyrene. Proc. Natl. Acad. Sci. U.S.A. **77**, 5933–5937.
42. Keller, G. M., and Jefcoate, C. R. (1983). Modulation of microsomal benzo[a]pyrene metabolism by DNA. Mol. Pharmacol. (in press).
43. Ryan, D. E., Thomas, P. E., Korzeniowski, D., and Levin, W. (1979). Separation and

characterization of highly purified forms of liver microsomal cytochrome P-450 from rats treated with polychlorinated biphenyls, phenobarbital, and 3-methylcholanthrene. *J. Biol. Chem.* **254,** 1365–1374.

44. Koop, D. R., Persson, A. V., and Coon, M. J. (1981). Properties of electrophoretically homogeneous constitutive forms of liver microsomal cytochrome P-450. *J. Biol. Chem.* **256,** 10704–10711.
45. Aoyama, T., Imai, Y., and Sato, R. (1982). Multiple forms of cytochrome P-450 from liver microsomes of drug-untreated rabbits: Purification and characterization. *In* "Microsomes, Drug Oxidations, and Drug Toxicity" (R. Sato and R. Kato, eds.), pp. 83–84. Wiley-Interscience, New York.
46. Lu, A. Y. H., and West, S. B. (1980). Multiplicity of mammalian microsomal cytochromes P-450. *Pharmacol. Rev.* **31,** 277–295.
47. Guengerich, F. P., Wang, P., Mason, P. S., and Mitchell, M. B. (1981). Immunological comparison of rat, rabbit, and human microsomal cytochromes P-450. *Biochemistry* **20,** 2370–2378.
47a. Dieter, H. H., and Johnson, E. F. (1982). Functional and structural polymorphism of rabbit microsomal cytochrome P-450 form 3b. *J. Biol. Chem.* **257,** 9315–9323.
48. Poland, A., Mak, I., and Glover, E. (1981). Species differences in responsiveness to 1,4-*bis*[2-(3,5-dichloropyridyloxy)]-benzene, a potent phenobarbital-like inducer of microsomal monooxygenase activity. *Mol. Pharmacol.* **20,** 442–450.
49. Nebert, D. W., Eisen, H. J., Negishi, M., Lang, M. A., and Hjelmeland, L. M. (1981). Genetic mechanisms controlling the induction of polysubstrate monooxygenase (P-450) activities. *Annu. Rev. Pharmacol. Toxicol.* **21,** 431–462.
50. Carlstedt-Duke, J. M. B. (1979). Tissue distribution of the receptor for 2,3,7,8-tetrachlorodibenzo-p-dioxin in the rat. *Cancer Res.* **39,** 3172–3176.
51. Gehly, E. B., Fahl, W. E., Jefcoate, C. R., and Heidelberger, C. (1979). The metabolism of benzo[a]pyrene by cytochrome P-450 in transformable and nontransformable C3H mouse fibroblasts. *J. Biol. Chem.* **254,** 5041–5048.
52. Wiebel, F. J., Selkirk, J. K., Gelboin, H.V., Haugen, D. A., van der Hoeven, T. A., and Coon, M. J. (1975). Position-specific oxygenation of benzo[a]pyrene by different forms of purified cytochrome P-450 from rabbit liver. *Proc. Natl. Acad. Sci. U.S.A.* **72,** 3917–3920.
53. Kaminsky, L. S., Fasco, M. J., and Guengerich, F. P. (1980). Comparison of different forms of purified cytochrome P-450 from rat liver by immunological inhibition of regio- and stereoselective metabolism of warfarin. *J. Biol. Chem.* **255,** 85–91.
54. Armstrong, R. N., Levin, W., Ryan, D. E., Thomas, P. E., Mah, H. D., and Jerina, D. M. (1981). Stereoselectivity of rat liver cytochrome P-450$_c$ on formation of benzo[a]pyrene 4,5-oxide. *Biochem. Biophys. Res. Commun.* **100,** 1077–1084.
55. Holder, G., Yagi, H., Dansette, P., Jerina, D. M., Levin, W., Lu, A. Y. H., and Conney, A. H. (1974). Effects of inducers and epoxide hydrase on the metabolism of benzo[a]pyrene by liver microsomes and reconstituted system: Analysis of high pressure liquid chromatography. *Proc. Natl. Acad. Sci. U.S.A.* **71,** 4356–4360.
56. Chou, M. W., Yang, S. K., Sydor, W., and Yang, C. S. (1981). Metabolism of 7,12-dimethylbenz[a]anthracene and 7-hydroxymethyl-12-methylbenz[a]anthracene by rat liver nuclei and microsomes. *Cancer Res.* **41,** 1559–1564.
57. Deutsch, J., Leutz, J. C., Yang, S. K., Gelboin, H. V., Chiang, Y. L., Vatsis, K. P., and Coon, M. J. (1978). Regio- and stereoselectivity of various forms of purified cytochrome P-450 in the metabolism of benzo[a]pyrene and (−)trans-7,8-dihydroxy-7,8-dihydrobenzo[a]pyrene as shown by product formation and binding to DNA. *Proc. Natl. Acad. Sci. U.S.A.* **75,** 3123–3127.

58. Wislocki, P. G., Kapitulnik, J., Levin, W., Lehr, R., Schaefer-Ridder, M., Karle, J. M., Jerina, D. M., and Conney, A. H. (1978). Exceptional carcinogenic activity of benz[a]anthracene 3,4-dihydrodiol in the newborn mouse and the bay region theory. *Cancer Res.* **38,** 693–696.
59. Thakker, D. R., Levin, W., Yagi, H., Ryan, D., Thomas, P. E., Karle, J. M., Lehr, R. E., Jerina, D. M., and Conney, A. H. (1979). Metabolism of benz[a]anthracene to its tumorigenic 3,4-dihydrodiol. *Mol. Pharmacol.* **15,** 138–153.
60. Wood, A. W., Chang, R. L., Levin, W., Thomas, P. E., Ryan, D., Stoming, T. A., Thakker, D. R., Jerina, D. M., and Conney, A. H. (1978). Metabolic activation of 3-methylcholanthrene and its metabolites to products metagenic to bacterial and mammalian cells. *Cancer Res.* **38,** 3398–3404.
61. Nordqvist, M., Thakker, D. R., Vyas, K. P., Yagi, H., Levin, W., Ryan, D. E., Thomas, P. E., Conney, A. H., and Jerina, D. M. (1981). Metabolism of chrysene and phenanthrene to bay-region diol epoxides by rat liver enzymes. *Mol. Pharmacol.* **19,** 168–178.
62. Nordqvist, M., Thakker, D. R., Levin, W., Yagi, H., Ryan, D. E., Thomas, P. E., Conney, A. H., and Jerina, D. M. (1979). The highly tumorigenic 3,4-dihydrodiol is a principal metabolite formed from dibenzo[a,h]anthracene by liver enzymes. *Mol. Pharmacol.* **16,** 643–655.
63. Metcalfe, S. A., Colley, P. J., and Neal, G. E. (1981). A comparison of the effects of pretreatment with phenobarbitone and 3-methylcholanthrene on the metabolism of aflatoxin B_1 by rat liver microsomes and isolated hepatocytes *in vitro*. *Chem.-Biol. Interact.* **35,** 145–157.
64. Johnson, E. F., Levitt, D. S., Müller-Eberhard, U., and Thorgeirrson, S. S. (1980). Catalysis of divergent pathways of 2-acetylaminofluorene metabolism by multiple forms of cytochrome *P*-450. *Cancer Res.* **40,** 4456–4459.
65. Enomoto, M., Miyake, M., and Sato, K. (1968). Carcinogenicity in the hamster of simultaneously administered 2-acetylaminofluorene and 3-methylcholanthrene. *Gann* **59,** 177–186.
66. Tu, Y. Y., Sonnenberg, J., Lewis, K. F., and Yang, C. S. (1981). Pyrazole-induced cytochrome *P*-450 in rat liver microsomes: An isozyme with high affinity for dimethylnitrosamine. *Biochem. Biophys. Res. Commun.* **103,** 905–912.
67. White, R. E., and Coon, M. J. (1980). Oxygen activation by cytochrome *P*-450. *Annu. Rev. Biochem.* **49,** 315–356.
68. Blake, R. C., II, and Coon, M. J. (1981). On the mechanism of action of cytochrome *P*-450: Evaluation of homolytic and heterolytic mechanisms of oxygen-oxygen bound cleavage during substrate hydroxylation by peroxidases. *J. Biol. Chem.* **256,** 12127–12133.
69. Ingelman-Sundberg, M., and Johansson, I. (1981). The mechanism of cytochrome *P*450-dependent oxidation of ethanol in reconstituted membrane vesicles. *J. Biol. Chem.* **256,** 6321–6326.
70. Nordblom, G. D., and Coon, M. J. (1977). Hydrogen peroxide formation and stoichiometry of hydroxylation reactions catalyzed by highly purified liver microsomal cytochrome *P*-450. *Arch. Biochem. Biophys.* **180,** 343–347.
71. Marnett, L. J., Reed, G. A., and Johnson, J. T. (1977). Prostaglandin synthetase dependent benzo[a]pyrene oxidation: Products of the oxidation and inhibition of their formation by antioxidants. *Biochem. Biophys. Res. Commun.* **79,** 569–576.
72. Sivarajah, K., Lasker, J. M., and Eling, T. E. (1981). Prostaglandin synthetase-dependent co-oxidation of (±)-benzo[a]pyrene 7,8-dihydrodiol by human lung and other mammalian tissues. *Cancer Res.* **14,** 1834–1839.

73. Zenser, T. V., Mattammal, M. B., Armbrecht, H. J., and Davis, B. B. (1980). Benzidine binding to nucleic acids mediated by the peroxidative activity of prostaglandin endoperoxide synthetase. *Cancer Res.* **40,** 2839–2845.
74. Bartsch, H., and Hecker, E. (1971). On the metabolic activation of N-hydroxy-N-2-acetylaminofluorene. III. Oxidation with horseradish peroxidase to yield 2-nitrofluorene and N-acetoxy-N-2-acetylaminofluorene. *Biochim. Biophys. Acta* **237,** 567–578.
75. Correia, M. A., and Mannering, G. J. (1973). Reduced diphosphopyridine nucleotide synergism of the reduced triphosphopyridine nucleotide-dependent mixed-function oxidase system of hepatic microsomes. II. Role of the type I drug-binding site of cytochrome P-450. *Mol. Pharmacol.* **9,** 470–485.
76. Brunström, A., and Ingelman-Sundberg, M. (1980). Benzo[a]pyrene metabolism by purified forms of rabbit liver microsomal cytochrome P-450, cytochrome b_5 and epoxide hydrase in reconstituted phospholipid vesicles. *Biochem. Biophys. Res. Commun.* **95,** 431–439.
77. Sugiyama, T., Miki, N., and Yamano, T. (1979). The obligatory requirement of cytochrome b_5 in the p-nitroanisole O-demethylation reaction catalyzed by cytochrome P-450 with a high affinity for cytochrome b_5. *Biochem. Biophys. Res. Commun.* **90,** 715–720.
78. Nastainczyk, W., Ahr, H. J., and Ullrich, V. (1982). The reductive metabolism of halogenated alkanes by liver microsomal cytochrome P-450. *Biochem. Pharmacol.* **31,** 391–396.
79. Testa, B., and Jenner, P. (1981). Inhibitors of cytochrome P-450s and their mechanism of action. *Drug Metab. Rev.* **12,** 1–117.
80. Friedman, M. A., and Woods, S. (1977). Effects of piperonyl butoxide on hydroxylation of 2-acetylaminofluorene in mouse liver. *Res. Commun. Chem. Pathol. Pharmacol.* **17,** 623–630.
81. Grasdalen, H., Bäckström, D., Eriksson, L. E. G., Ehrenberg, A., Moldeus, P., von Bahr, C., and Orrenius, S. (1975). Heterogeneity of cytochrome P-450 in rat liver microsomes: Selective interaction of metyrapone and SKF 525-A with different fractions of microsomal cytochrome P-450. *FEBS Lett.* **60,** 294–299.
82. Gurtoo, H. L., and Dahms, R. P. (1979). Effects of inducers and inhibitors on the metabolism of aflatoxin B_1 by rat and mouse. *Biochem. Pharmacol.* **28,** 3441–3449.
83. Huang, M.-T., Johnson, E. F., Müller-Eberhard, U., Koop, D. R., Coon, M. J., and Conney, A. H. (1981). Specificity in the activation and inhibition by flavonoids of benzo[a]pyrene hydroxylation by cytochrome P-450 isozymes from rabbit liver microsomes. *J. Biol. Chem.* **256,** 10897–10901.
84. Sardana, M. K., Sassa, S., and Kappas, A. (1980). Adrenalectomy enhances the induction of heme oxygenase and the degradation of cytochrome P-450 in liver. *J. Biol. Chem.* **255,** 11320–11323.
85. Gurantz, D., and Correia, M. A. (1981). Morphine-mediated effects on rat hepatic heme and cytochrome P-450 *in vivo*. *Biochem. Pharmacol.* **30,** 1529–1536.
86. Kamataki, T., and Neal, R. A. (1976). Metabolism of diethyl-p-nitrophenyl phosphorothionate (parathion) by a reconstituted mixed function oxidase system: Studies of covalent binding of the sulfur atom. *Mol. Pharmacol.* **12,** 933–944.
87. Ortiz de Montellano, P. R., Kunze, K. L., and Mico, B. A. (1980). Destruction of cytochrome P-450 by olefins: N-Alkylation of prosthetic heme. *Mol. Pharmacol.* **18,** 602–605.
88. Soda, D. M., and Levy, G. (1975). Inhibition of drug metabolism by hydroxylated metabolites: Cross-inhibition and specificity. *J. Pharma. Sci.* **64,** 1928–1931.

2. Integration of Xenobiotic Metabolism

89. Burke, M. D., Vadi, H., Jernström, B., and Orrenius, S. (1977). Metabolism of benzo[a]pyrene with isolated hepatocytes and the formation and degradation of DNA-binding derivatives. *J. Biol. Chem.* **252**, 6424–6431.
90. Shen, A. L., Fahl, W. E., and Jefcoate, C. R. (1980). Metabolism of benzo[a]pyrene by isolated hepatocytes and factors affecting covalent binding of benzo[a]pyrene metabolites to DNA in hepatocyte and microsomal systems. *Arch. Biochem. Biophys.* **204**, 511–523.
91. Thakker, D. R., Levin, W., Yagi, H., Tada, M., Ryan, D. E., Thomas, P. E., Conney, A. H., and Jerina, D. M. (1982). Stereoselective metabolism of the (+)- and (−)-enantiomers of trans-3,4-dihydroxy-3,4-dihydrobenz[a]anthracene by rat liver microsomes and by a purified and reconstituted cytochrome *P*-450 system. *J. Biol. Chem.* **257**, 5103–5110.
92. Nemoto, N., Takayama, S., and Gelboin, H. V. (1978). Sulfate conjugation of benzo[a]pyrene metabolites and derivatives. *Chem.-Biol. Interact.* **23**, 19–30.
93. Nemoto, N., and Gelboin, H. V. (1976). Enzymatic conjugation of benzo[a]pyrene oxides, phenols and dihydrodiols with UDPglucuronic acid. *Biochem. Pharmacol.* **25**, 1221–1226.
94. Oesch, F. (1980). Microsomal epoxide hydrolase. *In* "Enzymatic Basis of Detoxication" (W. B. Jakoby, ed.), Vol. 2, pp. 277–290. Academic Press, New York.
95. Ota, K., and Hammock, B. D. (1980). Cytosolic and microsomal epoxide hydrolases: Differential properties in mammalian liver. *Science* **207**, 1479–1481.
96. James, M. O., Fouts, J. R., and Bend, J. R. (1976). Hepatic and extrahepatic metabolism, *in vitro*, of an epoxide (8-^{14}C-styrene oxide) in the rabbit. *Biochem. Pharmacol.* **25**, 187–193.
97. Vaz, A. D., Fiorica, V. M., and Griffin, M. J. (1981). New heterocyclic stimulators of hepatic epoxide hydrolase. *Biochem. Pharmacol.* **30**, 651–656.
98. Seidegard, J. and DePierre, J. W. (1980). Benzil, a potent activator of microsomal epoxide hydrolase *in vitro*. *Eur. J. Biochem.* **112**, 643–648.
99. Fahl, W. E., Nesnow, S., and Jefcoate, C. R. (1977). Microsomal metabolism of benzo[a]pyrene: Multiple effects of epoxide hydratase inhibitors. *Arch. Biochem. Biophys.* **181**, 649–664.
100. Lu, A. Y. H., and Miwa, G. T. (1980). Molecular properties and biological functions of microsomal epoxide hydrase. *Annu. Rev. Pharmacol. Toxicol.* **20**, 513–531.
101. Seifried, H. E., Birkett, D. J., Levin, W., Lu, A. Y. H., Conney, A. H., and Jerina, D. M. (1977). Metabolism of benzo[a]pyrene: Effect of 3-methylcholanthrene pretreatment on metabolism by microsomes from lungs of genetically "responsive" and "nonresponsive" mice. *Arch. Biochem. Biophys.* **178**, 256–263.
102. Kaminsky, L. S., Kennedy, M. W., and Guengerich, F. P. (1981). Differences in the functional interaction of two purified cytochrome *P*-450 isozymes with epoxide hydrolase. *J. Biol. Chem.* **256**, 6359–6362.
103. Gehly, E. B., Fahl, W. E., Jefcoate, C. R., and Heidelberger, C. (1979). The metabolism of benzo[a]pyrene by cytochrome *P*-450 in transformable and nontransformable C3H mouse fibroblasts. *J. Biol. Chem.* **254**, 5041–5048.
104. Kauffman, F. C., Evans, R. K., and Thurman, R. G. (1977). Alterations in nicotinamide and adenine nucleotide systems during mixed-function oxidation of *p*-nitroanisole in perfused livers from normal and phenobarbital-treated rats. *Biochem. J.* **166**, 583–592.
104a. Sligar, S. G., Cinti, D. L., Gibson, G. G., and Schenkman, J. B. (1979). Spin state control of the hepatic cytochrome *P*-450 redox potential. *Biochem. Biophys. Res. Commun.* **90**, 925–932.
105. Moldeus, P., Grundin, R., Vadi, H., and Orrenius, S. (1974). A study of drug metabo-

lism linked to cytochrome P-450 in isolated rat-liver cells. *Eur. J. Biochem.* **46**, 351–360.
106. Seidegard, J., DePierre, J. W., and Ernster, L. (1980). Increased activity of the pentose pathway after administration of *trans*-stilbene oxide. *In* "Microsomes, Drug Oxidations and Chemical Carcinogenesis" (M. J. Coon, A. H. Conney, R. W. Estabrook, H. V. Gelboin, J. R. Gillette, and P. J. O'Brien, eds.), pp. 949–952. Academic Press, New York.
107. Sirica, A. E., and Pitot, H. C. (1980). Drug metabolism and effects of carcinogens in cultured hepatic cells. *Pharmacol. Rev.* **31**, 205–228.
108. Reinke, L. A., Kauffman, F. C., and Thurman, R. G. (1979). Stimulation of p-nitroanisole O-demethylation by ethanol in perfused livers from fasted rats. *J. Pharmacol. Exp. Ther.* **211**, 133–139.
109. Jones, D. P., Thor, H., Andersson, B., and Orrenius, S. (1978). Detoxification reactions in isolated hepatocytes: Role of glutathione peroxidase, catalase, and formaldehyde dehydrogenase in reactions relating to N-demethylation by the cytochrome P-450 system. *J. Biol. Chem.* **253**, 6031–6037.
110. Thurman, R. G., Reinke, L. A., Belinsky, S. A., and Kauffman, F. C. (1980). The influence of the nutritional state on rates of p-nitroanisole O-demethylation and p-nitrophenol conjugation in perfused rat livers. *In* "Microsomes, Drug Oxidations and Chemical Carcinogenesis" (M. J. Coon, A. H. Conney, R. W. Estabrook, H. V. Gelboin, J. R. Gillette, and P. J. O'Brien, eds.), Vol. 2, pp. 913–916. Academic Press, New York.
111. Seidegard, J., DePierr, J. W., and Ernster, L. (1980). Increased activity of hepatic UDPglucose dehydrogenase after treatment of rats with *trans*-stilbene oxide. *Acta Chem. Scand., Ser. B* **B34**, 382–384.
112. Moldeus, P., Andersson, B., and Norling, A. Interaction of ethanol oxidation with glucuronidation in isolated hepatocytes. *Biochem. Pharmacol.* **27**, 2583–2588.
113. Jones, D. P. (1981). Hypoxia and drug metabolism. *Biochem. Pharmacol.* **30**, 1019–1023.
114. Dutton, G. J. (1975). Control of UDPglucuronyltransferase activity. *Biochem. Pharmacol.* **24**, 1835–1841.
115. Wiebkin, P., Parker, G. L., Fry, J. R., and Bridges, J. W. (1979). Effect of various metabolic inhibitors on biphenyl metabolism in isolated rat hepatocytes. *Biochem. Pharmacol.* **28**, 3315–3321.
116. Moldeus, P., Andersson, B., and Gergely, V. (1979). Regulation of glucuronidation and sulfate conjugation in isolated hepatocytes. *Drug Metab. Dispos.* **7**, 416–419.
117. Mulder, G. J., and Keulemans, K. (1978). Metabolism of inorganic sulphate in the isolated perfused rat liver: Effect of sulphate concentration on the rate of sulphation by phenol sulphotransferase. *Biochem. J.* **176**, 959–965.
118. Meister, A. (1981). Metabolism and functions of glutathione. *Trends Biochem. Sci.* **6**, 231–234.
119. Vina, J., Hems, R., and Krebs, H. A. (1978). Maintenance of glutathione content in isolated hepatocytes. *Biochem. J.* **170**, 627–630.
120. Carlberg, I., DePierre, J. W., and Mannervik, B. (1981). Effect of inducers of drug-metabolizing enzymes on glutathione reductase and glutathione peroxidase in rat liver. *Biochim. Biophys. Acta* **677**, 140–145.
121. Wendel, A., and Feuerstein, S. (1981). Drug-induced lipid peroxidation in mice. I. Modulation by monooxygenase activity, glutathione and selenium status. *Biochem. Pharmacol.* **30**, 2513–2520.

2. Integration of Xenobiotic Metabolism

122. Fiala, S., Mohindru, A., Kettering, W. G., Fiala, A. E., and Morris, H. P. (1976). Glutathione and gamma glutamyl transpeptidase in rat liver during chemical carcinogenesis. *JNCI, J. Natl. Cancer Inst.* **57**, 591–598.
123. Sies, H., Graf, P., and Estrela, J. M. (1981). Hepatic calcium efflux during cytochrome P-450-dependent drug oxidations at the endoplasmic reticulum in intact liver. *Proc. Natl. Acad. Sci. U.S.A.* **78**, 3358–3362.
124. Kasper, C. B., and Henton, D. (1980). Glucuronidation. *In* "Enzymatic Basis of Detoxication" (W. B. Jakoby, ed.), Vol. 2, pp. 3–36. Academic Press, New York.
125. Jakoby, W. B., Sekura, R. D., Lyon, E. S., Marcus, C. J., and Wang, J.-L. (1980). Sulfotransferases. *In* "Enzymatic Basis of Detoxication" (W. B. Jakoby, ed.), Vol. 2, pp. 199–228. Academic Press, New York.
126. Wishart, G. J. (1978). Demonstration of functional heterogeneity of hepatic uridine diphosphate glucuronosyltransferase activities after administration of 3-methylcholanthrene and phenobarbital to rats. *Biochem. J.* **174**, 671–672.
127. Eletr, S., Zakim, D., and Vessey, D. A. (1973). A spin-label study of the role of phospholipids in the regulation of membrane-bound microsomal enzymes. *J. Mol. Biol.* **78**, 351–362.
128. Vessey, D. A., and Zakin, D. (1971). Regulation of microsomal enzymes by phospholipids. II. Activation of hepatic uridine diphosphate glucuronyl transferase. *J. Biol. Chem.* **246**, 4649–4656.
129. Zakin, D., and Vessey, D. A. (1977). Regulation of microsomal UDP-glucuronyltransferase. *Biochem. Pharmacol.* **26**, 129–131.
130. Jones, D. P., Sundby, G.-B., Ormstad, K., and Orrenius, S. (1979). Use of isolated kidney cells for study of drug metabolism. *Biochem. Pharmacol.* **28**, 929–935.
131. Benson, A. M., Hunkeler, M. J., and Talalay, P. (1980). Increase of NAD(P)H : quinone reductase by dietary antioxidants: Possible role in protection against carcinogenesis and toxicity. *Proc. Natl. Acad. Sci. U.S.A.* **77**, 5216–5220.
132. Shen, A. L., Fahl, W. E., Wrighton, S. A., and Jefcoate, C. R. (1979). Inhibition of benzo[a]pyrene and benzo[a]pyrene 7,8-dihydrodiol metabolism by benzo[a]pyrene quinones. *Cancer Res.* **39**, 4123–4129.
133. Jakoby, W. B., and Habig, W. H. (1980). Glutathione transferases. *In* "Enzymatic Basis of Detoxication" (W. B. Jakoby, ed.), Vol. 2, pp. 63–94. Academic Press, New York.
134. Hayakawa, T., Udenfriend, S., Yagi, H., and Jerina, D. M. (1975). Substrates and inhibitors of hepatic glutathione S-epoxide transferase. *Arch. Biochem. Biophys.* **170**, 438–451.
135. Morgenstern, R., Meijer, J., DePierre, J. W., and Ernster, L. (1980). Characterization of rat-liver microsomal glutathione S-transferase activity. *Eur. J. Biochem.* **104**, 167–174.
136. Morgenstern, R., Guthenberg, C., and DePierre, J. W. (1982). Purification of microsomal glutathione S-transferase. *Acta Chem. Scand.* **36**, 257–259.
137. Seidegard, J., Morgenstern, R., DePierre, J. W., and Ernster, L. (1979). *Trans*-stilbene oxide: A new type of inducer of drug-metabolizing enzymes. *Biochim. Biophys. Acta* **586**, 10–21.
138. Benson, A. M., Batzinger, R. P., Ou, S.-Y. L., Bueding, E., Cha, Y.-N., and Talalay, P. (1978). Elevation of hepatic glutathione S-transferase activities and protection against mutagenic metabolites of benzo[a]pyrene by dietary antioxidants. *Cancer Res.* **38**, 4486–4495.
139. Lam, L. K. T., Sparnins, V. L., Hochalter, J. B., and Wattenberg, L. W. (1981). Effects

of 2- and 3-*tert*-butyl-4-hydroxyanisole on glutathione S-transferase and epoxide hydrolase activities and sulfhydryl levels in liver and forestomach of mice. *Cancer Res.* **41,** 3940–3943.

140. Wattenberg, L. W. (1978). Inhibition of chemical carcinogenesis. *JNCI, J. Natl. Cancer Inst.* **60,** 11–18.

141. Smith, B. R., Plummer, J. L., and Bend, J. R. (1980). Pulmonary metabolism and excretion of arene oxides. *In* "Microsomes, Drug Oxidations and Chemical Carcinogenesis" (M. J. Coon, A. H. Conney, R. W. Estabrook, H. V. Gelboin, J. R. Gillette, and P. J. O'Brien, eds.), Vol. 2, pp. 683–686. Academic Press, New York.

142. Smith, B. R., and Bend, J. R. (1979). Metabolism and excretion of benzo[*a*]pyrene 4,5-oxide by the isolated perfused rat liver. *Cancer Res.* **39,** 2051–2056.

143. DePierre, J. W., Seidegard, J., Morgenstern, R., Balk, L., Meijer, J., and Aström, A. (1981). Induction of drug-metabolizing enzymes: A status report. *In* "Mitochondria and Microsomes" (C. P. Lee, G. Schatz, and G. Dallner, eds.), pp. 585–610. Addison-Wesley, Reading, Massachusetts.

144. Schwarz, L. R. (1980). Modulation of sulfation and glucuronidation of 1-naphthol in isolated rat liver cells. *Arch. Toxicol.* **44,** 137–145.

145. Meerman, J. H. N., van Doorn, A. B. D., and Mulder, G. J. (1980). Inhibition of sulfate conjugation of *N*-hydroxy-2-acetylaminofluorene in isolated perfused rat liver and in the rat *in vivo* by pentachlorophenol and low sulfate. *Cancer Res.* **40,** 3772–3779.

146. Kadlubar, F. F., Unruh, L. E., Flammang, T. J., Sparks, D., Mitchum, R. K., and Mulder, G. J. (1981). Alteration of urinary levels of the carcinogen, *N*-hydroxy-2-naphthylamine, and its *N*-glucuronide in the rat by control of urinary pH, inhibition of metabolic sulfation, and changes in biliary excretion. *Chem.-Biol. Interact.* **33,** 129–147.

147. Thor, H., Thorold, S., and Orrenius, S. (1980). The mechanism of cytochrome *P*-450-mediated cytotoxicity studied with isolated hepatocytes exposed to bromobenzene. *In* "Microsomes, Drug Oxidations and Chemical Carcionogenesis" (M. J. Coon, A. H. Conney, R. W. Estabrook, H. V. Gelboin, J. R. Gillette, and P. J. O'Brien, eds.), Vol. 2, pp. 907–911. Academic Press, New York.

148. Thor, H., and Orrenius, S. (1980). The mechanism of bromobenzene-induced cytotoxicity studied with isolated hepatocytes. *Arch. Toxicol.* **44,** 31–43.

149. Dybing, E., Soderlund, E., Haug, L. T., and Thorgeirsson, S. S. (1979). Metabolism and activation of 2-acetylaminofluorene in isolated rat hepatocytes. *Cancer Res.* **39,** 3268–3275.

150. Fahl, W. B., Jefcoate, C. R., and Kasper, C. B. (1978). Characteristics of benzo[*a*]pyrene metabolism and cytochrome *P*-450 heterogeneity in rat liver nuclear envelope and comparison to microsomal membrane. *J. Biol. Chem.* **253,** 3106–3113.

151. Sundheimer, D. W., White, R. D., Brendel, K., and Sipes, I. G. (1982). The bioactivation of 1,2-dibromethane in rat hepatocytes: Covalent binding to nucleic acids. *Carcinogenesis* **3,** 1129–1133.

152. Pezzuto, J. M., Lea, M. A., and Yang, C. S. (1977). The role of microsomes and nuclear envelope in the metabolic activation of benzo[*a*]pyrene leading to binding with nuclear macromolecules. *Cancer Res.* **37,** 3427–3433.

153. Gillette, J. R., Mitchell, J. R., and Brodie, B. B. (1974). Biochemical mechanisms of drug toxicity. *Annu. Rev. Pharmacol.* **14,** 271–288.

154. Shu, H. P., and Nichols, A. V. (1981). Uptake of lipophilic carcinogens by plasma lipoproteins: Structure–activity studies. *Biochim. Biophys. Acta* **665,** 376–384.

CHAPTER 3

Ontogenesis

Julian E. A. Leakey

I. Introduction 77
   A. Background and Scope 77
   B. Basic Concepts of Enzyme Ontogenesis 78
   C. Experimental Approach to Enzyme Ontogenesis 81
II. Developmental Profiles of Drug-Metabolizing Enzymes 83
   A. Cytochrome P-450 and Its Monooxygenase Activities 83
   B. Other Monooxygenases 86
   C. Epoxide Hydrolase 86
   D. UDPglycuronyltransferases 87
   E. Sulfotransferases 88
   F. Glutathione S-Transferases 89
   G. N-Acetyltransferases 90
   H. Overall Trends 90
III. Factors Influencing Drug-Metabolizing Enzyme Development . . 91
   A. Hormonal Influences 91
   B. Xenobiotic Effects 94
IV. Comments: The Biological and Clinical Consequences of Drug-Metabolizing Enzyme Ontogenesis 95
   References 97

I. INTRODUCTION

A. Background and Scope

Drug metabolism in the fetus and neonate has been extensively studied since the 1950s, when it became apparent that the human fetus and infant had an increased susceptibility to certain drugs and xenobiotics: toxic effects were produced at doses harmless to an adult.[1] The fetus and neonate are physiologically very different from the adult because of their relatively sheltered environment and their primary commitment to growth and development. Such differences result in altered drug disposition and impaired excretion of drug metabolites,[2,3] but the major cause of increased drug susceptibility appears to be the immaturity or absence of the hepatic enzymes that metabolize these compounds.

These enzymes of detoxication, which metabolize many classes of xenobiotics and potentially toxic endogenous compounds, have been described in previous volumes of this series and are alluded to in Chapters 1 and 2, this volume. This chapter describes the ontogenic profiles of the drug-metabolizing enzymes, as far as they are known, and reviews our current understanding of their developmental control. For the benefit of the nonspecialist in developmental enzymology, some of the basic concepts of enzyme ontogenesis are described, along with the experimental approaches used in its study.

B. Basic Concepts of Enzyme Ontogenesis

Enzymes may be ontogenically divided into two groups: (1) those responsible for primary functions in the undifferentiated cell that are present from the earliest stages of cytodifferentiation and do not show marked changes in activity during later development and (2) those that become active at later stages of development in a given tissue when their activities become necessary for the phenotypic expression of the tissue's characteristics. The latter group do not accumulate gradually but instead exhibit rapidly increasing activities over comparatively short periods. Moreover, the critical periods at which such enzymes develop are not randomly spread throughout an organ's development. Rather, "clusters" of different enzymes begin to develop together at a specific time. This applies to many different species, including human[4] and also to different tissues. However, the enzymes associated with the different clusters may not be the same among tissues and species.

Rat liver is probably the most thoroughly studied developing organ for which Greengard[5,6] has proposed three critical periods when such clusters of enzymes begin to appear. In addition, a fourth cluster may develop at puberty, when hepatic sexual differentiation begins. The major functional types of enzymes associated with these clusters are summarized in Table I.

The enzymes developing in these clusters appear, for the most part, to complement the physiological needs of the developing organism in its changing environment. This can best be illustrated with the enzymes controlling glucose homeostasis; at birth the perinate loses its constant, ma-

TABLE I

Enzyme Clusters in Rat Liver

Cluster	Age of onset	Functional type of enzyme developing	Examples[a]
Late fetal	17 days gestation	Glycogen synthesis Urea cycle Amino acid catabolism	Glycogen synthase Ornithine carbamoyltransferase Serine dehydratase
Neonatal	Birth	Gluconeogenesis Certain detoxicating enzymes	Phosphoenolpyruvate carboxylase UDPglucuronyltransferase toward bilirubin
Late suckling	16 days postpartum	Amino acid metabolism Fatty acid synthesis	Tryptophan 2,3-dioxygenase Cytosolic malate dehydrogenase
Pubertal	30–40 days postpartum	Steroid metabolism and associated enzymes showing sex differences	Steroid 5α-reductase

[a] Taken from Greengard[5] and Gustaffson et al.[9]

ternally derived supply of glucose and has to maintain its blood glucose levels from an intermittent supply of milk, which is rich in protein and fat but relatively low in carbohydrate. Later, at weaning, the rat pup must adapt to a fluctuating, predominantly carbohydrate solid-food diet.

As shown in Table I, hepatic enzymes develop to cope with these demands. For instance, the capacity to synthesize and store glycogen develops with the late fetal cluster, so that at birth perinatal rat liver contains sufficient glycogen for maintaining glucose homeostasis for the first few hours of postnatal life.[7] By this time, the neonatally developing enzymes of gluconeogenesis have reached activities sufficient to supply adequate glucose from the gluconeogenic precursors derived from milk. Later, during "late suckling," enzymes catalyzing interconversion of amino acids and fat synthesis develop in anticipation of the weanling's new diet.

As with the induction of drug-metabolizing enzymes by xenobiotics, enzyme development involves synthesis of new enzyme protein rather than activation of existing protein, although activation processes do play a secondary role in the development of some enzymes.[8]

Unlike induction by xenobiotics, however, enzyme ontogenesis is under more complex control mechanisms than a simple adaptation to cope with the presence of a new (xenobiotic) substrate, because in many instances enzymes develop *in anticipation* of their physiological need and *before* their substrates accumulate.[5] This is a logical consequence of the nature and timing of an organism's developmental changes, which, unlike changes in its xenobiotic environment, have remained relatively constant in the genome for sufficient time for anticipatory control mechanisms to evolve.

Enzyme ontogenesis appears to be under hormonal control. The beginning of each cluster coincides with changes in secretion of one or more hormones that stimulate the development of many enzymes.[5,10] Conversely, many developing enzymes respond to more than one hormone, and although most enzymes are primarily associated with specific clusters, their overall ontogenic profiles are complex and can differ significantly, even between enzymes of the same cluster.[6] This appears to be the case for a number of reasons: first, although a cluster of enzymes may all begin to surge in activity at a specific time, their individual rates of synthesis and their half-lives may vary, thereby allowing them to reach maximum activity over a range of time; second, the enzymes need not develop from zero to maximum activities in a given cluster but often merely increase quantitatively; and third, enzymes developing in one cluster may increase or even decrease their activities in later clusters.

The responsiveness of an enzyme to a hormonal stimulus may also vary with age. This is most important to the concept of enzyme clusters, as different enzymes can become competent to respond to the same hormone at different times. This is best illustrated with the glucocorticoid hormones: glycogen synthase, which develops in the late fetal cluster in rat liver, may be precociously induced *in utero* with glucocorticoids from midgestation,[8] whereas tyrosine aminotransferase (neonatal cluster) is not competent to respond to glucocorticoids until just after birth[11] and tryptophan 2,3-dioxygenase (late-suckling cluster) is not competent until the fourth postnatal day.[12] The mechanism by which competence to respond to glucocorticoids is successively evoked for these enzymes remains unclear, although other hormones possibly play a role.[10,13]

Finally, enzyme ontogenesis is complicated by the reverse phenomenon: hormones that evoke the development of an enzyme in a differentiating tissue are not usually required for the maintenance of that enzyme's total activity in the adult. It appears that once a hormonal stimulus has programmed a differentiating cell to produce a given enzyme, the later absence of that hormone does not necessarily allow the cell to revert back to its undifferentiated state.[5,6]

C. Experimental Approach to Enzyme Ontogenesis

1. Developmental Profiles

When investigating the developmental profile of a specific enzyme, it is necessary to consider several pitfalls that may lead to artifactual results. Some of these are outlined here.

It is useful to remember that the rate-limiting enzyme in a metabolic pathway in the mature animal need not be rate limiting during all stages of development, because of, for instance, the relative immaturity of other enzymes in the pathway or those supplying cofactors. The presence of *in vitro* enzyme activity does not therefore guarantee *in vivo* metabolic capacity.

Whether lack of enzyme activity is due to lack of enzyme protein or inactivity of an existing enzyme under the conditions of assay must be established. Optimal cell fractionation and assay conditions can vary with age, and this is particularly important for membrane-bound enzymes such as the microsomal drug-metabolizing enzymes.[13] In perinatal liver the endoplasmic reticulum matures qualitatively as well as quantitatively,[14] and this may lead to altered stability of membrane-dependent enzyme activities and altered activation characteristics of latent enzymes.[13,15] Such developmental changes are capable of altering as well the sedimentation characteristics of membranes[16] and their hepatic phospholipid content.[17]

Changes in the cellular makeup of an organ can lead to additional problems. For example, fetal rat liver contains a greater number of hematopoietic cells than hepatocytes at midgestation; they are lost during the late fetal cluster, so that at birth the liver contains predominantly hepatocytes.[5] However, because of their smaller size, this loss of hematopoietic cells results in only about a twofold enrichment of hepatocytes over this period when measured on a liver wet-weight basis. Hepatocyte enrichment will therefore contribute to the rise in late fetal cluster enzyme activities but is not its sole cause. In addition, this hepatocyte enrichment will decrease the DNA content relative to protein content or wet weight of the developing liver, as illustrated by Kistler.[18] Care must therefore be taken in the selection of parameters for use as the basis of measurement of enzyme activities during development.

When experimentally practical, the ideal method of determining an enzyme's developmental profile is to approach the problem on two levels: (1) assay using physiological methods, such as intact cells or perfusates, to establish how the overall metabolic pathway changes during development, and (2) assay of enzyme activity in cell fractions using optimized fractionation and assay conditions, coupled, when possible, with quantification by immunoprecipitation techniques to establish the changes in the intracellular concentrations of the rate-limiting enzyme. Care must be taken with the latter technique because of the possible differential development of immunospecific isoenzymes.[13,19] Finally, the enzyme should be assayed sufficiently frequently during ontogenesis to pinpoint the specific cluster in which it develops.

2. Control Mechanisms

Once the developmental profile of an enzyme has been established, the hormonal factors regulating its development may be investigated; the best candidates are those hormones whose concentrations change concurrently with enzyme activity. It is possible (1) to stimulate an enzyme's development precociously by administering that hormone before its plasma concentration normally rises *in vivo*, and (2) to retard the enzyme's development by inhibiting the hormone's endogenous production. The procedure may be reversed for a hormone suspected of repressing an enzyme activity.

The ease with which these techniques can be used depends upon the hormone under study. Most hormones have short biological half-lives, and for enzymes with relatively slow rates of change, including many drug-metabolizing enzymes, continued administration may be required. Hormone treatment of the fetus has the additional problem of passage through and metabolism by the placenta.[20] For some hormones, synthetic derivatives are available that readily cross the placenta,[21] whereas others must be injected directly into the fetus. Although the latter technique is relatively simple, it leads to maternal stress[22] and is not practical for repeated injections.

The secretion of a hormone may be inhibited by excising the responsible endocrine gland or by using drugs inhibiting its synthesis. One difficulty with the former approach is that many endocrine glands produce more than one type of hormone, and it is sometimes difficult to identify the hormone responsible, for example, corticosterone and adrenalin as neonatal inducers of tyrosine aminotransferase.[23] Chemical inhibitors are more specific, but such xenobiotics may themselves act as enzyme inducers by nonhormonal mechanisms.

Tissue culture is an attractive technique because the culture medium is devoid of hormones present *in vivo* if serum is not present in the medium. Furthermore, hormones added to the medium will effect the cultured tissue directly, and the mechanism of such effects is more accessible to investigation when transcription and translation inhibitors are used. Unfortunately, many drug-metabolizing enzymes are unstable during culture. Microsomal drug-metabolizing enzymes from postnatal mammalian liver lose much of their activity during primary culture.[13,24] Although a hormone may control such an enzyme's development *in vivo*, its stimulation of the enzyme's activity will not be seen in culture if factors maintaining that activity are missing.[13]

II. DEVELOPMENTAL PROFILES OF DRUG-METABOLIZING ENZYMES

Several aspects of the development of drug-metabolizing enzymes have been extensively reviewed over the last few years.[2,3,13,25-35] This section correlates the developmental profiles of such enzymes in rat liver with the established clusters outlined in the previous section and compares the profiles in rat liver with those in other species such as human.

A. Cytochrome *P*-450 and Its Monooxygenase Activities

The cytochrome *P*-450 complex and its resultant monooxygenase activities have long been known to be deficient perinatally in liver microsomes from most species of laboratory animals,[25,29,30] and are also low when measured in neonatal liver perfusates.[36]

The developmental profiles of cytochrome P-450 and its reductase in microsomes from Wistar rat liver are illustrated in Fig. 1. Cytochrome P-450 is detectable in the fetal liver but develops rapidly with the neonatal cluster from <10 to 50% of its concentration in adult male rat liver. This increase is associated with a concurrent increase in the microsomal concentration of a ~50,000-dalton, heme-staining polypeptide,[13,37] which implies de novo synthesis of cytochrome P-450. The microsomal concentration of cytochrome P-450 rises further with the late suckling cluster and again at puberty in the male, resulting in the well-known sex difference in cytochrome P-450 concentration. Microsomal cytochrome b_5 develops at essentially the same rate as cytochrome P-450,[38] whereas the neonatal development of cytochrome P-450 reductase is more rapid and is followed by a fall in activity, which increases again during late suckling.

Not all microsomal monooxygenases develop in parallel, however, and this is illustrated in Fig. 2. The developmental profile of aminopyrine N-demethylating activity closely follows that of cytochrome P-450. Those of 7-ethoxycoumarin O-deethylating and 7-ethoxyresorufin O-deethylating activity, however, show large peaks between late suckling and puberty; the latter is more active in adult female rats than in male rats.

It is possible that such differences in the developmental profiles of different activities can be due to the existence of the multiple forms of cytochrome P-450 in "uninduced" rat liver,[39] which may develop at different rates. However, the microsomal membrane is also developing both in hepatic concentration and in phospholipid content.[13,14] It is therefore likely that developmental changes in the specific forms of cytochrome P-450, in cytochrome P-450 reductase, and in their phospholipid environment all contribute to the differential development of monooxygenase activities. Such changes, moreover, would result in the observed developmental differences in monooxygenase kinetic values and in cytochrome P-450 binding spectra.[28,29]

Microsomal monooxygenase activity predominates in the smooth endoplasmic reticulum (SER), which proliferates in rat liver neonatally.[28] Certain work suggests that nuclear membrane monooxygenase activities develop in parallel with those of microsomes.[40] The postnatal increases in monooxygenase activities relative to liver weight are therefore understandably greater than the corresponding increases measured in microsomes.

Hepatic cytochrome P-450–dependent monooxygenases develop postnatally in other species of nonprimate mammals, including guinea pig, in which they predominantly form with the neonatal cluster, and swine, in which their development is slower.[25,28,29] Development of extrahepatic monooxygenases is less well studied, but available data suggest that ac-

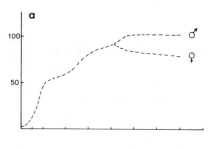

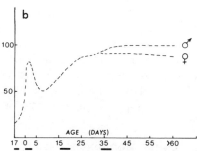

Fig. 1. Developmental profiles of (a) cytochrome P-450 (measured spectrophotometrically) and (b) cytochrome P-450 reductase (measured with cytochrome c) in Wistar rat liver microsomes. Activities are expressed semischematically as percentages of that in adult males. The periods of development of the four clusters are marked with a bar along the horizontal axis.

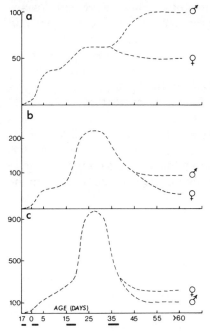

Fig. 2. Developmental profiles of (a) aminopyrine N-demethylation, (b) 7-ethoxycoumarin O-deethylation, and (c) 7-ethoxyresorufin O-deethylation. For details, see Fig. 1.

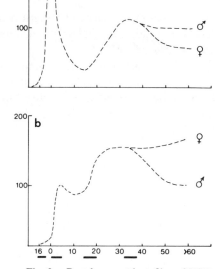

Fig. 3. Developmental profiles of UDP-glucuronyltransferase activity toward (a) o-aminophenol and (b) bilirubin. The activities were maximally activated by digitonin.[15] For details, see Fig. 1.

tivities are low in fetal kidney,[41] intestine,[42] and lung[25] and that their rates of development *post partum* are essentially similar to that in liver.

In primates, however, significant amounts of both SER and cytochrome P-450 are present in fetal liver from a period quite early in gestation.[2,28,31–33] Some monooxygenase activities (e.g., aminopyrine N-demethylation) are also present here, but others (e.g., benzopyrene hydroxylation are generally low or absent.[30] Monooxygenases active toward xenobiotics are also present in primate fetal adrenal gland and in placenta,[30,33,34] and it is possible that the primary function of fetal monooxygenases is steroid hydroxylation.[43]

By term, human fetal liver cytochrome P-450 and some monooxygenase activities may reach more than 50% of their values in adults.[31] However, *in vivo* studies show that the human infant demethylates aminopyrine at only 25% of the rate seen in adults.[44]

B. Other Monooxygenases

The microsomal flavin-containing monooxygenase develops concurrently with hepatic cytochrome P-450 reductase in rat liver (as shown in Fig. 1b), showing a large peak in activity neonatally.[45] In mouse liver, N-oxidation catalyzed by this enzyme also develops neonatally. In females, activity increases further at puberty, mature females having 30–40% more activity than mature males.[46] In the human fetus, microsomal N-oxidation, like cytochrome P-450, is present as early as 13 weeks' gestation.[47]

Microsomal heme oxygenase activity is present in late fetal rat liver at twice the adult rate and increases still further with the neonatal cluster. Hepatic heme oxygenase activity falls to adult values with the late suckling cluster, whereas activity in the spleen develops predominantly in the late suckling stage.[48]

C. Epoxide Hydrolase

Microsomal epoxide hydrolase activity toward styrene oxide is present at a very low range from 17 days' gestation in rat liver. Activity increases with the neonatal cluster to 70–80% of adult female values and then rises again later in male rat liver, resulting in a large sex difference.[49] Activity toward 3-methylcholanthrene 11,12-oxide develops similarly.[50]

Epoxide hydrolase levels are also low in fetal rabbit liver and intestine.[51] In liver it develops mainly between 20 and 30 days postpartum, whereas in intestine it does not increase until after puberty. However, pulmonary and renal activities are some 50% of the adult value in the late fetus and rise further neonatally, as do epoxide hydrolase activities in guinea pig liver and intestine.[51,52]

Epoxide hydrolase can also be present at up to 50% of adult values in human and primate fetal liver and is also present in human placenta and fetal adrenal.[13,31-33]

D. UDPglucuronyltransferases

Development of UDPglucuronyltransferase has been reviewed in detail.[13,32] Transferase activity in rat and mouse liver initially develops with either the late fetal or neonatal clusters, depending upon the substrate. Substrates for the late fetal cluster activities tend to have molecular structures that are smaller and more planar than those for the neonatal cluster. Substrate specificity has been studied using a series of substituted phenols, and the limiting configuration for a phenolic substrate to be accepted by the late fetal form(s) of the enzyme has been determined.[53]

The developmental profiles of UDPglucuronyltransferase acting on o-aminophenol and bilirubin are illustrated in Fig. 3. Activity toward o-aminophenol initially develops with the late fetal cluster, reaching a maximum over birth, whereas activity toward bilirubin develops neonatally. Both activities fall to some extent postnatally before increasing again with the late suckling cluster. At puberty, activity toward o-aminophenol decreases in the female and that toward bilirubin decreases in the male, thus resulting in a characteristic sex difference in adult rat liver.

The developmental profiles of UDPglucuronyltransferases toward o-aminophenol and 1-naphthol in rat liver homogenates, and in liver "snips,"[54] essentially follow those shown in Fig 3a.[13]

UDPglucuronyltransferase activities toward o-aminophenol and 1-naphthol are enhanced both by detergents and by UDP-N-acetylglucosamine (a proposed physiological activator of the enzyme)[55] in fetal and neonatal rat liver homogenates. However, perinatal liver is more sensitive to such perturbations, with lower detergent concentrations required for activation of the enzyme[15] and activation by UDP-N-acetylglucosamine may be lost during preparation or storage of microsomes.[13] However, activity toward o-aminophenol from rat liver is not stimulated by ketones (B. Burchell, personal communication), as is the adult activity,[56] which implies structural differences between the fetal and adult forms of UDPglucuronyltransferases for this substrate.

The neonatal cluster of UDPglucuronyltransferase activities toward estriol, testosterone, and androsterone increase similarly to the activities toward bilirubin (Fig. 3) immediately after birth[13,57] but differ in their subsequent development. For example, in Wistar rat liver microsomes, activity toward estrone shows a sex difference similar to that toward bilirubin in the adult but it does not increase over weaning, whereas activity toward testosterone does not show a sex difference in adult rats[58] (also R. Forrest and J. E. A. Leakey, unpublished).

Two populations of Wistar rat exist, which exhibit different developmental profiles of UDPglucuronyltransferase with androsterone as substrate; in one, activity remains low in both sexes from late suckling through adulthood; in the other, activity rises 10- to 15-fold over puberty in both sexes so that there is no sex difference in the adult.[57] The genetic expression of the latter population is autosomal dominant.[59] These findings imply that multiple forms of neonatally developing UDPglucuronyltransferase exist, as does evidence from purification studies showing that distinct forms of UDPglucuronyltransferase act on testosterone, estrone, and bilirubin; these forms are less specific for a range of xenobiotic substrates.[56] No absolute correlation exists between forms glucuronidating a specific substrate and the cluster in which activity develops, but many substrates of the late fetal cluster activity are preferentially glucuronidated by the form specific for testosterone. A better correlation does exist, however, between the cluster in which an activity develops and its phospholipid dependency. Activities toward o-aminophenol or p-nitrophenol (late fetal cluster) are stable throughout purification procedures, whereas activities toward bilirubin and testosterone (neonatal cluster) are inactivated upon purification unless phospholipid is added back.[13,56]

UDPglucuronyltransferases do not develop in similar clusters in all species. In human, for instance, activities toward o-aminophenol and bilirubin both surge neonatally,[60] although low activity toward bilirubin,[61] estriol,[62] and morphine[35] has been shown to be present in human fetal liver.

The development of other glycosyltransferases has been less well studied, but UDPglucosyltransferase and UDPxylosyltransferase activities toward bilirubin both develop with the neonatal cluster in rat liver,[13,63] and therefore do not replace UDPglucuronyltransferase in conjugating bilirubin in fetal rat liver.

E. Sulfotransferases

Both the hydroxysteroid sulfotransferases and the aryl sulfotransferases exist in rat liver in multiple forms with overlapping substrate specificities.[64,65] Hepatic aryl sulfotransferase activity with p-nitrophenol rises in both the neonatal and late suckling clusters in rats[57] and mice.[66] However, at puberty activity continues to rise in male rats, resulting in a threefold sex difference in this species, whereas in the mouse, activity falls in both sexes at puberty. Aryl sulfotransferase activity with N-hydroxy-2-acetylaminofluorene shows a sex difference similar to that of the p-nitrophenol activity in adult rat liver.[67]

Hepatic hydroxysteroid sulfotransferases are more active in adult female rats than in males. Activities toward cortisol, corticosterone,[65] and androsterone[57] all develop postnatally to maximum values in male rats at weaning but continue to rise in females. These changes in activity are associated with changes in the relative concentrations of the three forms of hydroxysteroid sulfotransferase.[65]

Hepatic hydroxysteroid sulfotransferase activity toward dehydroepiandrosterone shows a developmental profile in both male and female mice similar to that of the male rat. However, in humans, hydroxysteroid sulfotransferase appears to be present at almost adult activity in fetal liver,[13] an organ that is also active in aryl sulfation.[68]

F. Glutathione S-Transferases

Hepatic glutathione S-transferase also exists in multiple forms with overlapping substrate specificity. In adult rat liver, the major form present is "transferase B." This form is structurally similar to the cytosolic binding protein ligandin that binds many electrophilic and hydrophobic compounds.[69,69a]

Glutathione S-transferase activities toward several substrates have been shown to increase linearly from birth to puberty in rat liver, lung, and kidney.[70-72] Activities toward some substrates, for example, 1,2-dichloro-4-nitrobenzene,[73] show sex differences. The results of immunological studies are consistent with the doubling of glutathione S-transferase B concentration in rat liver with the neonatal cluster.[72] Its concentration rises again with the late suckling cluster and, in the female, also at puberty to produce higher concentrations in adult female than in adult male rat liver.[73]

Glutathione S-transferase activity toward styrene oxide and 1,2-dichloro-4-nitrobenzene is low in fetal rabbit liver but relatively high, compared with adult values, in fetal intestine, lung, and kidney.[51] Activity rises further postnatally in all four tissues. Glutathione S-transferase activity is also relatively high in human fetal liver and is also present in placenta[13]; thus, adequate amounts are probably available to cope with toxic metabolites produced by fetal monooxygenases.

The enzymes of mercapturic acid formation show differential development in rat liver: γ-glutamyltranstransferase is present at greater than adult activities in late fetal liver and falls postnatally, whereas hepatic cysteine N-acetyltransferase activity does not develop until puberty.[71] Renal γ-glutamyltransferase activity rises as the hepatic activity falls. Such large developmental changes in these three enzymes, however, had little effect on the mercapturic acid conjugation of 1,2-epoxy-3-(p-nitrophenoxy)propane, which was conjugated at similar rates in both adult and neonatal rats.[71]

G. N-Acetyltransferases

N-Acetyltransferase activity toward p-aminobenzoic acid increases with the late fetal cluster in rat liver but falls again neonatally. Activity is also present in rat placenta. In rabbits, it also develops with the late fetal cluster in liver and lung, but neonatally in intestine and kidney.[74] It is probable that neonatal rabbit liver contains isoenzyme(s) of N-acetyltransferase that are different from those in adult liver.[75] In human, hepatic N-acetyltransferase appears lower in the neonates than in adults.[13]

H. Overall Trends

In the rat, therefore, it is evident that the critical periods for development of the hepatic enzymes of detoxication are, for the most part, coincidental with the clusters established for enzymes of intermediary metabolism in this tissue. However, developmental profiles of enzyme activities are more complex when multiple forms of the enzymes are present with overlapping substrate specificity (e.g., glutathione S-transferase), and the future use of more specific immunological techniques to assay the development of individual isoenzymes would be of great value. The parallel development of different drug-metabolizing enzymes implies common control mechanisms (discussed in the next section).

It may be seen that, with the exception of UDPglucuronyltransferase and possibly N-acetyltransferase, human fetal liver appears adequately equipped with these enzymes. Unfortunately, the developmental profile of all the enzymes of detoxication is incomplete and is likely to remain so for ethical reasons;[76] for instance, little is known of how they develop postnatally. It is interesting that the drug-metabolizing enzyme developmental profiles in chick embryo liver resemble those of the human fetus in that cytochrome P-450, monooxygenase activity, and epoxide hydrolase activities are all present during late incubation, whereas UDPglucuronyltransferase develops only after hatching.[13]

III. FACTORS INFLUENCING DRUG-METABOLIZING ENZYME DEVELOPMENT

A. Hormonal Influences

1. Glucocorticoids

Glucocorticoid hormones influence the ontogenesis of many enzymes, including those metabolizing xenobiotics. In the rat, the predominant glucocorticoid is corticosterone, which begins to increase in concentration from day 16 of gestation in response to ACTH secreted by the fetal pituitary.[13,77] The total plasma corticosterone concentration reaches a maximum 2 days before birth and remains relatively low for the first 14 days postpartum[10,78] before increasing again with the late suckling cluster.[79] However, plasma transcortin falls after birth, so that the plasma concentration of unbound corticosterone (the form available for entry into the hepatocyte) initially rises at birth. Increases in corticosterone entering the hepatocyte can, therefore, be associated with the late fetal, neonatal, and late suckling clusters, and, as outlined in Section I,B, glucocorticoids may stimulate enzymes developing with all three clusters.

Corticosterone is now considered to be the hormone stimulating the late fetal development of UDPglucuronyltransferase and γ-glutamyltransferase in rat liver.[13,80] Maternal glucocorticoid treatment at 14–16 days gestation precociously stimulates the development of both enzymes *in utero*, and fetal hypothysectomy (i.e., ablation of ACTH to stimulate the fetal adrenal) inhibits the natural, late fetal development of UDPglucuronyltransferase.[80,81] This activity is also stimulated by glucocorticoids in organ cultures of fetal liver by a process dependent upon protein synthesis.[82]

Glucocorticoid treatment *in utero* fails to evoke precocious development of drug-metabolizing enzymes belonging to the neonatal cluster; included are cytochrome P-450, its reductase, and dependent monooxygenases, glutathione S-transferase, and the UDPglucuronyltransferase activities of the neonatal cluster.[13] After birth, however, glucocorticoids stimulate the development of all of these enzymes,[82–86] and the induction of cytochrome P-450 by them in postnatal rat liver has been studied in detail.[13,80] The form of cytochrome P-450 induced by glucocorticoids in neonatal rat liver appears chromatographically and electrophoretically similar to the predominant form that develops naturally here and appears distinct from those forms induced by barbiturates or polycyclic aromatic hydrocarbons.[13,37] Functional adrenal glands are necessary for development of hepatic cytochrome P-450 in neonatal or suckling rats,[13,83] but adrenalectomy in adults only evokes small decreases in hepatic cytochrome P-450 content.[87] Furthermore, glucocorticoids readily induce cytochrome P-450 only in immature rat liver; in adults significant induction of cytochrome P-450 does not occur, although cytochrome P-450 reductase and some monooxygenase activities do increase.[88]

From this evidence it appears likely that corticosterone plays a major role in the control of the postnatal development of cytochrome P-450 and its dependent monooxygenases, and possibly of glutathione S-transferase and UDPglucuronyltransferase as well. Possible mechanisms for glucocorticoid control have been outlined.[13,80] It has been proposed that these neonatally developing, drug-metabolizing enzymes (unlike late fetal cluster UDPglucuronyltransferase) are not competent to respond to glucocorticoids in fetal liver *in utero*, but that as with tyrosine aminotransferase (Section I,B), they become competent to respond at birth and then develop with the neonatal cluster in response to circulating (unbound) corticosterone. These enzyme activities increase further with the late suckling cluster in response to the increasing plasma corticosterone concentrations at that stage. The precise mechanism evoking competence remains uncertain, but repression *in utero* by hormones such as insulin, progesterone, or estrogens could participate.[13,80]

The time at which drug-metabolizing enzymes become competent in responding to glucocorticoid stimulation varies among species and tissues. For instance, N,N-dimethylaniline N-demethylating and N-oxidative activities in late fetal rabbit lung are already competent to respond to glucocorticoid,[89] as are cytochrome P-450, aminopyrine N-demethylase, and UDPglucuronyltransferase in chick embryo liver.[86]

In human, little is known about the effects of glucocorticoids on the development of these enzymes, but the human fetus is exposed to increasing glucocorticoid concentrations during the third trimester.[90] Plasma cortisol is high during natural birth but is relatively lower in naturally premature, cesarian-delivered, and "induced" infants.[13] During the first postnatal week plasma glucocorticoid concentrations fall, but they increase again between 3 and 12 months,[91] in a possible analogous way to late suckling in the rat. Significantly, the absence of glucocorticoid in the human infant has been associated with neonatal hyperbilirubinemia.[13]

2. Estrogen, Androgen, and Progesterone

Estrogen, androgen, and progesterone have all been implicated in the control of drug-metabolizing enzyme development, particularly at puberty.[87] Human and rat fetuses are exposed to relatively high progesterone concentrations *in utero*,[77,90] and postnatally, suckling infants may also be exposed to progesterone and its derivatives present in their mother's milk.[13,61] Both progesterone and estrogen have been reported to possess antiglucocorticoid activity[13] that could possibly control the competence of these enzymes in their response to glucocorticoid.[13,86] Both progestogen and estrogen inhibit some monooxygenases[13,92] and possibly UDP-glucuronyltransferase,[61] but this is probably the result of direct interaction with the microsomal membrane.[92]

The plasma concentration of testosterone in rat neonates is higher in males than in females,[93] and it programs the male pituitary and hypothalamus for later "maleness."[9] At puberty, plasma concentrations of androgen and estrogen rise in male and female rats, respectively, and evoke their characteristic sex differences in drug-metabolizing and other enzymes.[9,87,94]

Both estrogen and androgen are now thought to act on the liver indirectly via the pituitary.[9,94] However, a direct action is also possible because rat liver contains binding proteins for both estrogen and testosterone,[95] and the latter steroid increases the cytochrome P-450 concentration of cultured rat hepatocytes.[96]

3. Growth Hormone

Early work showed that growth hormone could suppress certain hepatic monooxygenases in adult male rats, leading to the proposal that high neonatal concentrations of growth hormone were responsible for their low monooxygenase activities.[97] However, the concentration of plasma growth hormone, although falling neonatally,[78] rises again at puberty in both sexes of the rat.[98]

It was later shown that growth hormone suppresses only those monooxygenase activities that are higher in male than in female rat liver, that is, it essentially converts male hepatic monooxygenase profiles into female profiles.[94] In this respect, growth hormone is functionally identical to feminotropin, the pituitary feminizing factor[9]; feminotropin is presently considered a derivative of growth hormone.[99]

The precise mechanism whereby female rat liver is "feminized" remains uncertain, but feminotropin[8] and growth hormone[94] are candidates for directly feminizing hepatic enzymes by interacting with the hepatic "lactogenic receptor,"[100] which is estrogen dependent and increases its concentration at puberty in female but not in male rats. Alternatively, estrogen may directly feminize hepatic enzymes by interacting with the specific growth hormone–dependent estrogen receptor that is present at similar concentrations in liver from intact male and female rats but at much lower concentrations in hypophysectomized rats of either sex.[95]

4. Other Hormones

Although glucagon and cyclic AMP precociously stimulate neonatally developing tyrosine aminotransferase *in utero*,[6] they have been ineffective in precociously stimulating development of cytochrome *P*-450 or neonatally developing UDPglucuronyltransferase activities in fetal rat liver.[13] In the neonate, glucagon treatment lowers the cytochrome *P*-450 concentration.[83]

Thyroid hormone administration stimulates cytochrome *P*-450 reductase and heme oxygenase activities in fetal, neonatal, and adult rat liver, while decreasing the cytochrome *P*-450 concentration and epoxide hydrolase activity.[21,87,100] Increases in plasma thyroid hormone concentrations at birth may thus be partly responsible for the rapid neonatal increase in cytochrome *P*-450 reductase activity.

B. Xenobiotic Effects

Xenobiotics themselves play a regulatory role because of their ability to induce some of the enzymes of detoxication. Although these effects are noted throughout this volume, certain aspects of the subject require mention here because of their importance during the perinatal period.

It is probable that at least some of the many xenobiotics that induce drug-metabolizing enzymes will be present in the developing organism's natural environment at sufficient concentrations to influence the developmental profiles of these enzymes, thereby allowing developed levels of the enzymes to be partly maintained by such xenobiotics. It must also be emphasized that the potency of xenobiotics in inducing the detoxication enzymes will change with age. Immature animals are usually more responsive than adults, largely because of the initially slower rates at which the inducer is metabolically inactivated; such other factors as increased concentrations of specific receptor proteins for the inducer can also contribute.[101] Fetal and pregnant animals generally have low sensitivity to induction, largely because of the repressive effects of hormones associated with pregnancy.[102]

However, some fetal livers completely lack the competence to respond to certain xenobiotics. For example, rat liver UDPglucuronyltransferase and cytochrome *P*-450–dependent monooxygenase do not respond to phenobarbital induction prenatally, whereas they are readily induced by 3-methylcholanthrene and pregnenolone-16α-carbonitrile, respectively.[80,103] Maternally administered phenobarbital, however, is able to reach the fetal hepatocyte and convert fetal rough endoplasmic reticulum to SER,[16,26] and induces both enzyme activities postnatally.

3. Ontogenesis

Many similarities exist between the induction of enzymes by xenobiotics and glucocorticoids. This is particularly so for the polycyclic aromatic hydrocarbons that reach the cell nucleus and stimulate transcription by a mechanism analogous to that used by the glucocorticoids.[80,104] As with the steroid hormones (Section I,B), different enzyme activities become competent in responding to polycyclic aromatic hydrocarbon inducers at different periods.[101] For instance, these compounds induce the "P_1-450" form of cytochrome P-450 in perinatal rat liver several days earlier than they begin to induce the "P-448" form; in developing rabbits, 3-methylcholanthrene induces cytochrome P-448 only after the tenth postnatal day and cytochrome P_1-450 only in preweanling rabbits.[105]

Similarly, 3-methylcholanthrene, which is hepatocarcinogenic only in young animals,[69] will stimulate hepatic γ-glutamyltransferase, tyrosine aminotransferase, and tryptophan 2,3-dioxygenase activities in preweanling but not adult rats.[80] These effects were associated with changes in the rat liver glucocorticoid receptor concentration and involve the adrenals. Significantly, the induction of cytochrome P-448–dependent monooxygenase activity in rat liver is also partly dependent upon the adrenals.[80] The teratogenic effects of polycyclic aromatic hydrocarbons and associated inducers could possibly also involve interactions with adrenal hormones because excess glucocorticoid produces similar teratogenic effects in fetal rats and mice.

IV. COMMENTS: THE BIOLOGICAL AND CLINICAL CONSEQUENCES OF DRUG-METABOLIZING ENZYME ONTOGENESIS

A healthy member of a successful species may be expected to have acquired, through natural selection, the enzymes necessary for the detoxication of any potentially toxic molecule it routinely encounters in its natural environment, whether such a metabolite is a naturally occurring xenobiotic or a waste product of its own metabolism. The mammalian fetus in its natural environment is largely protected from such toxic metabolites by the drug-metabolizing enzymes of the maternal liver and placenta,[30] and high fetal drug-metabolizing enzyme activities are thus generally not necessary for survival.

After birth the neonate must develop those enzyme activities necessary for the detoxication of its own potentially toxic waste products, and those of any xenobiotic sequested in the milk, before they accumulate to toxic concentrations. The young mammal, however, does not become fully exposed to its chemical environment until weaning, by which time it must possess a sufficient complement of detoxicating enzymes to cope with the influx of dietary xenobiotics. Puberty largely reflects changes in the organism's internal environment, and drug-metabolizing enzymes must adapt to cope with new concentrations of gonadal steroids. As Section II shows, the enzymes of detoxication of the rat and probably of other species adapt to meet these requirements.

In recent times, however, humans and animals have tended not to live in the ecological niches in which they evolved. Their chemical environment in particular has undergone major changes since the introduction of organic chemistry. Entirely new chemicals of bizarre structure have appeared, and others, although structurally similar to naturally occurring xenobiotics, may be present in much larger amounts than before.[106] Although enzyme induction will allow a large and possibly infinite[107] capacity to adapt to these environmental changes, xenobiotics not routinely encountered before will be metabolized according to their structural similarities with the natural substrates of the drug-metabolizing enzymes.[108] Such patterns of metabolism often produce a spectrum of metabolites and products, some of which are more toxic than the parent compound.[109,110]

Exposure to such unnatural xenobiotics can have more serious and unpredictable consequences for the developing organism than for the adult. A number of factors contribute. Should the xenobiotic form toxic metabolites, which are then inactivated by further metabolism, its degree of toxicity will vary with age because of the differential development of the various enzymes involved in its metabolism. In addition, the generally low drug-metabolizing enzyme activities in the neonate may result in a longer metabolic half-life for the xenobiotic and thus a greater risk because of toxic effects from the xenobiotic itself, or from accumulation of endogenous compounds competing with it for the same detoxicating pathways.[111] Such effects will be further exaggerated in the fetus if the relevant metabolizing enzymes are not yet competent to respond to its inducing effects.

Increased sensitivity to inducers in the infant and neonate (Section III,B) can cause additional problems. Hormone-metabolizing activities can be additionally induced, upsetting the organism's hormonal balance, which is at greater risk because of immaturity of pituitary feedback mechanisms.[79,98] Hormone–xenobiotic interactions of this sort in the developing organism may upset ordered development and increase risks of teratogeny, carcinogenesis, and imprinted abnormalities,[80] in addition to their immediate toxic effects.

If the healthy perinate is potentially at risk from changes in its chemical environment, the sick or premature infant is doubly so. Many pathological conditions additionally alter hormonal balance and the effective concentration of drug-metabolizing enzymes;[87] such changes are probably different in degree at this age than in the adult. Illness may require drug therapy, with consequent additional alteration to the infant's chemical environment. The latter is a major problem in modern pediatric pharmacology. Many drugs in routine use for the adult are unsuitable for the fetus and infant because safe dose regimens are not available.[76] Complete elucidation of the factors controlling ontogenesis of the enzymes of detoxication could therefore be of great value. It should then be possible not only to use hormone therapy in aid of the premature neonate's adaption to its new environment but also to predict which additional changes in the development of drug-metabolizing enzymes are likely to be associated with the pathological conditions of those infants requiring drug therapy.

REFERENCES

1. Gädeke, R. C. (1972). Unwanted effects of drugs in the neonate, premature and young child. *Drug-Induced Dis.* **4**, 585–616.
2. Yaffe, S. J., and Juchau, M. R. (1974). Perinatal pharmacology. *Annu. Rev. Pharmacol.* **14**, 219–238.
3. Yaffe, S. J., and Danish, M. (1978). Problems of drug administration in the paediatric patient. *Drug Metab. Rev.* **8**, 303–318.
4. Greengard, O. (1977). Enzymic differentiation of human liver: Comparison with the rat model. *Pediatr. Res.* **11**, 669–676.
5. Greengard, O. (1971). Enzymic differentiation in mammalian tissues. *Essays Biochem.* **7**, 159–205.
6. Greengard, O., and Bodanszky, H. (1981). Enhanced and delayed maturation of tissue enzyme patterns. *In* "Physiological and Biochemical Basis for Perinatal Medicine" (M. Monset-Couchard and A. Minkowski, eds.), pp. 217–226. Karger, Basel.
7. Girard, J. R., Pegorier, J. P., Leturque, A., and Ferre, P. (1981). Glucose homeostasis in the newborn rat. *In* "Physiological and Biochemical Basis for Perinatal Medicine" (M. Monset-Couchard and A. Minkowski, eds.), pp. 90–96. Karger, Basel.
8. Vanstapel, F., Dopere, F., and Stalmans, W. (1980). Role of glycogen synthase phosphatase in the glucocorticoid-induced desposition of glycogen in foetal rat liver. *Biochem. J.* **192**, 607–612.
9. Gustaffson, J. A., Mode, A., Norstedt, G., Hökfelt, T., Sonnenschein, C., Eneroth, P., and Skett, P. (1980). The hypothalamo–pituitary–liver axis: A new hormonal system in control of hepatic steroid and drug metabolism. *In* "Biochemical Actions of Hormones" (G. Litwack, ed.), Vol. 7, pp. 48–49. Academic Press, New York.
10. Greengard, O. (1975). Steroids and the maturation of rat tissues. *J. Steroid Biochem.* **6**, 639–642.
11. Cake, M. H., Ghisalberti, A. V., and Oliver, I. T. (1973). Cytoplasmic binding of dexamethasone and induction of tyrosine aminotransferase in neonatal rat liver. *Biochem. Biophys. Res. Commun.* **54**, 983–990.
12. Killewick, L. A., and Feigelson, P. (1977). Developmental control of messenger RNA for hepatic tryptophan-2,3-dioxygenase. *Proc. Natl. Acad. Sci. U.S.A.* **74**, 5392–5396.

13. Dutton, G. J., and Leakey, J. E. A. (1981). Perinatal development of drug metabolizing enzymes: What factors trigger their onset? *Prog. Drug Res.* **25**, 191–269.
14. Dallner, G., Siekevitz, P., and Palade, G. E. (1966). Biogenesis of endoplasmic reticulum membranes. *J. Cell Biol.* **30**, 73–96.
15. Dutton, G. J., Leakey, J. E. A., and Pollard, M. R. (1981). Assays for UDP-glucuronyltransferase activites. *In* "Methods in Enzymology" (W. G. Jakoby, ed.), Vol. 77, pp. 383–391. Academic Press, New York.
16. Cresteil, T., Flinois, J. P., Pfister, A., and Lerouk, J. P. (1979). Effect of microsomal preparations and induction of cytochrome P-450–dependent monooxygenase in foetal and neonatal rat liver. *Biochem. Pharmacol.* **28**, 2057–2063.
17. Abe, M., Endo, M., Nagai, H., and Shimojo, T. (1981). Phosphatidylcholine and glycogen in liver, lung and kidney of the developing rabbit. *IRCS Med. Sci.: Libr. Compend.* **9**, 315–316.
18. Kistler, A. (1979). Tissue specific changes in DNA, RNA and protein content during the late fetal and postnatal development in the rat. *Int. J. Biochem.* **10**, 975–980.
19. Dreyfus, J.-C., and Schapira, F. (1981). Evolution of enzymes and isoenzymes in the fetus and the neonate. *In* "Physiological and Biochemical Basis for Perinatal Medicine" (M. Monset-Couchard and A. Minkowski, eds.), pp. 205–216. Karger, Basel.
20. Wong, M. D., Thomson, M. J., and Burton, A. F. (1976). Metabolism of the natural and synthetic corticosteroids in relation to their effects on mouse fetuses. *Biol. Neonate* **28**, 12–17.
21. Kriz, B. M., Jones, A. L., and Jorgensen, E. C. (1978). The effects of a thyroid hormone analogue on fetal rat hepatocyte ultrastructure and microsomal function. *Endocrinology* **102**, 712–722.
22. Jacquot, R. L., Plas, C., and Nagel, J. (1973). Two examples of physiological maturations in rat fetal liver. *Enzyme* **15**, 296–303.
23. Ghisalberti, A. V., Steele, J. G., Cake, M. H., McGrath, M. C., and Oliver, I. T. (1980). Role of adrenaline and cyclic AMP in appearance of tyrosine aminotransferase in perinatal rat liver. *Biochem. J.* **190**, 685–690.
24. Bissell, D. M., and Guzelian, P. S. (1980). Phenotypic stability of adult rat hepatocytes in primary monolayer culture. *In* "Differentiation and Carcinogenesis in Liver Cell Cultures" (C. Borek and G. M. Williams, eds.), pp. 85–98. N.Y. Acad. Sci., New York.
25. Fouts, J. R. (1973). Microsomal mixed-function oxidases in the fetal and newborn rabbit. *In* "Fetal Pharmacology" (L. O. Boréus, ed.), pp. 305–320. Raven Press, New York.
26. Gillette, J. R., and Stripp, B. (1975). Pre- and postnatal enzyme capacity for drug metabolite production. *Fed. Proc., Fed. Am. Soc. Exp. Biol.* **34**, 172–179.
27. Hänninen, O. (1975). Age and exposure factors in drug metabolism. *Acta Pharmacol. Toxicol.* **36**, 1–20.
28. Short, C. R., Kinden, D. A.,and Stith, R. (1976). Fetal and neonatal development of the microsomal monooxygenase system. *Drug Metab. Rev.* **5**, 1–42.
29. Neims, A. H., Warner, M., Laughnan, P. M., and Arande, J. V. (1976). Developmental aspects of the hepatic cytochrome P-450 monooxygenase system. *Annu. Rev. Pharmacol.* **16**, 427–445.
30. Pelkonen, O. (1977). Transplacental transfer of foreign compounds and their metabolism by the foetus. *Prog. Drug Metab.* **2**, 119–161.
31. Henderson, P. T. (1978). Development and maturation of drug metabolizing enzymes. *Eur. J. Drug Metab. Pharmacokinet.* **1**, 1–14.

3. Ontogenesis

32. Dutton, G. J. (1978). Developmental aspects of drug conjugation, with special reference to glucuronidation. *Annu. Rev. Pharmacol. Toxicol.* **18,** 17–35.
33. Nau, H., and Neubert, D. (1978). Development of drug-metabolizing monooxygenase systems in various mammalian species including man. Its significance for transplacental toxicity. *In* "Role of Pharmacokinetics in Prenatal and Perinatal Toxicology" (D. Neubert, H.-J. Merker, H. Nau, and J. Langman, eds.), pp. 13–44. Thieme, Stuttgart.
34. Pelkonan, O. (1980). Developmental drug metabolism. "Concepts in Drug Metabolism" (P. Jenner and B. Testa, eds.), pp. 285–309. Dekker, New York.
35. Rane, A., and Tomson, G. (1980). Prenatal and neonatal drug metabolism in man. *Eur. J. Clin. Pharmacol.* **18,** 9–15.
36. Sonawane, B. R., Yaffe, S. J., Reinke, L. A., and Thurman, R. G. (1980). Postnatal development of mixed-function oxidation and conjugation in haemoglobin free perfused rat liver. *J. Pharmacol. Exp. Ther.* **216,** 473–478.
37. Leakey, J. E. A., Dutton, G. J., and Fouts, J. R. (1980). Partial purification and characterization of dexamethasone-induced cytochrome P-450 from neonatal rat liver. *Biochem. Soc. Trans.* **8,** 344–345.
38. Uehleke, H. (1978). Covalent binding of xenobiotics in the fetal and newborn rodent. *In* "Role of Pharmocokinetics in Prenatal and Perinatal Toxicology" (D. Neubert, H.-J. Merker, H. Nau, and J. Langman, eds.), pp. 225–234. Thieme, Stuttgart.
39. Kamataki, T., Maeda, K., Yamazoe, Y., Nagain, T., and Kato, R. (1981). Partial purification and characterization of cytochrome P-450 responsible for the occurence of sex difference in drug metabolism in the rat. *Biochem. Biophys. Res. Commun.* **103,** 1–7.
40. Nunnink, J. C., Chuang, A. H. L., and Bresnick, E. (1978). The ontogeny of nuclear aryl hydrocarbon hydroxylase. *Chem. Biol. Interact.* **22,** 225–230.
41. Goldsmith, P. K. (1980). Postnatal development of some membrane-bound enzymes of rat liver and kidney. *Biochim. Biophys. Acta* **672,** 45–56.
42. Tredger, J. M., and Chhabra, R. S. (1979). Factors affecting the properties of mixed function oxidases in the liver and small intestine of neonatal rabbits. *Pharmacol. Exp. Ther.* **8,** 16–21.
43. Telegdy, G. (1973). Possible influence of drugs on fetal steroid metabolism. *In* "Fetal Pharmacology" (L. Boréus, ed.), pp. 335–354. Raven Press, New York.
44. Jäger, E., Gregg, B., Knies, S., Helge, H., and Bochert, G. (1978). Postnatal development of human liver N-demethylation activity measured with the $^{13}CO_2$ breath test after application of ^{13}C-dimethylaminopyrine. *In* "Role of Pharmacokinetics in Prenatal and Perinatal Toxicology" (D. Neubert, H.-J. Merker, H. Nau, and J. Longman, eds.), pp. 211–214. Thieme, Stuttgart.
45. Uehleke, H., Reiner, O., and Hellmer, K. H. (1971). Perinatal development of tertiary amine N-oxidation and NADPH cytochrome c reduction in rat liver microsomes. *Res. Commun. Chem. Pathol. Pharmacol.* **2,** 793–805.
46. Wirth, P. J., and Thorgeirsson, S. S. (1977). Amine oxidase in mice–sex differences and developmental aspects. *Biochem. Pharmacol.* **27,** 601–603.
47. Rane, A. (1973). N-oxidation of a tertiary amine (N,N-dimethylaniline) by human fetal liver microsomes. *Clin. Pharmacol. Ther.* **15,** 32–38.
48. Thaler, M. M., Gemes, D. L., and Bakken, A. F. (1972). Enzymatic conversion of heme to bilirubin in normal and starved fetuses and newborn rats. *Pediatr. Res.* **6,** 197–201.
49. Oesch, F. (1976). Different control of rat microsomal "aryl hydrocarbon" monooxygenase and epoxide hydratase. *J. Biol. Chem.* **251,** 79–87.

50. Stoming, T. A., and Bresnick, E. (1974). Hepatic epoxide hydrase in neonatal and partially hepatectomized rats. *Cancer Res.* **34**, 2810–2813.
51. James, M. O., Foureman, G. L., Law, F. C., and Bend, J. R. (1977). The perinatal development of epoxide-metabolizing enzyme activities in liver and extrahepatic organs of guinea pig and rabbit. *Drug Metab. Dispos.* **5**, 19–28.
52. Bend, J. R., James, M. O., Deveraux, T. R., and Fouts, J. R. (1975). Toxication-detoxication systems in hepatic and extrahepatic tissues in the perinatal period. *In* "Basic and Therapeutic Aspects of Perinatal Pharmacology" (P. L. Morselli, S. Garattini, and F. Sereni, eds.), pp. 229–243. Raven Press, New York.
53. Wishart, G. J., and Campbell, M. T. (1978). Demonstration of two functionally heterogeneous groups within the activities of UDP-glucuronyltransferase towards a series of 4-alkyl-substituted phenols. *Biochem. J.* **178**, 443–447.
54. Pollard, M. R., and Dutton, G. J. (1982). Liver snips. *Biochem. J.* **202**, 469–473.
55. Dutton, G. J. (1980). "Glucuronidation of Drugs and Other Compounds." CRC Press, Boca Raton, Florida.
56. Burchell, B. (1981). Identification and purification of multiple forms of UDP-glucuronyltransferase. *Rev. Biochem. Toxicol.* **3**, 1–32.
57. Matsui, M., and Watanabe, H. K. (1982). Developmental alterations of hepatic UDP-glucuronosyltransferase and sulfotransferase towards androsterone and 4-nitrophenol in Wistar rats. *Biochem. J.* **204**, 441–447.
58. Fuchs, M., Rae, G. S., Rao, M. L., and Bieuer, H. (1977). Studies on the properties of an enzyme forming the glucuronide of pregnanediol and the pattern of development of steroid glucuronyltransferase in rat liver. *J. Steroid Biochem.* **7**, 235–240.
59. Matsui, M., and Watanabe, H. K. (1981). Classification and genetic expression of Wistar rats with high and low hepatic microsomal UDP-glucuronyltransferase activity towards androsterone. *Biochem. J.* **202**, 171–174.
60. Onishi, S., Kawade, N., Itoh, S., Isobe, K., and Sugiyama, S. (1979). Post-natal development of uridine diphosphate glucuronyltransferase activity towards bilirubin and 2-aminophenol in human liver. *Biochem. J.* **184**, 705–707.
61. Odell, G. B. (1980). "Neonatal Hyperbilirubinemia." Grune & Stratton, New York.
62. Burchell, B. (1974). UDP-glucuronyltransferase activity towards oestriol in fresh and cultured foetal tissues from man and other species. *J. Steroid Biochem.* **5**, 261–267.
63. Vaisman, S. L., Lee, K. G., and Gartner, L. M. (1976). Xylose, glucose and glucuronic acid conjugation of bilirubin in the newborn rat. *Pediatr. res.* **10**, 967–971.
64. Singer, S. S., Giera, D., Johnson, J., and Sylvester, S. (1976). Enzymatic sulfation of steroids. I. The enzymatic basis for the sex difference in cortisol sulfation by rat liver preparations. *Endocrinology* **98**, 963–974.
65. Seura, R. D., and Jakoby, W. B. (1979). Phenol sulphotransferases. *J. Biol. Chem.* **254**, 5658–5663.
66. Carroll, J., and Armstrong, L. M. (1976). Regulation of hepatic sulphotransferases. *Biochem. Soc. Trans.* **4**, 871–873.
67. Debaun, J. R., Miller, E. C., and Miller, J. A. (1970). N-hydroxy-2-acetylaminofluorene sulphotransferase: Its probable role in carcinogenesis and in protein-(methion-S-yl) binding in rat liver. *Cancer Res.* **30**, 577–595.
68. Rollins, D. E., Glaumann, H., Moldéus, P., and Rane, A. (1979). Acetaminophen: Potentially toxic metabolite formed by human fetal and adult liver microsomes and isolated fetal liver cells. *Science* **205**, 1414–1416.
69. Smith, G. J., Ohl, V. S., and Litwack, G. (1977). Ligandin, the glutathione S-transferases, and chemically induced hepatocarcinogenesis: A review. *Cancer Res.* **37**, 8–14.

69a. Kitahara, A., and Sato, K. (1981). Immunological relationships among subunits of glutathione S-transferases A, AA, B and ligandin and hybrid formation between AA and ligandin by guanidine hydrochloride. *Biochem. Biophys. Res. Commun.* **103,** 943–950.
70. Mukhtar, H., and Bresnick, E. (1976). Glutathione-S-epoxide transferase activity during development and the effect of partial hepatectomy. *Cancer Res.* **36,** 937–940.
71. James, S. P., and Pheasant, A. E. (1978). Glutathione conjugation and mercapturic acid formation in the developing rat. *Xenobiotica* **8,** 207–217.
72. Hales, B. F., and Neims, A. H. (1976). Developmental aspects of glutathione S-transferase B (ligandin) in rat liver. *Biochem. J.* **160,** 231–236.
73. Hales, B. F., and Neims, A. H. (1976). A sex difference in hepatic glutathione S-transferase B and the effect of hypophysectomy. *Biochem. J.* **160,** 223–229.
74. Sonawane, B. R., and Lucier, G. W. (1975). Hepatic and extrahepatic N-acetyl-transferase. Perinatal development using a new radioassay. *Biochim. Biophys. Acta* **411,** 97–105.
75. Cohen, S. N., Baumgartner, R., Steinberg, M. S., and Weber, W. W. (1973). Changes in the physicochemical characteristics of rabbit liver N-acetyl-transferase during post-natal development. *Biochim. Biophys. Acta* **304,** 473–481.
76. Yaffe, S. J. (1980). Pediatric pharmacology, overview and perspectives. *Pediatr. Pharmacol.* **1,** 3–5.
77. Martin, C. E., Cake, M. H., Hartmann, P. E., and Cook, I. F. (1977). Relationship between foetal corticosteroids, maternal progesterone and parturition in the rat. *Acta Endocrinol. (Copenhagen)* **84,** 167–176.
78. Poland, R. E., Weichsel, M. E., Jr., and Rubin, R. T. (1979). Postnatal maturation patterns of serum corticosterone and growth hormone in rats: Effect of chronic thyroxine administration. *Horm. Metab. Res.* **11,** 222–227.
79. Poland, R. E., Weichsel, M. E., Jr., and Rubin, R. T. (1981). Neonatal dexamethasone administration. I. Temporary delay of development of the circadian serum corticosterone rhythm in rats. *Endocrinology* **81,** 1049–1054.
80. Leakey, J. E. A., Wishart, G. J., and Dutton, G. J. (1982). Interrelationship of carcinogen and glucocorticoid in perinatal enzyme induction. *Int. J. Biol. Res. Pregnancy,* **3,** 108–113.
81. Wishart, G. J., and Dutton, G. J. (1977). Regulation of onset of development of UDP-glucuronyltransferase activity towards o-aminophenol by glucocorticoids in late-foetal rat liver *in utero. Biochem. J.* **168,** 507–511.
82. Wishart, G. J., Goheer, M. A., Leakey, J. E. A., and Dutton, G. J. (1977). Precocious development of uridine diphosphate glucuronosyltransferase activity during organ culture of foetal rat liver in the presence of glucocorticoids. *Biochem. J.* **166,** 249–253.
83. Leakey, J. E. A., and Fouts, J. R. (1979). Precocious development of cytochrome P-450 in neonatal rat liver after glucocorticoid treatment. *Biochem. J.* **182,** 233–235.
84. Mukhtar, H., Sahib, M. K., and Kidwai, J. R. (1973). Precocious induction of hepatic aniline hydroxylase and aminopyrine N-demethylase with hydrocortisone in neonatal rat. *Biochem. Pharmacol.* **23,** 345–349.
85. Mukhtar, H., Leakey, J. E. A., Elmamlouk, T. H., Fouts, J. R., and Bend, J. R. (1978). Precocious development of hepatic glutathione S-transferase activity with glucocorticoid administration in neonatal rat. *Biochem. Pharmacol.* **28,** 1801–1803.
86. Leakey, J. E. A., Dutton, G. J., and Wishart, G. J. (1979). Differential stimulation of hepatic monooxygenase and glucuronidating systems in chick embryo and neonatal rat by glucocorticoids. *Med. Biol.* **57,** 256–261.
87. Kato, R. (1977). Drug metabolism under pathological and abnormal physiological states in animals and man. *Xenobiotica* **7,** 25–92.

88. Tredger, J. M., Chakraborty, J., and Parke, D. V. (1976). Effect of natural and synthetic glucocorticoids on rat hepatic microsomal drug metabolism. *J. Steroid Biochem.* **7,** 351–356.
89. Devereux, T. R., and Fouts, J. R. (1977). Effect of dexamethasone treatment on *N,N*-dimethylaniline demethylation and N-oxidation in pulmonary microsomes from pregnant and fetal rabbits. *Biochem. Pharmacol.* **27,** 1007–1008.
90. Pettit, B. R., and Fry, D. E. (1978). Corticosteroids in amniotic fluid and their relationship to fetal lung maturation. *J. Steroid Biochem.* **9,** 1245–1249.
91. Sippell, W. G., Dorr, H. G., Bidlingmaier, F., and Knorr, D. (1980). Plasma levels of aldosterone, corticosterone, 11-deoxycorticosterone, progesterone, 17-hydroxyprogesterone, cortisol, and cortisone during infancy and childhood. *Pediatr. Res.* **14,** 39–46.
92. Leakey, J. E. A., Mukhtar, H., Fouts, J. R., and Bend, J. R. (1982). Thyroid hormone-induced changes in the hepatic monooxygenase system, heme oxygenase activity and epoxide hydrolase activity in adult male, female and immature rats. *Chem.-Biol. Interact.* **40,** 257–264.
93. Pang, S. F., Caggiula, A. R., Gay, V. L., Goodman, R. L., and Pang, C. S. F. (1979). Serum concentrations of testosterone, oestrogens, luteinizing hormone and follicle-stimulating hormone in male and female rats during the critical period of neural sexual differentiaton. *J. Endocrinol.* **80,** 103–110.
94. Colby, H. D. (1980). Regulation of hepatic drug and steroid metabolism by androgens and estrogens. *Adv. Sex Horm. Res.* **4,** 27–73.
95. Lucier, G. W., Slaughter, S. R., Thompson, C., Lamartiniere, C. A., and Powell-Jones, W. (1981). Selective actions of growth hormone on rat liver estrogen binding proteins. *Biochem. Biophys. Res. Commun.* **103,** 872–879.
96. Decad, G. M., Hieh, D. P. H., and Byard, J. L. (1977). Maintenance of cytochrome P-450 and metabolism of aflatoxin B_1 in primary hepatocyte cultures. *Biochem. Biophys. Res. Commun.* **78,** 279–287.
97. Wilson, J. T. (1970). Alteration of normal development of drug metabolism by injection of growth hormone. *Nature (London)* **225,** 861–863.
98. Eden, S. (1979). Age- and sex-related differences in episodic growth hormone secretion in the rat. *Endocrinology* **105,** 555–560.
99. Mode, A., Norstedt, G., Simic, B., Eneroth, P., and Gustafsson, J.-A. (1981). Continuous infusion of growth hormone feminizes hepatic steroid metabolism in the rat. *Endocrinology* **108,** 2103–2108.
100. Posner, B. I. Kelly, P. A., and Friesen, H. G. (1974). Induction of a lactogenic receptor in rat liver: Influence of estrogen and the pituitary. *Proc. Natl. Acad. Sci. U.S.A.* **71,** 2407–2410.
101. Kahl, G. F., Friederici, D. E., Bigelow, S. W., Okey, A. B., and Nebert, D. W. (1980). Ontogenetic expression of regulatory and structural gene products associated with the *Ah* locus. *Dev. Pharmacol. Ther.* **1,** 137–162.
102. Feuer, G. (1979). Action of pregnancy and various progesterones on hepatic microsomal activities. *Drug Metab. Rev.* **9,** 147–169.
103. Guenthner, T. M., and Mannering, G. J. (1976). Induction of hepatic monooxygenase systems in fetal and neonatal rats with phenobarbital, polycyclic hydrocarbons and other xenobiotics. *Biochem. Pharmacol.* **26,** 567–575.
104. Nebert, D. W., Eisen, H. J., Negishi, M., Lang, M. A., and Hjelmeland, L. M. (1981). Genetic mechanisms controlling the induction of polysubstrate monooxygenase (P-450) activities. *Annu. Rev. Pharmacol. Toxicol.* **21,** 431–462.
105. Atlas, S. A., Boobis, A. R., Felton, J. S., Thorgeirsson, S. S., and Nebert, D. W.

(1977). Ontogenetic expression of polycyclic aromatic compound-inducible monooxygenase activities and forms of cytochrome P-450 in rabbit. *J. Biol. Chem.* **252,** 4712–4721.
106. Jakoby, W. B. (1980). Introduction. *In* "Enzymatic Basis of Detoxication" (W. B. Jakoby, ed.), Vol. 1, pp. 1–6. Academic Press, New York.
107. Nebert, D. W. (1979). Multiple forms of inducible drug-metabolizing enzymes: A reasonable mechanism by which any organism can cope with adversity. *Mol. Cell. Biochem.* **27,** 27–46.
108. Dutton, G. J. (1982). Biochemical development of the foetus and neonate. *In* "The Neonate" (C. T. Jones, ed.), pp. 823–844. Elsevier-North Holland, Amsterdam.
109. Gillette, J. R., Mitchell, J. R., and Brodie, B. B. (1974). Biochemical mechanisms of drug toxicity. *Annu. Rev. Pharmacol. Toxicol.* **14,** 271–287.
110. Mulder, G. J. (1979). Detoxification or toxification? Modification of the toxicity of foreign compounds by conjugation in the liver. *Trends Biochem. Sci.* **4,** 86–90.
111. Stockley, I. (1980). The role of drug metabolism in the development of clinically significant adverse durg interactions. *Rev. Metab. Drug Interact.* **3,** 2–29.

CHAPTER 4

Intratissue Distribution of Activating and Detoxicating Enzymes*

Jeffrey Baron and Thomas T. Kawabata

I. Introduction 105
II. Distribution within the Liver 107
   A. Histological Basis for Heterogeneous Intrahepatic Distribution 107
   B. Heterogeneity of Xenobiotic Metabolism within the Liver Lobule 112
III. Distribution within the Lung 121
   A. Nonciliated Bronchiolar Epithelial Cells 122
   B. Type II Alveolar Epithelial Cells 123
   C. Pulmonary Alveolar Macrophages 124
IV. Distribution within the Skin 124
V. Comments 127
   References 128

I. INTRODUCTION

It is now evident that chemicals often exert relatively selective toxic actions within many tissues. That is, they frequently induce damage either in cells of a specific morphological type or in morphologically similar cells that are found within specific areas or regions of tissues. For example,

---

* Recent research in this laboratory was supported by PHS Grant Number 12675, awarded by the National Institute of General Medical Sciences, DHHS.

hepatotoxins such as acetaminophen,[1] bromobenzene and other halogenated aromatic hydrocarbons,[2,3] furosemide,[4] carbon tetrachloride,[5,6] and thioacetamide[7] all exert their toxic effects primarily upon those parenchymal cells that surround the central veins, whereas other hepatotoxins, such as 2-acetylaminofluorene[8,9] and both allyl formate[10] and its metabolite, allyl alcohol,[8] primarily damage those parenchymal cells that surround the portal triads. A similar picture of the relatively selective nature of chemically induced toxicity is seen in the lung, where the nonciliated bronchiolar epithelial (Clara) cell is the primary, if not the sole, target for toxins such as 4-ipomeanol[11] and carbon tetrachloride.[11,12] Factors involved in determining the basis for such selectivity in chemically induced toxicity undoubtedly include the intratissue distribution of the toxicant, differences in the uptake of the toxicant into specific cells, and differences in the intratissue distributions of the various enzymes and cofactors that are involved in the metabolism of the toxicant.

Although it is readily apparent that tissues are not composed of a single population of morphologically identical cells and that significant enzymatic differences exist among morphologically dissimilar cells, it has not always been appreciated that morphologically similar cells can and often do exhibit significant differences in the contents and activities of enzymes. Enzymatic differences existing among both morphologically dissimilar and similar cells may be of the utmost importance in determining the susceptibility of cells to toxicities induced by xenobiotics because chemically inert substances are very often enzymatically transformed into highly reactive, toxic, electrophilic metabolites that are capable of covalently interacting with cellular macromolecules, thereby initiating the cytotoxic process. The relative amounts and activities of bioactivating and detoxicating enzymes within individual cells may be the critical factor determining whether cytotoxic metabolites accumulate within specific cells that would thereby be susceptible to this type of chemically induced toxicity. For these reasons, a number of laboratories have initiated investigations on the intratissue distributions of enzymes that catalyze both the bioactivation and detoxication of cytotoxic chemicals.

Among those enzymes participating in the metabolic activation and detoxication of both endogenous and exogenous chemicals, the numerous isoenzymes of cytochrome P-450 play central roles. In addition to these hemoproteins, NADPH-cytochrome c (P-450) reductase, cytochrome b_5, epoxide hydrolase, and those enzymes catalyzing phase II (conjugation) reactions such as the glutathione S-transferases, UDP-glucuronyltransferases, and sulfotransferases all play important roles in the biotransformation of toxicants and other chemicals. Detailed descriptions of the properties, intracellular distributions, and roles of these

4. Intratissue Distribution of Enzymes

enzymes are beyond the scope of this chapter, and the reader is referred to two earlier volumes in this series[13] in which these aspects have been reviewed. No attempt will be made here to provide a comprehensive review of the localizations and distributions of xenobiotic activating and detoxicating enzymes within all tissues. Rather, attention will be focused upon three tissues in which the distributions of these enzymes have been most extensively studied: liver, lung, and skin. Particular emphasis will be placed on results obtained from studies in which immunohistochemical, microspectrophotometric (microdensitometric), and histochemical approaches have been employed. However, this review is not meant to be exhaustive. The intent is to introduce the reader to a concept—the heterogeneous distributions of xenobiotic-metabolizing enzymes within tissues—that should provide a more complete understanding of why many chemicals frequently exert their toxic effects upon only certain cells within tissues.

II. DISTRIBUTION WITHIN THE LIVER

A. Histological Basis for Heterogeneous Intrahepatic Distribution

To appreciate the phenomenon of enzymatic heterogeneity within the liver, a brief overview of the microscopic anatomy of this tissue and its relationship to metabolic function is necessary. At the light microscopic level, several major cell types can be identified: parenchymal cells (hepatocytes), which are arranged in a series of communicating cell plates and which have been estimated to comprise approximately 80% of the cell population within the human liver[14]; flattened, epithelial sinusoidal cells; phagocytic Kupffer cells occurring at points along the sinusoidal lining; bile duct cells; and cells associated with hepatic blood vessels. The parenchymal cell is the major site for the metabolic activation and detoxication of chemicals, although acetylation of a limited number of compounds occurs within Kupffer cells.[15]

Three different models have been developed to describe the histological and/or functional unit of the liver: the classic lobule, the liver acinus, and the portal lobule. With respect to chemically induced toxicity, most attention has been focused on the classic liver lobule and liver acinus models. The classic liver lobule is roughly hexagonal in shape. At the angles of the hexagon are the portal triads or canals in which the terminal branches of the hepatic artery, portal vein, and bile duct are found. At the center of the lobule, the sinusoids that separate the parenchymal cell plates con-

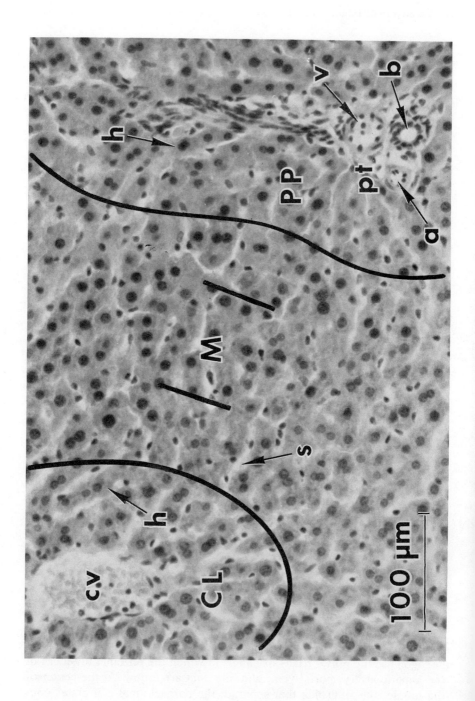

4. Intratissue Distribution of Enzymes

verge to form the central vein, also called the *terminal hepatic venule*. The classic liver lobule is considered to be composed of three separate regions: the centrilobular or pericentral region that surrounds the central vein; the periportal region that surrounds the portal triad; and the midzonal region, an area lying between a central vein and a portal triad. Loud[16] has estimated that centrilobular and periportal cells each account for approximately 20% of the parenchymal cells across a lobule in rat liver, with midzonal cells comprising the remaining 60%. Conservative estimates of the limits of the three regions of the classic liver lobule employed by this laboratory for semiquantitative immunohistochemical analyses of xenobiotic activating and detoxicating enzymes are indicated in Fig. 1.

Although many laboratories have utilized the classic liver lobule model, others have employed the liver acinus model introduced by Rappaport and his colleagues.[17–19] The simple liver acinus is a small, irregular mass of parenchymal cells arranged in three concentric zones around an axis consisting of a portal canal: zone I consists of those cells closest to the portal canal; zone III consists of those cells farthest from the portal canal; and zone II consists of those cells lying between zones I and III. In general, zone I cells of this model can be equated with periportal cells of the classic liver lobule, zone II cells with midzonal cells, and zone III (periacinar) cells with centrilobular cells.

Evidence for morphological heterogeneity among hepatic parenchymal cells has been provided by both cell density and morphometric analyses. Sedimentation velocity analysis of isolated hepatocytes demonstrated the existence of significant heterogeneity in both cell size and density.[20] Similar findings were obtained after isolated rat hepatic parenchymal cells were separated by isopycnic centrifugation, leading to the conclusion that centrilobular cells possess a significantly lower density than do periportal cells.[21,22] Morphometric analyses revealed significant differences in the contents of subcellular organelles within parenchymal cells in the three regions of the liver lobule.[16] Whereas rough endoplasmic reticulum was rather uniformly distributed throughout the lobule, smooth endoplasmic

---

Fig. 1. Portion of a lobule in the liver of an untreated male rat stained with hematoxylin and eosin. The arrows indicate the following: a, branch of a hepatic artery; b, branch of a bile duct; h, plate of hepatocytes (parenchymal cells); s, sinusoid; and v, branch of a portal vein. For semiquantitative immunohistochemical analyses of xenobiotic activating and detoxicating enzymes, centrilobular (CL) hepatocytes are considered to lie within the first five cell layers surrounding a central vein (cv), periportal (PP) hepatocytes are considered to lie within the first five cell layers surrounding a portal triad (pt), and midzonal (M) hepatocytes are considered to lie within three cell layers on either side of a line midway between a central vein and a portal triad.

reticulum was more abundant within centrilobular cells than within either midzonal or periportal cells. Differences in both the size and distribution of mitochondria were also demonstrated across the lobule, with centrilobular cells containing the smallest but most numerous mitochondria.[16]

Regardless of the histological model of the liver that is chosen, it must be kept in mind that, whereas bile flows from the parenchymal cells, where it is formed, through the bile canaliculi toward the interlobular bile ducts in the portal canals, blood flows from the terminal hepatic arterioles and the terminal portal venules in the portal triads through the sinusoids toward the central veins. Because of the directional flow of blood, it has been suggested that a gradient of oxygen and other nutrients is produced in sinusoidal blood between the periportal region and the centrilobular region, thereby creating different microenvironments across the liver lobule.[17-19] Because the metabolic function of a cell is dependent upon its supply of oxygen and other nutrients, the microcirculation of the liver has been proposed[17-19] as forming the basis for metabolic heterogeneity among hepatic parenchymal cells and, in addition, as explaining why centrilobular parenchymal cells are frequently much less resistant to chemically induced damage than are either midzonal or periportal parenchymal cells. Although oxygen tensions within periportal regions have been demonstrated to be greater than those within centrilobular regions,[18,23] the mere presence of such an oxygen gradient across the liver lobule does not in itself appear to be the major factor determining the greater susceptibility of centrilobular cells to damage after exposure to hepatotoxins. Moreover, because oxygen is required for monooxygenation reactions catalyzed by cytochrome P-450, a significantly lower oxygen tension within centrilobular cells might be expected to be mirrored by correspondingly lower rates of cytochrome P-450–catalyzed monooxygenations within this region, including those resulting in the formation of highly reactive, toxic, electrophilic metabolites from olefinic and aromatic substances. However, as discussed subsequently, centrilobular hepatocytes seem to possess the greatest capacity for catalyzing the oxidative metabolism of most substrates.

Substantial evidence for metabolic zonation within the liver lobule has been provided by both enzyme histochemical analyses and enzymatic analyses conducted on populations of parenchymal cells obtained from different regions of the lobule by means of microdissection techniques. Results of qualitative and quantitative enzyme histochemical analyses conducted during the past three decades have clearly revealed that many enzymes are not uniformly distributed throughout the liver lobule (Table I). It appears that cells lying within the periportal regions of the liver lobule have greater roles in gluconeogenesis and glycogen metabolism

4. Intratissue Distribution of Enzymes

TABLE I

Predominant Intrahepatic Localization of Some Enzymatic Activities Not Directly Involved in the Metabolic Activation and Detoxication of Xenobiotics

Enzyme	Region of greatest activity	References
Alkaline phosphatase	Periportal	24
Cytochrome oxidase	Periportal	25, 26
Glucose 6-phosphatase	Periportal	26, 27
Glucose 6-phosphate dehydrogenase	Centrilobular	26
Glucosephosphate isomerase	Periportal	24
Glutamate dehydrogenase	Centrilobular	24, 26, 28
Glutamic-pyruvic transaminase	Periportal	24, 29
β-Hydroxybutyrate dehydrogenase	Centrilobular	26
Isocitrate dehydrogenase	Centrilobular	24, 28
Lactate dehydrogenase	Periportal	24, 26, 28
Malate dehydrogenase	Periportal	24, 28
Succinate dehydrogenase	Periportal	25, 26, 28, 30

than do cells surrounding the central veins.[19,26,31,32] Cells located within the periportal region also possess greater activity associated with both the tricarboxylic acid cycle and mitochondrial respiration than do centrilobular cells. LeBouton[33] has provided evidence that the periportal region is also the primary site for both protein synthesis and protein degradation within the liver.

The activity of alcohol dehydrogenase has also been histochemically demonstrated to be greatest within the periportal region of the lobule.[10] Because allyl formate is metabolically converted via allyl alcohol to the cytotoxic substance, acrolein (allyl aldehyde), by alcohol dehydrogenase, this histochemical finding offers an explanation for the selective damage to periportal cells seen after the administration of allyl formate or allyl alcohol.[10]

In contrast to the functions of parenchymal cells found within the periportal region of the liver lobule, parenchymal cells found within the centrilobular region appear to be more active in glycolysis, glycogen storage, liponeogenesis, and pigment formation.[19,26,31,32] Centrilobular parenchymal cells also possess greater NADPH- and NADH-tetrazolium reductase activities than do cells surrounding the portal triads.[34,35] Moreover, the generation of NADPH, the primary source of reducing equivalents for monooxygenation reactions catalyzed by cytochrome P-450, proceeds at a significantly greater rate within the cytosol of centrilobular cells than within the cytosol of periportal cells.[36] However, it has been suggested that the supply of NADPH may limit the rates of cytochrome

P-450–catalyzed monooxygenations within centrilobular but not periportal regions of the liver lobule.[37]

B. Heterogeneity of Xenobiotic Metabolism within the Liver Lobule

It should be clear that there is a definite zonation of metabolic function within the liver lobule, especially with respect to carbohydrate homeostasis. It is also possible, therefore, that xenobiotics and other chemicals do not undergo bioactivation and detoxication uniformly within the liver lobule. Such metabolic zonation or nonuniform distribution of bioactivating and detoxicating enzymes could contribute greatly to the relatively selective damage to parenchymal cells frequently observed after exposure to a variety of hepatotoxins. A number of laboratories have undertaken intensive examinations of this possibility, the results of which are summarized here. In this discussion, emphasis will be placed on results obtained using untreated animals, although the effects of enzyme induction will be considered where data are available.

1. Intralobular Distributions of Xenobiotic Activating and Detoxicating Enzymes

a. Cytochrome P-450. Of all the enzymes involved in the bioactivation and detoxication of exogenous and endogenous substances in the liver, cytochrome P-450 has received the greatest attention. By means of a microspectrophotometric (microdensitometric) technique with which cytochrome P-450 can be directly detected and measured within hepatic parenchymal cells in unfixed, cryostat tissue sections, livers of rats pretreated with phenobarbital were found to contain more than twice as much cytochrome P-450 than did livers of untreated rats.[38] However, the regions within the liver lobule from which the microspectrophotometric measurements were obtained were not identified. Subsequent extensions of this study[39–41] led to the conclusion that the concentration of cytochrome P-450 within livers of both untreated and phenobarbital-treated rats varied significantly across the lobule: the cytochrome P-450 concentration within hepatocytes decreased fairly steadily from the central vein toward the portal triad. In livers of untreated rats, the concentration of cytochrome P-450 within centrilobular cells was about 1.6 times greater than that within periportal cells. After treatment with phenobarbital, cytochrome P-450 was primarily induced within centrilobular cells, which contained between 2.8 and 4.0 times as much cytochrome P-450 as did periportal cells. More conventional spectrophotometric analyses, using microsomes prepared from isolated subpopulations of hepatic paren-

4. Intratissue Distribution of Enzymes

chymal cells, revealed that phenobarbital produced a three- to fivefold increase in the cytochrome P-450 content within centrilobular cells without significantly altering the content of the hemoprotein within periportal cells.[42]

Although the microspectrophotometric technique allows for the quantitative investigation of cytochrome P-450 across the liver lobule, the sensitivity of the method is fairly low.[40] Moreover, it does not discriminate among the various isoenzymes of the hemoprotein, that is, it detects all forms of cytochrome P-450 associated with the endoplasmic reticulum, the nucleus, and mitochondria. Although different isoenzymes of hepatic microsomal and nuclear cytochrome P-450 exhibit overlapping substrate specificities, a metabolite formed under the influence of one isoenzyme may be a highly reactive and toxic electrophile, whereas another metabolite of the same parent compound formed under the influence of a different isoenzyme may be much less reactive and toxic. Thus, differences in the intralobular distributions of the several cytochrome P-450 isoenzymes may be of critical importance in determining both the locations and the severities of specific chemically induced hepatotoxicities.

In order to determine whether different cytochrome P-450 isoenzymes are similarly distributed within the liver lobule, Baron and colleagues[43-48] initiated qualitative and semiquantitative immunohistochemical investigations on two forms of the hemoprotein, cytochromes P-450 PB-B and MC-B. These protein species represent the major forms of cytochrome P-450 isolated from hepatic microsomes of rats treated with phenobarbital and 3-methylcholanthrene, respectively.[49,50] Whereas the content of both cytochrome P-450 species were found to be greatest within centrilobular parenchymal cells in livers of untreated rats, they exhibited significantly different patterns of intralobular distribution (Fig. 2).[43-46] The content of cytochrome P-450 PB-B decreased fairly steadily from the central vein toward the portal triad: centrilobular cells contained approximately 1.3 times as much cytochrome P-450 PB-B as did midzonal cells, whereas cells within the centrilobular region contained approximately 1.6 times as much cytochrome P-450 PB-B as did periportal cells. This immunohistochemical finding is, therefore, quite similar to the difference determined for cytochrome P-450 content between centrilobular and periportal cells in livers of untreated rats by means of the microspectrophotometric technique.[39-41] In contrast, the content of cytochrome P-450 MC-B was similar within midzonal and periportal hepatocytes; these cells contained between 20 and 30% less cytochrome P-450 MC-B than did centrilobular parenchymal cells. Thus, the two isoenzymes of cytochrome P-450 exhibit different patterns of intralobular distribution within the livers of untreated animals.

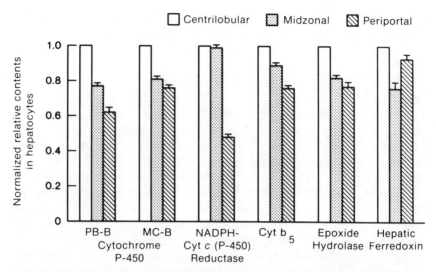

Fig. 2. Intralobular distributions of monooxygenase components and epoxide hydrolase within livers of untreated male rats. Each bar represents the mean ± SE normalized content of enzyme within hepatocytes in the specified lobular region. The data were derived from semiquantitative immunohistochemical determinations of the relative extents of antibody binding to centrilobular, midzonal, and periportal hepatocytes[44] and were normalized by assigning a value of 1.0 for the extent of antibody binding to centrilobular hepatocytes.

Because inductions of the different cytochrome P-450 isoenzymes within the different regions of the liver lobule could profoundly modify both the locations and severities of certain chemically induced hepatotoxicities, the effects of a number of xenobiotics on cytochromes P-450 PB-B and MC-B within the liver were also investigated by immunohistochemical techniques.[46–48] The administration of phenobarbital to rats did not result in any apparent alteration in hepatic cytochrome P-450 MC-B. In contrast, both 3-methylcholanthrene and β-naphthoflavone induced this isoenzyme within parenchymal cells throughout the lobule. However, whereas cytochrome P-450 MC-B was induced by a similar extent within midzonal and periportal hepatocytes following the administration of each polycyclic aromatic hydrocarbon, significantly less induction was detected within centrilobular hepatocytes.[46,47] The induction of cytochrome P-450 PB-B by both phenobarbital and trans-stilbene oxide within the liver lobule was also found to be nonuniform. The administration of phenobarbital to rats resulted in similar degrees of induction of cytochrome P-450 PB-B within midzonal and periportal hepatocytes, but significantly less induction of this isoenzyme occurred within centrilobular hepatocytes.[48] In contrast to the effects of phenobarbital, a gradient of induction of cytochrome P-450 PB-B was

4. Intratissue Distribution of Enzymes

produced across the lobule after rats had been treated with *trans*-stilbene oxide; the extent of induction was greatest within periportal and least within centrilobular hepatocytes.[48] Treatment of rats with β-naphthoflavone had no apparent effect on hepatic cytochrome *P*-450 PB-B,[48] although its content decreased within centrilobular cells after administration of 3-methylcholanthrene.[47,48] Thus, not only do cytochromes *P*-450 PB-B and MC-B normally exhibit nonuniform patterns of intralobular distribution, but they are not affected uniformly across the lobule following the administration of xenobiotics.

b. NADPH-Cytochrome c (P-450)Reductase. NADPH-cytochrome *c* (*P*-450) reductase is capable of catalyzing the NADPH-dependent reduction of tetrazolium dyes, and the observation that histochemical staining for NADPH-tetrazolium reductase activity was most pronounced within centrilobular hepatocytes[34,35] suggested that the reductase was predominantly localized within the centrilobular region of the liver lobule. However, other hepatic enzymes, including the NADPH-dependent flavoprotein dehydrogenase (hepatic ferredoxin reductase) associated with mitochondrial cytochrome *P*-450 and certain NADH-linked dehydrogenases that would be supplied with NADH via mitochondrial transhydrogenase activity, could also participate in the histochemical staining reaction for NADPH-tetrazolium reductase activity. Moreover, these enzymatic activities may not be uniformly distributed across the liver lobule. For this reason, the intralobular distribution of NADPH-cytochrome *c* (*P*-450) reductase has been investigated by qualitative and semiquantitative immunohistochemical techniques.[44,47,51–54] These studies demonstrated that the reductase is not uniformly distributed throughout the lobule in livers of untreated animals: centrilobular and midzonal hepatocytes contained similar amounts of NADPH-cytochrome *c* (*P*-450) reductase, whereas hepatocytes within periportal regions of the lobule contained only about half as much enzyme (Fig. 2).[44,47,52,53] Induction within the liver lobule by xenobiotics has also been investigated. Although treatment of rats with 3-methylcholanthrene did not result in any apparent alteration in hepatic NADPH-cytochrome *c* (*P*-450) reductase,[47,53] both phenobarbital and pregnenolone-16α-carbonitrile induced the enzyme within hepatocytes throughout the lobule. However, although phenobarbital induced the reductase uniformly across the lobule,[47,53] pregnenolone-16α-carbonitrile produced significantly different amounts of induction within the three regions of the liver lobule; the greatest degree of induction occurred within periportal hepatocytes and the least within midzonal hepatocytes.[53] Thus, not only does NADPH-cytochrome *c* (*P*-450) reductase normally exhibit a nonuniform pattern of intralobular distribution (one that differs significantly from those exhibited by cytochromes P-450 PB-B and MC-B), but it is affected quite differently within the three regions of the liver lobule following the administration of specific xenobiotics.

c. Cytochrome b_5. The intralobular distribution of cytochrome b_5 has been studied only by an immunohistochemical approach and only within livers of untreated rats.[44,55,56] Although immunohistochemical staining for cytochrome b_5 was initially reported to be limited to parenchymal cells surrounding the central veins,[56] results of a more recent study demonstrated that the hemoprotein was present within hepatocytes throughout the liver lobule.[44] However, cytochrome b_5 was not distributed uniformly across the lobule: its concentration decreased from the central vein toward the portal triad (Fig. 2). The pattern of intralobular distribution of cytochrome b_5 within livers of untreated rats, therefore, resembles that of cytochrome P-450 PB-B, although the difference in cytochrome b_5 content between centrilobular and periportal hepatocytes was only about half as great as that determined for cytochrome P-450 PB-B.

d. Epoxide Hydrolase. The intralobular distribution of epoxide hydrolase (EC 3.3.2.3; formerly referred to as epoxide hydrase or epoxide hydratase) has been studied by means of immunohistochemical approaches within livers of both untreated and xenobiotic-treated rats. In liver of untreated rats, immunohistochemical staining for the enzyme was detected within parenchymal cells throughout the lobule, although centrilobular hepatocytes were more intensely stained for this enzyme than were either midzonal or periportal hepatocytes.[44,57–60] Semiquantitative immunohistochemical analyses revealed that the intralobular distribution of epoxide hydrolase within livers of untreated rats was essentially identical to that determined for cytochrome P-450 MC-B, that is, midzonal and periportal hepatocytes contained similar amounts of the hydrolase, whereas between 20 and 30% more enzyme was present within centrilobular hepatocytes (Fig. 2).[44,58] Upon administration of the hepatocarcinogen 2-acetylaminofluorene, a substance that initially produces selective damage to cells lying within the periportal regions of the liver lobule,[8,9] the intensity of immunohistochemical staining for epoxide hydrolase increased, especially within midzonal and periportal regions.[59,60] This observation suggested that 2-acetylaminofluorene induced the enzyme best within midzonal and periportal hepatocytes. Semiquantitative immunohistochemical analyses demonstrated that *trans*-stilbene oxide also induced epoxide hydrolase to a similar extent within midzonal and

4. Intratissue Distribution of Enzymes

periportal hepatocytes but to a significantly lesser degree within centrilobular hepatocytes.[61] 3-Methylcholanthrene exerted only a relatively weak inductive action upon epoxide hydrolase; the enzyme was similarly affected within the three regions of the liver lobule. In contrast, after the administration of phenobarbital, epoxide hydrolase was induced to the greatest extent within midzonal hepatocytes and to the least extent within periportal hepatocytes.[61] Epoxide hydrolase, therefore, also normally has a nonuniform pattern of distribution across the liver lobule and is affected in markedly different ways within the three regions of the liver lobule following the administration of several different xenobiotics.

e. Hepatic Ferredoxin. Hepatic mitochondria have been reported to be capable of catalyzing the bioactivation of the hepatocarcinogen, aflatoxin B_1.[62] One of the electron transport components required for monooxygenations catalyzed by hepatic mitochondrial cytochrome P-450 is the iron–sulfur protein, hepatic ferredoxin (hepatoredoxin).[63-66] This protein has been immunohistochemically demonstrated as present within parenchymal cells throughout the lobule in livers of untreated rats.[44,66] Semiquantitative immunohistochemical analyses, however, revealed that hepatic ferredoxin is not uniformly distributed across the liver lobule; although centrilobular and periportal hepatocytes contained similar amounts of this protein, significantly less ferredoxin was present within midzonal hepatocytes (Fig. 2).[44] This pattern of intralobular distribution of hepatic ferredoxin differs markedly from those determined for cytochromes P-450, NADPH-cytochrome c (P-450) reductase, cytochrome b_5, and epoxide hydrolase for untreated rats. Moreover, hepatic ferredoxin and NADPH-cytochrome c (P-450) reductase represent two enzymes participating in the bioactivations of exogenous and endogenous substances that are not preferentially concentrated within centrilobular parenchymal cells.

f. Glutathione S-Transferases. The glutathione S-transferases are a family of detoxicating enzymes that catalyze the conjugation of the sulfhydryl group of reduced glutathione with a wide variety of electrophilic substances, the first step leading to the formation of water-soluble mercapturic acid derivatives that are readily excreted via the kidney.[13,67] The distribution of glutathione S-transferase B (also referred to as *ligandin* and *azo dye-binding protein*) within the liver lobule has been investigated immunohistochemically, although conflicting findings have been reported. Initially, it was observed that centrilobular parenchymal cells in livers of untreated rats were much more intensely stained for glutathione S-transferase B than were those hepatocytes surrounding the portal triad.[68] Subsequently, parenchymal cells throughout the lobule were reported to be uniformly stained for this transferase in both rat[69] and human liver.[70] Later studies have demonstrated that whereas parenchymal cells throughout the lobule in livers of control rats were immunohistochemically stained for transferase B and two other glutathione S-transferase isoenzymes, transferases C and E,[67] the intensity of staining for each enzyme was more pronounced within centrilobular cells than within periportal cells.[71] Although the discrepancies in these immunohistochemical observations remain to be resolved, it will be appreciated that the role glutathione S-transferases play in the detoxication of chemicals in the liver is dependent, in large measure, upon the content of reduced glutathione within hepatic parenchymal cells. The distribution of reduced glutathione across the liver lobule has been investigated by histochemical means, although conflicting findings have been obtained. Histochemical staining for reduced glutathione was relatively uniformly dispersed throughout the lobule in livers of untreated rats,[72,73] but quantitative cytochemical analyses revealed that reduced glutathione is not evenly distributed across the liver lobule; centrilobular hepatocytes contained much less glutathione than did those hepatocytes lying within the midzonal and periportal regions of the lobule.[74] This observation may explain, at least in part, the frequently greater susceptibility of centrilobular hepatocytes to chemically induced damage resulting from the generation of reactive electrophiles that are inactivated by conjugation with reduced glutathione.

g. UDPglucuronyltransferases. The intralobular distribution of UDPglucuronyltransferases, enzymes that catalyze the transfer of the glucuronyl moiety of UDPglucuronic acid to a variety of endogenous and exogenous compounds and thereby render these substances more water soluble and more readily excreted via the kidney, has not received a great deal of attention. Rappaport[19] suggested that glucuronidation is most active within the periportal regions of the liver lobule. On the basis of the effects of several hepatotoxins that cause relatively selective regional damage within the liver lobule, UDPglucuronyltransferase activity was concluded to be predominantly localized within periportal and midzonal parenchymal cells.[75] In contrast, the results of studies on the pharmacokinetics of harmol (7-hydroxy-1-methyl-9H-pyrido[3,4b]indole) in perfused rat liver preparations have led to the conclusion that UDPglucuronyltransferases are localized primarily within hepatocytes that are distal to the portal triad.[76] Although this discrepancy remains to be resolved, it is possible that different UDPglucuronyltransferases exhibit distinct patterns of intralobular distribution.

4. Intratissue Distribution of Enzymes

h. Aryl Sulfotransferases. The intralobular distributions of the several aryl sulfotransferases have also not received a great deal of attention, although indirect evidence provided by studies on the bioactivation of 2-acetylaminofluorene[8,9,77] and on the sulfation of drugs in perfused liver preparations[76,78] indicates that these enzymes are predominantly localized within the periportal regions of the liver lobule. Again, however, it is possible that different aryl sulfotransferase isoenzymes exhibit different patterns of intralobular distribution.

2. Intralobular Distributions of Xenobiotic Monooxygenase Activities

a. Covalent Binding of Xenobiotic Metabolites to Hepatic Parenchymal Cells. It is now well established that highly reactive, toxic, electrophilic metabolites are quite frequently generated from relatively inert chemicals under the influence of cytochromes P-450 (see Chapter 2, this volume). Once formed, these metabolites can attack and covalently bind to nucleophilic sites on cellular macromolecules and thereby elicit cytotoxicity that is manifested as necrosis, mutagenesis, and carcinogenesis. A number of laboratories have utilized autoradiographic techniques in an attempt to determine the precise location(s) within the liver lobule for the generation of reactive metabolites *in vivo* after the administration of radiolabeled hepatotoxins to animals. Results of these investigations have revealed that reactive metabolites of hepatotoxins such as acetaminophen[79] and bromobenzene[2,3] preferentially bind to parenchymal cells lying within those regions of the liver lobule that are selectively damaged by the xenobiotics. Moreover, the degree of covalent binding of reactive, electrophilic metabolites to hepatocytes was found (1) to be both dose dependent and directly proportional to the severity of the chemically induced damage (e.g., necrosis); (2) to be reduced by the prior administration of substances that inhibit either the synthesis or activity of cytochromes P-450; and (3) in general, to be increased by the prior administration of substances that induce the hepatic cytochrome P-450 species that participate in the bioactivations of the hepatotoxins.[2,3,79] The results of these autoradiographic analyses imply that the rates of activation exceed the rates of detoxication of hepatotoxins within those regions of the liver lobule that are preferentially damaged by the xenobiotics and are consistent with the finding that enzymes participating in xenobiotic bioactivation and detoxication are not uniformly distributed within the liver lobule. Support for this concept (summarized in Chapter 4) is provided by results of histochemical analyses of a number of cytochrome P-450–catalyzed xenobiotic monooxygenations.

b. Benzo[a]pyrene Hydroxylase Activity. The pioneering histochemical studies conducted 2 decades ago by Wattenberg and Leong[80] provided the first direct proof for the nonuniform distribution of a xenobiotic monooxygenase activity within the liver lobule. Employing a fluorescence histochemical technique, these investigators were able to demonstrate that several polycyclic aromatic hydrocarbons, including benzo[a]pyrene, were hydroxylated most extensively within centrilobular regions in livers of untreated rats. Aryl hydrocarbon hydroxylase activity was also reported to be enhanced to the greatest degree within the central portion of the liver lobule following the treatment of rats with 3-methylcholanthrene.[80] Thus, not only did this study demonstrate that xenobiotics such as polycyclic aromatic hydrocarbons do not undergo oxidative metabolism uniformly across the liver lobule, but also that the oxidative metabolism of this group of compounds is not induced equally within all regions of the liver lobule.

c. Aniline Hydroxylase Activity. Using a histochemical technique, Gangolli and Wright[81] found that p-aminophenol, the hydroxylated metabolite of aniline, was produced within parenchymal cells throughout the lobule in livers of untreated rats. Periportal hepatocytes, however, were observed to exhibit the greatest aniline hydroxylase activity. On the basis of analyses of enzymatic activities catalyzed *in vitro* by hepatic microsomes isolated from rats treated with various hepatotoxins, aniline hydroxylase activity was also concluded to be concentrated within the midzonal and periportal regions of the liver lobule.[75] Although histochemical staining for aniline hydroxylase activity was enhanced within all hepatic parenchymal cells following treatment with phenobarbital and 3-methylcholanthrene,[81] there was no indication of the degree to which this monooxygenase activity was induced within the three regions of the liver lobule.

d. 7-Ethoxycoumarin O-Deethylase Activity. A microfluorometric technique with which the formation of 7-hydroxycoumarin (umbelliferone) from 7-ethoxycoumarin can be determined has been described.[37] Employing micro-light guides, the relative rates of 7-hydroxycoumarin formation were measured within centrilobular and periportal regions in perfused rat livers, and it was found that 7-ethoxycoumarin was O-deethylated within centrilobular regions at a rate that was approximately twice as great as that within periportal regions.[37]

e. Other Xenobiotic Monooxygenase Activities. The distributions of a limited number of other xenobiotic monooxygenase activities within

4. Intratissue Distribution of Enzymes

the liver lobule have been investigated only indirectly. Examination of the metabolism of phenacetin in perfused rat livers led to the conclusion that the O-deethylation of this xenobiotic occurs predominantly within centrilobular hepatocytes.[78] On the basis of the results of enzymatic analyses of microsomal preparations derived from livers of rats treated with various hepatotoxins, it was concluded that centrilobular hepatocytes also possess greater aminopyrine N-demethylase activity than do periportal hepatocytes.[75] A similar conclusion was reached after examination of the *in vivo* metabolism of aminopyrine in rats treated with hepatotoxins.[82] In contrast, the distribution of p-nitroanisole N-demethylase activity within the liver lobule appears to resemble that determined for aniline hydroxylase activity, that is, p-nitroanisole N-demethylase activity is greatest within the midzonal and periportal regions of the liver lobule.[75]

III. DISTRIBUTION WITHIN THE LUNG

The lung is one of the major portals for entry of xenobiotics into the body. It is also continuously exposed to those chemicals that are present in the general circulation and represents another organ in which many chemicals exert relatively selective toxicities. The ability of the lung to metabolize drugs, carcinogens, and other xenobiotics has received a great deal of attention, and we now know that these substances can undergo both bioactivation and detoxication within this organ.[83-86] During the past several years, cytochrome P-450,[87-92] NADPH-cytochrome c (P-450) reductase,[87,89-91] cytochrome b_5,[87,89,90] epoxide hydrolase,[93-95] glutathione S-transferase,[96,97] UDPglucuronyltransferase,[98,99] and sulfotransferase[85,100] have all been shown to be present within the mammalian lung. However, the exact cellular localizations of these enzymes within the lung are largely unknown, primarily because of the complex and heterogeneous nature of the organ: at last 40 distinctive cell types are present,[101] some of which cannot be unequivocally identified at the light microscopic level. Moreover, attempts to localize xenobiotic-metabolizing activities within the respiratory tract have often yielded conflicting observations. For example, although Wattenberg and Leong[102] were unable to detect histochemically benzo[*a*]pyrene hydroxylase activity within tracheal or bronchial mucosa, others[103-106] have provided biochemical evidence for the oxidative metabolism of this polycyclic aromatic hydrocarbon within both. Moreover, a considerable degree of histochemical staining for aniline hydroxylase activity has been observed within bronchial epithelial cells,[107] and cytochrome P-450[108] and NADPH-cytochrome c

(P-450) reductase[54] have both been immunohistochemically detected in bronchi.

Despite the conflicting observations for the metabolism of benzo[a]pyrene in the trachea and bronchi, the results of many studies indicate that xenobiotics can be metabolized within bronchiolar and alveolar cells, more specifically, within nonciliated bronchiolar epithelial (Clara) cells and type II alveolar epithelial cells. Results of a number of other studies further indicate that pulmonary alveolar macrophages are also capable of metabolizing a variety of xenobiotics. Evidence for the participation of these three pulmonary cells in the metabolism of xenobiotics is summarized below.

A. Nonciliated Bronchiolar Epithelial Cells

Two major epithelial cell types, one ciliated and the other nonciliated, line the bronchioles. Although much less numerous than the ciliated bronchiolar epithelial cell, the nonciliated bronchiolar epithelial cell (commonly referred to as the Clara cell) appears to be one of the most metabolically active cells in the lung.[109] It has also been shown to be the target for many pulmonary toxins that require metabolic activation, including 4-ipomeanol,[11,12,110] carbon tetrachloride,[11,12] 3-methylfuran,[11] naphthalene,[111] and N-nitrosamines.[112] In contrast to the ciliated bronchiolar epithelial cell, the Clara cell contains an unusual abundance of smooth endoplasmic reticulum,[109] membranes with which cytochromes P-450 and many other xenobiotic-metabolizing enzymes are associated. Although cytochrome P-450 has been immunohistochemically detected within Clara cells,[108,113,114] immunohistochemical staining was not detected within ciliated bronchiolar epithelial cells.[113] Immunohistochemical staining for NADPH-cytochrome c (P-450) reductase has also been detected within Clara cells, although other cell types within the bronchioles were also reported to be immunohistochemically stained for the reductase; these other cell types, however, were not identified.[54]

Clara cells are capable of oxidatively metabolizing xenobiotics. Indirect evidence for this was provided by *in vitro* enzymatic analyses conducted on microsomal preparations isolated from the lungs of animals treated with pulmonary toxins that require metabolic activation. The administration of carbon tetrachloride to rats and mice resulted in both necrosis of Clara cells and a marked depression in pulmonary benzphetamine N-demethylase activity.[12] Similarly, naphthalene-induced necrosis of Clara cells in mouse lung is associated with significant reductions in pulmonary benzo[a]pyrene hydroxylase and 7-ethoxyresorufin O-deethylase

4. Intratissue Distribution of Enzymes

activities.[111] More direct evidence for the participation of Clara cells in the bioactivations of pulmonary toxins has been provided by the results of autoradiographic analyses conducted by Boyd and his colleagues.[11,110,115] After the administration of radiolabeled 4-ipomeanol to animals, Clara cells were found to be specifically alkylated,[110] although some covalent binding to ciliated bronchiolar epithelial cells was also noted, albeit to a much lesser extent.[11] The selective alkylation of Clara cells was greatly reduced by the prior administration of inhibitors of the cytochrome P-450–mediated metabolic activation of the pulmonary toxin.[110] Incubation of freshly isolated, intact Clara cells with 4-ipomeanol also resulted in the alkylation of these cells that was almost totally prevented by an inhibitor of cytochrome P-450.[115] Moreover, the incubation of sonicated Clara cells with 4-ipomeanol, NADPH, and reduced glutathione resulted in the formation of glutathione conjugates,[115] an observation suggestive of the presence of the glutathione S-transferases within these cells.

B. Type II Alveolar Epithelial Cells

Two major epithelial cell types are also present in the alveolar wall: type I cells (also referred to as squamous alveolar epithelial cells, membranous pneumocytes, small alveolar cells, and respiratory cells) and type II cells (also referred to as great alveolar cells, granular pneumocytes, large alveolar cells, and septal cells). The type II alveolar epithelial cell possesses considerable metabolic activity,[109] contains a large amount of endoplasmic reticulum,[109] and is damaged by toxins that require metabolic activation, such as carbon tetrachloride.[11,12,116] Immunohistochemical staining for cytochrome P-450[108,114] and NADPH-cytochrome c (P-450) reductase[54] has been detected within alveolar epithelial cells, and both benzo[a]pyrene[102] and aniline[107] hydroxylase activities have been histochemically demonstrated within the alveolar epithelium. Cytochromes P-450, NADPH-cytochrome c (P-450) reductase, and cytochrome b_5 have each been detected within isolated type II alveolar epithelial cells.[114,117] Although the levels of these enzymes within type II cells are considerably lower than those within Clara cells,[117] isolated type II cells are capable of supporting the O-deethylation of 7-ethoxycoumarin,[117,118] the N-oxidation of N,N-dimethylaniline,[117] and the hydroxylation of benzo[a]pyrene.[117,119] Moreover, the oxidative metabolism of benzo[a]pyrene by isolated type II cells produced metabolites that bound covalently to nuclear and cytoplasmic fractions within these cells.[119] Similarly, the metabolism of 4-ipomeanol by isolated type II cells resulted in the alkylation of these cells, albiet to a significantly lesser extent than were Clara cells, and this

was prevented by incubating cells with an inhibitor of the cytochrome
P-450–dependent metabolism of 4-ipomeanol.[115] When sonicated type II
cells were incubated with 4-ipomeanol, NADPH, and reduced glutathione, glutathione conjugates were formed,[115] suggesting that type II
alveolar epithelial cells also possess glutathione S-transferase activity.

C. Pulmonary Alveolar Macrophages

The pulmonary alveolar macrophage residing in the wall and lumen of
the pulmonary alveoli is the only other pulmonary cell for which there is
information regarding xenobiotic metabolism. Investigations on the ability
of these cells to metabolize xenobiotics, however, have frequently yielded
conflicting observations. For example, although neither Devereux et al.[114]
nor Hook et al.[120] were able to detect cytochrome P-450 within pulmonary alveolar macrophages, Fisher et al.[121] reported that microsomes
isolated from these cells do contain this protein. However, alveolar macrophage cytochrome P-450 comprises only a small fraction of the total
pulmonary content of this hemoprotein. Although NADPH-cytochrome
c (P-450) reductase and cytochrome b_5 are both present within pulmonary
alveolar macrophages,[120] a number of investigations have indicated that
these cells either cannot oxidatively metabolize xenobiotics[115,118,120,122] or
exhibit only very low activities.[120,121] In contrast, pulmonary alveolar
macrophages have been reported to possess aryl hydrocarbon hydroxylase activity[105,123,124] which is inducible.[123] Moreover, the incubation of
alveolar macrophages with benzo[a]pyrene was found to result in the
formation of a dihydrodiol metabolite,[124] suggesting the presence of both
cytochrome P-450 and epoxide hydrolase. Although neither glutathione
S-transferase nor UDPglucuronyltransferase activity was detected within
pulmonary alveolar macrophages,[120] these cells are capable of acetylating
xenobiotics.[122] Although it appears that the pulmonary alveolar macrophage is capable of metabolizing xenobiotics, the specific role that this
cell plays in the pulmonary metabolism of chemicals remains to be
clarified.

IV. DISTRIBUTION WITHIN THE SKIN

The skin is another common portal of entry of xenobiotics. Not only is it
continuously exposed to a great many environmental chemicals, but it is
also exposed to many therapeutic agents, and this exposure will increase
as transdermal medication—the delivery of drugs through the unbroken
skin—becomes more popular. The skin is also a target for chemically

4. Intratissue Distribution of Enzymes

induced toxicity, especially carcinogenesis. In fact, chemically induced neoplasia has probably been studied to a greater extent in the skin than in any other organ.

Although the skin is often considered as merely acting as a passive barrier to the entrance of chemicals, it is clearly capable of metabolizing endogenous substances as well as drugs and other exogenous chemicals.[125,126] Attempts at examining xenobiotic activating and detoxicating enzymes within the skin have frequently met with little or no success, primarily because of technical difficulties attendant on the homogenization and subsequent preparation of subcellular fractions from skin and because of the relatively low levels of the cutaneous enzymes. Despite the low activity of these enzymes in skin, it must be appreciated that the skin is the largest organ in the body, comprising between 4 and 6% of total body weight.[127] Thus, the skin could significantly contribute to the animal's total metabolism of chemicals. Although the indicated difficulties have hampered their study, cytochrome P-450,[128,129] NADPH-cytochrome c (P-450) reductase,[129,130] epoxide hydrolase,[93,131,132] glutathione S-transferase,[132,133] UDPglucuronyltransferase,[134,135] and aryl sulfotransferase[136] have all been detected within the skin, and microsomal preparations isolated from skin are capable of catalyzing aryl hydrocarbon hydroxylase,[128–130,132] aniline hydroxylase,[129] 7-ethoxycoumarin O-deethylase,[129,132] aminopyrine N-demethylase,[137] and p-nitroanisole O-demethylase[138] activities. These cutaneous enzymes and monooxygenase activities can be induced by a variety of xenobiotics.[125,126,128–130,132,135]

Little is known of the specific sites for xenobiotic biotransformations within this tissue. The skin is a relatively complex and heterogeneous organ, consisting of an epidermis and a dermis (Fig. 3). The epidermis is a stratified, squamous, keratinizing epithelium composed of a variable number of cell layers, whereas the dermis is a thick, dense connective tissue layer extending from the epidermis to the fatty, areolar subcutaneous tissue. The dermis is rich in vasculature and nerves and contains hair follicles and sebaceous glands. The hair follicle is a cylindrical formation, derived from the epidermis, that grows down into the dermal connective tissue; the outermost layer of the hair follicle, the outer root sheath, is continuous with the epidermis. Sebaceous glands are also derived from the epidermis and are usually associated with hair follicles, discharging their secretions into the upper part of the hair follicle via a short duct.

By means of autoradiographic techniques, Nakai and Shubik[139] found that an appreciable amount of radioactivity was concentrated within the epidermis, the upper part of hair follicles, and sebaceous glands after the topical application of tritiated 7,12-dimethylbenz[a]anthracene to mouse skin. Isolated hair follicles have been reported to be capable of metaboliz-

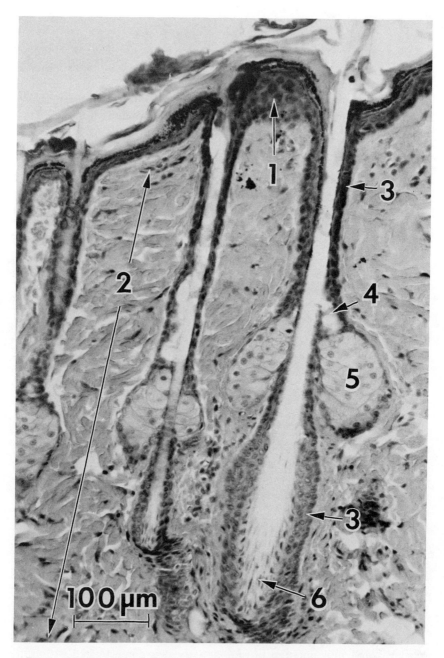

Fig. 3. Skin of an untreated female rat stained with hematoxylin and eosin. 1. Epidermis. 2. Upper portion of the dermis. 3. Outer root sheath of a hair follicle shown in longitudinal section. 4. Excretory duct of a sebaceous gland. 5. Sebaceous gland. 6. Hair cortex.

4. Intratissue Distribution of Enzymes

ing benzo[a]pyrene to yield dihydrodiols, the formation of which was prevented in the presence of an inhibitor of epoxide hydrolase activity.[140] The epidermis, sebaceous glands, and epithelium of hair follicles in skin of untreated rats have each been observed to be histochemically stained for aniline hydroxylase activity, sebaceous glands being the most intensely stained.[107] In contrast, although benzo[a]pyrene hydroxylase activity was not histochemically demonstrable within either epidermis or hair follicles in skin of untreated mice, aryl hydrocarbon hydroxylase activity increased within sebaceous glands and became apparent within the epidermis and the upper portion of the hair follicle following the topical application of β-naphthoflavone to mouse skin.[102] Results of immunohistochemical investigations[71,141] are consistent with the observations indicating that the epidermis, sebaceous glands, and hair follicles are the sites of xenobiotic activation and detoxication within the skin: cells of the epidermis, the sebaceous glands, and the outer root sheath of hair follicles in skin of untreated rats and mice were all found to be immunohistochemically stained for three isoenzymes of cytochrome P-450 (PB-B, MC-B, and BNF-B),[141] NADPH-cytochrome c (P-450) reductase,[141] epoxide hydrolase,[141] and glutathione S-transferases B, C, and E.[41]

V. COMMENTS

It will be apparent that enzymes participating in both bioactivation and detoxication of drugs, carcinogens, mutagens, and other xenobiotics are not uniformly distributed within tissues such as the liver, lung, and skin. Although distinctive cell types may be the only cells in which chemicals can be metabolically activated and detoxicated within a given tissue, it is equally clear that morphologically similar cells can frequently exhibit significant differences in their ability to metabolize xenobiotics. Metabolic differences can often be accentuated when the xenobiotic-metabolizing enzymes are induced.

Differences in the concentration and activity of xenobiotic activating and detoxicating enzymes within either discrete cell types or morphologically similar cells undoubtedly account, in large measure, for the relatively selective nature of many chemically induced toxicities. Within the next few years, we can anticipate a significant increase in our knowledge of the localizations and distributions of these important enzymes within many other mammalian tissues, thereby increasing our understanding of the biological basis for the selective damage frequently induced within either specific cells or tissue regions after the exposure to specific xenobiotics.

REFERENCES

1. Mitchell, J. R., Jollow, D. J., Potter, W. Z., Davis, D. C., Gillette, J. R., and Brodie, B. B. (1973). Acetoaminophen-induced hepatic necrosis. I. Role of drug metabolism. *J. Pharmacol. Exp. Ther.* **187**, 185-194.
2. Brodie, B. B., Reid, W. D., Cho, A. K., Sipes, G., Krishna, G., and Gillette, J. R. (1971). Possible mechanism of liver necrosis caused by aromatic organic compounds. *Proc. Natl. Acad. Sci. U.S.A.* **68**, 160-164.
3. Reid, W. D., and Krishna, G. (1973). Centrolobular necrosis related to covalent binding of metabolites of halogenated aromatic hydrocarbons. *Exp. Mol. Pathol.* **18**, 80-99.
4. Mitchell, J. R., Nelson, W. L., Potter, W. Z., Sasame, H. A., and Jollow, D. J. (1976). Metabolic activation of furosemide to a chemically reactive, hepatotoxic metabolite. *J. Pharmacol. Exp. Ther.* **199**, 41-52.
5. Vorne M., and Alavaikko, M. (1971). Effect of carbon tetrachloride induced progressive liver damage on the metabolism of hexobarbital and bilirubin *in vivo*. *Acta Pharmacol. Toxicol.* **29**, 402-416.
6. Ferreyra, E. C., De Fenos, O. M., Bernacchi, A. S., De Castro, C. R., and Castro, J. A. (1977). Treatment of carbon tetrachloride-induced liver necrosis with chemical compounds. *Toxicol. Appl. Pharmacol.* **42**, 513-521.
7. Hunter, A. L., Holscher, M. A., and Neal, R. A. (1977). Thioacetamide-induced hepatic necrosis. I. Involvement of the mixed-function oxidase enzyme system. *J. Pharmacol. Exp. Ther.* **200**, 439-448.
8. Thorgeirsson, S. S., Mitchell, J. R., Sasame, H. A., and Potter, W. Z. (1976). Biochemical changes after hepatic injury by allyl alcohol and N-hydroxy-2-acetylaminofluorene. *Chem.-Biol. Interact.* **15**, 139-147.
9. Meerman, J. H. N., and Mulder, G. J. (1981). Prevention of the hepatotoxic action of N-hydroxy-2-acetylaminofluorene in the rat by inhibition of N-O-sulfation by pentachlorophenol. *Life Sci.* **28**, 2361-2365.
10. Rees, K. R., and Tarlow, M. J. (1967). The hepatotoxic action of allyl formate. *Biochem. J.* **104**, 757-761.
11. Boyd, M. R. (1980). Biochemical mechanisms in chemical-induced lung injury: Roles of metabolic activation. *CRC Crit. Rev. Toxicol.* **7**, 103-176.
12. Boyd, M. R., Statham, C. N., and Longo, N. S. (1980). The pulmonary Clara cell as a target for toxic chemicals requiring metabolic activation; studies with carbon tetrachloride. *J. Pharmacol. Exp. Ther.* **212**, 109-114.
13. Jakoby, W. B., ed. (1980). "Enzymatic Basis of Detoxication," Vols. 1 and 2. Academic Press, New York.
14. Gates, G. A., Henley, K. S., Pollard, H. M., Schmidt, E., and Schmidt, F. W. (1961). The cell population of human liver. *J. Lab. Clin. Med.* **57**, 182-201.
15. Grovier, W. C. (1965). Reticuloendothelial cells as the site of sulfanilamide acetylation in the rabbit. *J. Pharmacol. Exp. Ther.* **150**, 305-308.
16. Loud, A. V. (1968). A quantitative stereological description of the ultrastructure of normal rat liver parenchymal cells. *J. Cell Biol.* **37**, 27-46.
17. Rappaport, A. M., Borowy, Z. J., Lougheed, W. M., and Lotto, W. N. (1954). Subdivision of hexagonal liver lobules into a structural and functional unit. Role in hepatic physiology and pathology. *Anat. Res.* **119**, 11-27.
18. Rappaport, A. M. (1976). The microcirculatory acinar concept of normal and pathological hepatic structure. *Beitr. Pathol. Anat.* **157**, 215-243.
19. Rappaport, A. M. (1979). Physioanatomical basis of toxic liver injury. *In* "Toxic Injury

4. Intratissue Distribution of Enzymes

of the Liver" (E. Farber and M. M. Fisher, eds.), Part A, pp. 1–57. Dekker, New York.
20. Sweeney, G. D., Garfield, R. E., Jones, K. G., and Latham, A. N. (1978). Studies using sedimentation velocity on heterogeneity of size and function of hepatocytes from mature male rats. *J. Lab. Clin. Med.* **91,** 432–443.
21. Drochmans, P., Wanson, J.-C., and Mosselmans, R. (1975). Isolation and subfractionation on Ficoll gradients of adult rat hepatocytes. Size, morphology, and biochemical characteristics of cell fractions. *J. Cell Biol.* **66,** 1–22.
22. Wanson, J.-C., Drochmans, P., May, C., Penasse, W., and Popowski, A. (1975). Isolation of centrolobular and perilobular hepatocytes after phenobarbital treatment. *J. Cell Biol.* **66,** 23–41.
23. Ji, S., Lemasters, J. J., and Thurman, R. G. (1980). A non-invasive method to study metabolic events within sublobular regions of hemoglobin-free perfused liver. *FEBS Lett.* **113,** 37–41.
24. Shank, R. E., Morrison, G., Cheng, C. H., Karl, I., and Schwartz, R. (1959). Cell heterogeneity within the hepatic lobule (quantitative histochemistry). *J. Histochem. Cytochem.* **7,** 237–239.
25. Schumacher, H.-H. (1957). Histochemical distribution pattern of respiratory enzymes in the liver lobule. *Science* **125,** 501–503.
26. Novikoff, A. B. (1959). Cell heterogeneity within the hepatic lobule of the rat (staining reactions). *J. Histochem. Cytochem.* **7,** 240–244.
27. Chiquoine, A. D. (1953). The distribution of glucose-6-phosphatase in the liver and kidney of the mouse. *J. Histochem. Cytochem.* **1,** 429–435.
28. Wimmer, M., and Pette, D. (1979). Microphotometric studies on intraacinar enzyme distribution in rat liver. *Histochemistry* **64,** 23–33.
29. Welsh, F. A. (1972). Changes in distribution of enzymes within the liver lobule during adaptive increases. *J. Histochem. Cytochem.* **20,** 107–111.
30. Seligman, A. M., and Rutenburg, A. M. (1951). The histochemical demonstration of succinic dehydrogenase. *Science* **113,** 317–320.
31. Novikoff, A. B., and Essner, E. (1960). The liver cell. Some new approaches to its study. *Am. J. Med.* **29,** 102–131.
32. Jungermann, K., and Sasse, D. (1978). Heterogeneity of liver parenchymal cells. *Trends Biochem. Sci.* **3,** 198–202.
33. LeBouton, A. B. (1968). Heterogeneity of protein metabolism between liver cells as studied by autoradiography. *Curr. Med. Biol.* **2,** 111–114.
34. Wachstein, M. (1959). Enzymatic histochemistry of the liver. *Gastroenterology* **37,** 525–537.
35. Koudstaal, J., and Hardonk, M. J. (1969). Histochemical demonstration of enzymes related to NADPH-dependent hydroxylating systems in rat liver after phenobarbital treatment. *Histochemie* **20,** 68–77.
36. Smith, M. T., and Wills, E. D. (1981). The effects of dietary lipid and phenobarbitone on the production and utilization of NADPH in the liver. A combined biochemical and quantitative cytochemical study. *Biochem. J.* **200,** 691–699.
37. Ji, S., Lemasters, J. J., and Thurman, R. G. (1981). A fluorometric method to measure sublobular rates of mixed-function oxidation in the hemoglobin-free perfused rat liver. *Mol. Pharmacol.* **19,** 513–516.
38. Altman, F. P., Moore, D. S., and Chayen, J. (1975). The direct measurement of cytochrome P-450 in unfixed tissue sections. *Histochemistry* **41,** 227–232.
39. Chayen, J., Bitensky, L., Johnstone, J. J., Gooding, P. E., and Slater, T. F. (1979). The application of microspectrophotometry to the measurement of cytochrome P-450. *In*

"Quantitative Cytochemistry and its Applications" (J. R. Pattison, L. Bitensky, and J. Chayen, eds.), pp. 129-137. Academic Press, New York.
40. Gooding, P. E., Chayen, J., Sawyer, B., and Slater, T. F. (1978). Cytochrome P-450 distribution in rat liver and the effect of sodium phenobarbitone administration. Chem.-Biol. Interact. **20**, 299-310.
41. Smith, M. T., and Wills, E. D. (1981). Effects of dietary lipid and phenobarbitone on the distribution and concentration of cytochrome P-450 in the liver studied by quantitative cytochemistry. FEBS Lett. **127**, 33-36.
42. Gumucio, J. J., DeMason, L. J., Miller, D. L., Krezoski, S. O., and Keener, M. (1978). Induction of cytochrome P-450 in a selective subpopulation of hepatocytes. Am. J. Physiol. **234**, C102-C109.
43. Baron, J., Redick, J. A., and Guengerich, F. P. (1978). Immunohistochemical localizations of cytochromes P-450 in rat liver. Life Sci. **23**, 2627-2632.
44. Redick, J. A., Kawabata, T. T., Guengerich, F. P., Krieter, P. A., Shires, T. K., and Baron, J. (1980). Distributions of monooxygenase components and epoxide hydratase within the livers of untreated male rats. Life Sci. **27**, 2465-2470.
45. Baron, J., Redick, J. A., and Guengerich, F. P. (1981). An immunohistochemical study on the localizations and distributions of phenobarbital- and 3-methylcholanthrene-inducible cytochromes P-450 within the livers of untreated rats. J. Biol. Chem. **256**, 5931-5937.
46. Baron, J., Redick, J. A., and Guengerich, F. P. (1982). Effects of 3-methylcholanthrene, β-naphthoflavone, and phenobarbital on the 3-methylcholanthrene-inducible isozyme of cytochrome P-450 within centrilobular, midzonal, and periportal hepatocytes. J. Biol. Chem. **257**, 953-957.
47. Baron, J., Taira, Y., Redick, J. A., Greenspan, P., Kapke, G. F., and Guengerich, F. P. (1980). Effects of xenobiotics on the distributions of monooxygenase components in liver. In "Microsomes, Drug Oxidations and Chemical Carcinogenesis" (M. J. Coon, A. H. Conney, R. W. Estabrook, H. V. Gelboin, J. R. Gillette, and P. J. O'Brien, eds.), pp. 501-504. Academic Press, New York.
48. Baron, J. Redick, J. A., and Guengerich, F. P. (1983). Effects of phenobarbital, 3-methylcholanthrene, β-naphthoflavone, and *trans*-stilbene oxide on the phenobarbital-inducible isozyme of cytochrome P-450 within centrilobular, midzonal, and periportal hepatocytes. J. Biol. Chem. (in press).
49. Guengerich, F. P. (1977). Separation and purification of multiple forms of microsomal cytochrome P-450. Activities of different forms of cytochrome P-450 towards several compounds of environmental interest. J. Biol. Chem. **252**, 3970-3979.
50. Guengerich, F. P. (1978). Separation and purification of multiple forms of microsomal cytochrome P-450. Partial characterization of three apparently homogeneous cytochromes P-450 prepared from livers of phenobarbital- and 3-methylcholanthrene-treated rats. J. Biol. Chem. **253**, 7931-7939.
51. Baron, J., Redick, J. A., Greenspan, P., and Taira, Y. (1978). Immunohistochemical localization of NADPH-cytochrome c reductase in rat liver. Life Sci. **22**, 1097-1102.
52. Taira, Y., Redick, J. A., and Baron, J. (1980). An immunohistochemical study on the localization and distribution of NADPH-cytochrome c (P-450) reductase in rat liver. Mol. Pharmacol. **17**, 374-381.
53. Taira, Y., Greenspan, P., Kapke, G. F., Redick, J. A., and Baron, J. (1980). Effects of phenobarbital, pregnenolone-16α-carbonitrile, and 3-methylcholanthrene pretreatments on the distribution of NADPH-cytochrome c (P-450) reductase within the liver lobule. Mol. Pharmacol. **18**, 304-312.
54. Dees, J. H., Coe, L. D., Yasukochi, Y., and Masters, B. S. S. (1980). Immunofluores-

cence of NADPH-cytochrome c (P-450) reductase in rat and minipig tissues injected with phenobarbital. *Science* **208,** 1473–1475.
55. Müller-Eberhard, U., Yam, L., Tavassoli, M., Cox, K., and Ozols, J. (1974). Immunohistochemical demonstration of cytochrome b_5 and hemopexin in rat liver parenchymal cells using horseradish peroxidase. *Biochem. Biophys. Res. Commun.* **61,** 983–988.
56. Tavassoli, M., Ozols, J., Sugimoto, G., Cox, K. H., and Müller-Eberhard, U. (1976). Localization of cytochrome b_5 in rat organs and tissues by immunohistochemistry. *Biochem. Biophys. Res. Commun.* **72,** 281–287.
57. Baron, J., Redick, J. A., and Guengerich, F. P. (1980). Immunohistochemical localization of epoxide hydratase in rat liver. *Life Sci.* **26,** 489–493.
58. Kawabata, T. T., Guengerich, F. P., and Baron, J. (1981). An immunohistochemical study on the localization and distribution of epoxide hydrolase within livers of untreated rats. *Mol. Pharmacol.* **20,** 709–714.
59. Bentley, P., Waechter, F., Oesch, F., and Staubli, W. (1979). Immunochemical localization of epoxide hydratase in rat liver: Effects of 2-acetylaminofluorene. *Biochem. Biophys. Res. Commun.* **91,** 1101–1108.
60. Enomoto, K., Ying, T. S., Griffin, M. J., and Farber, E. (1981). Immunohistochemical study of epoxide hydrolase during experimental liver carcinogenesis. *Cancer Res.* **41,** 3281–3287.
61. Kawabata, T. T., Guengerich, F. P., and Baron, J. (1983). Effects of phenobarbital, 3-methylcholanthrene, and *trans*-stilbene oxide on epoxide hydrolase within centrilobular, midzonal, and periportal hepatocytes. *J. Biol. Chem.* (in press).
62. Niranjan, B. G., and Avadhani, N. G. (1980). Activation of aflatoxin B_1 by a monooxygenase system localized in rat liver mitochondria. *J. Biol. Chem.* **255,** 6575–6578.
63. Sato, R., Atsuta, Y., Imai, Y., Taniguchi, S., and Okuda, K. (1977). Hepatic mitochondrial cytochrome P-450: Isolation and functional characterization. *Proc. Natl. Acad. Sci. U.S.A.* **74,** 5477–5481.
64. Pedersen, J. I. (1978). Rat liver mitochondrial cytochrome P-450 active in a reconstituted steroid hydroxylation reaction. *FEBS Lett.* **85,** 35–39.
65. Ohashi, M., and Omura, T. (1978). Presence of the NADPH-cytochrome P-450 reductase system in liver and kidney mitochondria. *J. Biochem. (Tokyo)* **83,** 249–260.
66. Kapke, G. F., Redick, J. A., and Baron, J. (1978). Immunohistochemical demonstration of an adrenal ferredoxin-like iron-sulfur protein in rat hepatic mitochondria. *J. Biol. Chem.* **253,** 8604–8608.
67. Jakoby, W. B. (1978). The glutathione S-transferases: A group of multifunctional detoxification proteins. *Adv. Enzymol.* **46,** 383–414.
68. Bannikov, G. A., Guelstein, V. I., and Tchipsheva, T. A. (1973). Distribution of basic azo-dye-binding protein in normal rat tissues and carcinogen-induced liver tumors. *Int. J. Cancer* **11,** 398–411.
69. Fleischner, G. M., Robbins, J. B., and Arias, I. M. (1977). Cellular localization of ligandin in rat, hamster, and man. *Biochem. Biophys. Res. Commun.* **74,** 992–1000.
70. Campbell, J. A. H., Bass, N. M., and Kirsch, R. E. (1980). Immunohistological localization of ligandin in human tissues. *Cancer* **45,** 503–510.
71. Redick, J. A., Jakoby, W. B., and Baron, J. (1982). Immunohistochemical localization of glutathione S-transferases B, C, and E in liver and skin. *Fed. Proc., Fed. Am. Soc. Exp. Biol.* **41,** 1573.
72. Asghar, K., Reddy, B. G., and Krishna, G. (1975). Histochemical localization of glutathione in tissues. *J. Histochem. Cytochem.* **23,** 774–779.
73. Deml, E., and Oesterle, D. (1980). Histochemical demonstration of enhanced

glutathione content in enzyme altered islands induced by carcinogens in rat liver. *Cancer Res.* **40,** 490-491.
74. Smith, M. T., Loveridge, N., Wills, E. D., and Chayen, J. (1979). The distribution of glutathione in the rat liver lobule. *Biochem. J.* **182,** 103-108.
75. James, R., Desmond, P., Kupfer, A., Schenker, S., and Branch, R. A. (1981). The differential localization of various drug metabolizing systems within the rat liver lobule as determined by the hepatotoxins allyl alcohol, carbon tetrachloride and bromobenzene. *J. Pharmacol. Exp. Ther.* **217,** 127-132.
76. Pang, K. S., Koster, H., Halsema, I. C. M., Scholtens, E., and Mulder, G. J. (1981). Aberrant pharmacokinetics of harmol in the perfused rat liver preparation: Sulfate and glucuronide conjugations. *J. Pharmacol. Exp. Ther.* **219,** 134-140.
77. De Baun, J. R., Smith, J. Y. R., Miller, E. C., and Miller, J. A. (1970). Reactivity *in vivo* of the carcinogen N-hydroxy-2-acetylaminofluorene: Increase by sulfate ions. *Science* **167,** 184-186.
78. Pang, K. S., and Terrell, J. A. (1981). Retrograde perfusion to probe the heterogeneous distribution of hepatic drug metabolizing enzymes in rats. *J. Pharmacol. Exp. Ther.* **216,** 339-346.
79. Jollow, D. J., Mitchell, J. R., Potter, W. Z., Davis, D. C., Gillette, J. R., and Brodie, B. B. (1973). Acetaminophen-induced hepatic necrosis. II. Role of covalent binding *in vivo*. *J. Pharmacol. Exp. Ther.* **187,** 185-202.
80. Wattenberg, L. W., and Leong, J. L. (1962). Histochemical demonstration of reduced pyridine nucleotide dependent polycyclic hydrocarbon metabolizing systems. *J. Histochem. Cytochem.* **10,** 412-420.
81. Gangolli, S., and Wright, M. (1971). The histochemical demonstration of aniline hydroxylase activity in rat liver. *Histochem. J.* **3,** 107-116.
82. Willson, R. A., and Hart, J. R. (1981). *In vivo* drug metabolism and liver lobule heterogeneity in the rat. *Gastroenterology* **81,** 563-569.
83. Gram, T. E. (1973). Comparative aspects of mixed function oxidation by lung and liver of rabbits. *Drug Metab. Rev.* **2,** 1-32.
84. Brown, E. A. B. (1974). The localization, metabolism, and effects of drugs and toxicants in lung. *Drug Metab. Rev.* **3,** 33-87.
85. Hook, G. E. R., and Bend, J. R. (1976). Pulmonary metabolism of xenobiotics. *Life Sci.* **18,** 279-290.
86. Gram, T. E. (1980). The metabolism of xenobiotics by mammalian lung. *In* "Extrahepatic Metabolism of Drugs and Other Foreign Compounds" (T. E. Gram, ed.), pp. 159-209. Spectrum Publ., New York.
87. Matsubara, T., and Tochino, Y. (1971). Electron transport systems of lung microsomes and their physiological functions. I. Intracellular distribution of oxidative enzymes in lung cells. *J. Biochem. (Tokyo)* **70,** 981-991.
88. Matsubara, T., Prough, R. A., Burke, M. D., and Estabrook, R. W. (1974). The preparation of microsomal fractions of rodent respiratory tract and their characterization. *Cancer Res.* **34,** 2196-2203.
89. Philpot, R. M., Arinc, E., and Fouts, J. R. (1975). Reconstitution of the rabbit pulmonary microsomal mixed-function oxidase system from solubilized components. *Drug Metab. Dispos.* **3,** 118-126.
90. Arinc, E., and Philpot, R. M. (1976). Preparation and properties of partially purified pulmonary cytochrome P-450 from rabbits. *J. Biol. Chem.* **251,** 3213-3220.
91. Guengerich, F. P. (1977). Preparation and properties of highly purified cytochrome P-450 and NADPH-cytochrome P-450 reductase from pulmonary microsomes of untreated rabbits. *Mol. Pharmacol.* **13,** 911-923.
92. Franklin, M. R., Wolf, C. R., Serabjit-Singh, C., and Philpot, R. M. (1980). Quantita-

tion of two forms of pulmonary cytochrome P-450 in microsomes, using substrate specificities. *Mol. Pharmacol.* **17**, 415–420.
93. Oesch, F., Glatt, H., and Schmassmann, H. (1977). The apparent ubiquity of epoxide hydratase in rat organs. *Biochem. Pharmacol.* **26**, 603–607.
94. Smith, B. R., Maguire, J. H., Ball, L. M., and Bend, J. R. (1978). Pulmonary metabolism of epoxides. *Fed. Proc., Fed. Am. Soc. Exp. Biol.* **37**, 2480–2484.
95. Gill, S. S., and Hammock, B. D. (1980). Distribution and properties of a mammalian soluble epoxide hydrase. *Biochem. Pharmacol.* **29**, 389–395.
96. Mukhtar, H., and Bresnick, E. (1976). Mouse liver and lung glutathione S-epoxide transferase: Effects of phenobarbital and 3-methylcholanthrene administration. *Chem.-Biol. Interact.* **15**, 59–67.
97. Kraus, P., and Kloft, H.-D. (1980). The activity of glutathione S-transferases in various organs of the rat. *Enzyme* **25**, 158–160.
98. Aitio, A. (1973). Glucuronide synthesis in the rat and guinea pig lung. *Xenobiotica* **3**, 13–22.
99. Gram, T. E., Litterst, C. L., and Mimnaugh, E. G. (1974). Enzymatic conjugation of foreign chemical compounds by rabbit lung and liver. *Drug Metab. Dispos.* **2**, 254–258.
100. Cohen, G. M., Haws, S. M., Moore, B. P., and Bridges, J. W. (1976). Benzo[a]pyrene-3-yl hydrogen sulphate, a major ethyl acetate-extractable metabolite of benzo[a]pyrene in human, hamster, and rat lung culture. *Biochem. Pharmacol.* **25**, 2561–2570.
101. Sorokin, S. P. (1970). The cells of the lung. *In* "Morphology of Experimental Respiratory Carcinogenesis" (P. Nettesheim, M. G. Hanna, Jr., and J. W. Deatherage, eds.), pp. 3–43. U.S. At. Energy Comm., Oak Ridge, Tennessee.
102. Wattenberg, L. W., and Leong, J. L. (1971). Tissue distribution studies of polycyclic hydrocarbon hydroxylase activity. *Handb. Exp. Pharmakol.* **28**, Part A, 422–430.
103. Cohen, G. M., and Moore, B. P. (1976). Metabolism of [^{3}H]benzo[a]pyrene by different portions of the respiratory tract. *Biochem. Pharmacol.* **25**, 1623–1629.
104. Mass, M. J., and Kaufman, D. G. (1978). [^{3}H]Benzo[a]pyrene metabolism in tracheal epithelial microsomes and tracheal organ cultures. *Cancer Res.* **38**, 3861–3866.
105. Autrup, H., Harris, C. C., Stoner, G. D., Selkirk, J. K., Schafer, P. W., and Trump, B. F. (1978). Metabolism of [^{3}H]benzo[a]pyrene by cultured human bronchus and cultured human pulmonary alveolar macrophages. *Lab. Invest.* **38**, 217–224.
106. Kahng, M. W., Smith, M. W., and Trump, B. F. (1981). Aryl hydrocarbon hydroxylase in human bronchial epithelium and blood monocyte. *J. Natl. Cancer Inst.* **66**, 227–232.
107. Grasso, P., Williams, M., Hodgson, R., Wright, M. G., and Gangolli, S. D. (1971). The histochemical distribution of aniline hydroxylase in rat tissues. *Histochem. J.* **3**, 117–126.
108. Masters, B. S. S., Yasukochi, Y., Okita, R. T., Parkhill, L. K., Taniguchi, H., and Dees, J. H. (1980). Laurate hydroxylation and drug metabolism in pig liver and kidney microsomes and in reconstituted systems from pig liver and kidney. *In* "Microsomes, Drug Oxidations and Chemical Carcinogenesis" (M. J. Coon, A. H. Conney, R. W. Estabrook, H. V. Gelboin, J. R. Gillette, and P. J. O'Brien, eds.), pp. 709–719. Academic Press, New York.
109. Sorokin, S. P. (1977). The respiratory system. *In* "Histology" (L. Weiss and R. O. Greep, eds.), pp. 765–830. McGraw-Hill, New York.
110. Boyd, M. R. (1977). Evidence for the Clara cell as a site of cytochrome P-450–dependent mixed-function oxidase activity in lung. *Nature (London)* **269**, 713–715.
111. Tong, S. S., Hirokata, Y., Trush, M. A., Mimnaugh, E. G., Ginsburg, E., Lowe, M. C.,

and Gram, T. E. (1981). Clara cell damage and inhibition of pulmonary mixed-function oxidase activity by naphthalene. *Biochem. Biophys. Res. Commun.* **100,** 944–950.
112. Reznik-Schuller, H., and Hague, B. F. (1980). A morphometric study of the pulmonary Clara cell in normal and nitrosoheptamethyleneimine-treated European hamsters. *Exp. Pathol.* **18,** 366–371.
113. Serabjit-Singh, C. J., Wolf, C. R., Philpot, R. M., and Plopper, C. G. (1980). Cytochrome P-450: Localization in rabbit lung. *Science* **207,** 1469–1470.
114. Devereux, T. R., Serabjit-Singh, C. J., Slaughter, S. R., Wolf, C. R., Philpot, R. M., and Fouts, J. R. (1981). Identification of cytochrome P-450 isozymes in nonciliated bronchiolar epithelial (Clara) and alveolar type II cells isolated from rabbit lung. *Exp. Lung Res.* **2,** 221–230.
115. Devereux, T. R., Jones, K. G., Bend, J. R., Fouts, J. R., Statham, C. N., and Boyd, M. R. (1982). *In vitro* metabolic activation of the pulmonary toxin, 4-ipomeanol, in nonciliated bronchiolar epithelial (Clara) and alveolar type II cells isolated from rabbit lung. *J. Pharmacol. Exp. Ther.* **220,** 223–227.
116. Chen, W.-J., Chi, E. Y., and Smuckler, E. A. (1977). Carbon tetrachloride-induced changes in mixed function oxidases and microsomal cytochromes in the rat lung. *Lab. Invest.* **36,** 388–394.
117. Devereux, T. R., and Fouts, J. R. (1981). Xenobiotic metabolism by alveolar type II cells isolated from rabbit lung. *Biochem. Pharmacol.* **30,** 1231–1237.
118. Devereux, T. R., Hook, G. E. R., and Fouts, J. R. (1979). Foreign compound metabolism by isolated cells from rabbit lung. *Drug Metab. Dispos.* **7,** 70–75.
119. Teel, R. W., and Douglas, W. H. J. (1980). Aryl hydrocarbon hydroxylase activity in type II alveolar lung cells. *Experientia* **36,** 107.
120. Hook, G. E. R., Bend, J. R., and Fouts, J. R. (1972). Mixed-function oxidases and the alveolar macrophage. *Biochem. Pharmacol.* **21,** 3267–3277.
121. Fisher, A. B., Huber, G. A., and Furia, L. (1977). Cytochrome P-450 content and mixed-function oxidation by microsomes from rabbit alveolar macrophages. *J. Lab. Clin. Med.* **90,** 101–108.
122. Reid, W. D., Glick, J. M., and Krishna, G. (1972). Metabolism of foreign compounds by alveolar macrophages of rabbits. *Biochem. Biophys. Res. Commun.* **49,** 626–634.
123. McLemore, T. L., Martin, R. R., Toppell, K. L., Busbee, D. L., and Cantrell, E. T. (1977). Comparison of aryl hydrocarbon hydroxylase induction in cultured blood lymphocytes and pulmonary macrophages. *J. Clin. Invest.* **60,** 1017–1024.
124. Harris, C. C., Hsu, I. C., Stoner, G. D., Trump, B. F., and Selkirk, J. K. (1978). Human pulmonary alveolar macrophages metabolise benzo[*a*]pyrene to proximate and ultimate mutagens. *Nature (London)* **272,** 633–634.
125. Pannatier, A., Jenner, P., Testa, B., and Etter, J. C. (1978). The skin as a drug-metabolizing organ. *Drug Metab. Rev.* **8,** 319–343.
126. Bickers, D. R., and Kappas, A. (1980). The skin as a site of chemical metabolism. In "Extrahepatic Metabolism of Drugs and Other Foreign Compounds" (T. E. Gram, ed.), pp. 295–318. Spectrum Publ., New York.
127. Leider, M., and Buncke, C. M. (1954). Physical dimensions of the skin. Determination of the specific gravity of skin, hair, and nail. *Arch. Dermatol.* **69,** 563–569.
128. Bickers, D. R., Kappas, A., and Alvares, A. P. (1974). Differences in the inducibility of cutaneous and hepatic drug metabolizing enzymes and cytochrome P-450 by polychlorinated biphenyls and 1,1,1-trichloro-2,2-bis(p-chlorophenyl)ethane (DDT). *J. Pharmacol. Exp. Ther.* **188,** 300–309.
129. Pohl, A. J., Philpot, R. M., and Fouts, J. R. (1976). Cytochrome P-450 content and mixed-function oxidase activity in microsomes isolated from mouse skin. *Drug Metab. Dispos.* **4,** 442–450.

4. Intratissue Distribution of Enzymes

130. Vizethum, W., Ruzicka, T., and Goerz, G. (1980). Inducibility of drug-metabolizing enzymes in the rat skin. *Chem.-Biol. Interact.* **31**, 215–219.
131. Pyerin, W. G., and Hecker, E. (1975). Epoxide hydrase activity in mouse skin epidermis. *Z. Krebsforsch.* **83**, 81–83.
132. Mukhtar, H., and Bickers, D. B. (1981). Drug metabolism in skin. Comparative activity of mixed-function oxidases, epoxide hydratase, and glutathione S-transferase in liver and skin of the neonatal rat. *Drug Metab. Dispos.* **9**, 311–314.
133. Mukhtar, H., and Bresnick, E. (1976). Glutathione S-epoxidetransferase in mouse skin and human foreskin. *J. Invest. Dermatol.* **66**, 161–164.
134. Harper, K. H., and Calcutt, G. (1960). Conjugation of 3:4-benzopyrenols in mouse skin. *Nature (London)* **186**, 80–81.
135. Dutton, G. J., and Stevenson, I. H. (1962). The stimulation by 3,4-benzpyrene of glucuronide synthesis in the skin. *Biochim. Biophys. Acta* **58**, 633–634.
136. Berliner, D. L., Pasqualine, J. R., and Gallegos, A. J. (1968). The formation of water soluble steroids by human skin. *J. Invest. Dermatol.* **50**, 220–224.
137. Mukhtar, H., and Bickers, D. R. (1981). Aminopyrine N-demethylase activity in neonatal rat skin. *Biochem. Pharmacol.* **30**, 3257–3260.
138. Pannatier, A., Testa, B., and Etter, J.-C. (1981). Aryl ether O-dealkylase activity in the skin of untreated mice *in vitro*. *Xenobiotica* **11**, 345–350.
139. Nakai, T., and Shubik, P. (1964). Autoradiographic localization of tissue-bound tritiated 7,12-dimethylbenz[a]anthracene in mouse skin 24 and 48 hours after single application. *J. Natl. Cancer Inst.* **33**, 887–891.
140. Vermorken, A. J. M., Goos, C. M. A. A., Roelofs, H. M. J., Henderson, P. T., and Bloemendal, H. (1979). Metabolism of benzo[a]pyrene in isolated human scalp hair follicles. *Toxicology* **14**, 109–116.
141. Baron, J., Richards, J. L., Redick, J. A., and Guengerich, F. P. (1982). Localizations of cutaneous xenobiotic-metabolizing enzymes. *Fed. Proc., Fed. Am. Soc. Exp. Biol.* **41**, 1573.

CHAPTER 5

Nonenzymatic Biotransformation

Bernard Testa

I.	Introduction	137
II.	Reactions with Macromolecules as Borderline Cases	139
III.	Reactions with Endogenous Nucleophiles	141
IV.	Reactions with Endogenous Electrophiles	144
V.	Breakdown and Rearrangement Reactions of Xenobiotics and Prodrugs in Acidic and Neutral Aqueous Media	145
VI.	Reactions between Two Xenobiotics	146
VII.	Conclusion	148
	References	149

I. INTRODUCTION

Two facts dominate past and current aspects of research in xenobiotic metabolism: (1) that xenobiotics in a biological environment can experience a variety of chemical reactions and (2) that enzymes are the chemical operators of the biosphere. These facts imply that xenobiotic metabolism must result essentially from enzymatic reactions, as confirmed in innumerable studies.

Granting enzymatic reactions, there is a growing awareness that xenobiotics can be biotransformed by other mechanisms. Indeed, a number of situations can be recognized in the pathways leading from a xenobiotic to a metabolite (Fig. 1). The most obvious possibility is for the xenobiotic to undergo an enzymatic reaction, as is discussed elsewhere in this volume. Another simple and as yet poorly recognized situation in-

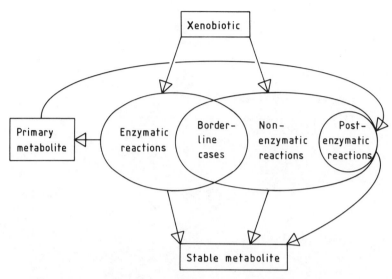

Fig. 1. Enzymatic and nonenzymatic modes of xenobiotic metabolism.

volves a nonenzymatic reaction directly generating a stable metabolite. This chapter is concerned mainly with such situations, presenting representative evidence and attempting to delineate those metabolic pathways in which nonenzymatic reactions are of significance. The topic has already been reviewed in a somewhat different form.[1]

Aside from the two conceptually simple situations of enzymatic and nonenzymatic reactions, other more complex cases exist. Thus, we will give some attention in Section II to borderline cases (Fig. 1) that are difficult to categorize as genuinely enzymatic or nonenzymatic because they involve macromolecules that are not enzymes, yet display an enzyme-like behavior. Again different are postenzymatic contributions to xenobiotic metabolism. Here, enzymatically generated primary metabolites display various degrees of instability and undergo spontaneous or nonenzymatically catalyzed breakdown (postenzymatic reactions) to stable metabolites.

Metabolic pathways including an enzymatic reaction followed by a postenzymatic breakdown or rearrangement have long been recognized and will not be considered in detail here. The interested reader is referred to two remarkably well-documented reviews.[2,3] One particularly important aspect of postenzymatic chemistry is that involving various activated forms of oxygen; this topic is extensively and lucidly reviewed by Trager[4] and in Chapter 7.

5. Nonenzymatic Biotransformation

II. REACTIONS WITH MACROMOLECULES AS BORDERLINE CASES

Reactions of xenobiotics with macromolecules imply more than the straightforward case in which these macromolecules are enzymes. Reactive metabolites can undergo, for example, a postenzymatic reaction (adduct formation) with such macromolecular systems as nucleic acids[5] and hemoglobin.[6,7] Relevant to the present context is the case of a chemical reaction catalyzed by nonenzymatic macromolecules. Some proteins, not categorized as enzymes, may display an enzyme-like behavior, a situation that must be regarded as a borderline case (Fig. 1) and the source of some semantic confusion. As an example, bovine serum albumin has been shown to catalyze the oxygenation of benzo[a]pyrene.[8] By stirring benzo[a]pyrene with albumin in water in the dark at 4°C for 5 days, a free radical was produced; it was identified as the 6-phenoxy radical (I), and not the 1- or 3-phenoxy radical or a semiquinone radical. Such conditions and results may or may not have physiological relevance. But the fact that

the reaction occurs at all, as well as its high regioselectivity, points to unexpected and challenging properties of serum albumin.

Esterasic activity of human serum albumin has been demonstrated[9] for a number of phenyl esters, including acetylsalicylic acid (II: X = o-COOH, R = COCH$_3$). Rate constants of hydrolysis are compared in Table I revealing some interesting facts. It is apparent that the rate constant of spontaneous hydrolysis (k_0) for the phenyl acetates is between one (acetylsalicylic acid) and two (p-nitrophenyl acetate) orders of magnitude smaller than the rate constant of an albumin-catalyzed hydrolysis. Both rate constants for the phenyl acetates are controlled by the electron-withdrawing power of the ring substituent; there are indeed very good linear correlations between $\log k$ and the pK_a of the corresponding phenol.

$$\log k_{cat} = -0.685 \mathrm{p}K_a + 4.98 \quad (1)$$
$$n = 6, \quad r^2 = 0.989$$

$$\log k_0 = -0.412 \mathrm{p}K_a + 0.435 \quad (2)$$
$$n = 6, \quad r^2 = 0.972$$

TABLE I

Rate Constants for the Hydrolysis of Phenyl Esters in the Presence (k_{cat}) or Absence (k_o) of Human Serum Albumin[a]

Substituent	k_{cat} (sec^{-1})	k_o (sec^{-1})
Phenyl acetates[b]		
o-COOH	4.00×10^{-4}	2.10×10^{-5}
p-OCH$_3$	7.30×10^{-3}	1.23×10^{-4}
p-Cl	2.97×10^{-2}	3.87×10^{-4}
m-Cl	4.42×10^{-2}	7.22×10^{-4}
m-NO$_2$	2.65×10^{-1}	1.18×10^{-3}
p-NO$_2$	1.24	2.25×10^{-3}
p-Nitrophenyl esters[c]		
Acetate	2.93×10^{-2}	—
Propionate	7.15×10^{-2}	—
Butyrate	1.02×10^{-2}	—
Valerate	2.58×10^{-3}	—
Capronate	2.22×10^{-3}	—
Isobutyrate	1.08×10^{-2}	—
Isovalerate	1.47×10^{-3}	—
Trimethyl acetate	6.42×10^{-4}	—

[a] Data from Kurono et al.[9]
[b] Substituents given for X (II: R = COCH$_3$; pH 9.9, 25°C).
[c] Substituents given for R (II: X = p-NO$_2$; pH 7.0, 25°C).

It is clear that electron withdrawal from the ester linkage facilitates hydrolysis and that albumin, like the hydroxyl anion, behaves as a nucleophilic reagent. The protein, however, is a weaker nucleophile, as shown by the different slopes in Eqs. (1) and (2). The overall greater activity of albumin as compared to H_2O/OH^- is due to the good binding of the substrates,[9] a process that favors subsequent hydrolysis and a role that albumin performs well *in vivo*.

The above mechanistic model is substantiated by the finding[10] that the primary reactive site of albumin contains a tyrosine residue. Because the esterase activity of albumin is inhibited by acetic anhydride,[11] it would appear that the tyrosine phenolic group is the nucleophilic reagent. In regard to the hydrolysis of p-nitrophenyl esters (Table I), structure–activity relationships are less apparent. A plot of these results reveals a parabolic relationship between rate of hydrolysis and size of the acyl group; activity decreases for acyl groups with more than three carbon atoms. Branching also decreases hydrolysis, but the number of observations is too small to derive regression equations.

5. Nonenzymatic Biotransformation

These studies thus show that serum albumin, although not regarded as an enzyme, displays a genuine enzyme-like behavior. Indeed, it contains a site able both to bind and hydrolyze phenyl esters, these processes being highly dependent upon the structure of the substrates and subject to competitive or noncompetitive inhibition.[10,11]

III. REACTIONS WITH ENDOGENOUS NUCLEOPHILES

This section discusses smaller nucleophilic compounds, as opposed to the macromolecules considered in the previous section. An interesting reaction is the nonenzymatic reduction of nitrosobenzene mediated by either NADPH or NADH.[12] In aqueous buffered solutions, quantitative conversion to phenylhydroxylamine was observed. The rate constants of reduction are practically identical for both reagents, and they are the same at pH 5.7 and 7.4 and are independent of buffer concentration, the reaction not being subject to general acid–base catalysis. These results are all compatible with the reagents acting by transfer of a nucleophilic hydride anion, as is the case with the NAD(P)H-catalyzed reduction of carbonyl groups. Such results open promising perspectives and may lead to a broader understanding of bioreduction processes.

Most nonenzymatic biotransformations involving endogenous nucleophiles other than water are mediated by thiols, particularly glutathione (GSH). Throughout the biological literature are a number of isolated observations showing that GSH conjugates can be formed nonenzymatically from several substrates. Ketterer[13] has discussed these observations in terms of two factors, namely, the susceptibility of an electrophile to nucleophilic attack by GSH and the catalytic effectiveness of the GSH transferases. Some combinations of these two factors are illustrated by the reaction of 1,2-disubstituted ethanes with GSH to yield ethylene. The most probable reaction is shown in Fig. 2; the rates of enzymatic and

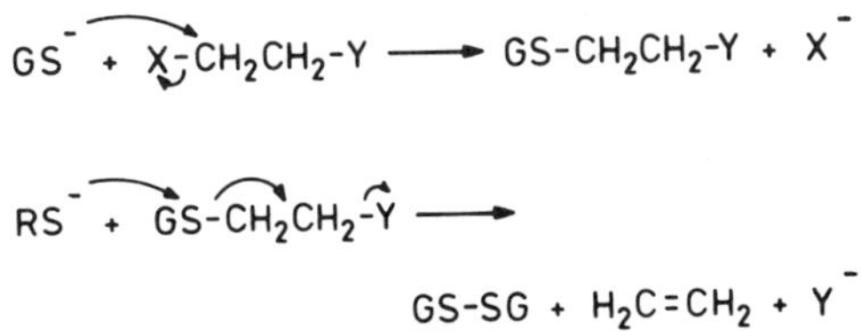

$$GS^- + X-CH_2CH_2-Y \longrightarrow GS-CH_2CH_2-Y + X^-$$

$$RS^- + GS-CH_2CH_2-Y \longrightarrow$$

$$GS-SG + H_2C=CH_2 + Y^-$$

Fig. 2. Proposed mechanism for the reaction of 1,2-disubstituted ethanes with GSH.[14]

TABLE II

Metabolism of 1,2-Disubstituted Ethanes with GSH to Yield Ethylene[a]

Substrate	Ethylene formed (pmol/min)[b]	
	Nonenzymatic[c]	Enzymatic[d]
1,2-Dibromoethane	323.5 ± 161.2	305.1 ± 87.6
1-Bromo-2-chloroethane	5.9 ± 4.9	137.7 ± 55.6
1,2-Dichloroethane	0.6 ± 0.2	26.8 ± 8.5
2-Chloroethyl methanesulfonate	6.2 ± 1.1	13.6 ± 4.7
Ethylene bis(methanesulfonate)	11.9 ± 1.0	8.3 ± 1.8

[a] Data from Liversey and Anders.[14]
[b] Substrate and GSH concentrations 25 and 10 mM, respectively.
[c] In buffer.
[d] Enzymatic reaction is the difference between total (in rat liver cytosol) and nonenzymatic reactions.

nonenzymatic reactions are compared in Table II. It is apparent that for one substrate, both reactions proceed very efficiently. For other substrates, both reactions proceed at moderate rates, and for still others, the nonenzymatic reaction is far less efficient than the enzymatic one. However, rationalizing the relative importance of these two contributions in terms of the structure of substrates appears difficult with available data.

Certain reactions of GSH with arylnitroso compounds appear of interest in a toxicological context. These xenobiotics react nonenzymatically with GSH to yield arylhydroxylamines and arylamines,[15] as shown in a simplified way in Fig. 3. The reactions appear of particular significance in the biotransformation of nitrosobenzene in red cells.[16] Although the

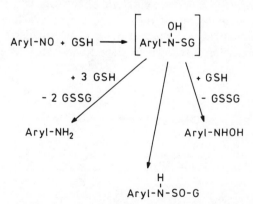

Fig. 3. Redox reactions of arylnitroso compounds with GSH.[15]

5. Nonenzymatic Biotransformation

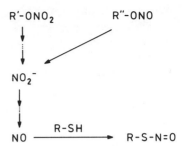

Fig. 4. Proposed mechanisms for the generation of S-nitrosothiols from organic nitrates and nitrites, nitrite anion, and nitric oxide.[17]

GSH-mediated reduction of nitroso compounds is clearly a detoxication process, its benefits are overcome by the rapid reoxidation taking place in red cells.[16]

Cysteine has been shown to react with nitric oxide (NO) or sodium nitrite ($NaNO_2$) to yield S-nitrosocysteine.[17] The latter compound and other S-nitrosothiols are strong activators of coronary arterial guanylate cyclase, an activation leading to vascular smooth-muscle relaxation. The same study[17] has shown that organic nitrates (e.g., glyceryl trinitrate) and organic nitrites (e.g., amyl nitrite) also react with cysteine to yield S-nitrocysteine and that the presence of cysteine or other thiols is required for these smooth-muscle relaxants to activate guanylate cyclase. The scheme presented in Fig. 4 has thus been proposed in order to explain the formation of a common mediator, S-nitrosothiol, from nitrates and nitrites. A number of probably nonenzymatic reductive steps are involved in the formation of nitric oxide, which then undergoes nonenzymatic attack by thiols. However, the exact mechanism of this key reaction has not been detailed.[17]

IV. REACTIONS WITH ENDOGENOUS ELECTROPHILES

Biogenic aldehydes and ketones such as formaldehyde, acetaldehyde, acetone, pyruvic acid, α-ketoglutaric acid, and acetoacetic acid form a group of endogenous electrophiles able to react nonenzymatically with a variety of amino derivatives, forming Schiff bases. O'Donnell[18] has extensively discussed the reaction of amines with carbonyls from three points of view, namely, the mechanistic aspects of the reaction, with particular emphasis on the type of catalysis depending upon the basicity of the nucleophile; the stability of the products; and the significance of the reaction in the nonenzymatic metabolism of xenobiotics.

The reactions of carbonyls with a number of medicinal hydrazines and hydrazides are especially well studied, because of their significance in biotransformation processes. Hydralazine (**III**) is particularly illustrative in this respect, and several studies have indicated its readiness to form hydrazones under physiological conditions of pH and temperature. Thus, characterized hydrazones result from the reaction with formaldehyde (**IV**), acetaldehyde (**V**), pyruvic acid (**VI**), acetone (**VII**), and *α*-ketoglutaric acid (**VIII**).[19,20] The hydrazones tend to rearrange, yielding either triazolo[3,4-*a*]phthalazine derivatives[19–21] (**IX**) or phthalazinone[20] (**X**).

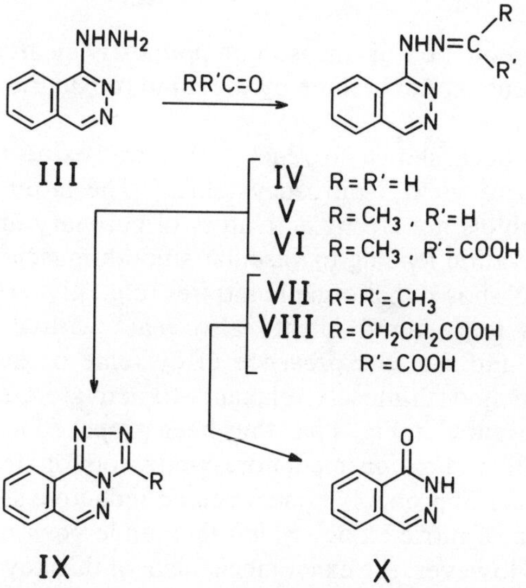
NHNH2
N
N
RR'C=O
NHN=C
R
R'
III
IV R=R'=H
V R=CH3, R'=H
VI R=CH3, R'=COOH
VII R=R'=CH3
VIII R=CH2CH2COOH
R'=COOH
N—N
N
R
O
NH
IX
X

Other examples exist that are chemically less well understood. Xylidide local anesthetics yield a number of cyclic metabolites, conceivably generated postenzymatically or nonenzymatically, the carbon atom that operates the ring closure either belonging to the molecule or not. Thus, tocainide (**XI**) in rats yields the hydantoin derivative (**XII**).[22] The reaction, although not fully understood, is believed to be nonenzymatic. The involvement of carbamyl phosphate has been postulated,[22] leading to the carbamylation and subsequent cyclization of tocainide. Small amounts of **XII** were also formed from tocainide and CO_2.[22] Whatever the actual mechanism, formation of the hydantoin, **XII**, is one more feature of the complex metabolic behavior of xylidides.

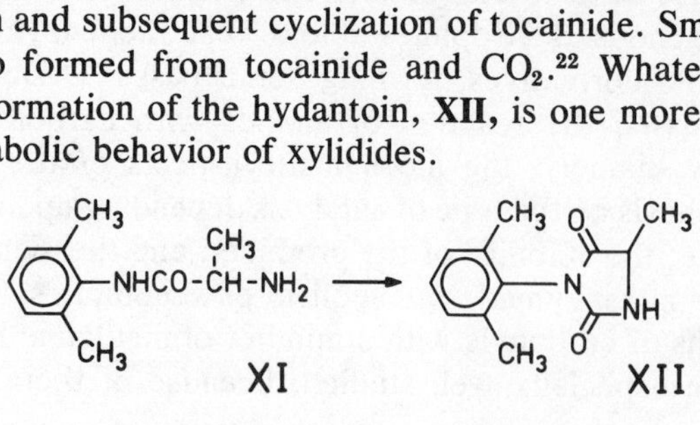
CH3
CH3
NHCO-CH-NH2
CH3
XI
CH3
O
CH3
N
NH
O
CH3
XII

V. BREAKDOWN AND REARRANGEMENT REACTIONS OF XENOBIOTICS AND PRODRUGS IN ACIDIC AND NEUTRAL AQUEOUS MEDIA

Many xenobiotics contain functional groups that display relatively limited stability under physiological conditions. Thus, the hydrolytic breakdown of acid-sensitive penicillins under gastric conditions of acidity is a well-known phenomenon. Breakdown reactions in neutral media, particularly in blood plasma, occur for a number of drugs and other xenobiotics and may even play a significant metabolic role. For example, the two ester bonds of cocaine (**XIII**) differ markedly in a metabolic sense.[23] Both liver and serum esterases hydrolyze the benzoyl ester bond to yield ecgonine methyl ester (**XIV**). In contrast, the methyl ester bond is resistant to enzymatic hydrolysis but is readily hydrolyzed nonenzymatically (~40% after 24 hr at pH 7.4 and 37°C) to benzoylecgonine (**XV**). The latter is generally considered to be the major metabolite of cocaine in humans, thereby indicating the significance of the nonenzymatic reaction of hydrolysis.

In a study on the carcinogenicity of 6-substituted benzo[a]pyrene (BP) derivatives,[24] it was found that activity resides in those compounds capable of conversion to 6-hydroxymethyl-BP. The 6-chloromethyl and 6-bromomethyl derivatives were found to be good precursors of 6-hydroxymethyl-BP. Interestingly, the reaction of nucleophilic substitution was proven to be nonenzymatic; under physiological conditions in the absence of enzyme, 80–90% of the incubated amount was transformed in 1 hr.[24]

An actively studied class of medicinal compounds is that of prodrugs, compounds that can be converted to their active metabolite(s) either enzymatically or nonenzymatically. The compound WR-2721 (**XVI**) is a pro-

$$R-CONH-CH_2-NR'R'' \rightleftharpoons RCO\bar{N}H + CH_2=\overset{+}{N}R'R'' \underset{H_2O}{\rightleftharpoons}$$

$$RCONH_2 + HO-CH_2-NR'R'' \rightleftharpoons RCONH_2 + CH_2O + HNR'R''$$

Fig. 5. Proposed mechanism for the decomposition of N-Mannich bases.[27]

drug of the corresponding free thiol N-2-mercaptoethyl-1,3-diaminopropane, which is an active mucolytic agent.[25] The prodrug is enzymatically hydrolyzed by intestinal and hepatic esterases, but its poor stability under acidic conditions (quantitative hydrolysis in 6 hr at pH 3.4 and 20°C) suggests that part of the free thiol generated *in vivo* is produced nonenzymatically in the stomach.

Many prodrugs of amides, imines, and amines have been prepared, as extensively reviewed by Pitman.[26] Important representative groups include ring-opened derivatives of cyclic drugs, for example, benzodiazepines, N-hydroxymethylated amines, and N-Mannich bases, all of which break down nonenzymatically. Many N-Mannich bases have been studied for their stability,[27] and their rate of decomposition is shown to be heavily dependent upon the substituents, which act through steric and electronic effects, i.e., by modulation of pK_a. It appears that the rate constants of decomposition of N-Mannich bases are two to three orders of magnitude larger for the neutral than for the protonated forms of the bases; the proposed mechanism of decomposition[27] is shown in Fig. 5.

VI. REACTIONS BETWEEN TWO XENOBIOTICS

A few instances are known of a xenobiotic reacting nonenzymatically under biological conditions with another xenobiotic or with a metabolite. Often, such reactions involve adduct formation, although at least one example of a redox reaction is known and will be discussed.

Quite unexpected was the finding that coadministration of acetylsalicylic acid (**XVII**) and acetaminophen (**XVIII**) to rats or rabbits leads to the formation of salicylic acid (**XIX**) and N,O-diacetyl-p-aminophenol (**XX**).[28] Transacetylation of this sort is likely to be nonenzymatic because phenol acetylation is not a known pathway of xenobiotic metabolism and because **XX** is not a metabolite when acetaminophen is administered alone.

Under the acid conditions likely to exist in the stomach, norethindrone was shown to react with isoniazid to form the hydrazone **XXI**.[29] The optimum pH is 2.5 and the reaction is essentially complete in 6 min at

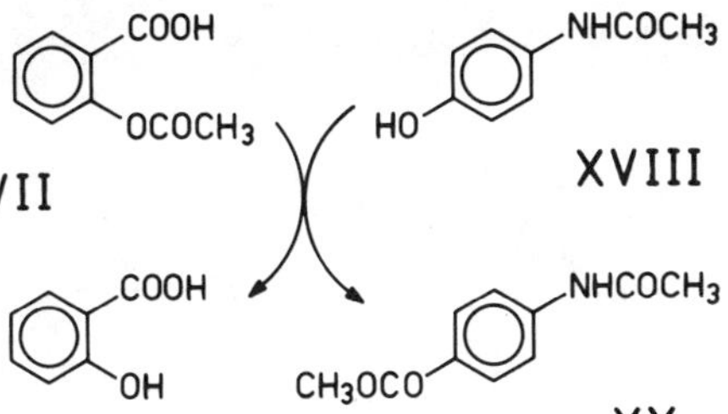

5. Nonenzymatic Biotransformation

XVII (COOH, OCOCH₃) + HO-C₆H₄-NHCOCH₃ (XVIII) → XIX (COOH, OH) + CH₃OCO-C₆H₄-NHCOCH₃ (XX)

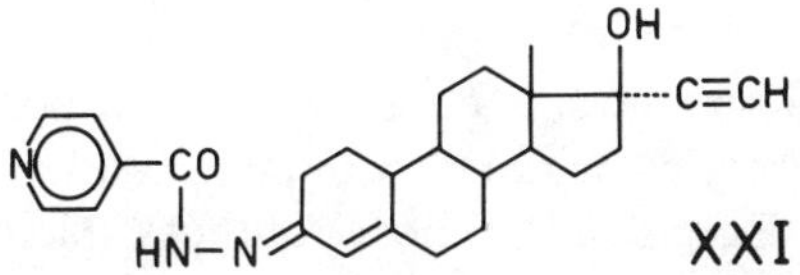

XXI

37°C. When administered orally to guinea pigs, the hydrazone was detected in plasma, thus proving its ability to be absorbed intact from the gastrointestinal tract. The same, however, was not found in rats, probably because of faster cleavage in the liver of the hydrazone, as shown by *in vitro* experiments. Such a finding suggests the possibility of interference with the contraceptive effect of norethindrone in women.

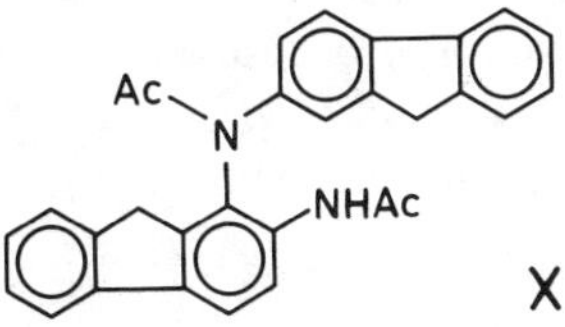

XXII

When a xenobiotic reacts with one of its own metabolites, or with another xenobiotic metabolite, the situation pertains to postenzymatic as much as to nonenzymatic metabolism. This is documented by a number of examples that have been reviewed.[1] An interesting illustration is that of the dimer **XXII** that is formed as an *in vitro* metabolite of *N*-hydroxy-2-acetylaminofluorene (*N*-OH-2-AAF).[30] The sulfation of *N*-OH-2-AAF to its *N,O*-sulfate ester is a necessary step in the formation of the dimer, suggesting the involvement of a nitrenium ion. The latter then is trapped by 2-AAF (added or formed by *N*-OH-2-AAF reduction); such a dimer has been shown to be far less mutagenic than 2-AAF or its *N*-oxygenated metabolites. The formation of dimer **XXII** must be viewed as a genuine process of detoxication.[30]

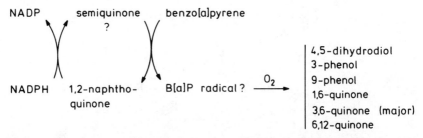

Fig. 6. A postulated pathway for the nonenzymatic oxidation of benzo[a]pyrene mediated by NADPH plus 1,2-naphthoquinone.[1,31] Reproduced with the permission of Marcel Dekker, Inc.

An interesting although isolated example of a nonenzymatic reaction between two xenobiotics without adduct formation has been reported by Nesnow and Bergman.[31] These authors have shown that 1,2-naphthoquinone can function as an electron transport component in the NADPH-dependent oxidation of benzo[a]pyrene. The postulated mechanism is shown in Fig. 6, with the semiquinone as the most likely intermediate formed in the reaction. The reduced quinone can conceivably act by reducing either benzo[a]pyrene or O_2, the former mechanism appearing more likely.[31] Other quinones were tested in this system: 1,4-naphthoquinone and 9,10-anthraquinone were inactive; menadione and vitamin K were moderately active; and 1,4-benzoquinone and phenanthrene-9,10-quinone were markedly active.[31] This nonenzymatic pathway deserves continued study because a number of quinones can be postulated as active redox agents catalyzing the oxidation of their parent aromatic hydrocarbon or of other aromatic hydrocarbons.

VII. CONCLUSION

The present chapter has tried to show that the biotransformation of xenobiotics is not restricted to enzymatic reactions. Nonenzymatic reactions, whether direct or postenzymatic, occur and are documented. However, their importance in xenobiotic metabolism is difficult to assess. *In vitro* systems, in general, vary from poor to good in their relevance to *in vivo* situations, but for nonenzymatic reactions their degree of relevance is usually unknown. As a consequence, one can only speculate on the relevance of most of the previously described examples to *in vivo* biotransformation. Because the concept of nonenzymatic metabolism is not yet familiar to every biologist, valuable findings could have escaped unaware investigators. It is hoped that in the near future, a more fully documented view will be provided of nonenzymatic reactions in the biotransformation and toxication–detoxication of xenobiotics.

ACKNOWLEDGMENTS

The author expresses his appreciation to Prof. Edward Bresnick for drawing his attention to Nagata et al.[8]

REFERENCES

1. Testa, B. (1982). Nonenzymatic contributions to xenobiotic metabolism. *Drug Metab. Rev.* **13,** 25–50.
2. Hathway, D. E. (1980). The importance of (non-enzymic) chemical reaction processes to the fate of foreign compounds in mammals. *Chem. Soc. Rev.* **9,** 63–89.
3. Lindeke, B. (1982). The non- and postenzymatic chemistry of N-oxygenated molecules. *Drug Metab. Rev.* **13,** 71–121.
4. Trager, W. F. (1982). The postenzymatic chemistry of activated oxygen. *Drug Metab. Rev.* **13,** 51–70.
5. Hathway, D. E., and Kolar, G. F. (1980). Mechanisms of reaction between ultimate chemical carcinogens and nucleic acid. *Chem. Soc. Rev.* **9,** 241–258.
6. Osterman-Golkar, S., Ehrenberg, L. E., Segerbäck, D., and Hällström, I. (1976). Evaluation of genetic risks of alkylating agents. II. Haemoglobin as a dose monitor. *Mutat. Res.* **34,** 1–10.
7. Pereira, M. A., and Chang, L. W. (1981). Binding of chemical carcinogens and mutagens to rat hemoglobin. *Chem.-Biol. Interact.* **33,** 301–305.
8. Nagata, C., Inomata, M., Kodama, M., and Tagashira, Y. (1968). Electron spin resonance study on the interaction between the chemical carcinogens and tissue components. III. Determination of the structure of the free radical produced either by stirring 3,4-benzopyrene with albumin or incubating it with liver homogenates. *Gann* **59,** 289–298.
9. Kurono, Y., Maki, T., Yotsuyanagi, T., and Ikeda, K. (1979). Esterase-like activity of human serum albumin: Structure-activity relationships for the reactions with phenyl acetates and p-nitrophenyl esters. *Chem. Pharm. Bull.* **27,** 2781–2786.
10. Ozeki, Y., Kurono, Y., Yotsuyanagi, T., and Ikeda, K. (1980). Effects of drug binding on the esterase activity of human serum albumin: Inhibition modes and binding sites of anionic drugs. *Chem. Pharm. Bull.* **28,** 535–540.
11. Morikawa, M., Inoue, M., Tsuboi, M., and Sugiura, M. (1979). Studies on aspirin esterase of human serum. *Jpn. J. Pharmacol.* **29,** 581–586.
12. Becker, A. R., and Sternson, L. A. (1980). Nonenzymatic reduction of nitrosobenzene to phenylhydroxylamine by NAD(P)H. *Bioorg. Chem.* **9,** 305–312.
13. Ketterer, B. (1982). Xenobiotic metabolism by nonenzymatic reactions of glutathione. *Drug Metab. Rev.* **13,** 161–187.
14. Liversey, J. C., and Anders, M. W. (1979). In vitro metabolism of 1,2-dihaloethanes to ethylene. *Drug Metab. Dispos.* **7,** 199–203.
15. Dölle, B., Töpner, W., and Neumann, H. G. (1980). Reaction of arylnitroso compounds with mercaptans. *Xenobiotica* **10,** 527–536.
16. Eyer, P., and Lierheimer, E. (1980). Biotransformation of nitrosobenzene in the red cell and the role of glutathione. *Xenobiotica* **10,** 517–526.

17. Ignarro, L. J., Lippton, J., Edwards, J. C., Baricos, W. H., Hymon, A. L., Kadowitz, P. J., and Gruetter, C. A. (1981). Mechanism of vascular smooth muscle relaxation by organic nitrates, nitrites, nitroprusside and nitric oxide: Evidence for the involvement of S-nitrosothiols as active intermediates. *J. Pharmacol. Exp. Ther.* **218,** 739–749.
18. O'Donnell, J. P. (1982). The reaction of amines with carbonyls: Its significance in the nonenzymatic metabolism of xenobiotics. *Drug Metab. Rev.* **13,** 123–159.
19. O'Donnell, J. P., Proveaux, W. J., and Ma, J. K. H. (1979). HPLC studies of reaction of hydralazine with biogenic aldehydes and ketones. *J. Pharm. Sci.* **68,** 1524–1526.
20. Nakano, M., Tomitsuka, T., and Juni, K. (1980). Identification of the decomposition products of hydrazine hydrazones with three endogenous ketones and kinetic study of the formation of 3-methyl-s-triazolo[3,4-a]phthalazine from hydralazine and pyruvic acid. *Chem. Pharm. Bull.* **28,** 3407–3411.
21. Talseth, T., Haegele, K. D., and McNay, J. L. (1980). Studies in the rabbit on 3-methyltriazolophthalazine, an acetylated metabolite of hydralazine. Evidence for an alternative route of formation. *Drug Metab. Dispos.* **8,** 73–76.
22. Venkataramanan, R., and Axelson, J. E. (1981). 3-(2,6-Xylyl)-5-methylhydantoin—A metabolite or a metabonate of tocainide in rats? *Xenobiotica* **11,** 259–265.
23. Stewart, D. J., Inaba, T., Lucassen, M., and Kalow, W. (1979). Cocaine metabolism: Cocaine and norcocaine hydrolysis by liver and serum esterases. *Clin. Pharmacol. Ther.* **25,** 464–468.
24. Sydnor, K. L., Bergo, C. H., and Flesher, J. W. (1980). Effect of various substituents in the 6-position on the relative carcinogenic activity of a series of benzo[a]pyrene derivatives. *Chem.-Biol. Interact.* **29,** 159–167.
25. Tabachnik, N. F., Peterson, C. M., and Cerami, A. (1980). Studies on the reduction of sputum viscosity in cystic fibrosis using an orally absorbed protected thiol. *J. Pharmacol. Exp. Ther.* **214,** 246–249.
26. Pitman, I. H. (1981). Pro-drugs of amides, imines and amines. *Medic. Res. Rev.* **1,** 189–214.
27. Bundgaard, H., and Johansen, M. (1981). Hydrolysis of N-Mannich bases and its consequences for the biological testing of such agents. *Int. J. Pharm.* **9,** 7–16.
28. Kyo, Y. N., Muranaka, Y., Hikichi, N., and Niwa, H. (1980). N,O-Diacetyl-p-aminophenol formation by combined use of aspirin with phenacetin and acetaminophen *in vivo*. *J. Pharm. Soc. Jpn.* **100,** 1043–1047.
29. Watanabe, H., Menzies, J. A., and Loo, J. C. K. (1981). An investigation of the interaction between isoniazid and the contraceptive steroid norethindrone in vivo. *Experientia* **37,** 883–884.
30. Andrews, L. S., Pohl, L. R., Hinson, J. A., Fisk, C. L., and Gillette, J. R. (1979). Production of a dimer of 2-acetylaminofluorene during the sulfation of N-hydroxy-2-acetylaminofluorene in vitro. *Drug Metab. Dispos.* **7,** 296–300.
31. Nesnow, S., and Bergman, H. (1979). 1,2-Naphthoquinone: A mediator of nonenzymatic benzo[a]pyrene oxidation. *Life Sci.* **25,** 2099–2104.

CHAPTER 6

Unmetabolized Compounds

A. G. Renwick

I.	Introduction	151
II.	Criteria for Classification of a Foreign Compound as Being Unmetabolized	152
	A. Analytical Methodology	152
	B. Dose Selection and Purity	153
	C. Species	155
	D. Route of Administration	156
	E. Effect of Chronic Administration	157
	F. *In Vivo* and *in Vitro* Data	157
III.	Properties of Unmetabolized Compounds	158
	A. Highly Polar Compounds	158
	B. Volatile Compounds	168
	C. Nonpolar Compounds	169
	D. Metabolites as Unmetabolized Compounds	169
	E. Nonabsorbed Compounds	173
IV.	Toxicological Implications	173
	A. Mechanistic Implications	173
	B. Interpretation of Animal Data	174
V.	Comments	175
	References	175

I. INTRODUCTION

Compounds that pass through the body without undergoing metabolic transformation have been considered briefly in reviews of drug metabolism.[1,2] The finding that a compound is excreted unchanged usually reduces its interest to researchers in the fields of drug metabolism, biochemical pharmacology, and toxicology. Accordingly, most reviews have considered unmetabolized compounds in a cursory manner, and many aspects and implications have been overlooked. This chapter presents a critical account of the criteria for classification of a compound as "unmetabolized," followed by an assessment of the nature of those compounds that have been reported to be unmetabolized, with comparison to the fate of metabolized structural analogs. Finally, some toxicological implications of the absence of metabolism are considered briefly.

II. CRITERIA FOR CLASSIFICATION OF A FOREIGN COMPOUND AS BEING UNMETABOLIZED

Unmetabolized implies an absolute quality, the absence of metabolism, which can never be proved experimentally. Although it is easy to demonstrate that a thing exists, the converse (i.e., its absence) is a relative value that is limited by both the ability of the observer and the conditions under which the observation was made. Thus, it is important to realize that there is no truly unmetabolized compound, but merely that there are a number of compounds that are eliminated from the body largely in unchanged form, and for which no metabolites have been detected. However, if this description were used, future improvements in the sensitivity and resolution of analytical techniques would soon render this chapter both out of date and incorrect, despite the continuing validity of many of the broader conclusions reached. With these considerations in mind, an unmetabolized compound should be regarded as one in which the vast majority of the dose is either retained in the body and/or eliminated as the unchanged chemical and for which metabolism is of negligible importance in the overall elimination of the compound. In the context of this chapter, a balance of more than 95% unchanged and less than 5% metabolism will be used arbitrarily to separate unmetabolized from metabolized chemicals.

A. Analytical Methodology

Early studies on the metabolism of foreign compounds concentrated on the detection and characterization of excretory products. General conclusions with respect to pathways of drug metabolism resulting from such pioneering studies were reviewed in the classic text of R. T. Williams.[3] Most of the early work was qualitative, and therefore, reports of the excretion of unchanged compounds cannot be evaluated properly. The

6. Unmetabolized Compounds

application of quantitative methods, initially colorimetry and gravimetric analysis, allowed an estimation of the percentage of drug excreted. However, it was not until the application of separation techniques such as chromatography and electrophoresis, combined with spray reagents for visualization, that it was possible to conclude that the material present in excreta was the unchanged xenobiotic. A further major advance in the detection of metabolites came with the incorporation of a radiolabeled atom into the foreign compound. This allowed both detection and quantification of the metabolites, as well as measurement of the percentage of the dose recovered in excreta. Other advances have come from refinements in separation, such as high-performance liquid chromatography (hplc), which have allowed increased separation and resolution of structurally related chemicals.

A wide range of techniques of varying credibility and validity have been applied to compounds reported to be excreted unchanged, which will be discussed in detail subsequently. An attempt (Table I) has been made to rate these in order of validity, and these ratings are then applied to subsequent tables. It should be realized, however, that frequently few or no details of the methods used have been reported. On the basis of the dates of such publications and the techniques available, these have either been assigned a rating as given in Table I or are indicated by a question mark (see Tables II–VII).

B. Dose Selection and Purity

1. Dose Level

The total elimination or clearance of a compound from the body (Cl_{tot}) is the sum of the various contributory pathways, such as renal clearance (Cl_R), biliary clearance (Cl_B), metabolism (Cl_m), excretion via the lungs, and so on (Eq. 1) (see also Chapters 8 and 9, this volume).

$$Cl_{tot} = Cl_R + Cl_B + Cl_m + \cdots \qquad (1)$$

Each pathway may consist of both nonsaturable first-order processes, such as diffusion or spontaneous decomposition, which are independent of dose, and saturable first-order processes, such as renal tubular secretion or enzyme-mediated biotransformation, which will decrease in importance when high doses are given. These saturable processes may show a high or low affinity and a high or low capacity for the foreign compound. Because the products detected in excreta reflect the balance of all contributory elimination processes, it is possible for the dose selected to alter dramatically the extent of metabolism. Two such situations can be envisaged.

TABLE I

Validity of Analytical Techniques[a]

Techniques	Information available	Validity	Rating
Detection of drug and possible metabolites by spray reagents, color tests, and isolation from urine	Qualitative detection of parent compound but not metabolites	Inadequate	*
As above, but with quantitation of the parent compound in excreta (95%+) using a derivatization technique and colorimetry, glc, or isotope dilution. Such methods are usually of questionable specificity with respect to structural analogs such as metabolites	Quantitative detection of parent compound but not metabolites	Adequate	**
Use of a radiolabeled compound with quantitative recovery of radioactivity (95%+) in excreta and carcass. No metabolites separable by techniques such as paper and thin-layer chromatography or electrophoresis. Analysis usually restricted to one or two systems, and limits of detection not given. Usually only one animal species studied.	Quantitative detection of parent compound and a reasonable likelihood of detecting structural analogs	Good	***
As above, but with the use of multiple solvent systems or high-resolution techniques such as hplc. Metabolic fate studied at multiple dose levels, in various animal species, and with different routes of administration and duration of exposure	Quantitative detection of parent compound and a high likelihood of detecting structural analogs. Variables such as dose are investigated	Very good	****

[a] The rating system given above (and used in Tables II–VII) reflects the validity of concluding from the published data that the compound is not metabolized and does not reflect the validity of the data with respect to any other criteria or conclusions reached by the authors.

6. Unmetabolized Compounds

a. Dose Too High. If a metabolic process has a low capacity for the foreign compound, then the pathway may be saturated at quite low doses, even if the process has a high affinity. As the dose administered is increased, the excess is eliminated by nonmetabolic pathways such as renal elimination of the unchanged compound. An example is found in the conjugation of thiodipropionic acid,[4] for which the extent of conjugation decreases from 97% at 3 mg/kg to 20% at 627 mg/kg. It seems probable that in some cases, compounds are classified as unmetabolized simply because a high dose has been used in the metabolism study to aid detection and characterization of the metabolites.

b. Dose Too Low. If an enzyme has a low affinity for the foreign compound, then the pathway may be undetectable at low doses, as most of the material will be eliminated by high-affinity processes such as renal tubular excretion. However, as the dose is increased, the high-affinity process may become saturated and the resultant buildup in plasma and tissue levels of the compound may allow the low-affinity metabolic process to be detected. An example is seen with the fate of 2,4,5-trichlorophenoxyacetic acid (2,4,5-T),[5] which is excreted unchanged at low doses but undergoes detectable metabolism at doses sufficient to saturate the renal clearance of the parent compound. Such phenomena are frequently of importance in high-dose animal toxicity studies.[6]

2. Dose Purity

The presence of impurities in the dose material is likely to result in a truly unmetabolized chemical appearing to produce metabolites. If in a metabolism study the amount of any metabolite detected in excreta is less than the impurity content of the dose material, then the "metabolite" need not be a biotransformation product of the test compound. The synthetic sweetener saccharin provides an example. Early studies, reviewed by Williams,[3] indicated that saccharin was excreted unchanged. However, with the advent of radiochemical techniques, studies in the early 1970s suggested that up to 1% of the dose was metabolized by heterocyclic ring fission to 2-sulfamoylbenzoic acid and 2-sulfobenzoic acid (see the later discussion). These findings could not be reproduced by other workers using highly purified, radiolabeled saccharin, and it appears likely that the [^{14}C]saccharin used in the early studies contained a radiolabeled impurity (such as benz[*d*]isothiazoline 1,1-dioxide) that produced the reported "metabolites."[7]

C. Species

It is now recognized that there occur marked species differences in the extent of metabolism of some xenobiotics.[1,2,8] These differences are usu-

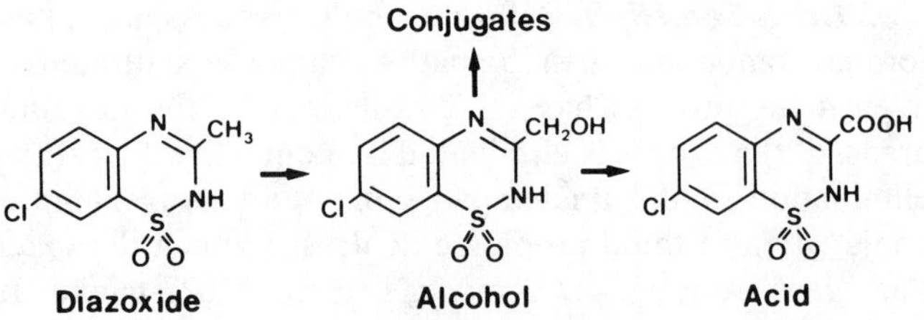

Conjugates

Diazoxide → Alcohol → Acid

		Diazoxide	Alcohol	Acid
Human	P.O.	31	21	23
Monkey	I.V.	70	13	10
Dog	P.O.	70	30	
Rat	I.P.	98		<2
Guinea Pig	I.P.	98		<2

Fig. 1. Species differences in the fate of diazoxide.[9] The numbers given refer to the proportion of radioactivity present in urine. In humans this represented about 90% of the dose, in monkey 8–63%, in dog 13%, in rat 38–68%, and in guinea pig 67–77%.

ally in the extent to which a drug or metabolite utilizes alternative pathways and results in different proportions of urinary metabolites in different species. However, for a compound that has only one major pathway of metabolism, species differences may lead to an apparent absence of metabolism. Thus, diazoxide (Fig. 1) undergoes extensive metabolism in the human, the monkey, and the dog but is excreted essentially unchanged in the rat and the guinea pig.[9] Conversely, cyclohexylamine, a strong base, undergoes negligible metabolism in the human (2%) but is subject to extensive deamination and aliphatic hydroxylation in the rat (5%), the guinea pig (5%), and especially the rabbit (27%).[10] A similar species difference is seen in the conjugation of 4-nitrophenylacetic acid, which is negligible in the human and the monkey but extensive (61%) in the rat.[11] It should be apparent that metabolism studies in a single species may incorrectly suggest that a compound is not subjected to biotransformation.

D. Route of Administration

The number of metabolic barriers that a compound must cross prior to entering the blood, with the consequent possibility of renal elimination unchanged, is dependent on the route of administration[6] (see Chapters 8 and 9, this volume). Parenteral administration such as intravenous, intraperitoneal, or subcutaneous injection allows the compound to reach the systemic circulation without passing any enzyme-rich tissue other than the lung. However, after oral administration, the compound must pass the metabolic hurdles of the gut contents and its enzymes, both host and microbial, the gut wall with its high capacity for conjugation, the high

6. Unmetabolized Compounds

metabolic activity of the liver, and finally the lung. The nature and extent of first-pass metabolism of foreign compounds in these tissues have been reviewed[12] (see also Chapters 9 and 10, this volume). Because the balance of unchanged compound to metabolites in the excreta must reflect, in part, the extent to which the compound is presented to the organs of excretion and metabolism, the proportion of unchanged chemical is usually higher after parenteral administration. Indeed, for compounds such as amygdalin,[12] isoprenaline,[13] methyldopa,[14] and cyclamate,[15] which are metabolized in the intestinal wall or contents, this difference may be so large that the compound is excreted essentially unchanged after parenteral administration, but largely as metabolites after oral administration.

E. Effect of Chronic Administration

Certain enzyme-mediated reactions are readily inducible by exposure of the animal to suitable environmental stimuli. When the environmental stimulus is the test chemical itself, there is the potential for the induction of metabolism during chronic administration. Under such circumstances, the fate of the compound will be different if tested prior to and during chronic administration. An excellent example of this is the nonnutritive sweetener cyclamate, which is excreted unchanged in normal animals but is metabolized extensively, but variably, to cyclohexylamine when given after a period of chronic administration.[15] A further possible example is seen in the fate of certain nonmetabolized organochlorine compounds that are retained in the body and that can act as potent inducers of the cytochrome P-450 system.[16]

F. *In Vivo* and *in Vitro* Data

Although *in vitro* techniques have provided invaluable tools for detailed analysis of mechanisms of enzyme-mediated reactions and cell toxicity, they have a limited value in determining whether or not a compound is unmetabolized (as defined earlier). *In vitro* investigations may be misleading in two ways:

1. The selection of inappropriate sources of enzyme, incubation conditions, or concentrations of substrate or cofactors may suggest that a compound is not capable of undergoing metabolism despite extensive biotransformation *in vivo*.

2. With the exception of biliary elimination in the isolated perfused liver, *in vitro* techniques cannot take any account of alternative nonmetabolic pathways of elimination, and the chemical is trapped with the

enzyme for the duration of the study. Under such circumstances, the compound may undergo measurable biotransformation, despite the fact that it is eliminated largely unchanged *in vivo*. An excellent example of this is the aerosol propellant trichlorofluoromethane, which is converted to dichlorofluoromethane when incubated anaerobically in a sealed tube with liver microsomes[17] but excreted almost totally unchanged in the expired air *in vivo*.[18]

For these reasons, the presence or absence of metabolism during *in vitro* studies has been regarded as insufficient evidence for classifying a compound as metabolized or unmetabolized for the purposes of this discussion.

III. PROPERTIES OF UNMETABOLIZED COMPOUNDS

A wide variety of chemical structures are eliminated without significant enzyme-mediated transformation. From the diversity of chemical and physical properties, a number of groups emerge with a major subdivision separating compounds that are chemically reactive and those that are stable under physiological conditions. The former are frequently of high toxicity and are not excreted as the parent compound (considered in Chapter 10, this volume). The latter are the subject of this chapter and may be divided into a number of groups.

A. Highly Polar Compounds

Most therapeutic drugs and many anutrients encountered in the food and external environment contain ionizable functional groups such as $COOH$, NH_2, and SO_3H. Such groups exist in equilibrium between the ionized, water-soluble and un-ionized, lipid-soluble forms. The extent of ionization in aqueous solution will depend on the polarity of the compound, that is, its tendency to lose or gain a proton.

Because it is the lipid-soluble, un-ionized form that is primarily responsible for transfer of a chemical across cell membranes, highly polar molecules frequently show only a slow rate of diffusion into cells with high metabolic capacity. Such a diffusion barrier does not exist at the glomerulus or for compounds that undergo active renal tubular secretion. The presence, therefore, of a highly ionized group, combined with the absence of an obvious target for metabolism, will predispose a compound to elimination without metabolism.

6. Unmetabolized Compounds

1. Acids

a. Carboxylic Acids. The principal metabolic reactions of carboxylic acids are conjugation with glucuronic acid on the endoplasmic reticulum or with an amino acid within the mitochondria. Caldwell[8] has concluded that carboxylic acid groups may be excreted unchanged if they have a low pK_a, have a carboxyl group that is sterically hindered, or are more readily metabolized at other sites. The first two characteristics are of key importance to the present discussion but appear inadequate to explain all of the examples of unmetabolized carboxylic acids given in Fig. 2 and Table II.

Carboxylic acids may be classified according to the number of acid groups present on the molecule, as shown in Fig. 2.

Monocarboxylic acids are usually subjected to extensive metabolism. If the carboxylic acid group is attached to an aromatic ring, it will undergo a conjugation reaction, as previously described. If the acid group is attached to an alkyl side chain, this will usually undergo β-oxidation, ultimately yielding an aromatic or arylacetic acid. There are very few exam-

TABLE II

Carboxylic Acids

Compound	Metabolism (%)	Dose[a] (%) recovered	Species	Validity[b]
Monocarboxylic acids				
4-Sulfamoylbenzoic	0	105	Rat[19]	***
2-Nitrobenzoic	18[c]	72	Rat[20]	
Triphenylacetic	0	?	Dog, rabbit[3]	?
Dicarboxylic acids				
Adipic	100	70[d]	Rat[21]	
β-Ethyladipic	20 (?)	80	Dog[3]	
Phthalic	0	?	Dog[3]	?
Methylenedisalicylic	0	85	Various[22]	****
Cromoglycate	0	85+	Rat,[23] human[24]	****
Tricarboxylic acids				
Nitrilotriacetic	0	90+	Rat,[25] human[26]	****
Aurin tricarboxylic	0	10[e]	Rat[3]	**
Tetracarboxylic acids				
Ethylenediaminetetra-acetic	0	96+	Rat[27]	***

[a] As parent compound plus any metabolites.
[b] For unmetabolized compounds (see Table I).
[c] Nitroreduction plus some conjugation (?).
[d] As CO_2.
[e] Extensive retention in body.

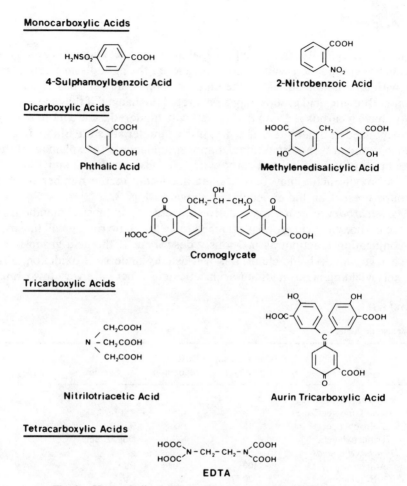

Fig. 2. Unmetabolized carboxylic acids considered in Table II.

ples of unmetabolized monocarboxylic acids and these may be explained adequately by the following:

1. The presence of an electron-withdrawing group such as a sulfonamide group (4-sulfamoylbenzoic acid[19] and 2-sulfamoylbenzoic acid) (see later discussion) or a nitro group (4-nitrophenylacetic acid[11] or 2-nitrobenzoic acid[8], although one study[20] indicates both conjugation and nitroreduction in the rat). Such electron-withdrawing groups tend to deactivate the aromatic ring, stabilize the carboxylate anion, and thus decrease the pK_a of the acid (4-sulfamoylbenzoic acid, pK_a 3.5[19]; 2-nitrobenzoic acid, pK_a 2.2[8])

2. The presence of a sterically hindered carboxylic acid group, as is found in triphenylacetic acid, which is excreted unchanged.[8]

Dicarboxylic acids are a group that shows a pronounced tendency to elimination as the parent compound. Simple aliphatic dicarboxylic acids are common substrates for oxidative intermediary metabolism, with little excreted in the urine unchanged. However, the presence of chemical substituents may dramatically increase the elimination of unchanged compound. Adipic acid is almost entirely oxidized,[21] but β-alkyl derivatives of adipic acid could be recovered from the urine in up to 85% yield.[3] Aromatic dicarboxylic acids represent a group of compounds with a high probability of avoiding biotransformation. In part, this may be due to their low pK_a (phthalic acid,[3] pK_a 3.0; methylenedisalicylic acid,[22] pK_a 3.5; sodium cromoglycate,[23,24] pK_a 1.9), and to the fact that the un-ionized form may have low lipid solubility.[23] However, comparison of the absence of metabolism of methylenedisalicylic acid with the extensive conjugation of salicylic acid suggests a further important determinant, which is that the organic acid must contain both a carboxylic acid and a hydrophobic center for extensive conjugation. A similar scheme has been suggested[8] specifically for the amino acid conjugation of arylacetic acids, but the data presented here suggest a more general requirement for a hydrophobic center. Indeed, the absence of such a center in 4-sulfamoylbenzoic acid appears to be a more likely explanation of its lack of metabolism than does its pK_a (3.5), which is similar to those of aromatic acids that undergo extensive conjugation.[8]

Tricarboxylic acids such as nitrilotriacetic acid[25,26] and aurin tricarboxylic acid,[3] and *tetracarboxylic acids* such as ethylenediaminetetraacetic acid (EDTA),[27] are not subject to significant metabolism *in vivo*. These findings support the suggestion made previously that a hydrophobic center is a requisite for the conjugation of carboxylic acid groups.

b. Sulfonic and Sulfamic Acids.

These functional groups are strong organic acids with pK_a values of 2 or less[3] and, as such, would predispose the associated molecules to elimination without metabolism. This is true for simple molecules such as methanesulfonic acid,[3] iodomethanesulfonic acid,[3] sulfoacetic acid,[3] ethanesulfonic acid,[3] benzenesulfonic acid,[3] 2-naphthylamine-1- and -6-sulfonic acids,[28] and single doses of cyclohexylsulfamic acid (cyclamate)[15] (Fig. 3; Table III). The high polarity of these acids results in their incomplete absorption from the gut, which provides the possibility of induction of metabolism by the intestinal microflora. This was found for cyclamate,[15] and the enzyme that was induced could hydrolyze other simple sulfamic acids.[29] A number of azo dyes

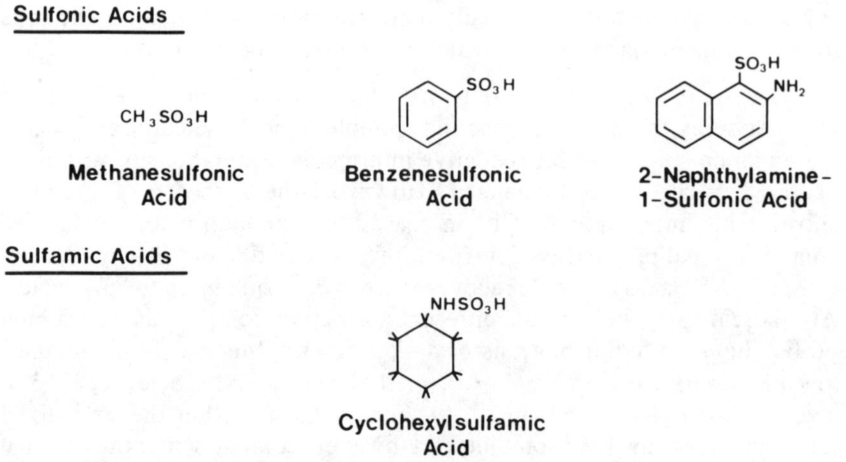

Sulfonic Acids

CH₃SO₃H

Methanesulfonic Acid

Benzenesulfonic Acid

2-Naphthylamine-1-Sulfonic Acid

Sulfamic Acids

Cyclohexylsulfamic Acid

Fig. 3. Unmetabolized sulfonic and sulfamic acids considered in Table III.

contain sulfonic acid groups and are subjected to metabolism, but because of intestinal microbial azo reduction and not by host tissues.

c. Sulfonamides and Their Heterocyclic Analogs.

The sulfonamide group is a weakly acidic functional group that is not a prime site of metabolism, although the SO_2NH group has been reported to undergo acetylation and reduction in the dog[3] and glucuronidation in humans.[1]

TABLE III

Sulfonic and Sulfamic Acids[a]

Compound	Metabolism (%)	Dose recovered (%)	Species	Validity
Methanesulfonic	0[b]	96[c]	Rat[3]	**
Iodomethanesulfonic	0[b]	96[c]	Rat[3]	**
Benzenesulfonic	0	?	Dog[3]	?
4-Hydroxybenzenesulfonic	0	?	Rabbit[3]	?
2-Naphylamine-1-sulfonic	0	84+	Dog[28]	***
2-Naphylamine-6-sulfonic	0	84+	Dog[28]	***
Cyclohexylsulfamic	0[d]	90+	Various[15]	***

[a] For other details, see Tables I and II.
[b] No inorganic sulfate detected; 70+% of dose isolated and characterized.
[c] As neutral S in urine.
[d] After a single dose (see text). Some cyclohexylamine found in one human subject in later urine samples.

6. Unmetabolized Compounds

Simple sulfonamides such as toluene-4-sulfonamide and homosulfanilamide (see Fig. 8) or sulfanilamide undergo metabolism, usually at a site other than the sulfonamide group, for example, C-oxidation or N-acetylation. However, an interesting group of unmetabolized compounds (Fig. 4; Table IV) is found when the sulfonamide group is part of a delocalized heterocyclic ring system.

i. Benz[d]isothiazoline 1,1-Dioxide and Derivatives. The parent compound is oxidized extensively at the CH_2 group (see Fig. 8) to yield saccharin and 2-sulfamoylbenzoic acid.[30] However, the presence of a

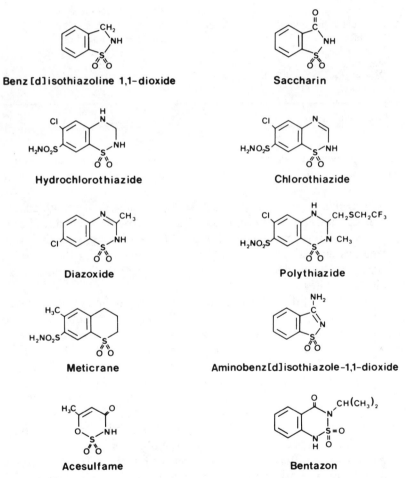

Fig. 4. Unmetabolized cyclic sulfonamides and related metabolized analogs considered in Table IV.

TABLE IV

N,S-Heterocycles

Compound	Metabolism (%)	Dose recovered (%)	Species	Validity
Benz[d]isothiazoline 1,1-dioxide	90+	74–94	Rat, human[30]	
Saccharin	0	90+	Various[7,31]	****
5-Chlorosaccharin	0	89	Rat[32]	**
3-Aminobenz[d]isothiazole 1,1-dioxide	1	95	Rat[32]	***
Chlorothiazide	0	95	Human[33]	***[a]
Hydrochlorothiazide	0	90	Human[34]	***(*)
Polythiazide	80	100 (?)	Rat, dog[35]	
Diazepoxide	0–44	8–92	Various[9]	
Meticrane	?	?	Rat[36]	?[a]
Acesulfame	0	?	Various[37]	***(?)[a]
Bentazon	3	91	Rat[38]	

[a] No details given.

π-bonded system, as in saccharin,[31] 5-chlorosaccharin, and 3-aminobenz[d]isothiazole,[32] results in a system that is resistant to metabolism.

ii. Benzothiadiazine and Related Analogs. Simple benzothiadiazines such as the diuretics, chlorothiazide,[33] and hydrochlorothiazide,[34] are excreted in the urine unchanged. However, the addition of side chains, as in the analogs diazoxide[9] (see Fig. 1) and polythiazide,[35] or the omission of the heterocyclic nitrogen atoms, as in meticrane,[36] allows extensive oxidation at these sites.

The presence of the π-bonded system adjacent to the cyclic sulfonamide groups seen in many of these unmetabolized aromatic heterocyclic compounds and in the monocyclic nonnutritive sweetener acesulfame-K[37] results in a marked increase in the tendency of SO_2NH to lose a proton, which lowers the pK_a; for example, saccharin has a pK_a of 2. In addition, there would be marked deactivation of the associated benzene ring. However, comparison of the structure of hydrochlorothiazide with benz-[d]isothiazoline 1,1-dioxide suggests that the former should also undergo extensive oxidation of the methylene group. The reason for this anomaly is not a species difference because both compounds were studied in humans. The difference possibly arises from the further deactivation introduced by the noncyclic sulfonamide group in hydrochlorothiazide. Bentazon is a structural analog of saccharin in which the N atom of the

6. Unmetabolized Compounds

CONHSO$_2$ system in alkylated; this compound undergoes slight metabolism, although the site was not identified.[38]

2. Bases

a. Amines. Previous reviews[1-3] have recognized that primary (RNH$_2$), secondary (R$_2$NH), and tertiary (R$_3$N) amines usually undergo extensive biotransformation, and "almost unmetabolized" examples are restricted to highly polar primary and secondary amines such as cyclohexylamine[10] and morpholine.[39] Because quaternary amines are fixed in the charged form, they are highly polar and, after parenteral injection, may be eliminated from the body without metabolism, either in bile or in the urine by active renal tubular secretion (Fig. 5; Table V). The presence or absence of metabolism cannot be related to the number of quaternary groups but rather to the presence within the molecule of a readily available site of metabolism. Thus, simple mono- and diquaternary amines such as tetraethylammonium,[40] hexamethonium,[41] decamethonium,[42] and bretylium[43] are excreted unchanged. The extensive metabolism of emepronium[44] is probably related to the lipophilic diphenyl substituent, because dibenzyl- and tribenzylmethylammonium undergo only slight metabolism, whereas the more lipid-soluble cetyltrimethylammonium is metabolized more extensively.[40] The absence of metabolism of the complex monoquaternary compound, stercuronium,[45] is in contrast to the extensive biotransformation of the diquaternary neuromuscular blocker,

TABLE V

Quaternary Amines

Compound	Metabolism (%)	Dose recovered (%)	Species	Validity
Tetraethylammonium	0	45–76	Various[40]	****
Hexamethonium	0	95–103	Mouse[41]	**(*)
Sulfonium analog	0	?	Mouse[41]	?
Decamethonium	0	80	Rabbit[42]	***
Bretylium	0	74	Cat[43]	**(*)
Emepronium	23^a	—	Dog[44]	
Stercuronium	0	70–80	Rat[45]	***
Pancuronium	28	79	Cat[46]	
Gallamine	0	95	Dog[47]	***
d-Tubocurarine	0	30–82	Various[48]	****
Paraquat	0	88	Rat[49]	***

^{a} Based on clearance values, and agrees well with electrophoresis data.

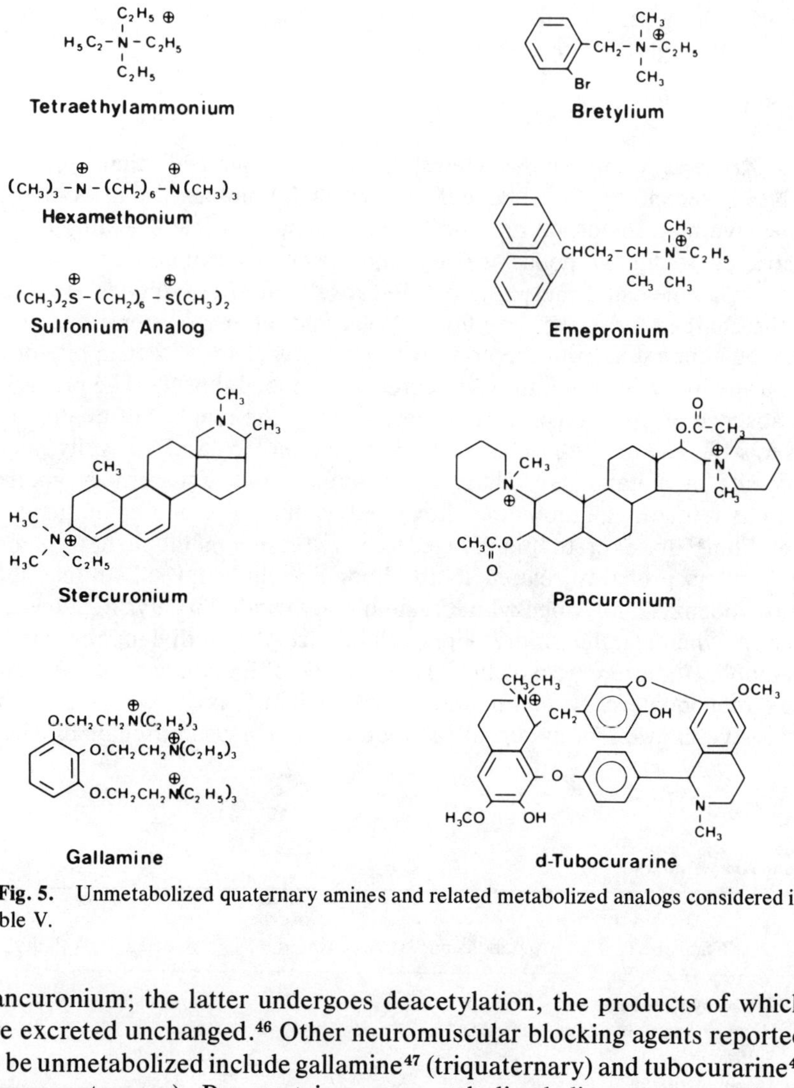

Fig. 5. Unmetabolized quaternary amines and related metabolized analogs considered in Table V.

pancuronium; the latter undergoes deacetylation, the products of which are excreted unchanged.[46] Other neuromuscular blocking agents reported to be unmetabolized include gallamine[47] (triquaternary) and tubocurarine[48] (monoquaternary). Paraquat is an unmetabolized diquaternary amine[49] used as an herbicide and has a high toxic potential (see the following discussion and Chapter 14, this volume).

b. Guanidines. Because the guanidine group is strongly basic, with a pK_a of 13.6, little metabolism would be expected at this site. Guanidine itself and simple analogs such as methylguanidine are excreted entirely unchanged.[3]

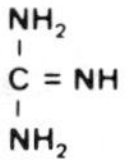

Guanidine

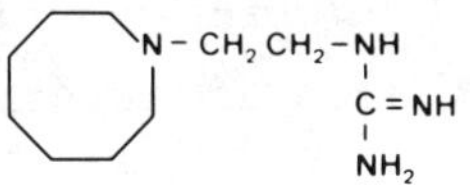

Guanethidine

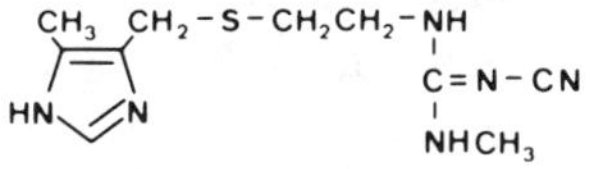

Cimetidine

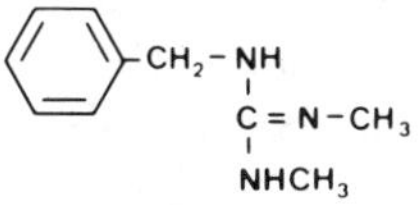

Bethanidine

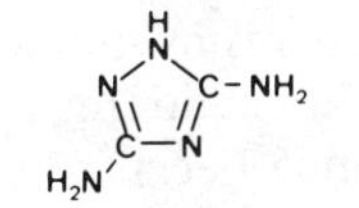

Guanazole

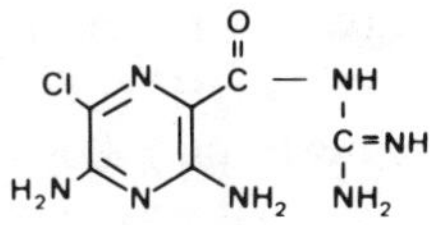

Amiloride

Fig. 6. Unmetabolized guanidines and related metabolized analogs considered in Table VI.

More complex molecules (Fig. 6; Table VI) may undergo extensive metabolism if a suitable site is present, that is, N-oxidation in guanethidine[50] or S-oxidation in cimetidine,[51] or they may be excreted unchanged despite the presence of possible sites of metabolism, such as the benzyl group of bethanidine,[52] and the amino substituents of amiloride[53] and guanazole.[54] However, all current studies support the suggestion[3] that the guanidine moiety itself is resistant to biotransformation.

TABLE VI

Guanidines

Compound	Metabolism (%)	Dose recovered (%)	Species	Validity
Guanidine	0	80	Hen[3]	?
Guanethidine	39–64	77–89	Human[50]	
Cimetidine	17–40	58–72	Various[51]	
Bethanidine	0	30–75	Cat, rat[52]	***
Amiloride	0	93	Human[53]	***
Guanazole	0[a]	95	Rat, human[54]	*(*)

[a] Based on incubation of radiolabeled drug with isolated perfused rat liver and quantitation of unlabeled drug in human urine by colorimetry.

3. Un-Ionized Polar Molecules

Carbohydrates normally undergo complete digestion and metabolism. However, they are a group of highly polar molecules and as such may be eliminated in urine, providing that they are not substrates for host tissue enzymes or active reuptake from urine, such as glucose. Thus, glycosides such as amygdalin are excreted unchanged if given by injection, which avoids metabolism by the gut flora.[12] The proposed new sweeteners, L-sugars, may also undergo little tissue metabolism.

B. Volatile Compounds

Humans may be exposed to volatile compounds either unintentionally, as a result of their use in industry, or intentionally when they are given as general anesthetics. There are two main mechanisms by which such chemicals may be removed: metabolism and elimination in the expired air. The latter is potentially of greater importance because of the large surface area of the lungs and the fact that it receives the complete cardiac output, not a fraction of the blood circulation, as does the liver.

Volatile compounds demonstrate clearly that the extent of metabolism *in vivo* is a balance between competing processes of elimination. *In vivo* and *in vitro* comparisons with trichlorofluoromethane and dichlorodifluoromethane have been discussed. Studies with dichloromethane in rats have shown that, after a single intraperitoneal dose, 93% was recovered unchanged in the expired air,[55] but 76% was converted to CO and CO_2 in a closed rebreathing system.[56] Similarly, properties of general

TABLE VII

General Anesthetics

Compound	Metabolism (%)[a]	Boiling point (°C)[b]	Blood/gas[c]
Cyclopropane	0.5	<20	0.5
Aliflurane	1	?	1.7
Enflurane	2.5	57	1.9
Ether	5–10	35	15.0
Fluroxene	10	43	1.9
Halothane	12–25	50	2.3
Methoxyflurane	40	105	13.0

[a] Data obtained from various sources and mainly apply to rat and humans. Quantitation has utilized glc analysis of expired air and measurement of urinary organic and inorganic fluoride, where appropriate.

[b] <20 indicates a gas at room temperature.

[c] Blood/gas partition coefficient.

anesthetics that predispose them toward tissue retention and extensive metabolism include a low volatility as indicated by the boiling point, and a high affinity for the circulation as indicated by the blood/gas partition coefficient (Table VII).

C. Nonpolar Compounds

Nonpolar compounds cannot be eliminated rapidly because of their reabsorption within the organs of elimination. Therefore, normally they are subjected to extensive metabolism initially in the lipoid environment of the endoplasmic reticulum, as a consequence of which lipid solubility and hence reabsorption are reduced. However, it has long been recognized[3] that certain molecules, despite their high lipid solubility and ability to reach the cytochrome P-450 of the endoplasmic reticulum, do not possess a suitable site for metabolism and are resistant to biotransformation. Halogenated, unmetabolized, lipophilic compounds have been reviewed,[57] and the following conclusions may be reached:

1. Fecal elimination of the unchanged compound is the most important route of elimination but ceases after a few days.

2. The urine usually contains only trace amounts of the dose, present as hydroxylated metabolites.

3. There is retention of up to 80% of the dose, principally in adipose tissue. This material is retained as a permanent body burden unless there is a dramatically increased utilization of, and release from, adipose tissue, as occurs during starvation when large amounts (up to 50%) may be eliminated.

4. There are pronounced species differences in the extent of metabolism and, therefore, in the classification of what is or is not unmetabolized.

5. Aromatic halogenated compounds are metabolized, presumably via epoxidation, to phenols if there are two adjacent unsubstituted carbon atoms. Conversely, compounds lacking this characteristic are usually unmetabolized and retained in the body. Thus, in Fig. 7, 4-mono-, 4,4'-di-, and 2,5,2',5'-tetrachlorobiphenyl are metabolized, whereas 2,4,5,2',4',5'-hexachlorobiphenyl is not. Similarly, dibenzo-p-dioxin is metabolized completely to polar metabolites,[38] whereas 2,3,7,8-tetrachlorodibenzo-p-dioxin is unmetabolized in most species.[57]

6. The term "unmetabolized" is appropriate for these compounds because metabolism does not contribute to their elimination.

D. Metabolites as Unmetabolized Compounds

Because the end products of drug metabolism represent suitable molecules for elimination, it might be expected that, if administered, they

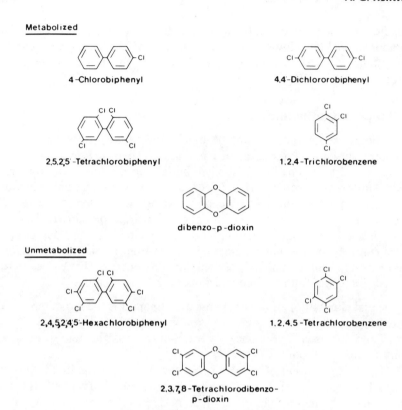

Fig. 7. Chlorinated hydrocarbons.

would be excreted unchanged. This should be particularly true of the polar products of conjugation reactions.[3] However, it must be realized that these conjugates may undergo hydrolysis, either in the tissues or by the gut flora which have the capacity to metabolize glucuronides, sulfates (to a limited extent), amides (amino acid and acetyl conjugates), and glutathione conjugates.[12] Consideration of the nature of oxidation, reduction, and hydrolysis reactions suggests that the formation of polar, and possibly unmetabolized, compounds is most likely to result from oxidation, particularly of an alkyl group to a carboxylic acid, and from hydrolysis, again with the formation of a polar carboxylic acid or amine.

Despite, or perhaps because of, the logic of this concept, surprisingly few studies have investigated its validity (Fig. 8). An early study[57] showed that about 90% of an oral dose of phenylsulfate was recovered gravimetrically as ethereal sulfate, and no extra glucuronide was excreted. However, an oral dose of phenylglucuronide resulted in the excretion of in-

6. Unmetabolized Compounds

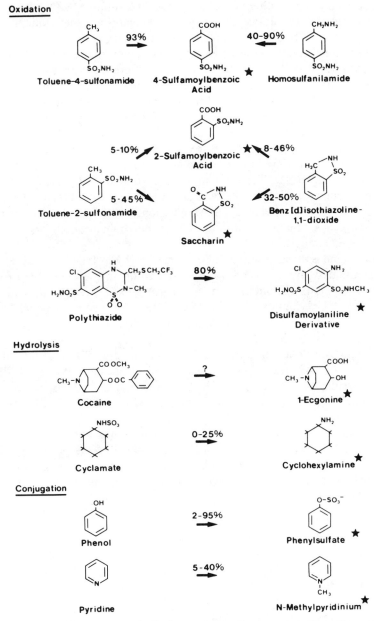

Fig. 8. Metabolites as unmetabolized compounds. Compounds indicated by a star are excreted unchanged after administration to experimental animals.

creased amounts of both glucuronide (35%) and sulfate (18%). These data suggest that sulfate but not glucuronide conjugates are metabolically stable. The site of hydrolysis was the gut because no extra sulfate was detected after subcutaneous injection of the glucuronide that was recovered quantitatively (97%) in the urine. This difference probably originates in the activities of the bacterial β-glucuronidase and arylsulfatases.[12]

Studies on the metabolism of the impurities found in commercial saccharin have shown that both toluene-2-sulfonamide[58] and benz[d]-isothiazoline 1,1-dioxide[30] are oxidized to 2-sulfamoylbenzoic acid and saccharin (Fig. 8), which are excreted unchanged, principally in the urine[59] (see Table IV). The isomeric 4-sulfamoylbenzoic acid is also excreted unchanged (see Table II), and it is formed as a metabolite of both toluene-4-sulfonamide[19] and homosulfanilamide.[3] Polythiazide (Fig. 8) undergoes extensive heterocyclic ring fission, but the disulfamoyl aniline metabolite is eliminated without further metabolism[35] (50–80% recovered; validity, good).

1-Ecgonine is an acidic metabolite formed by hydrolysis of cocaine, and it is eliminated largely in the urine within 24 hr, in the unchanged form, after intravenous administration of the rat[60] (99% recovered; validity, good). Cyclohexylamine is a polar, basic hydrolysis product of cyclamate, formed by the gut flora and excreted with little metabolism in humans (see the preceding discussion).

A final, well-documented example of an unmetabolized metabolite is the product of the N-methylation of pyridine. When N-[^{14}C]-methylpyridinium was given by intraperitoneal injection to rats and guinea pigs, no metabolites were detected in urine[61] (51–81% recovered; validity, good).

Although the preceding is not a comprehensive review, there is a comparative paucity of data supporting the hypothesis. This probably arises from the lack of suitably pure radiolabeled reference metabolites for administration to animals. However, the following tentative conclusions may be drawn.

1. The products of oxidative, reductive, and hydrolysis reactions usually undergo further enzyme action, either by oxidation or by conjugation. For example, phenols produced by oxidation with the cytochrome P-450 system are only weakly acidic and are conjugated prior to elimination, whereas alcohols usually undergo further oxidation to a carboxylic acid group, which normally either undergoes β-oxidation or is conjugated with glucuronic acid or an amino acid.

2. The products of conjugation are likely to be eliminated unchanged if given parenterally. If given orally, however, or if eliminated in the bile, considerable deconjugation by the gut flora may be anticipated.

E. Nonabsorbed Compounds

Compounds that are not absorbed from the gut cannot undergo metabolism by the host tissues but may still be a substrate for host enzymes secreted into the gut lumen or for the intestinal microflora, as occurs with cyclamate. Such compounds may be divided into two classes.

1. Low Modular Weight Compounds

Highly polar acids such as sodium cromoglycate or bases such as amiloride are only partially absorbed from the gut, and a large fraction may be recovered in the feces. In such cases, the absence of metabolism and the incomplete absorption originate in the same physicochemical property, that is, a low lipid solubility. By contrast, most lipid-soluble compounds are both readily absorbed from the gut and metabolized. Interesting exceptions are the lipid-soluble, sterically hindered, phenolic antioxidants, of which <1% may be recovered in the urine as metabolites and the remainder eliminated unchanged in the feces.[62]

2. Undigested Polymers

Sulfated glycoproteins, which have antiulcerogenic and antipeptic properties, undergo negligible absorption in the rat, and in radiolabeled studies only a small amount of ^{35}S (0.5%) was recovered in urine as low molecular weight organic sulfate.[63] Other polymers that are not substrates for digestive enzymes (e.g., microcrystalline cellulose[38]) are also recovered quantitatively in the feces unchanged after oral administration.

IV. TOXICOLOGICAL IMPLICATIONS

A. Mechanistic Implications

An obvious consequence of lack of metabolism of a compound is that any biological effects seen must arise from the chemical itself, either directly on the system affected or indirectly. Such a conclusion is of obvious mechanistic importance and has been the stimulus for major research efforts on compounds such as paraquat and saccharin. Paraquat is a diquaternary amine that is used as an herbicide, and accidental poisoning is associated with severe toxicity to the lung, liver, and kidney. The compound is poorly absorbed from the gut and is eliminated without undergoing detectable metabolism.[49] Unlike many other toxins, it does not bind covalently to tissue macromolecules[64] and must exert its effects in a less direct manner. Various other nonmetabolic mechanisms have been studied, and the most probable is the generation of superoxide radicals via a redox system[65] (see Chapter 1, this volume). Similarly, saccharin is unmetabolized even after prolonged feeding[31] and does not bind covalently to DNA; yet when fed to rats for two generations, this inert chemical produces an increase in the number of bladder tumors. A large number of studies have concentrated on possible mechanisms of this effect, seen only at very high dietary levels (5% or more), so far without notable success. However, it has been demonstrated that the feeding of such high levels does not reveal a dose-dependent metabolism of saccharin (see the previous discussion) despite the saturation of renal clearance.[6] The absence of metabolism suggests that the anionic saccharin molecule cannot act as a classic electrophilic carcinogen,[31] and this may affect profoundly the interpretation of the animal data, the extrapolation of risk to humans, and even the regulation for the use of this compound.

B. Interpretation of Animal Data

In the extrapolation of animal toxicity data to a possible risk for humans, information is needed on the fate of the compound in both the test species and in humans because of the possibility of large interspecies differences in metabolism.[1-3,8] The absence of metabolism in both species greatly simplifies such transspecies extrapolations and leaves only the problems associated with extrapolation down the dose–response curve, possible pharmacokinetic differences between species, and the general applicability of the mechanism of action.

Differences in toxicity between test species and between structural analogs frequently can be related to the presence or absence of metabolism. The absence of hepatotoxicity of trichlorofluoromethane compared with other halogenated hydrocarbons originates in its lack of metabolic activation because of its rapid elimination unchanged in the expired air in all species studied. Conversely, 2,3,7,8-tetrachlorodibenzo-p-dioxin (TCDD) is extremely toxic to most species in which it is not metabolized but is retained in the body. However, in the hamster[66] it is much less toxic and is more rapidly eliminated as metabolites in the urine. Studies in monkey and mouse have suggested that TCDD may be less resistant to enzymatic action than previously believed.

It is, therefore, apparent that the presence or absence of metabolism may significantly affect how we view a particular compound and assess the likely impact of its addition to the human environment. The absence of biotransformation is frequently a stimulus to an intensification of investigations in a search for the mechanism of toxicity.

V. COMMENTS

Compounds that are eliminated or retained without undergoing biotransformation represent a small minority of the chemicals encountered in the environment. Of this minority, very few have been studied in a range of species, at different doses, and using modern analytical techniques. Many unmetabolized examples presented in this chapter may be shown to undergo biotransformation in future studies.

The major determinant of the extent of metabolism is the nature of the chemical, with highly polar chemicals representing the most important group of unmetabolized compounds. Such compounds have a low lipid solubility and poor absorption from the gut. Thus, most examples are compounds either intended not to have any systemic effect in humans, such as food additives, herbicides, and other environmental chemicals, or have to be given parenterally by injection or inhalation to obtain a therapeutic effect. Compounds that are active therapeutically after oral administration and that are eliminated without undergoing biotransformation represent a very small proportion of the clinical armamentarium. Such compounds (chlorothiazide, hydrochlorothiazide, bethanidine, and amiloride) represent a combination of marked polarity, which results in slow and incomplete absorption from the gut, and the absence of a suitable site for enzyme action, so that competing nonmetabolic processes are responsible for removing the drug from the body. The other major group of unmetabolized compounds includes those in which the lack of biotransformation is dependent solely on the absence of a suitable site of enzyme attack. Such compounds are highly lipid soluble, accumulate in the body, and may show a high toxic potential. The absence of metabolism of a compound may greatly simplify the investigation of any associated pharmacological or toxicological problem; alternatively, it may provide a stimulus for even greater research activity in attempting to define the mechanism of action.

REFERENCES

1. Williams, R. T. (1974). Inter-species variations in the metabolism of xenobiotics. *Biochem. Soc. Trans.* **2**, 359–377.
2. Williams, R. T., and Millburn, P. (1975). Detoxication mechanisms—The biochemistry of foreign compounds. *In* "Physiological and Pharmacological Biochemistry" (H. K. F. Blaschko, ed.), pp. 211–266. Butterworth, London.
3. Williams, R. T. (1959). "Detoxication Mechanisms." Chapman & Hall, London.

4. Reynolds, R. C., Astill, B. P., and Fassett, D. W. (1974). The fate of [^{14}C]thiodipropionates in rats. *Toxicol. Appl. Pharmacol.* **28**, 133–141.
5. Gehring, P. J., and Young, J. D. (1978). Application of pharmacokinetic principles in practice. *In* "Proceedings of the First International Congress on Toxicology: Toxicology as a Predictive Science" (G. L. Plaa and W. A. M. Duncan, eds.), pp. 119–141. Academic Press, New York.
6. Renwick, A. G. (1982). Pharmacokinetics in toxicology. *In* "Methods in Toxicology" (A. W. Hayes, ed.), pp. 659–710. Raven Press, New York.
7. Renwick, A. G. (1979). The metabolism, distribution and elimination of non-nutritive sweeteners. *In* "Health and Sugar Sustitutes" (B. Guggenheim, ed.), pp. 41–47. Karger, Basel.
8. Caldwell, J. (1982). Conjugation of xenobiotic carboxylic acids. *In* "Metabolic Basis of Detoxication" (W. B. Jakoby, J. R. Bend, and J. Caldwell, eds.), pp. 271–290. Academic Press, New York.
9. Pruitt, A. W., Faraj, B. A. and Dayton, P. G. (1974). Metabolism of diazoxide in man and experimental animals. *J. Pharmacol. Exp. Ther.* **188**, 248–256.
10. Renwick, A. G., and Williams, R. T. (1972). The metabolites of cyclohexylamine in man and certain animals. *Biochem. J.* **129**, 857–867.
11. James, M. O., Smith, R. L., and Williams, R. T. (1972). The conjugation of 4-chloro- and 4-nitro-phenylacetic acids in man, monkey and rat. *Xenobiotica* **2**, 499–506.
12. George, C. F., Shand, D. G., and Renwick, A. G., eds. (1982). "Presystemic Drug Elimination." Butterworth, London.
13. Connolly, M. E., Davies, D. S., Dollery, C. T., Morgan, C. D., Paterson, J. W., and Sandler, M. (1972). Metabolism of isoprenaline in dog and man. *Br. J. Pharmacol.* **46**, 458–472.
14. Kwan, K. C., Foltz, E. L., Breault, G. O., Baer, J. E., and Totaro, J. A. (1976). Pharmacokinetics of methyldopa in man. *J. Pharmacol. Exp. Ther.* **198**, 264–277.
15. Renwick, A. G., and Williams, R. T. (1972). The fate of cyclamate in man and other species. *Biochem. J.* **129**, 869–879.
16. Bickel, M. H., and Muehlebach, S. (1980). Pharmacokinetics and ecodisposition of polyhalogenated hydrocarbons: Apsects and concepts. *Drug Metab. Rev.* **11**, 149–190.
17. Wolf, C. R., King, L. J., and Parke, D. V. (1975). Anaerobic dechlorination of trichlorofluoromethane by liver microsomal preparations *in vitro*. *Biochem. Soc. Trans.* **3**, 175–177.
18. Blake, D. A., and Mergner, G. W. (1974). Inhalation studies on the biotransformation and elimination of [^{14}C]trichlorofluoromethane and [^{14}C]dichlorodifluoromethane in beagles. *Toxicol. Appl. Pharmacol.* **30**, 396–407.
19. Ball, L. M., Williams, R. T., and Renwick, A. G. (1978). The fate of saccharin impurities. The excretion and metabolism of [^{14}C]toluene-4-sulphonamide and 4-sulphamoylbenzoic acid in the rat. *Xenobiotica* **8**, 183–190.
20. Gardner, D. M., and Renwick, A. G. (1978). The reduction of nitrobenzoic acids in the rat. *Xenobiotica* **8**, 679–690.
21. Rusoff, I. I., Baldwin, R. R., Dominques, F. J., Mander, C., Ohan, W. J., and Thiessen, R. (1960). Intermediary metabolism of adipic acid. *Toxicol. Appl. Pharmacol.* **2**, 316–330.
22. Davison, C., and Williams, R. T. (1968). The metabolism of 5,5'-methylenedisalicylic acid in various species. *J. Pharm. Pharmacol.* **20**, 12–18.
23. Moss, G. F., Jones, K. M., Ritchie, J. T., and Cox, J. S. G. (1970). Distribution and metabolism of disodium cromoglycate in rats. *Toxicol. Appl. Pharmacol.* **17**, 691–698.

6. Unmetabolized Compounds

24. Walker, S. R., Evans, M. E., Richards, A. J., and Paterson, J. W. (1971). The fate of [^{14}C]disodium cromoglycate in man. *J. Pharm. Pharmacol.* **24,** 525–531.
25. Michael, W. R., and Wakim, J. M. (1971). Metabolism of nitrilotriacetic acid (NTA). *Toxicol. Appl. Pharmacol.* **18,** 407–416.
26. Budny, J. A., and Arnold, J. D. (1973). Nitrilotriacetate (NTA): Human metabolism and its importance in the total safety evaluation program. *Toxicol. Appl. Pharmacol.* **25,** 48–53.
27. Foreman, H., Vier, M., and Magee, M. (1953). The metabolism of ^{14}C-labelled ethylenediamine-tetraacetic acid in the rat. *J. Biol. Chem.* **203,** 1045–1053.
28. Batten, B. L. (1979). Metabolism of 2-naphthylamine sulfphonic acids. *Toxicol. Appl. Pharmacol.* **48,** A171.
29. Renwick, A. G. (1977). Microbial metabolism of drugs. *In* "Drug metabolism—From Microbe to Man" (D. V. Parke and R. L. Smith, eds.), pp. 169–189. Taylor & Francis, London.
30. Renwick, A. G., and Williams, R. T. (1978). The fate of saccharin impurities: The excretion and metabolism of [3-^{14}C]benz[*d*]isothiazoline-1,1-dioxide (BIT) in man and rat. *Xenobiotica* **8,** 475–486.
31. Sweatman, T. W., and Renwick, A. G. (1979). Saccharin metabolism and tumorigenicity. *Science* **205,** 1019–1020.
32. Renwick, A. G. (1978). The fate of saccharin impurities: The metabolism and excretion of 3-amino[3-^{14}C]benz[*d*]isothiazole-1,1-dioxide and 5-chlorosaccharin in the rat. *Xenobiotica* **8,** 487–495.
33. Brettell, H. R., Aikawa, J. K., and Gordon, G. S. (1960). Studies with chlorothiazide tagged with radioactive carbon (^{14}C) in human beings. *Arch. Intern. Med.* **106,** 57–63.
34. Beermann, B., Groschinsky-Grind, M., and Rosen, A. (1976). Absorption, metabolism and excretion of hydrochlorothiazide. *Clin. Pharmacol. Ther.* **19,** 531–537.
35. Wiseman, E. H., Schreiber, E. C., and Pinson, R. (1962). Studies of N-dealkylation of some aromatic sulphonamides. *Biochem. Pharmacol.* **11,** 881–886.
36. Hathway, D. E. (1972). Biotransformations. *In* "Foreign Compound Metabolism in Mammals" (D. E. Hathway, ed.), Vol. 2, pp. 163–327. Chemical Society, London.
37. Arpe, H. J. (1979). Acesulfame-K, a new noncaloric sweetener. *In* "Health and Sugar Substitutes" (B. Guggenheim, ed.), pp. 178–183. Karger, Basel.
38. Hathway, D. E. (1975). Biotransformations. *In* "Foreign Compound Metabolism in Mammals" (D. E. Hathway, ed.), Vol. 3, pp. 201–448. Chemical Society, London.
39. Rhodes, C., and Case, D. E. (1977). Non-metabolite residues of ICI 58,834 (viloxazine). Studies with [^{14}C]morpholine, [^{14}C]ethanolamine and [^{14}C]glyoxylate. *Xenobiotica* **7,** 112.
40. Hughes, R. D., Millburn, P., and Williams, R. T. (1973). Molecular weight as a factor in the excretion of monoquaternary cations in the bile of the rat, rabbit and guinea pig. *Biochem. J.* **136,** 967–978.
41. Levine, R. (1960). The physiological disposition of hexamethonium and related compounds. *J. Pharmacol. Exp. Ther.* **129,** 296–304.
42. Christensen, C. B. (1972). The biological fate of decamethonium. *In vivo* and *in vitro* studies. *Acta Pharmacol. Exp. Ther.* **31,** Suppl. III.
43. Duncombe, W. G., and McCoubrey, A. (1960). The excretion and stability to metabolism of bretylium. *Br. J. Pharmacol.* **15,** 260–264.
44. Hallen, B., Sundwall, A., Elwin, C. E., and Vassman, J. (1979). Renal, biliary and intestinal clearance of a quaternary ammonium compound Emepronium, in the dog. *Acta Pharmacol. Toxicol.* **44,** 43–59.

45. Hespe, W., and Wieriks, J. (1971). Metabolic fate of the short-acting peripheral neuromuscular blocking agent stercuronium in the rat as related to its long action. *Biochem. Pharmacol.* **20**, 1213–1224.
46. Agoston, S., Kersten, U. W., and Meyer, D. K. F. (1973). The fate of pancuronium in the cat. *Acta Anaesthiol. Scand.* **17**, 129–135.
47. Feldman, S. A., Cohen, E. N., and Golling, R. C. (1969). The excretion of gallamine in the dog. *Anesthiology* **30**, 593–598.
48. Hughes, R. D., Millburn, P., and Williams, R. T. (1973). Biliary excretion of some diquaternary ammonium cations in the rat, guinea pig and rabbit. *Biochem. J.* **136**, 979–984.
49. Murray, R. E., and Gibson, J. E. (1974). Paraquat disposition in rats, guinea pigs and monkeys. *Toxicol. Appl. Pharmacol.* **27**, 283–291.
50. Lukas, G. (1973). Metabolism and biochemical pharmacology of guanethidine and related compounds. *Drug Metab. Rev.* **2**, 101–116.
51. Taylor, D. C., and Cresswell, P. R. (1975). The metabolism of cimetidine in the rat, dog and man. *Biochem. Soc. Trans.* **3**, 884–885.
52. Baura, A. L. A., Duncombe, W. G., Robson, R. D., and McCoubrey, A. (1962). The distribution and excretion by cats of a new hypotensive drug, N-benzyl-N',N''-dimethylguanidine. *J. Pharm. Pharmacol.* **14**, 722–726.
53. Weiss, P., Hersey, R. M., Dojovne, C. A., and Bianchine, J. R. (1969). The metabolism of amiloride hydrochloride in man. *Clin. Pharmacol. Ther.* **10**, 401–406.
54. Gerber, N., Seibert, R., Desiderio, D., Thompson, R. M., and Lane, M. (1973). Pharmacokinetics of guanazole in man. *Clin. Pharmacol. Ther.* **14**, 264–270.
55. DiVincenzo, G. D., and Hamilton, M. L. (1975). Fate and disposition of [^{14}C]methylene chloride. *Toxicol. Appl. Pharmacol.* **32**, 385–393.
56. Rodkey, F. L., and Collison, H. A. (1971). Biological oxidation of [^{14}C]methylene chloride to carbon monoxide and carbon dioxide by the rat. *Toxicol. Appl. Pharmacol.* **40**, 33–38.
57. Garton, G. A., and Williams, R. T. (1949). Studies in Detoxication. 26. The fates of phenol, phenylsulphuric acid and phenylglucuronide in the rabbit in relation to the metabolism of benzene. *Biochem. J.* **45**, 158–163.
58. Renwick, A. G., Ball, L. M., Corina, D. L., and Williams, R. T. (1978). The fate of saccharin impurities: The excretion and metabolism of toluene-2-sulphonamide in man and rat. *Xenobiotica* **8**, 461–474.
59. Ball, L. M., Renwick, A. G., and Williams, R. T. (1977). The fate of [^{14}C]saccharin in man, rat and rabbit and of 2-sulphamoyl[^{14}C]benzoic acid in the rat. *Xenobiotica* **7**, 189–203.
60. Misra, A. L., Vadlamani, N. L., Bloch, R., Nayak, P. K., and Mule, S. J. (1974). Physiological disposition and metabolism of [^{3}H]ecgonine (cocaine metabolite) in the rat. *Res. Commun. Chem. Pathol. Pharmacol.* **8**, 55–63.
61. D'Souza, J., Caldwell, J., and Smith, R. L. (1980). Species variations in the N-methylation and quaternisation of [^{14}C]-pyridine. *Xenobiotica* **10**, 151–157.
62. Wright, A. S., Crowne, R. S., and Potter, D. (1972). The fate of 1,3,5-*tri*(3,5-*di-tert*-butyl-4-hydroxybenzyl)-2,4,6-trimethylbenzene (IONOX 330) in rats fed the compound over a prolonged period. *Xenobiotica* **2**, 7–23.
63. Chasseaud, L. F., Fry, B. J., Saggers, V. H., Sword, I. P., and Hathway, D. E. (1972). Studies on the possible absorption of sulphated glycopeptide (GLPS) in relation to its mode of action. *Biochem. Pharmacol.* **21**, 3121–3130.
64. Ilett, K. F., Stripp, B., Menard, R. H., Reid, W. D., and Gillette, J. R. (1974). Studies on the mechanisms of the lung toxicity of paraquat: Comparison of tissue distribution and

6. Unmetabolized Compounds

some biochemical parameters in rats and rabbits. *Toxicol. Appl. Pharmacol.* **28,** 216–226.
65. Bus, J. S. Cagen, S. Z., Olgaard, M., and Gibson, J. E. (1976). A mechanism of paraquat toxicity in mice and rats. *Toxicol. Appl. Pharmacol.* **35,** 501–503.
66. Olsen, J. R., Gasiewicz, T. A., and Neal, R. A. (1980). Tissue distribution, excretion and metabolism of 2,3,7,8-tetrachlorodibenzo-*p*-dioxin (TCDD) in the golden Syrian hamster. *Toxicol. Appl. Pharmacol.* **56,** 78–85.

CHAPTER 7

Biological Basis of Detoxication of Oxygen Free Radicals*

Helmut Sies and Enrique Cadenas

I.	Introduction	182
II.	Oxygen Free Radicals	182
	A. Superoxide Anion Radical and Hydrogen Peroxide	182
	B. Hydroxyl Radical	184
	C. Singlet Molecular Oxygen	184
	D. Alkoxy and Peroxy Radicals	186
III.	Defense Mechanisms: Enzymatic and Nonenzymatic	187
	A. Superoxide Dismutase	187
	B. Catalase	188
	C. Glutathione Peroxidases	188
	D. Vitamins E and A	189
IV.	Factors Influencing Defense Mechanisms and the Production of Oxygen Free Radicals	190
	A. Dietary Influences on the Cellular Level of Antioxidant Defenses	190
	B. Glutathione Concentrations: Cellular and Subcellular	191
	C. Level of Oxygenation	194
V.	Biological Systems Associated with Increased Oxygen Free Radical Production	195
	A. Toxic Drug Effects and Chemotherapy	195
	B. Inflammation and Microcirculation	197
	C. Radiobiology, Carcinogenesis, and Aging	200
VI.	Comments	201
	References	201

* Supported in part by Deutsche Forschungsgemeinschaft, Schwerpunktprogramm "Mechanismen toxischer Wirkungen von Fremdstoffen," and by Ministerium für Wissenschaft und Forschung, Nordrhein-Westfalen. E. C. was the recipient of a fellowship from the Alexander-v. Humboldt-Stiftung.

I. INTRODUCTION

The field of oxygen free radicals in biology received a major stimulus with the discovery of superoxide dismutase in 1969.[1] The scope has enlarged considerably, so that interest in oxygen free radicals now encompasses the superoxide anion radical, O_2^-, hydrogen peroxide, H_2O_2, and the hydroxyl radical, $OH\cdot$, as well as oxygen-centered radicals of polyunsaturated fatty acids of the general structure of alkoxy or peroxy radicals, $RO\cdot$ and $ROO\cdot$, respectively. Major interest in the latter group is focused on arachidonate metabolites arising from lipoxygenase activity.

The current state of knowledge in this area has been summarized in symposia on oxygen free radicals and tissue damage,[2] oxygen and oxyradicals in chemistry and biology,[3] and chemical, biochemical,[4] biological, and clinical[4a] aspects of superoxide and superoxide dismutase, as well as in the series on free radicals in biology.[5] The enzymatic basis of detoxication of oxygen free radicals has been covered in part in this series in articles on superoxide dismutase[6] and glutathione peroxidase,[7] but catalase does not seem to have been viewed in this context. Although not a radical itself, H_2O_2 is intimately related to cytoxicity and therefore deserves attention here. The metabolic basis of detoxication of organic hydroperoxides has also been treated in this series,[8] and the metabolism of hydroperoxides in mammalian organs has been discussed more generally.[9]

What may be said about the biological basis of detoxication of oxygen free radicals? This is an emerging field for which substantial gaps in knowledge exist. Nevertheless, the biological factors influencing detoxication can be described and extended to the level of oxygenation (hyperoxia and hypoxia), membrane status and dietary factors, inflammation, and microcirculation, as well as compartmental aspects on the intraorgan and interorgan levels.

II. OXYGEN FREE RADICALS

A. Superoxide Anion Radical and Hydrogen Peroxide

The superoxide anion radical, O_2^-, is the product of one-electron reduction of molecular oxygen. It is produced in the autooxidation of a multitude of reduced biomolecules, in enzymatic reactions, during electron flow in the respiratory chain, and as a consequence of environmental

7. Biological Basis of Detoxication of Oxygen Free Radicals

factors such as radiation and toxic chemicals. Table I summarizes known sources of O_2^-.

H_2O_2 can be formed subsequently by the dismutation reaction from O_2^-, but also by two-electron (nonradical) reduction of oxygen by a number of oxidases, including acyl-CoA oxidase, NADH oxidase, monoamine

TABLE I.

Sources of Superoxide Anion Radical[a]

Autoxidation reactions (including "redox cycling")
　Flavins ($FADH_2$, $FMNH_2$)
　Quinones
　Aromatic nitro compounds, aromatic hydroxylamines
　Redox dyes (e.g., paraquat)
　Melanin
　Thiols
　Tetrahydropteridines
　Iron chelates

Enzymatic reactions and proteins
　Aldehyde oxidase
　Cytochrome P-450
　Ferredoxin
　Hemoglobin
　Indoleamine dioxygenase
　NADH–cytochrome b_5 reductase
　NADPH–cytochrome P-450 reductase
　NADPH oxidase
　Peroxidase
　Tryptophan dioxygenase
　Xanthine oxidase

Cellular sources
　Mitochondrial electron transport chain
　Microsomal electron transport chain
　Chloroplast photosystem I
　Leukocytes and macrophages during bactericidal activity
　　(plasma membrane)
　Bacterial electron transport chain

Environmental factors
　Ultraviolet light
　Ultrasonic sound
　X-rays
　γ-rays
　Toxic chemicals
　Metal ions

[a] Modified from Halliwell[10] and Kappus and Sies[11].

oxidases, uricase (urate oxidase), L-gulonolactone oxidase, and glutathione oxidase.

The steady-state concentrations in liver have been estimated to be 10^{-12}–10^{-11} M for O_2^- [12] and 10^{-9}–10^{-7} M for H_2O_2.[13] In general, major contributions to the total cellular production of these oxygen metabolites come from membrane-bound enzymes.

B. Hydroxyl Radical

The hydroxyl radical OH· is a potent oxidant capable of reacting with a multitude of molecules through H abstraction, addition, or electron transfer. Because of its high reactivity, the range of diffusion of OH· from the site of its generation is generally considered very limited.

The use of scavengers of OH·, chelators of iron, and, more recently, specific traps combined with ESR techniques, in a system generally sensitive to superoxide dismutase and catalase, have helped to demonstrate the presence of OH·. The iron-catalyzed Haber–Weiss reaction[14–17] is usually quoted as responsible for transmitting cytotoxic effects of oxygen free radicals. However, it was reported that H_2O_2, in the presence of an adequate electron donor, can function as an OH·-like oxidant.[18] The sources of generation of OH· are diverse, including enzymatic sources such as xanthine oxidase,[19] the microsomal electron transfer system,[20] the oxidation pathway to ethanol,[21] and a number of biological phenomena including inflammation,[22] phagocytosis,[23] and the activity of some antineoplastic drugs.[24]

C. Singlet Molecular Oxygen

Singlet molecular oxygen[25] ($^1O_2$) is an excited form of molecular oxygen that can arise, for example, from the reactions shown in Fig. 1. Although formed in comparatively low amounts, it has become of interest because of its involvement at a certain stage of lipid peroxidation and its ready detection by photon counting in intact biological systems.[26,27]

Sources of singlet oxygen within the frame of lipid peroxidation are illustrated in Fig. 1. The self-reaction of secondary lipid peroxy radicals could be a source of either singlet oxygen or excited carbonyl groups, decaying to the ground state and emitting light at about 634 and 703 nm (singlet oxygen dimole emission) and between 340 and 460 nm, respectively. Singlet oxygen can react with unsaturated double bonds of fatty acids and through a dioxetane intermediate yield excited carbonyl groups.[29]

Singlet oxygen can also be formed in the interaction of O_2^- and

7. Biological Basis of Detoxication of Oxygen Free Radicals

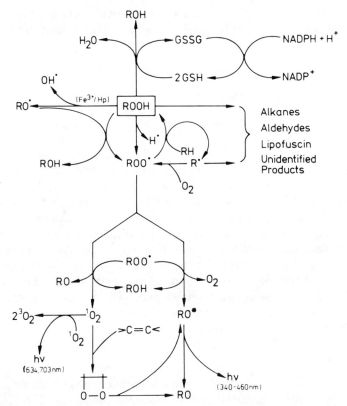

Fig. 1. Scheme of decomposition of lipid hydroperoxides. Lipid hydroperoxides (ROOH) are formed during the process of lipid peroxidation. The nonradical decomposition of ROOH, an activity that has been suggested for glutathione peroxidase with formation of the corresponding alcohol (ROH), is shown at the top. However, this activity has not been demonstrated for membrane-bound ROOH. The lower part represents the radical decomposition of lipid hydroperoxides with formation of lipid peroxy radicals (ROO·) and alkoxy (RO·) radicals. The former are also produced in the reaction of lipid radicals (R·) with oxygen and alkoxy radicals with lipid hydroperoxides. The production of alkanes (ethane, pentane, etc), aldehydes (malondialdehyde, alkenals such as 4-hydroxynonenal), and lipofuscin, as well as other unidentified products from lipid peroxides, can involve free radical intermediates during the lipid peroxidation process. Singlet O_2 formation, as well as the bimolecular decay reaction to triplet (ground state) O_2 with emission of low-level chemiluminescence, is shown at the lower left. Excited carbonyls RO* can also be formed (lower right). (Modified from Cadenas et al.[28]).

H_2O_2.[30–32] This is of particular interest because both of these reactants are formed during such processes as bactericidal activity, inflammation, redox cycling of drugs of quinone structure, cytostatic agents and aromatic nitro compounds, and enzymatic reactions such as xanthine oxidase. Although the nonenzymatic dismutation of O_2^- has been considered a source of singlet oxygen,[31,32] recent evidence against this has been brought forward.[33]

D. Alkoxy and Peroxy Radicals

Alkoxy and peroxy radicals, $RO\cdot$ and $ROO\cdot$, are formed in the decomposition of lipid hydroperoxides and are important in the maintenance of radical chain reactions in addition to the lipid radical $R\cdot$ (Fig. 1). Because the polyunsaturated fatty acids that undergo the process of lipid peroxidation are integral membrane components, much of the membrane pathology of oxygen free radicals is mediated by the balance of formation and decomposition of such alkoxy or peroxy radicals. Powerful protection systems (see the following discussion) control the localized occurrence of these radicals, but it now seems that there are factors that can export the process of lipid peroxidation from one site to another over sizable distances, even between different organs. Apart from the well-known products of lipid peroxidation (malondialdehyde, alkanes such as ethane and pentane, and lipofuscin), the potentially harmful effects of diffusible products[34,35] such as 4-hydroxynonenal[36] have been pointed out.

The formation of alkoxy and peroxy radicals can be partilly envisaged as follows.

1. Fragmentation of peroxides into alkoxy and peroxy radicals[37,38] (Eq. 1). However, the concentration of hydroperoxides required to yield free radical products through reaction (1) is extremely high, of the order 0.2 M.

$$2\ ROOH \rightarrow RO\cdot + ROO\cdot + H_2O \tag{1}$$

2. Hydroperoxides are stable in metal-free systems, and their interaction with metal ions will result in the formation of alkoxy radicals by a Fenton-type reaction[17] (Eq. 2). The reaction is favored over reaction (1) because of the presence of a variety of oxidizable substances that can react with hydroperoxides in the cell.[39]

$$Me^n + ROOH \rightarrow Me^{n+1} + RO\cdot + OH^- \tag{2}$$

3. Reaction of hydroperoxides with O_2^-.[40] Although *tert*-butyl hydroperoxide and linoleic acid hydroperoxide are thought not to react with O_2^-,[41] more reactive lipid hydroperoxides, such as linolenic or arachidonic hydroperoxides, might be able to undergo such a reaction (Eq. 3). However, the reaction between O_2^- and linoleic acid[40] remains feasible under the conditions that have been described,[40] where possible metal-catalyzed reactions are blocked by a chelator.

7. Biological Basis of Detoxication of Oxygen Free Radicals

$$ROOH + O_2^- \rightarrow RO\cdot + OH^- + O_2 \quad (3)$$

4. Induced decomposition of hydroperoxides by endogenous radicals ($Q\cdot$) present in the system (propagation reactions)[38] (Eq. 4).

$$Q\cdot + ROOH \rightarrow ROO\cdot + QH \quad (4)$$

5. The propagation reactions of an autoxidation chain of unsaturated fatty acids also yields $ROO\cdot$ (Eq. 5).

$$R\cdot + O_2 \rightarrow ROO\cdot \quad (5)$$

III. DEFENSE MECHANISMS: ENZYMATIC AND NONENZYMATIC

This section provides a brief introduction to relevant enzymes: superoxide dismutase, catalase, and glutathione peroxidase. It includes comments on nonenzymatic antioxidant systems, the biological capacity for induction of these enzymes, and relationships with trace metal biochemistry.

A. Superoxide Dismutase

There are three different types of superoxide dismutase[6]: An Mn enzyme is present in the matrix space of mitochondria of eukaryotic cells and is insensitive to cyanide. An iron-containing enzyme occurs in bacteria, apparently in the periplasmic space. A Cu- and Zn-containing enzyme is present in the cytosol of eukaryotic cells and is sensitive to cyanide ($K_i = 20 \ \mu M \ CN^-$)[42]; the enzyme is in the mitochondrial intermembrane space and in erythrocytes. Dismutase activity is also displayed by Cu^{2+} complexes[43] and by millimolar concentrations of Mn^{2+}.[44] The latter finding solves a problem that arose from observations of bacteria lacking the enzyme but retaining a defense against oxidative stress.

The total amount of the Cu-Zn enzyme in human was estimated to be close to 4 g,[45] liver having the highest content and fatty tissue the lowest. The Cu-Zn enzyme is present in human liver at about one-half the amount of the Mn enzyme, whereas the Cu-Zn enzyme accounts for almost all of the activity in the cerebrospinal fluid. In serum, the Cu-Zn enzyme is about 0.3 mg/liter; other metalloproteins present in serum, including ceruloplasmin, transferrin, and ferritin, also display small dismutase activity. Ceruloplasmin, for example, has an activity about 40,000-fold lower than that of the Cu-Zn enzyme.[46]

B. Catalase

Catalase[47] is present in almost all mammalian cells and in many it is compartmentalized in peroxisomes or microperoxisomes. By means of the catalatic reaction, two molecules of H_2O_2 are reduced to water and ground state molecular oxygen. The characteristic intermediate is catalase Compound I, which has been useful for the detection of H_2O_2 in intact tissues by virtue of its spectroscopic properties.[13,48,49] In the presence of suitable hydrogen donors, the enzyme can also catalyze peroxidatic reactions. The partitioning between the catalatic pathway and the peroxidatic pathway can be calculated from the concentrations of the reactants and the rate constants, as has been done extensively for the enzyme from rat liver.[50] It may be noteworthy in the present context that there is no "K_m" for the enzyme, so that even at very low enzyme levels catalase will effectively remove H_2O_2.

Work on cytotoxicity has established that catalase is efficient in protecting against oxygen free radical attack, thereby implicating the formation of H_2O_2 as a crucial feature in mediating cytotoxicity; this will be discussed in Sections IV,C and V,B.

C. Glutathione Peroxidases

The reaction shown at the top of Fig. 1 is catalyzed by the selenoenzyme glutathione (GSH) peroxidase[7,51,52] and by the nonselenium-dependent activity exhibited by some glutathione S-transferases. Whereas it is specific for GSH as physiological hydrogen donor, the Se enzyme accepts a wide variety of hydroperoxide substrates including H_2O_2; the non-Se activity is displayed only with organic hydroperoxides and not with H_2O_2. It should be noted that these activities have not yet been demonstrated for membrane-bound lipid hydroperoxides, and it is possible that, because of a problem of inaccessibility, only those lipid hydroperoxides that are released from membrane phospholipids by phospholipases serve as substrate for these peroxidases. Thus, the protective and repair roles for these enzymes with respect to damaged membranes remain to be demonstrated.

Some reports mention additional factors conferring protection. One such factor was characterized by a cytosolic glutathione-dependent enzyme protecting against lipid peroxidation in the NADPH–microsomal lipid peroxidation system; it was shown to be due, at least in part, to some of the glutathione S-transferases.[53] In a similar study, a factor, although also demonstrated, was shown to be due neither to Se–glutathione peroxidase nor to the S-transferase.[54] However, both studies emphasize that this activity prevents biological membranes from peroxidation rather than reducing lipid hydroperoxides once they are formed within membranes.

Glutathione peroxidase (the Se-enzyme) activity is higher in the perivenous zone of the rat liver lobule than in the pericentral zone,[55] similar to the distribution of its substrate, glutathione.[56]

The importance of glutathione in detoxication reactions of oxygen free radicals stems largely from its reaction with the various glutathione peroxidases and glutathione S-transferases. However, nonenzymatic reactions involving the thiyl radical are also possible.[38] The "glutathione status" denotes the different amounts of glutathione equivalents present as GSH, as GSSG, or as mixed disulfides with proteins or low molecular weight compounds.[57] The last are of increasing interest in biochemical pharmacology because they may result from redox cycling,[11,58] and regulatory effects on intermediary metabolism are entirely possible. Several enzymes of carbohydrate metabolism are known to be affected by the formation of mixed disulfides.[59] One example of an increase in hepatic mixed disulfides by redox cycling is given with paraquat[60] (see Section IV,B).

D. Vitamins E and A

The several tocopherols show similar vitamin E or antioxidant activity *in vitro*. However, absorption and retention of α-tocopherol are higher than for the other tocopherols, accounting for its higher activity *in vivo*[61] (see also Chapter 11, this volume). The antioxidant effects of vitamin E were demonstrated *in vivo* using pentane expiration as an index of lipid peroxidation.[62]

Vitamin E is a two-electron donor; the reactivity of the hydrogen atom from the OH group in the benzene ring has been shown to be higher than that of other possible electron donors in the molecule. Vitamin E free radicals, of low reactivity even within the frame of a radical–radical reaction, have been identified by ESR.[63] The hydroxyl group of the benzene ring of vitamin E (Vit-EOH) acts as a reductant, upon reaction with free radicals, and generates the vitamin E free radical (Vit-EO·) (Eq. 6).

$$ROO· + Vit\text{-}EOH \rightarrow ROOH + Vit\text{-}EO· \qquad (6)$$

This agrees with the hypothesis that antioxidants in general function by scavenging lipid-derived, oxygen-centered radicals. The efficiency of vitamin E in interrupting free radical reactions potentially damaging to biological membranes can be explained partially as follows:

1. Although the ratio of unsaturated fatty acids to vitamin E in a membrane could be as much as 1000 to 1, free radical processes of unsaturated fatty acids could proceed as a single kinetic chain[64] (Eq. 7) until the final peroxy radical is formed; thus, the possibility of vitamin E blocking this single kinetic free radical chain is enhanced.

$$R \cdot + R' \cdot \rightarrow R + R' \cdot \qquad (7)$$

2. The reaction of vitamin E with ROO· ($8 \times 10^4/M$/sec) is several orders of magnitude more effective than that of unsaturated fatty acids with ROO· ($50/M$/sec).[64]

The position of vitamin E on the membrane surface is important in its role as antioxidant[63] because it can both intercept free radical initiators and make possible their reduction by polar donors such as ascorbic acid (AH_2) (reaction 8)[64a] or the self-regenerating GSH system involving GS· (Eq. 9).[38]

$$\text{Vit-EO} \cdot + AH_2 \rightarrow AH \cdot + \text{Vit-EOH} \qquad (8)$$

$$\text{Vit-EO} \cdot + GSH \rightarrow GS \cdot + \text{Vit-EOH} \qquad (9)$$

The antioxidant activity exerted by carotenoids (vitamin A) has been related to their capacity to quench oxidant species such as singlet molecular oxygen; the mechanism seems to be a physical quenching reaction that does not affect chemically the structure of the pigment.[30] This ability of β-carotene is the basis for its protective role against damage initiated by visible light, as well as its possible use in the treatment of certain photosensitive diseases.[30] A protective effect of β-carotene against oxidative attack by polymorphonuclear leukocytes was observed in a β-carotene–containing bacterial mutant, whereas the carotenoid-lacking mutant was lysed.[65] It has been suggested that dietary β-carotene could have protective influence against cancer development.[66]

IV. FACTORS INFLUENCING DEFENSE MECHANISMS AND THE PRODUCTION OF OXYGEN FREE RADICALS

A. Dietary Influences on the Cellular Level of Antioxidant Defenses

Deficiencies in Cu and Zn, Mn, or Se will result in a perturbation in the activity of the enzymes mentioned in Section III (see also Chapter 10, this volume). Iron represents a special case because of its ability to catalyze free radical reactions, especially when not bound and sequestered in safe places. A prooxidative action of iron will be manifested when the diet is lacking antioxidants such as Se and vitamin E.[67] Diets deficient in Cu and Mn cause a decrease in the cytoplasmic superoxide dismutase activity in brain of neonatal rats and mitochondrial superoxide dismutase, respectively.[68,69]

Most of the reported symptoms of Se deficiency are attributable to the absence of glutathione peroxidase activity because this activity is directly related to the availability of selenium.[70] Se deficiency is one of the factors responsible for Keshan disease, a congestive cardiomyopathy affecting persons living in rural selenium-deficient areas of China.[71] Administration of sodium selenite was shown to prevent the disease. A case of cardiomyopathy in a patient with a diet-induced selenium deficiency was described.[72] In addition to glutathione peroxidase, there are other selenoproteins that may become of interest. The peroxidase was even viewed as a storage form for selenium that could supply the element to other proteins.[73]

Vitamin E deficiency has been shown to produce perturbations of the enzymes that afford protection against peroxidative stress, that is, the glutathione peroxidase system[74] and catalase.[75]

Polyunsaturated fatty acids are prone to undergo peroxidation reactions. Because their high intake in the diet decreases the tissue content of vitamin E,[76] dietary vitamin E needs to be raised. An increase in glutathione peroxidase after ingestion of peroxidized lipids has been reported.[77] A diet rich in polyunsaturated fatty acids exerts an increase in the content of cytochrome P-450 and the rates of oxidative metabolism.[78] In contrast, the activity of rat liver glucose-6-phosphate dehydrogenase was found to decrease in animals fed unsaturated fatty acids when compared to those fed a fat-free diet.[79] Fat-free diets increase the utilization of NADPH by peripheral hepatocytes for fatty acid synthesis and by centrilobular hepatocytes for mixed-function oxidase activity. Fatty acid–rich diets (corn oil) divert NADPH utilization in the periportal hepatocytes from fatty acid synthesis to mixed-function oxidase activity; the pathway of NADPH oxidation was not altered in the centrilobular regions.

B. Glutathione Concentrations: Cellular and Subcellular

The concentration of glutathione obviously may become critical when it is below a certain threshold. There is an ample literature on the lowering of glutathione concentrations during conditions of toxicological interest.[80,81] In general, it appears that down to about 20–30% of residual glutathione levels, protection remains. It should be noted that, in addition to glutathione peroxidase, there are many other detoxication functions of glutathione, notably alkylation reactions catalyzed by glutathione transferases.[82] It is clear that not only are the direct detoxication functions of glutathione sensitive to manipulations of the cellular level of glutathione but also its metabolic functions that indirectly contribute to the overall response of the organism. For example, normal GSH levels are necessary for the transduction of the activation signal from the exterior to the interior of the polymorphonuclear leukocyte,[83] or for the formation of leukotrienes C_4, the slow-reacting substances of anaphylaxis,[84] and of the products of the 12-lipoxygenase pathways in platelets.[85]

Mammalian cells contain glutathione at relatively high concentration[86] for which a pronounced circadian rhythm has been found in liver.[87] Upon starvation, there are substantial decreases in hepatic and intestinal levels, which have been termed *cyst(e)ine reservoirs*.[87a] Muscle GSH levels, in contrast, decrease only slightly, and erythrocyte GSH remains unchanged.[87b] Like liver,[87] heart[88] has been shown to exhibit a concomitant rise in mixed disulfides upon starvation. Cellular concentrations are maintained by *de novo* synthesis of GSH from the constituent amino acids. GSH, as such, does not penetrate such organs as liver[89] but was shown to exert a protective effect against acetaminophen toxicity when administered entrapped within liposomes to mice.[90]

In contrast, glutathione may be lost from cells in several ways, both reversibly and irreversibly[57,91] (Fig. 2). Normal hepatic glutathione turnover is accounted for by a GSH efflux of 12 nmol/min per gram of liver.[92] This efflux occurs predominantly across the sinusoidal plasma membrane. In the intact animal there is also a slight biliary GSH efflux of 2 nmol/min per gram of liver.[93] The decrease in hepatic glutathione levels observed during so-called oxidative stress is the result of a loss of glutathione disulfide (GSSG).[94] This occurs selectively across the biliary canalicular membrane,[95] and its rate is linearly related to the intracellular GSSG content.[96] The mechanism of delivery of oxidizing equivalents in the form of GSSG that may arise from oxygen free radicals, for example, by redox cycling, may be related to that of glutathione S-conjugates. In fact, competition of GSSG and S-conjugates for biliary transport has been observed.[97] A rise of intracellular GSSG can be counteracted by glutathione reductase, by which mechanism the bulk of GSH is actually regenerated from GSSG. However, removal from the cell, as an alternative to reduction of GSSG, fulfills a useful purpose because rises in cellular GSSG may be deleterious. Glutathione disulfide inhibits protein synthesis, and may influence tubulin polymerization and transmission of neuronal reflexes.[57]

Glutathione is compartmentalized at the subcellular level, so that the

7. Biological Basis of Detoxication of Oxygen Free Radicals

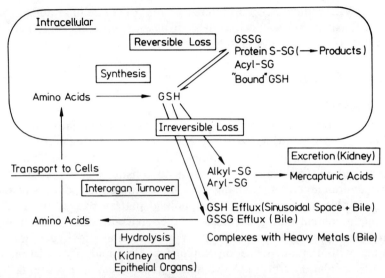

Fig. 2. Processes affecting the intracellular and extracellular glutathione status. (The scheme is based on Kosower and Kosower[57] and Sies[91].)

threshold for protection of about 20–30% of normal glutathione may mean almost complete abolition of localized but critical pools of glutathione. The mitochondrial matrix pool of GSH is 13–15% of the total,[98,99] and there are indications that it is intimately related to the transmission of deleterious effects of reactive oxygen species. In oxidative stress, mitochondrial lipoamide-dependent oxidation of ketoacids such as pyruvate or 2-oxoglutarate is impaired,[100] CoASH levels decrease while CoASSG levels increase,[101] and Ca^{2+} [102–104] and Mg^{2+} ions[102] are lost from the cell.

Although the liver is probably the organ central to metabolic disposition of noxious compounds, oxidative stress and formation of oxygen free radicals can occur in other organs as well, notably those exposed to high O_2 concentrations (lung, skin, eye, and the blood cells) or to high concentrations of xenobiotics (stomach and intestine). The concentration of glutathione in some cell types appears to be critical, as demonstrated in gastric cytoprotection.[105,106] Although a detailed presentation of the role of glutathione levels in all these different sites is not attempted here, some aspects are mentioned throughout the chapter.

The systemic application of glutathione has been suggested as having beneficial effects in clinical medicine (e.g., in the amelioration of shock[107,108]) and has a protective effect in an animal model of a hepatoma elicited by epoxide formation of aflatoxin B_1.[108] These and other studies

(Section V,B) indicate current interest in the role of cellular defense systems against oxygen radical damage in medical problems. Methods for experimental depletion of glutathione have been reviewed.[110,111]

C. Level of Oxygenation

1. Hyperoxia

During hyperoxia, either under normobaric or hyperbaric conditions, the identification of primary toxic species generated in the early stages of oxygen toxicity is difficult. Generation of H_2O_2[112] and O_2^-[113] and inhibition of reversed electron transport[114] appear to constitute early events at the mitochondrial level. Other changes include peroxidation of mitochondrial phospholipids, changes in the phospholipid pattern caused by a fall in cardolipin content, and a decrease in GSH concentration.[113] Brain levels of H_2O_2 and lipid hydroperoxides increased in parallel by hyperoxia, correlating with the incidence of convulsions.[115] Lipid peroxidation and NADPH oxidation constitute the initial steps of damage resulting from hyperbaric oxygenation in perfused liver and lung.[116] Low-level chemiluminescence was also found to be increased in perfused and *in situ* liver and in perfused lung under conditions of hyperbaric oxygenation[26,27] or under conditions that mimic hyperbaric oxygenation.[117]

In general, the cellular levels of enzymes protective against peroxidative stress (catalase, glutathione peroxidase, and superoxide dismutase) and the enzymes that maintain GSH in the reduced state are found to be increased during hyperoxia.[113,118,119] Perturbation of thiols seems to play a key role,[120] and it was early noted that most of the enzymes susceptible to hyperoxic toxicity contained thiol groups.[121] Dietary vitamin E seems to have an important role in preventing or decreasing the symptoms of oxygen toxicity; most of the effects studied are amplified by vitamin E-deficient diets or, conversely, quenched by vitamin E-rich diets. Vitamin E increases the preconvulsion time of animals exposed to hyperbaric oxygen and decreases lipid peroxidation.[122]

In organs essentially devoid of peroxisomes (brain, lung, and heart), the defense against hyperbaric oxygenation relies on systems other than catalase. In the case of brain, where glutathione peroxidase activity is about 20-fold lower than that of liver[123] and the ratio of vitamin E to polyunsaturated fatty acids is also very low,[124] toxic manifestations of hyperoxia would thus be expected to appear early.

2. Hypoxia

In hypoxia, the toxic effect of some drugs is potentiated.[11,125] For example, CCl_4-induced lipid peroxidation (measured as malondialdehyde ac-

7. Biological Basis of Detoxication of Oxygen Free Radicals

cumulated *in vitro* or as ethane expired *in vivo*)[126,127] and reductive hepatotoxic metabolism of halothane[128,129] are enhanced. The latter is of clinical relevance during anesthesia. These and other examples are of importance because the liver, the site of activation of many xenobiotics, has different oxygen concentrations in different areas.

Lower oxygen concentrations could modulate the toxicity of drugs undergoing redox cycling.[11] For example, partial protection against paraquat toxicity in lung was obtained with animals kept under hypoxia as compared with normal controls.[130] In these experiments, however, mortality of paraquat-treated animals kept under hypoxia was increased drastically after a subsequent brief exposure to normoxia. Presumably, an accumulation of paraquat radicals during hypoxia leads to a greater production of oxygen radicals when normal conditions are restored.

Rats exposed to hypoxia develop an enhanced activity of Mn–superoxide dismutase that permits them to survive subsequently under hyperoxia.[131] Conversely, during hyperoxia both the Cu–Zn- and Mn-superoxide dismutases are induced. Mitochondrial redox cycling of quinols and flavins may serve as a better source of O_2^- under hypoxia, leading to induction of the Mn enzyme. Such a concept may be considered as a basis for the free radical pathology observed during hypoxia or ischemia also in the central nervous system.[132]

V. BIOLOGICAL SYSTEMS ASSOCIATED WITH INCREASED OXYGEN FREE RADICAL PRODUCTION

A. Toxic Drug Effects and Chemotherapy

Metabolism of drugs can result in either a therapeutic or a toxic response[11] both of which may coexist. The mechanism of action of several antitumor drugs and that of therapeutic agents against infections with protozoa involve intracellular redox cycling with the formation of oxygen free radicals upon autoxidation of the drug free radicals, ultimately leading to deleterious effects on DNA, lipids, and proteins in target cells (scheme in Fig. 3).

1. Antitumor Agents

Active anticancer agents include those of quinones such as anthracycline antibiotics (adriamycin and daunorubicin), heterocyclic antibiotics (streptonigrin and mitomycin C), and others of benzoquinonelike structure (lapachol).[133] These have been classified by their selective toxicities

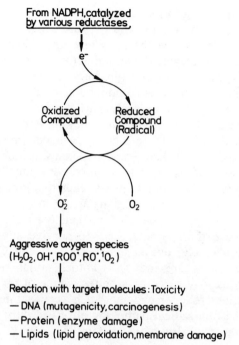

Fig. 3. Formation of reactive oxygen species upon redox cycling and some biological effects. (The scheme is based on Kappus and Sies[11].)

toward oxygenated and hypoxic tumor cells.[134] Although this establishes a new therapeutic approach, mainly for solid tumors with a large mass of hypoxic cells, the molecular mechanism that makes some antineoplastic agents more effective under hypoxic conditions remains to be clarified. Not only drugs but also endogenous compounds may operate in this way. This has been discussed for dopamine and related compounds in relation to Parkinson's disease.[135]

Reductive activation occurs in the endoplasmic reticulum, mitochondrial membranes, and nuclei of different tissues.[136] Oxygen is required for the cytotoxicity observed by antineoplastic drugs, as well as other substances that undergo redox cycling; the radical form of such drugs is unlikely to be responsible directly for cytotoxicity.[137] O_2^- and H_2O_2 are necessary to the degradation of DNA by bleomycin, thereby accounting for the greater toxicity of the drug in tumor cells that have higher O_2^- levels than do normal cells,[24] presumably because of a loss in superoxide dismutase activity in many tumors. DNA may bind Fe^{2+} and catalyze the decomposition of H_2O_2, with formation of $OH\cdot$.[138,139]

The cytotoxicity of anticancer drugs has been related to the varying glutathione peroxidase/superoxide dismutase ratio in different tumor cells[140,141]; these values were 1.0, 0.6, and 0.3 in liver, Yoshida tumor cells, and Ehrlich tumor cells, respectively. The major role of glutathione peroxidase is seen in the fact that catalase is essentially absent from tumor cells.[140,142,143] The cardiotoxicity of adriamycin is thought to be related to glutathione peroxidase, as the drug is more toxic in glutathione-deficient animals.[144] Superoxide dismutase is decreased, with some exceptions, in most experimental tumors.[145] Although higher susceptibility to lipid peroxidation might be assumed, altered membrane lipid composition seems to make tumor membrane cells less vulnerable to NADPH-induced lipid peroxidation.

2. Antiprotozoic Agents

One example of the role of oxygen-free radicals in host–invader relations are the compounds used to treat trypanosomiasis by redox cycling. Naphthoquinone[146] and nitrofuran[147] derivatives have been shown to lead to O_2^- and H_2O_2 production, the latter being the most effective drug used in the treatment of acute Chagas' disease. Because *Trypanosoma cruzi* is deficient in metabolizing H_2O_2,[148] the deleterious effect seems to be selective for the invader rather than the host. However, treatment with nitrofuran derivatives leads to increased hepatic redox cycling and, consequently, to increased GSSG release.[96,149] An alternative and complementary strategy would be to deplete the infecting cells of GSH, as shown with buthionine sulfoximine in *T. brucei*.[150]

B. Inflammation and Microcirculation

As a consequence of tissue damage, polymorphonuclear leukocytes and macrophages migrate to the site of injury. The defense mechanisms of these cells rest, in part, on the production of oxygen radicals.[151] NADPH oxidase-dependent formation of O_2^- in leukocytes and its conversion to H_2O_2 by superoxide dismutase, along with the reactions that deal with the formed H_2O_2 (myeloperoxidase, catalase, and glutathione peroxidase), determine the internal equilibrium of the phagocyte. Impairment of microbicidal activity occurs when enzymes are deficient or their cofactors are in short supply.

However, activated oxygen species are also released from the leukocyte to the extracellular milieu, possibly accounting for the tissue damage that accompanies the inflammatory process,[152] and also for alterations in microcirculation caused by changes in permeability. A simplified scheme of the phagocyte response related to the formation of O_2 metabolites and their involvement in inflammation is shown in Fig. 4.

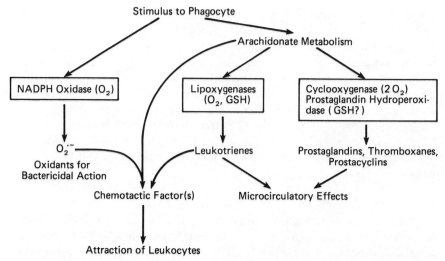

Fig. 4. Simplified scheme of phagocyte responses related to metabolites of oxygen.

$O_2^{\cdot-}$ released into the unprotected microvascular environment appears to serve as a signal for summoning leukocytes to the inflammation site by the formation of an $O_2^{\cdot-}$-dependent chemotactic factor[153] of albumin–lipid complex nature. Thus, the antiinflammatory role of superoxide dismutase could prevent the formation of this chemotactic factor. Oxygen radicals formed during inflammation could be related to chromosomal instability by activation of a clastogenic factor, as shown for autoimmune diseases.[154]

Increased permeability, macromolecular degradation, and bactericidal killing appear to be dependent on $OH\cdot$ or $OH\cdot$-derived products, rather than on $O_2^{\cdot-}$ itself.[155]

The flux through cyclooxygenase and lipoxygenase reactions at the site of inflammation may lead to an as yet unidentified oxygen radical (O_x) responsible for inhibition of enzymes, among them cyclooxygenase and prostacyclin isomerase.[156] However, work using optical antipodes of a radical quencher, MK477, has led to a more reserved view on the role of O_x.[157]

H_2O_2 has been identified as the mediator of cytotoxicity upon target molecules when released from experimentally activated macrophages[158] or generated enzymatically,[159] and on platelet function when released from latex-activated polymorphonuclear leukocytes.[160] The inhibition of GSH redox cycling in tumor cells led to a higher efficiency of the H_2O_2-mediated bactericidal activity of polymorphonuclear leukocytes.[160a]

1. In vitro and in Vivo Model Systems of Inflammation

Degradation of connective tissue as a characteristic feature of inflammatory arthritis was also related to the formation of oxygen free radicals by *in vitro* studies on hyaluronate and collagen molecules. The oxygen free radicals formed from the xanthine–xanthine oxidase system prevented gelation of soluble collagen and decreased the viscosity of a hyaluronate solution[161]; copper–penicillamine, catalase, superoxide dismutase, and mannitol were protective. Replacement of the xanthine-xanthine oxidase system by polymorphonuclear leukocytes led to similar observations.

Enzymatically generated oxygen radicals result in alterations in vascular permeability that might be associated with edema observed in inflammatory processes.[155] By making use of a hamster pouch model system and a permeability marker, the permeability changes in microcirculation that follow the exposure to a flux of oxygen radicals were assessed. The number of leakage sites per square centimeter was substantially decreased when scavengers of O_2^-, H_2O_2, or $OH\cdot$ were present. Superoxide dismutase added either topically or by intravenous infusion (as such or as a polyethyleneglycol complex) abolished the permeability changes completely. Interestingly, leukotrienes were potent in eliciting permeability changes.[162]

2. Drugs Against Inflammation

Although systemic application of superoxide dismutase as an antiinflammatory agent is limited,[163] there are interesting results on topical application of superoxide dismutase as a drug in patients with rheumatoid arthritis affecting the knee. A double-blind study demonstrated the higher efficiency of intrasynovial applications of superoxide dismutase as compared to aspirin in reducing the clinical symptoms of arthritis.[164]

The molecular mechanism of some types of therapeutic drugs against inflammation (i.e., steroid, nonsteroid, SH-containing and Cu-containing compounds, as well as superoxide dismutase) could partially rest on their inhibition of the formation of free radicals or their scavenging effect. Nonsteroidal antiinflammatory drugs may interfere with prostaglandin biosynthesis and also display a scavenging effect against generated free radicals: salicylate and copper–salicylate complexes act as O_2^- scavengers, and the latter can also enhance dismutation of O_2^-.[43] Indomethacin and mefenamic acid also show O_2^--scavenging activity, the extent of which depends on the oxygen radical generating system utilized.[165,166]

The antiinflammatory activity of corticosteroids seems more closely related to the inhibition of certain functions of polymorphonuclear leukocytes, such as O_2^- and H_2O_2 generation.[167,168] Corticosteroids at doses corresponding to *in vivo* therapeutic plasma concentrations inhibit O_2^- and H_2O_2 production from latex-activated polymorphonuclear leukocytes and, in the case of methylprednisolone, chemiluminescence as well.[169]

The SH group of compounds such as penicillamine, cysteine, cysteamine, and glutathione could account for their antioxidant and antiinflammatory activity through a reaction with free radicals or, for glutathione, also as a reductant of hydroperoxides in the glutathione peroxidase system. The beneficial effect of vitamin E on inflammation[170] may be due to its radical scavenger activity; it has no effect on prostaglandin biosynthesis.[171]

3. Systemic Response to Inflammation

As part of a systemic response to inflammation in the acute phase, the liver shows signs of oxidative stress characterized by a decrease in the concentration of reduced glutathione and ascorbic acid content, and of glutathione S-transferase and catalase activities.[172-174] Another aspect is the enhancement of ceruloplasmin which, in its capacity for storing copper, would display an antioxidant activity, act as a scavenger of O_2^-,[46] and may be responsible, in part, for the serum antioxidant activity.[175] Ceruloplasmin could also be synthesized locally at the site of inflammation.[176]

C. Radiobiology, Carcinogenesis, and Aging

Oxygen free radicals have been implicated in the manifestation of numerous biological processes. A detailed discussion of these fields is beyond the scope of this chapter. However, *ionizing radiation, radiobiology,* and *irradiation* are terms that, considering their biological effects, may even be interchangeable with the term *oxygen free radicals*.[177] Interest in the relationships between redox processes in radiation biology and cancer[178] and drugs used as radiosensitizers[179,180] involves oxygen free radicals. It is of interest that glutathione was shown to form not only a thiyl radical but also a carbon-centered radical after reaction with hydroxyl radicals.[181] Glutathione was also instrumental in the repair of DNA single-stranded breaks obtained after aerobic X radiation.[182]

Finally, much is known of the relationship between free radicals and aging,[10,183] and correlations have been made between superoxide dismutase levels and both life span[184] and the formation of catalytically defective forms of the enzyme with age.[185] Interestingly, it was hypothesized that uric acid provides an antioxidant defense against oxidant- and radical-caused aging and cancer[186] and that uric acid and ascorbate may have similar functions in humans.[187]

VI. COMMENTS

Oxygen free radicals mediate a wide range of biological effects, and potent detoxication systems are available for their control. It is of particular interest that nature not only evolved protection systems but, in specific cases, employed the otherwise harmful production of reactive oxygen species for useful purposes. The biological basis for detoxication of oxygen radicals arises from modulation of the available activities for the superoxide dismutase, catalase, and glutathione peroxidase reactions and of the availability of nonenzymatic antioxidant systems, notably vitamins A, C, and E. Because of the almost universal presence of oxygen-dependent toxic effects, their fine control in particular instances may be enormously variable, depending on localized conditions.

REFERENCES

1. McCord, J. M., and Fridovich, I. (1969). Superoxide dismutase: An enzymic function for erythrocuprein (hemocuprein). *J. Biol. Chem.* **244**, 6049–6055.
2. Ciba Foundation (1979). "Oxygen Free Radicals and Tissue Damage," Ciba Found Symp. No. 65 (new ser.). Excerpta Medica, Amsterdam.
3. Rodgers, M. A. J., and Powers, E. L., eds. (1981). "Oxygen and Oxy-Radicals in Chemistry and Biology." Academic Press, New York.
4. Bannister, J. V., and Hill, H. A. O., eds. (1980). "Chemical and Biochemical Aspects of Superoxide and Superoxide Dismutase," Vol. 11A. Elsevier/North Holland, New York.
4a. Bannister, W. H., and Bannister, J. V., eds. (1980). "Biological and Clinical Aspects of Superoxide and Superoxide Dismutase," Vol. 11B. Elsevier/North Holland, New York.
5. Pryor, W. A., ed. (1982). "Free Radicals in Biology," Vol. 5. Academic Press, New York.
6. Hassan, H. M., and Fridovich, I. (1980). Superoxide dismutases: Detoxication of a free radical. *In* "Enzymatic Basis of Detoxication" (W. B. Jakoby, ed.), Vol. 1, pp. 311–326. Academic Press, New York.
7. Wendel, A. (1980). Glutathione peroxidase. *In* "Enzymatic Basis of Detoxication" (W. B. Jakoby, ed.), Vol. 1, pp. 333–348. Academic Press, New York.
8. Sies, H., Wendel, A., and Bors, W. (1982). Metabolism of organic hydroperoxides. *In* "Metabolic Basis of Detoxication" (W. B. Jakoby, J. R. Bend, and J. Caldwell, eds.), pp. 307–321. Academic Press, New York.
9. Chance, B., Sies, H., and Boveris, A. (1979). Hydroperoxide metabolism in mammalian organs. *Physiol. Rev.* **59**, 527–605.
10. Halliwell, B. (1981). Free radicals, oxygen toxicity and aging. *In* "Age Pigments" (R. S. Sohal, ed.), pp. 1–62. Elsevier/North Holland, Biomedical Press, Amsterdam.
11. Kappus, H., and Sies, H. (1981). Toxic drug effects associated with oxygen metabolism: Redox cycling and lipid peroxidation. *Experientia* **37**, 1233–1241.
12. Tyler, D. D. (1975). Polarographic assay and intracellular distribution of superoxide dismutase in rat liver. *Biochem. J.* **147**, 493–504.
13. Oshino, N., Chance, B., Sies, H., and Bücher, T. (1973). The role of H_2O_2 generation in

perfused rat liver and the reaction of catalase compound I and hydrogen donors. *Arch. Biochem. Biophys.* **154,** 117–131.
14. Haber, F., and Weiss, J. (1934). The catalytic decomposition of hydrogen peroxide by iron salts. *Proc. R. Soc. London, Ser. A* **147,** 332–351.
15. McCord, J. M., and Day, E. M. (1977). Superoxide dependent production of hydroxyl radicals catalyzed by iron-EDTA complex. *FEBS Lett.* **86,** 139–142.
16. Halliwell, B. (1978). Superoxide dependent formation of hydroxyl radical in the presence of iron chelates. *FEBS Lett.* **92,** 321–326.
17. Walling, C. (1975). Fenton's reagent revisited. *Acc. Chem. Res.* **8,** 125–131.
18. Elstner, E. F., Osswald, W., and Konze, I. R. (1980). Reactive oxygen species: Electron donor–hydrogen peroxide complex instead of OH radicals? *FEBS Lett.* **121,** 219–221.
19. Beauchamp, C. O., and Fridovich, I. (1970). A mechanism for the production of ethylene from methional. *J. Biol. Chem.* **245,** 4641–4646.
20. Cohen, G., and Cederbaum, A. I. (1979). Chemical evidence for the production of hydroxyl radicals during microsomal electron transfer. *Science* **204,** 66–68.
21. Ohnishi, K., and Lieber, C. S. (1978). Respective role of superoxide anion and hydroxyl radical in the activity of the reconstituted microsomal ethanol oxidizing system. *Arch. Biochem. Biophys.* **191,** 798–803.
22. Salin, M. L., and McCord, J. M. (1977). Free radicals in leukocyte metabolism and inflammation. *In* "Superoxide and Superoxide Dismutases" (A. M. Michelson, J. M. McCord, and I. Fridovich, eds.), pp. 257–270. Academic Press, New York.
23. Weiss, S. J., King, G. W., and LoBuglio, A. F. (1977). Evidence for hydroxyl radical generation by human monocytes. *J. Clin. Invest.* **60,** 370–373.
24. Oberley, L. W., and Buettner, G. R. (1979). Role of superoxide dismutase in cancer. The production of hydroxyl radical by bleomycin and iron. II. *FEBS Lett.* **97,** 47–49.
25. Singh, A., and Petkau, A., eds. (1978). Singlet oxygen and related species in chemistry and biology. *Photochem. Photobiol.* **28,** 429–934.
26. Boveris, A., Cadenas, E., Reiter, R., Filipkowsky, M., Nakase, Y., and Chance, B. (1979). Organ chemiluminescence: Noninvasive assay for oxidative radical reactions. *Proc. Natl. Acad. Sci. U.S.A.* **77,** 347–351.
27. Cadenas, E., Arad, I. D., Boveris, A., Fisher, A. B., and Chance, B. (1980). Partial spectral analysis of the hydroperoxide-induced chemiluminescence of the perfused lung. *FEBS Lett.* **111,** 413–418.
28. Cadenas, E., Wefers, H., and Sies, H. (1981). Low level chemiluminescence of isolated hepatocytes. *Eur. J. Biochem.* **119,** 531–536.
29. Foote, C. S. (1976). Photosensitized oxidation and singlet oxygen: Consequences in biological systems. *In* "Free Radicals in Biology" (W A. Pryor, ed.), Vol. 2, pp. 85–133. Academic Press, New York.
30. Krinsky, N. I. (1977). Singlet oxygen in biological systems. *Trends Biochem. Sci.* **2,** 35–38.
31. Koppenol, W. H. (1976). Reactions involving singlet oxygen and superoxide anion. *Nature (London)* **262,** 420–421.
32. Khan, A. U. (1976). Singlet molecular oxygen. A new kind of oxygen. *J. Phys. Chem.* **80,** 2219–2227.
33. Foote, C. S., Shook, F. C., and Akaberli, R. B. (1980). Chemistry of superoxide anion. 4. Singlet oxygen is not a major product of dismutation. *J. Am. Chem. Soc.* **102,** 2503–2504.
34. Roders, M. K., Glende, E. A., and Recknagel, R. O. (1977). Prelytic damage of red cells in filtrates from peroxidizing microsomes. *Science* **196,** 1221–1222.

35. Benedetti, A., Casini, A. F., Ferrali, M., and Comporti, M. (1979). Effects of diffusible products of peroxidation of rat liver microsomal lipids. *Biochem. J.* **180,** 303–312.
36. Slater, T. F., and Benedetto, C. (1981). Free radical reactions in relation to lipid peroxidation, inflammation and prostaglandin metabolism. *In* "The Prostaglandin System" (F. Bert and G. P. Velo, eds.), pp. 109–126. Plenum, New York.
37. Tappel, A. L. (1973). Lipid peroxidation damage to cell components. *Fed. Proc., Fed. Am. Soc. Exp. Biol.* **32,** 1870–1874.
38. Pryor, W. A. (1976). The role of free radical reactions in biological systems. *In* "Free Radicals in Biology" (W. A. Pryor, ed.), Vol. 1, pp. 1–49. Academic Press, New York.
39. Mead, J. F. (1976). Free radical mechanisms of lipid damage and consequences for cellular membranes. *In* "Free Radicals in Biology" (W. A. Pryor, ed.), Vol. 1, pp. 51–68. Academic Press, New York.
40. Thomas, M. J., Mehl, K. S., and Pryor, W. A. (1978). The role of superoxide anion in the xanthine oxidase-induced autoxidation of linoleic acid. *Biochem. Biophys. Res. Commun.* **83,** 927–932.
41. Bors, W., Michel, C., and Saran, M. (1979). Superoxide anion do not react with hydroperoxides. *FEBS Lett.* **107,** 403–406.
42. Rotilio, G., Bray, R. C., and Fielden, E. M. (1972). A pulse radiolysis study of superoxide dismutase. *Biochim. Biophys. Acta* **268,** 605–609.
43. Brigelius, R., Spöttl, R., Bors, W., Lengfelder, E., Saran, M., and Weser, U. (1974). Superoxide dismutase activity of low molecular weight Cu^{2+}-chelates studied by pulse radiolysis. *FEBS Lett.* **47,** 72–75.
44. Archibald, F. S., and Fridovich, I. (1981). Manganese, superoxide dismutase, and oxygen tolerance in some lactic acid bacteria. *J. Bacteriol.* **146,** 928–936.
45. Marklund, S. (1980). Distribution of Cu–Zn superoxide dismutase and Mn superoxide dismutase in human tissues and extracellular fluids. *Acta Physiol. Scand., Suppl.* **492,** 19–23.
46. Goldstein, I. M., Kaplan, H. B., Edelson, H. B., and Weissmann, G. (1979). Ceruloplasmin, a scavenger of superoxide anion radicals. *J. Biol. Chem.* **254,** 4040–4045.
47. Chance, B. (1947). An intermediate compound in the catalase–hydrogen peroxide reaction. *Acta Chem. Scand.* **1,** 236–267.
48. Sies, H., and Chance, B. (1970). The steady state level of catalase compound I in isolated hemoglobin free perfused rat liver. *FEBS Lett.* **11,** 172–176.
49. Sies, H. (1981). Measurement of hydrogen peroxide formation *in situ*. *In* "Methods in Enzymology" (W. B. Jakoby, ed.), Vol. 77, pp. 15–20. Academic Press, New York.
50. Oshino, N., Oshino, R., and Chance, B. (1973). The properties of catalase peroxidatic reaction and its relationship to microsomal ethanol oxidation. *Biochem. J.* **131,** 555–567.
51. Lawrence, R. A., and Burk, R. F. (1976). Glutathione peroxidase activity in selenium-deficient rat liver. *Biochem. Biophys. Res. Commun.* **71,** 952–958.
52. Flohé, L. (1982). Glutathione peroxidase brought into focus. *In* "Free Radicals in Biology" (W. A. Pryor, ed.), Vol. 5, pp. 223–254. Academic Press, New York.
53. Burk, R. F., Trumble, M. J., and Lawrence, R. A. (1980). Rat hepatic cytosolic glutathione-dependent enzyme protection against lipid peroxidation in the NADPH-microsomal lipid peroxidation system. *Biochim. Biophys. Acta* **618,** 35–41.
54. Gibson, D. D., Hornbrook, K. R., and McCay, P. B. (1980). Glutathione-dependent inhibition of lipid peroxidation by a soluble, heat-labile factor in animal tissues. *Biochim. Biophys. Acta* **620,** 572–582.
55. Yoshimura, S., Komatsu, N., and Watanabe, K. (1980). Purification and immunohistochemical localization of rat liver glutathione peroxidase. *Biochim. Biophys. Acta* **621,** 130–137.

56. Smith, M. T., Loveridge, N., Wills, E. D., and Chayen, J. (1979). The distribution of glutathione in rat liver lobule. *Biochem. J.* **182**, 103–108.
57. Kosower, N. S., and Kosower, E. M. (1978). The glutathione status of the cells. *Int. Rev. Cytol.* **54**, 109–160.
58. Holtzman, J. L. (1982). Role of reactive oxygen and metabolite binding in drug toxicity. *Life Sci.* **30**, 1–9.
59. Mannervik, B., and Axelsson, K. (1980). Role of cytoplasmic thioltransferase in cellular regulation by thiol disulfide interchange. *Biochem. J.* **190**, 125–130.
60. Brigelius, R., Lenzen, R., and Sies, H. (1982). Increase in hepatic mixed disulfide and glutathione disulfide levels elicited by paraquat. *Biochem. Pharmacol.* **31**, 1637–1641.
61. Witting, L. A. (1980). Vitamin E and lipid antioxidants in free radical initiated reactions. *In* "Free Radicals in Biology" (W. A. Pryor, ed.), Vol. 4, pp. 295–319. Academic Press, New York.
62. Tappel, A. L. (1980). Measurement of and protection from in vivo lipid peroxidation. *In* "Free Radicals in Biology" (W. A. Pryor, ed.), Vol. 4, pp. 1–47. Academic Press, New York.
63. Simic, M. G. (1981). Vitamin E radicals. *In* "Oxygen and Oxy-Radicals in Chemistry and Biology" (M. A. J. Rodgers and E. L. Powers, eds.), pp. 109–118. Academic Press, New York.
64. Patterson, L. K. (1981). Studies of radiation-induced peroxidation in fatty acid micelles. *In* "Oxygen and Oxy-Radicals in Chemistry and Biology" (M. A. J. Rodgers and E. L. Powers, eds.), pp. 89–95. Academic Press, New York.
64a. Packer, J. E., Slater, T. F. and Willson, R. L. (1979). Direct Observation of a free radical interaction between vitamin E and vitamin C. *Nature (London)* **278**, 737–738.
65. Krinsky, N. I. (1974). Singlet excited oxygen as a mediator of the antibacterial action of leukocytes. *Science* **186**, 363–365.
66. Peto, R., Doll, R., Buckley, J. D., and Sporn, M. B. (1981). Can dietary beta-carotene materially reduce human cancer rates? *Nature (London)* **290**, 201–208.
67. Diplock, A. T. (1981). The role of vitamin E and selenium in the prevention of oxygen-induced tissue damage. *In* "Selenium in Biology and Medicine" (J. E. Spallholz, J. L. Martin, and H. E. Ganther, eds.), pp. 303–316. Avi Publ., Westport, Connecticut.
68. Prohaska, J. R., and Wells, W. W. (1974). Copper deficiency in the developing of rat brain: A possible model for Menkes steeley hair disease. *J. Neurochem.* **23**, 91–98.
69. De Rosa, G., Leach, R. M., and Hurley, L. S. (1978). Influence of dietary Mn^{++} on the activity of mitochondrial superoxide dismutase. *Fed. Proc., Fed. Am. Soc. Exp. Biol.* **37**, 594A.
70. Chow, C. K., and Tappel, A. L. (1974). Response of glutathione peroxidase to dietary selenium in rats. *J. Nutr.* **104**, 444–451.
71. Keshan Disease Research Group of the Chinese Academy of Medical Sciences (1979). Observations on effect of sodium selenite in prevention of Keshan disease. *Chin. Med. J. (Peking, Engl. Ed.)* **92**, 477–482.
72. Johnson, R. A., Baker, S. S., Fallon, J. T., Maynard, E. P., Ruskin, J. N., Wen, Z., Ge, K., and Cohen, H. J. (1981). An occidental case of cardiomyopathy and selenium deficiency. *N. Engl. J. Med.* **304**, 1210–1212.
73. Burk, R. F., and Gregory, P. E. (1982). Some characteristics of ^{75}Se-P, a selenoprotein found in rat liver and plasma, and comparison of it with selenoglutathione peroxidase. *Arch. Biochem. Biophys.* **213**, 73–80.
74. Chow, C. K., Reddy, K., and Tappel, A. L. (1973). Effect of dietary vitamin E on the activity of the glutathione peroxidase system in rat tissues. *J. Nutr.* **103**, 618–624.
75. Hauswirth, J. W., and Nair, P. P. (1975). Effects of different vitamin E deficient basal

diets on hepatic catalase and microsomal P_{450} and b_5 in rats. *Am. J. Clin. Nutr.* **28**, 1087–1094.
76. Bieri, J. G., Thorp, S. L., and Tolliver, T. J. (1978). Effect of dietary polyunsaturated fatty acids on tissue vitamin E status. *J. Nutr.* **108**, 392–398.
77. Reddy, J. K., and Tappel, A. L. (1974). Effect of dietary selenium and autooxidized lipids on the glutathione peroxidase system of gastrointestinal tract and other tissues in the rat. *J. Nutr.* **104**, 1069–1078.
78. Hammer, C. T., and Wills, E. D. (1979). Effect of dietary fats on the composition of the liver endoplasmic reticulum and oxidative drug metabolism. *Br. J. Ntur.* **41**, 464–475.
79. Smith, M. T., and Wills, E. D. (1981). The effects of dietary lipid and phenobarbitone on the production and utilization of NADPH in the liver. A combined biochemical and cytochemical study. *Biochem. J.* **200**, 691–699.
80. Gillette, J. R., Mitchell, J. R., and Brodie, B. B. (1974). Biochemical mechanisms of drug toxicity. *Annu. Rev. Pharmacol.* **14**, 271–288.
81. Reed, D. J., and Beatty, P. W. (1980). Biosynthesis and regulation of glutathione: Toxicological implications. *Rev. Biochem. Toxicol.* **2**, 213–241.
82. Jakoby, W. B., and Habig, W. H. (1980). Glutathione transferases. *In* "Enzymatic Basis of Detoxication" (W. B. Jakoby, ed.), Vol. 2, pp. 63–94. Academic Press, New York.
83. Wedner, H. J., Simchowitz, L., Stenson, W. F., and Fischman, C. M. (1981). Inhibition of human polymorphonuclear leukocyte function by 2-cyclohexene-1-one. *J. Clin. Invest.* **68**, 535–543.
84. Parker, C. W., Fischman, C. M., and Wedner, H. J. (1980). Relationship of biosynthesis of slow reacting substance to intracellular glutathione concentrations. *Proc. Natl. Acad. Sci. U.S.A.* **77**, 6870–6873.
85. Chang, W.-C., Nakao, J., Orimo, H., and Murota, S. (1982). Effects of reduced glutathione on the 12-lipoxygenase pathways in rat platelets. *Biochem. J.* **202**, 771–776.
86. Akerboom, T. P. M., and Sies, H. (1981). Assay of glutathione, glutathione disulfide and glutathione mixed disulfides in biological samples. *In* "Methods in Enzymology" (W. B. Jakoby, ed.), Vol. 77, pp. 373–382. Academic Press, New York.
87. Isaacs, J., and Binkley, F. (1977). Glutathione dependent control of protein disulfide–sulfhydryl content by subcellular fractions of hepatic tissue. *Biochim. Biophys. Acta* **497**, 192–204.
87a. Higashi, T., Tateishi, N., Naruse, A., and Sakamoto, Y. (1977). A novel physiological role of liver glutathione as a reservoir of L-cysteine. *J. Biochem. (Tokyo)* **82**, 117–124.
87b. Cho, E. S., Sahyoun, N., and Stegink, L. D. (1981). Tissue glutathione as a cyst(e)ine reservoir during fasting and refeeding of rats. *J. Nutr.* **111**, 914–922.
88. Harisch, G., and Mahmoud, M. F. (1980). The glutathione status in the liver and cardiac muscle of rats after starvation. *Hoppe-Seyler's Z. Physiol. Chem.* **361**, 1859–1862.
89. Hahn, R., Wendel, A., and Flohé, L. (1978). The fate of extrcellular glutathione in the rat. *Biochim. Biophys. Acta* **539**, 324–337.
90. Wendel, A., Jaeschke, H., and Gloger, M. (1982). Drug-induced lipid peroxidation in mice. II. Protection against paracetamol-induced liver necrosis by intravenous liposomally entrapped glutathione. *Biochem. Pharmacol.* **31**, 3601–3605.
91. Sies H. (1983). Reduced and oxidized glutathione efflux from liver. *In* "Glutathione—Storage, Transport and Turnover in Mammals" (Y. Sakamoto, T. Higashi, and N. Tateishi, eds.). Jpn. Sci. Soc. press, Tokyo (in press).
92. Bartoli, G. M., and Sies, H. (1978). Reduced and oxidized glutathione efflux from liver. *FEBS Lett.* **86**, 89–91.

93. Sies, H., Koch, O. R., Martino, E., and Boveris, A. (1979). Increased biliary glutathione disulfide (GSSG) release in chronically ethanol-treated rats. *FEBS Lett.* **103**, 287–290.
94. Sies, H., Gerstenecker, C., Menzel, H., and Flohé, L. (1972). Oxidation in the NADP-System and release of GSSG from hemoglobin-free perfused rat liver during peroxidatic oxidation of glutathione by hydroperoxides. *FEBS Lett.* **27**, 171–175.
95. Sies, H., Wahlländer, A., and Waydhas, C. (1978). Properties of glutathione disulfide (GSSG) and glutathione-S-conjugate release from perfused rat liver. In "Functions of Glutathione in Liver and Kidney" (H. Sies and A. Wendel, eds.), pp. 120–126. Springer-Verlag, Berlin/New York.
96. Akerboom, T. P. M., Bilzer, M., and Sies, H. (1982). The relationship of biliary glutathione disulfide (GSSG) efflux and intracellular GSSG content in perfused rat liver. *J. Biol. Chem.* **237**, 4248–4252.
97. Akerboom, T. P. M., Bilzer, M., and Sies, H. (1982). Competition between transport of glutathione disulfide (GSSG) and glutathione-S-conjugates from perfused rat liver into bile. *FEBS Lett.* **140**, 73–76.
98. Wahlländer, A., Soboll, S., and Sies, H. (1979). Hepatic mitochondrial and cytosolic glutathione content and the subcellular distribution of GSH-S-transferases. *FEBS Lett.* **97**, 138–140.
99. Meredith, M. J., and Reed, D. J. (1982). Status of the mitochondrial pool of glutathione in the isolated hepatocyte. *J. Biol. Chem.* **257**, 3747–3753.
100. Sies, H., and Moss, K. M. (1978). A role of mitochondrial glutathione peroxidase in modulating mitochondrial oxidations in liver. *Eur. J. Biochem.* **84**, 377–383.
101. Crane, D., Häussinger, D., and Sies, H. (1982). Rise of coenzyme A-glutathione mixed disulfide during hydroperoxide metabolism in perfused rat liver. *Eur. J. Biochem.* **127**, 575–578.
102. Siliprandi, N., Siliprandi, D., Bindoli, A., and Toninello, A. (1978). Effect of oxidation of glutathione and membrane thiol groups on mitochondrial functions. In "Functions of Glutathione in Liver and Kidney" (H. Sies and A. Wendel, eds.), pp. 139–147. Springer-Verlag, Berlin/New York.
103. Lötscher, H. R., Winterhalter, K. H., Carafoli, E., and Richter, C. (1979). Hydroperoxides can modulate the redox state of pyridine nucleotides and the calcium balance in rat liver mitochondria. *Proc. Natl. Acad. Sci. U.S.A.* **76**, 4340–4344.
104. Sies, H., Graf, P., and Estrela, J. M. (1981). Hepatic calcium efflux during cytochrome P-450–dependent drug oxidations at the endoplasmic reticulum in intact liver. *Proc. Natl. Acad. Sci. U.S.A.* **76**, 3358–3362.
105. Boyd, S. C., Sasame, H. A., and Boyd, M. R. (1979). High concentrations of glutathione in glandular stomach: Possible implications for carcinogenesis. *Science* **205**, 1010–1012.
106. Szabo, S., Trier, J. S., and Frankel, P. W. (1981). Sulfhydryl compounds may mediate gastric cytoprotection. *Science* **214**, 200–202.
107. Reichard, S. M., Bailey, N. M., and Galvin, M. J. (1980). Alterations in tissue glutathione levels following traumatic shock. *Adv. Shock Res.* **5**, 37–41.
108. Sumida, S., and Yagi, H. (1981). Experimental study on the inhibition of kinin release in endotoxin shock by glutathione, proteinase inhibitors, hydrocortisone and hyperbaric oxygen. *Jpn. Circ. J.* **45**, 1364–1368.
109. Novi, A. M. (1981). Regression of aflatoxin B_1-induced hepatocellular carcinomas by reduced glutathione. *Science* **212**, 541–542.
110. Plummer, J. L., Smith, B. R., Sies, H., and Bend, J. R. (1981). Chemical depletion of

7. Biological Basis of Detoxication of Oxygen Free Radicals

glutathione in vivo. *In* "Methods in Enzymology" (W. B. Jakoby, ed.), Vol. 77, pp. 50–59. Academic Press, New York.
111. Griffith, W. W. (1981). Depletion of glutathione by inhibition of biosynthesis. *In* "Methods in Enzymology" (W. B. Jakoby, ed.), Vol. 77, pp. 59–63. Academic Press, New York.
112. Boveris, A., and Chance, B. (1973). The mitochondrial generation of hydrogen peroxide. *Biochem. J.* **134**, 707–716.
113. Nohl, H., Hegner, D., and Summer, K. H. (1981). The mechanism of toxic action of hyperbaric oxygenation on the mitochondria of rat-heart cells. *Biochem. Pharmacol.* **30**, 1753–1757.
114. Chance, B., Jamieson, D., and Coles, H. (1965). Energy-linked pyridine nucleotide reduction: Inhibitory effects of hyperbaric oxygen in vitro and in vivo. *Nature (London)* **206**, 257–263.
115. Jerret, S. A., Jefferson, D., and Mengel, C. E. (1973). Seizures, hydrogen peroxide formation and lipid peroxide in brain during exposure to oxygen under high pressure. *Aerosp. Med.* **44**, 40–44.
116. Nishiki, K., Jamieson, D., Oshino, N., and Chance, B. (1976). Oxygen toxicity in the perfused liver and lung under hyperbaric oxygen conditions. *Biochem. J.* **160**, 343–345.
117. Boveris, A., Cadenas, E., and Chance, B. (1981). Ultraweak chemiluminescence: A sensitive assay for oxidative radical reactions. *Fed. Proc., Fed. Am. Soc. Exp. Biol.* **40**, 195–198.
118. Deneke, S. M., and Farnburg, B. L. (1980). Involvement of glutathione enzymes in oxygen tolerance development by diethyldithiocarbamate. *Biochem. Pharmacol.* **29**, 1367–1373.
119. Stevens, J. B., and Autor, A. P. (1981). Proposed mechanism for neonatal rat tolerance to normobaric hyperoxia. *Fed. Proc., Fed. Am. Soc. Exp. Biol.* **39**, 3138–3143.
120. Armstrong, D. A., and Buchanan, J. D. (1978). Reactions of superoxide anion, hydrogen peroxide and other oxidants with sulfhydryl enzymes. *Photochem. Photobiol.* **28**, 743–755.
121. Haugaard, N. (1968). Cellular mechanisms of oxygen toxicity. *Physiol. Rev.* **48**, 311–373.
122. Mengel, C. E. (1972). The effects of hyperoxia on red cells as related to tocopherol deficiency. *Ann. N.Y. Acad. Sci.* **203**, 163–171.
123. Lawrence, R. A., and Burk, R. F. (1978). Species, tissue and subcellular distribution of non-Se–dependent glutathione peroxidase activity. *J. Nutr.* **108**, 211–215.
124. Kornbrust, D. J., and Mavis, R. D. (1979). Relative susceptibility of microsomes from lung, heart, liver, kidney, brain and testes to lipid peroxidation: Correlation with vitamin E content. *Lipids* **15**, 315–322.
125. Jones, D. P. (1981). Hypoxia and drug metabolism. *Biochem. Pharmacol.* **30**, 1019–1023.
126. Kieczka, H., and Kappus, H. (1980). Oxygen dependence of CCl_4-induced lipid peroxidation in vitro and in vivo. *Toxicol. Lett.* **5**, 191–196.
127. Stacey, N. H., Ottenwälder, H., and Kappus, H. (1982). CCl_4-induced lipid peroxidation in isolated rat hepatocytes with different oxygen concentrations. *Toxicol. Appl. Pharmacol.* **62**, 421–427.
128. Nastainczyk, W., Ullrich, V., and Sies, H. (1978). Effect of oxygen concentration on the reaction of halothane with cythcorome P-450 in liver microsomes and isolated perfused rat liver. *Biochem. Pharmacol.* **27**, 387–392.
129. Jee, R. C., Sipes, I. G., Gandolfi, A. J., and Brown, B. R. (1980). Factors influencing halothane hepatotoxicity in rat hypoxic model. *Toxicol. Appl. Pharmacol.* **52**, 267–277.

130. Rhodes, M. L., Zavala, D. C., and Brown, D. (1977). Hypoxic protection in paraquat poisoning. *Lab. Invest.* **35,** 496–500.
131. Sjostrom, K., and Crapo, J. D. (1981). Adaptation to oxygen by preexposure to hypoxia: Enhanced activity of mangani superoxide dismutase. *Clin. Respir. Physiol.* **17,** Suppl., 111–116.
132. Demopoulos, H. B., Flamm, E. S., Pietronigro, D. D., and Seligman, M. L. (1980). The free radical pathology and the microcirculation in the major central nervous system disorders. *Acta Physiol. Scand.* **492,** 91–119.
133. Bachur, R. R., Gordon, S., and Gee, M. (1978). A general mechanism for microsomal activation of quinone anticancer agents to free radicals. *Cancer Res.* **38,** 1745–1750.
134. Teicher, B. A., Lazo, J. S., and Sartorelli, A. C. (1981). Classification of antineoplastic agents by their selective toxicities toward oxygenated and hypoxic tumors. *Cancer Res.* **41,** 73–81.
135. Graham, D. G., Tiffany, S. M., Bell, W. R., Jr., and Gutknecht, W. F. (1978). Autoxidation versus covalent binding of quinones as the mechanism of toxicity of dopamine, 6-hydroxydopamine, and related compounds toward C1300 Neuroblastoma cells in vitro. *Mol. Pharmacol.* **14,** 644–653.
136. Bachur, N. R., Gordon, S., Gee, M., and Kon, H. (1979). NADPH-cytochrome P_{450} reductase activation of quinone anticancer agents to free radicals. *Proc. Natl. Acad. Sci. U.S.A.* **76,** 954–957.
137. Bozzi, A., Mavelli, I., Mondovi, B., Strom, R., and Rotilio, G. (1981). Differential cytotoxicity of daunomycin in tumor cells is related to glutathione-dependent hydrogen peroxide metabolism. *Biochem. J.* **194,** 369–372.
138. Lown, J. W., and Sim, S. (1977). The mechanism of bleomycin-induced cleavage of DNA. *Biochem. Biophys. Res. Commun.* **77,** 1150–1157.
139. Floyd, R. A. (1981). DNA-ferrous iron-catalyzed hydroxyl free radical formation from hydrogen peroxide. *Biochem. Biophys. Res. Commun.* **99,** 1209–1215.
140. Bozzi, A., Mavelli, I., Finazzi-Agró, A., Strom, R., Wolf, A. M., Mondovi, B., and Rotilio, G. (1976). Enzyme defense against reactive oxygen derivatives. II. Erythrocytes and tumor cells. *Mol. Cell. Biochem.* **10,** 11–16.
141. Bartoli, G. M., and Galeotti, T. (1979). Growth-related lipid peroxidation in tumor microsomal membranes and mitochondria. *Biochim. Biophys. Acta* **574,** 537–541.
142. Peskin, A. B., Koen, Y. M., Zbarski, I. B., and Konstantinov, A. (1977). Superoxide dismutase and glutathione peroxidase activities in tumors. *FEBS Lett.* **78,** 41–45.
143. Greenstein, J. P. (1954). "Biochemistry of Cancer." Academic Press, New York.
144. Doroshow, J. H., Locker, G. Y., and Myers, C. E. (1980). Enzymatic defenses of the mouse heart against reactive oxygen metabolites. Alterations produced by doxorubicin. *J. Clin. Invest.* **65,** 128–135.
145. Galeotti, T., Bartoli, G. M., Santini, S., Bartoli, S., Neri, G., Vernole, P., Massoti, L., and Zannoni, C. (1981). Growth-related changes in tumor superoxide dismutase content. *In* "Oxygen and Oxy-Radicals in Chemistry and Biology" (M. A. J. Rodgers and E. L. Powers, eds.), pp. 641–644. Academic Press, New York.
146. Boveris, A., Docampo, R., Turrens, J. F., and Stoppani, A. O. M. (1978). Effect of β-lapachone on superoxide anion and hydrogen peroxide production in Trypanosoma cruzi. *Biochem. J.* **175,** 431–439.
147. Docampo, R., and Stoppani, A. O. M. (1979). Generation of superoxide anion and hydrogen peroxide induced by nifurtimox in Trypanosoma cruzi. *Arch. Biochem. Biophys.* **197,** 317–321.
148. Boveris, A., Sies, H., Martino, E. E., Docampo, R., Turrens, J. F., and Stoppani,

7. Biological Basis of Detoxication of Oxygen Free Radicals

A. O. M. (1980). Deficient metabolic utilization of hydrogen peroxide in Trypanosoma cruzi. *Biochem. J.* **188**, 643–648.
149. Dubin, M., Moreno, S. N. J., Martino, E. E., Docampo, R., and Stoppani, A. O. M. (1983). Increased biliary secretion and loss of hepatic glutathione in rat liver after nifurtimox treatment. *Biochem. Pharmacol.* (in press).
150. Bradley, A. A., Griffith, O. W., and Cerami, A. (1981). Inhibition of glutathione synthesis as a chemotherapeutic strategy for trypanosomiasis. *J. Exp. Med.* **153**, 720–725.
151. Babior, B. M. (1978). Oxygen-dependent microbial killing by phagocytes. Part I. *N. Engl. J. Med.* **298**, 659–668.
152. McCord, J. M. (1974). Free radicals and inflammation: Protection of synovial fluid by superoxide dismutase. *Science* **185**, 529–531.
153. Petrone, W. F., English, D. K., Wong, K., and McCord, J. M. (1980). Free radicals and inflammation: Superoxide-dependent activation of a neutrophil chemotactic factor in plasma. *Proc. Natl. Acad. Sci. U.S.A.* **77**, 1159–1163.
154. Emerit, I., and Michelson, A. M. (1981). Mechanism of photosensitivity in systemic lupus erythematosus patients. *Proc. Natl. Acad. Sci. U.S.A.* **78**, 2537–2540.
155. Del Maestro, R. F., Thaw, H. H., Björk, J., Planker, M., and Arfors, K. E. (1980). Free radicals as mediators of tissue injury. *Acta Physiol. Scand.* **492**, Suppl., 43–47.
156. Kuehl, F. A., Jr., and Egan, R. W. (1980). Prostaglandins, arachidonic acid and inflammation. *Science* **210**, 978–984; Kuehl, F. A., Jr., Humes, J. L., Ham, E. A., Egan, R. W., and Doughterty, H. W. (1980), Inflammation: The role of peroxidase-derived products. *Adv. Prostaglandin Thromboxane Res.* **6**, 77–86.
157. Payne, T. G., Dewald, B., Siegl, H., Gubler, H. U., Ott, H., and Baggiolini, M. (1982). Radical scavenging and stimulation of prostaglandin synthesis not anti-inflammatory? *Nature (London)* **296**, 160–162.
158. Nathan, C. F., Silverstein, S. C., Brukner, L. H., and Cohn, Z. A. (1979). Extracellular cytolysis by activated macrophages and granulocytes. II. Hydrogen peroxide as a mediator of cytotoxicity. *J. Exp. Med.* **149**, 100–113.
159. Simon, R. H., Scoggin, C. H., and Patterson, D. (1981). Hydrogen peroxide causes the fatal injury to human fibroblasts exposed to oxygen radicals. *J. Biol. Chem.* **256**, 7181–7186.
160. Krinsky, N. I., Sladdin, D. G., and Levine, P. H. (1981). Effect of oxygen radicals on platelet functions. *In* "Oxygen and Oxy-Radicals in Chemistry and Biology" (M. A. J. Rodgers and E. L. Powers, eds.), pp. 153–160. Academic Press, New York.
160a. Arrick, B. A., Nathan, C. F., Griffith, O. W., and Cohn, Z. A. (1982). Glutathione depletion sensitizes tumor cells to oxidative cytolysis. *J. Biol. Chem.* **257**, 1231–1237.
161. Greenwald, R. A. (1981). Effect of oxygen derived free radicals on connective tissue macromolecules: Inhibition by copper-penicillamine complex. *J. Rheumatol.* **8**, 9–13.
162. Dahlen, S.-E., Björk, J., Hedqvist, P., Arfors, K. E., Hammarström, S., Lindgren, J.-A., and Samuelsson, B. (1981). Leukotrienes promote plasma leakage and leukocyte adhesion in postcapillary venules: In vivo effects with relevance to the acute inflammatory response. *Proc. Natl. Acad. Sci. U.S.A.* **78**, 3887–3891.
163. McCord, J. M., Stokes, S. H., and Wong, K. (1979). *Adv. Inflammation Res.* **1**, 273–280.
164. Goebel, K.-M., Storck, U., and Neurath, F. (1981). Intrasynovial orgotein therapy in rheumtoid arthritis. *Lancet* **1**, 1015–1017.
165. Puig-Parallada, P., and Planas, J. M. (1978). Synovial fluid degradation induced by free radicals. In vitro action of several free radical scavengers and antiinflammatory drugs. *Biochem. Pharmacol.* **27**, 535–537.

166. Oyanagui, Y. (1976). Inhibition of superoxide anion production in macrophages by antiinflammatory drugs. *Biochem. Pharmacol.* **25**, 1473–1480.
167. Mandell, G. L., Rubin, W., and Hook, E. W. (1970). The effect of an NADH-oxidase inhibitor (hydrocortisone) on polymorphonuclear leukocytes bactericidal activity. *J. Clin. Invest.* **49**, 1381–1388.
168. Goldstein, I. M., Ross, D., Weissman, G., and Kaplan, H. B. (1976). Influence of corticosteroids on human polymorphonuclear leukocytes function in vitro. Reduction of lysosomal enzyme release and superoxide production. *Inflammation* **1**, 305–312.
169. Levine, P. H., Hardin, J. C., Scoon, K. L., and Krinsky, H. I. (1981). Effect of corticosteroids in the production of superoxide and hydrogen peroxide and the appearance of chemiluminescence by phagocytosing polymorphonuclear leukocytes. *Inflammation* **5**, 19–27.
170. Stuyvesant, V. W., and Yolley, W. B. (1967). Antiinflammatory activity of d-α-tocopherol (vitamin E) and linoleic acid. *Nature (London)* **216**, 585–586.
171. Panganamala, R. V., Miller, J. S., Gwebu, E. T., Sharma, H. M., and Cornwell, D. G. (1977). Differential inhibitory effects of vitamin E and other antioxidants on prostaglandin synthetase, platelets aggregation and lipoxydase. *Prostaglandins* **14**, 261–271.
172. Bragt, P. C., Bansberg, J. I., and Bonta, I. L. (1980). Depletion of hepatic antioxidants during granulomatous inflammation in the rat and local antiinflammatory effects of free radical scavengers. *Agents Actions* **7**, Suppl., 246–249.
173. Canonico, P. G., Bill, W., and Ayala, E. (1977). Effects of inflammation in peroxisomal enzyme activities, catalase synthesis and lipid metabolism. *Lab. Invest.* **37**, 459–468.
174. Fuhijara, E., Sandeman, V. A., and Whitehouse, M. W. (1979). Pathobiodynamics: Reduction of hepatic and intestinal ligandin (glutathione-S-transferase) levels in rats with severe and acute inflammation. *Biochem. Med.* **22**, 175–191.
175. Granfield, L. M., Gollan, J. L., White, A. G., and Dormandy, T. L. (1979). Serum antioxidant activity in normal and abnormal subjects. *Ann. Clin. Biochem.* **16**, 299–306.
176. Gitlin, J. D., Gitlin, J. I., and Gitlin, D. (1977). Localization of C-reactive protein in synovium of patients with rheumatoid arthritis. *Arthritis Rheum.* **20**, 1491–1499.
177. Willson, R. L. (1979). Hydroxyl radicals and biological damage in vitro: What relevance in vivo? *Ciba Found. Symp.* **65** (new ser.), 19–42.
178. Greenstock, C. L. (1981). Redox processes in radiation biology and cancer. *Radiat. Res.* **86**, 196–211.
179. Biaglow, J. E. (1981). Cellular electron transfer and radical mechanisms for drug metabolism. *Radiat. Res.* **86**, 212–242.
180. Mason, R. P. (1982). Free-radical intermediates in the metabolism of toxic chemicals. *In* "Free Radicals in Biology" (W. A. Pryor, ed.), Vol. 5, pp. 161–222. Academic Press, New York.
181. Sjöberg, L., Eriksen, T. E., and Révész, L. (1982). The reaction of the hydroxyl radical with glutathione in neutral and alkaline aqueous solution. *Radiat. Res.* **89**, 255–263.
182. Edgren, M., Révész, L., and Larsson, A. (1981). Induction and repair of single-strand DNA breaks after X-irradiation of human fibroblasts deficient in glutathione. *Int. J. Radiat. Biol.* **40**, 355–363.
183. Harman, D. (1982). The free-radical theory of aging. *In* "Free Radicals in Biology" (W. A. Pryor, ed.), Vol. 5, pp. 255–275. Academic Press, New York.
184. Tolmasoff, J. M., Ono, T., and Cutler, R. G. (1980). Superoxide dismutase: Correlation with life-span and specific metabolic rate in primate species. *Proc. Natl. Acad. Sci. U.S.A.* **77**, 2777–2781.
185. Glass, G. A., and Gershon, D. (1981). Enzymatic changes in rat erythrocytes with increasing cell and donor age: Loss of superoxide dismutase activity associated with

increases in catalytically defective forms. *Biochem. Biophys. Res. Commun.* **4,** 1245–1253.
186. Ames, B. N., Cathcart, R., Schwiers, E., and Hochstein, P. (1981). Uric acid provides an antioxidant defense in humans against oxidant- and radical-caused aging and cancer: A hypothesis. *Proc. Natl. Acad. Sci. U.S.A.* **78,** 6858–6862.
187. Proctor, P. (1970). Similar functions of uric acid and ascorbate in man? *Nature (London)* **228,** 868.

CHAPTER 8

Fate of Xenobiotics: Physiologic and Kinetic Considerations*

K. Sandy Pang

I. Introduction 214
II. The Xenobiotic 215
   A. Physicochemical Properties 215
   B. Interactions with Vascular Components 217
   C. Tissue Binding 221
   D. Substrate Specificity 223
III. Hemodynamics 223
IV. Xenobiotic Metabolism 225
   A. Concept of Organ Clearance and Total Body Clearance . . . 225
   B. Extraction Ratio 227
   C. Intrinsic Clearance 228
   D. Role of Organ Blood Flow in Clearance 230
   E. Role of Xenobiotic Binding to Vascular Components in Organ Clearance 230
   F. Role of Organ Intrinsic Clearance in Organ Clearance . . . 231
   G. Role of the Diffusional Barrier in Organ Clearance 232
   H. Role of Substrate Concentration in Organ Clearance 233
   I. Role of Spatial Organization of Organ and Enzyme Localization . 233
V. Xenobiotic Excretion 237
   A. The Kidneys 237
   B. The Liver 239

* This work was supported by USPHS Grant #GM-27323 and Research Career Development Award from the National Institute of Arthritis, Diabetes, and Digestive and Kidney Diseases AM-01028.

VI. Organs of Elimination 239
   A. The Liver 239
   B. The Intestine 240
   C. The Lung 241
   D. Sequential Organs: The Intestine, the Liver, and the Lung . . 242
   E. Effect of Route of Administration 244
VII. Comments 246
   References 246

I. INTRODUCTION

The biological fate of a xenobiotic is often defined in an *in vitro* system. Many methods, from purified enzyme or reconstituted enzyme systems through perfused organs, have been employed to probe the fate of many exogenously administered compounds. The information provided has laid the groundwork for drug metabolism and drug-mediated toxicity. The bulk of the information obtained from such *in vitro* systems, however, does not always agree with *in vivo* findings. The basis for the difference lies in the dynamic interplay of the physiological processes in the living organism that may contrast to the carefully controlled conditions of an *in vitro* system. Nevertheless, invaluable information has been gained from such *in vitro* studies.

The importance of the *in vitro* findings in an *in vivo* setting, however, may be placed in better perspective when the physiological processes in the body are integrated. For example, the manner in which a xenobiotic gains access, that is, the circulation that ultimately transports a xenobiotic to tissues and organs, requires consideration. The processes of absorption and distribution are key determinants that influence the rate of delivery of a xenobiotic to sites of metabolism and excretion as well as to sites of pharmacological or toxicological effect. The manner in which a xenobiotic is metabolized and excreted will, in turn, be dependent on the capacities of both metabolic and excretory systems and the rates at which these organs receive and eliminate the compound. The overall significance of a metabolic or excretory pathway, therefore, relies on its eliminatory rate relative to the sum total of all elimination rates in the body.

This chapter reviews the interactions between xenobiotic and body constituents, the physiological processes that govern the delivery of substrates to their organs of excretion and metabolism, and highlights the competition among eliminating organs that ultimately determines the fate of a xenobiotic. Awareness of the role and importance of each of the processes will aid in the interpretation of the differences between observa-

8. Fate of Xenobiotics

tions made with *in vitro* and *in vivo* systems. Major considerations will focus on the xenobiotic itself, body hemodynamics, the eliminating organ, and the anatomical positions of the eliminating organs *in vivo* with regard to different modes of administration.

II. THE XENOBIOTIC

A. Physicochemical Properties*

The chemical properties of the xenobiotic are the predominant factors dictating the rate of individual processes that determine the duration of a xenobiotic *in vivo*. Physicochemical properties such as molecular size, lipophilicity, and pK_a strongly influence the rate of delivery from the site of administration to the site of elimination or action. As shown in Fig. 1, these characteristics influence strongly the absorptive and distributive properties (routes 1 to 3) in the transport scheme of a xenobiotic.

At the site of administration (oral, intraperitoneal, and subcutaneous), molecules of small molecular size may gain entry into the cellular components of a membrane through small pores (70 Å) that intersperse the continuum. Inasmuch as xenobiotics gain entry mainly through membranous structures that are lipoidal, most lipophilic compounds that share some aqueous solubility will readily enter cellular components. A xenobiotic must be in its soluble form before it can be transported through lipoidal membranes. Solubility depends on chemical, electronic, and structural effects that influence the mutual interactions between the solute and the solvent. The lipophilic character of a xenobiotic plays an equally important role in determining its passage through lipid barriers. For ex-

---

* *Symbols used in this chapter:* C, total (bound and unbound) substrate concentration; subscripts P and B, plasma and blood concentrations; subscripts P,u and u, unbound plasma and unbound organ concentrations; subscripts In and Out, concentrations of substrate entering and leaving an eliminating organ; subscripts A, V, and U, arterial, venous, and urinary substrate concentrations. D, substrate. *Dose* denotes the administered dose; AUC, area under the curve; superscripts ia, ipl, iv, im, po, pia, and sc, intraarterial, intraportal, intravenous, intramuscular, oral, paraintraarterial, and subcutaneous routes of administration. Cl, organ clearance; Cl_T, total body or systemic clearance; Cl_R, renal clearance; Cl_u, clearance of an organ based on unbound concentration; Cl_{dif}, diffusional clearance to an organ; Cl_{eff}, effective clearance for organs arranged in sequence; Cl_{int}, intrinsic clearance of an organ. E, organ extraction ratio. F, availability of an eliminating organ; subscript sys denotes systemic availability. f, unbound fraction; Q, organ blood flow; subscripts I, H, and L, intestine, liver (total), and lung. V_{max} and K_m, maximal enzyme velocity and the Michaelis–Menten constant. M_I and M_{II}, primary and secondary metabolites. V_U, volume of urine. v, rate of loss of compound by an eliminating organ; subscripts I, H, and L, intestine, liver, and lung. Subscript ss denotes steady state.

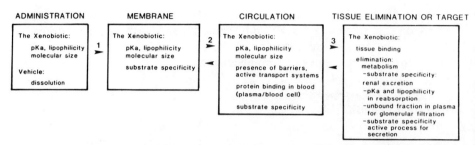

Fig.1. A schematic interaction between a xenobiotic and the organism.

ample, as the chain length of an aliphatic xenobiotic increases, lipid solubility increases with a concomitant decrease in water solubility, that is, partition coefficient correlates directly with chain length. Branched compounds are found to possess higher aqueous solubilities and lower partition coefficients than their straight-chain analogs. The partition coefficient has been shown to have a parabolic effect on absorption rate, a situation analogous to its effect on biological activity of the compound.[1] If the partition coefficient between lipid and water approaches zero, the compound will be so insoluble in the fat as to prevent rapid crossing of lipid membranes, and it will tend to remain localized in the first aqueous phase it contacts. Conversely, as the partition coefficient approaches infinity, the xenobiotic will be so insoluble in water that it will tend to remain localized within the lipoidal membrane. As transport entails a "random walk" of a solute in a water phase to a lipid phase, and a lipid phase to a water phase, the progress of a roving molecule through aqueous and lipophilic phases is heavily dependent on its hydrophilic–lipophilic balance. The driving force for transport by passive diffusion is the concentration gradient between the two phases that exist on both sides of a biological membrane.

Neutral compounds are considered to pass through the membrane more readily than their ionized and hence charged counterparts; the ratio of neutral to ionized molecules depends on the pH of the environment and the acid strength of the xenobiotic, or the pK_a. For weak bases, this acid strength is expressed in terms of the strength of the conjugate acid. Because most drugs are weak electrolytes and exist in solution as a mixture of the dissociated (ionized) and undissociated (un-ionized) forms, the proportion of a drug in the undissociated form depends on the pK_a as well as on the pH of the medium according to the Henderson–Hasselbach equation (Eqs. 1 and 2).[2] The un-ionized form is absorbed more rapidly than its ionized form.

8. Fate of Xenobiotics

For weak acids: $\quad \text{pH} = \text{p}K_a + \log [\text{ionized}]/[\text{un-ionized}] \quad (1)$

For weak bases: $\quad \text{pH} = \text{p}K_a + \log [\text{un-ionized}]/[\text{ionized}] \quad (2)$

With orally administered xenobiotics that are weak acids, for example, the undissociated state is more favorable for absorption from an acidic environment such as the stomach. The converse may hold for weak bases that may be more undissociated in a less acidic environment such as the small intestine. But most weak acids and weak bases, when given orally, are absorbed mostly in the small intestine. The small intestinal mucosa has an extremely large surface area relative to the volume of intestinal contents, allowing for greater absorption. The large surface is due to the mucosal villi, which are composed of epithelial cells covered by a brush border of microvilli (1 μm long and about 0.1 μm wide). It has been estimated that the presence of villi results in an increase of seven- or eightfold in the surface of the mucous membrane of the intestine. The microvilli increase the surface area by another 20-fold.[3] Because of water resorption and decreasing surface area at the terminal ileum, absorption of most xenobiotics occurs mainly at the upper end of the small intestine.

Operative in drug transport during the absorption phase are mechanisms such as facilitated diffusion as well as active processes. For facilitated diffusion, carriers that transport drug molecules across a membrane are implicated. The drug-carrier complex migrates through the membrane and relinquishes the drug molecule at the opposite side of the membrane, freeing the carrier to repeat the process. Active transport differs from the facilitated mechanism in that it operates against an electrochemical gradient at the expense of energy. Both processes can become saturated, and both are competitively inhibited by substrates that utilize the same mechanism.

B. Interactions with Vascular Components

1. Plasma Proteins

As a xenobiotic gains access to the blood, interactions with blood components occur almost instantly. Xenobiotics bind reversibly to blood constituents such as plasma proteins and red blood cells by means that include hydrophobic, induced-dipole, ionic bonding, electrostatic, and van der Waals forces. It is assumed that macromolecules, and therefore drugs bound to macromolecules, cross the biological membranes only with difficulty; it is the unbound drug that is considered to be the species that diffuses freely. Once the unbound form has penetrated into tissues, binding to cellular components or tissue binding may result.

The extent to which xenobiotics bind to blood proteins and cellular components depends on the specific macromolecule as well as the foreign compound. Albumin (69,000 daltons) is the major plasma protein (59%) and can exist extravascularly. Albumin is the predominant macromolecule involved in the binding of most compounds, especially anions, in the circulation. β-Lipoproteins have been implicated as assisting albumin in the binding of basic or cationic compounds,[4] with α- and γ-globulins responsible for other examples of binding.[5]

The quantitation of xenobiotic plasma protein binding usually assumes a lack of cooperativity, that is, groups on a protein molecule capable of interacting have identical affinities for the drug molecules. It also assumes that the affinity of any group is unaffected by the combination of drug molecules at binding sites. Because binding is an equilibrium reaction, the following scheme can be formulated.

$$X_u + P_u \underset{k_2}{\overset{k_1}{\rightleftharpoons}} XP \qquad (3)$$

Equation (3) denotes the equilibrium between the unbound xenobiotic in plasma X_u, and the unbound protein in plasma P_u, and the xenobiotic–protein complex XP. The molar concentrations at equilibrium (in square brackets) can be expressed in terms of the equilibrium constant K_A or the association constant of the xenobiotic–protein complex.

$$K_A = [XP]/[X_u][P_u] = k_1/k_2 \qquad (4)$$

For n, such equivalent binding sites on each protein molecule of total molar concentration $[P_T]$, $n[P_T]$ represents the total concentration of the binding sites and equals the sum of the bound and unbound protein concentration, $([XP] + [P_u])$. On substitution of $[P_u] = n[P_T] - [XP]$ into Eq. (4), and upon rearrangement, the ratio, r, that denotes the molar concentration of xenobiotic bound per molar concentration of total protein, is defined by Eq. (5).[6]

$$r = [XP]/[P_T] = nK_A[X_u]/(1 + K_A[X_u]) \qquad (5)$$

Alternatively, the fraction of total xenobiotic that is bound in plasma, that is, the ratio of the molar concentration of the bound xenobiotic in plasma [XP] and the molar concentration of the total (bound and unbound) in plasma $[X_T]$[7] is described by Eq. (6), and follows a nonlinear relationship with both the total plasma protein concentration and the xenobiotic concentration.

$$\text{Fraction bound} = [XP]/[X_T] = \frac{1}{1 + (1/nK_A[P_T]) + [X_u]/[P_T]} \qquad (6)$$

8. Fate of Xenobiotics

It will be clear that the degree of xenobiotic binding in plasma depends on the protein concentration, the xenobiotic concentration, the number of binding sites, and the association constant of the drug–protein complex. For multiple classes of binding sites with the number of binding sites of each class $n_1, n_2, n_3, \ldots$, and with the association constants for each class of $K_1, K_2, K_3, \ldots$, respectively, the total number of moles of xenobiotic bound per mole of protein (r_{total}) is given by Eq. (7).[8]

$$r_{\text{total}} = r_1 + r_2 + r_3 + \cdots r_i$$
$$= \frac{n_1 K_1 [X_u]}{1 + K_1 [X_u]} + \frac{n_2 K_2 [X_u]}{1 + K_2 [X_u]} + \frac{n_3 K_3 [X_u]}{1 + K_3 [X_u]} + \cdots \frac{n_i K_i [X_u]}{1 + K_i [X_u]} \quad (7)$$

The usefulness of this relationship has been translated into a linear form known as the Scatchard plot,[6] in which $r/[X_u]$ is plotted against r (Eq. 8).

$$r/[X_u] = nK_A - rK_A \quad (8)$$

The plot provides a positive intercept that equals nK_A, and a negative slope of K_A, and allows the number of binding sites to be calculated. Should binding involve more than one class of binding sites, curve-stripping procedures aided by computer analyses may resolve the several binding components, each representing a class of binding site. Usually, the different classes of binding site for a xenobiotic will surface only when the concentration range of the xenobiotic is varied extensively. It must be emphasized that strict interpretation of Scatchard plots requires that the assumption of reversibility be met experimentally.

The respective binding constants K_A strongly influence the degree of binding of a xenobiotic in the circulation.[8] For xenobiotics that have tight coupling with plasma proteins ($K_A > 10^4 \, M^{-1}$), xenobiotic–protein complex formation occurs readily at a low concentration of the xenobiotic, and the unbound fraction (unbound concentration/total concentration) of the xenobiotic in plasma may be very small. At an increasing body burden of the xenobiotic, however, the finite number of binding sites on the protein molecules will eventually become fully occupied. Further increases in the amount of xenobiotic bound to the body will lead to large increases in the unbound portion. This shift occurs over a small increment of xenobiotic when the binding sites on the protein are fully occupied. By contrast, for compounds of less tight coupling with plasma proteins ($K_A < 10^4 \, M^{-1}$), the xenobiotic tends not to bind despite an excess of binding sites that are present. The unbound fraction remains quite high. As loading with such a compound increases, excess binding sites will accommodate some of the additional drug, and the unbound fraction will increase only gradually. Thus, only a gradual rather than an abrupt shift occurs in the unbound fraction with an increase in the body burden of the xenobi-

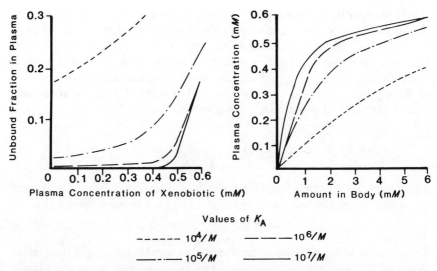

Fig. 2. The effect of the association constant, K_A, on (A) the fraction unbound in plasma and (B) plasma concentrations on varying body burdens of the xenobiotic. [Reprinted by permission from *Nature* **207**, 274–276. Copyright © 1965 McMillan Journals Limited.]

otic (Fig. 2). Therefore, with the tightly bound anticoagulent warfarin (>99% bound),[10] there may be an abrupt change upon binding of this drug, whereas with acetaminophen (20–30% bound),[11] only a gradual change would be expected upon alteration of the concentration.

2. Blood Cell

The binding between xenobiotic and plasma protein is not unique, as similar interactions between a xenobiotic and other components of the blood, namely the red blood cells, also occur. Little is known about the interaction with other blood components such as white blood cells and fibrinogen. The partitioning or binding of a xenobiotic to erythrocyte components, unlike plasma protein, achieves equilibrium rather slowly. Although several compounds may be bound by phospholipids in the membranes of the red cells, hemoglobin may be the binding species for others, and the rate of equilibration with the unbound drug in plasma water is impeded by the erythrocyte membrane.[12,13] Nonetheless, the same general treatment of the data for plasma protein binding can be applied to erythrocyte binding. The effects of xenobiotic–erythrocyte binding on the ultimate delivery of a xenobiotic to target sites (see Chapter 13, this volume) or organs of elimination (see Chapter 10, this volume) can also be interpreted analogously to those provided for protein binding.

8. Fate of Xenobiotics

3. Interrelationship between Plasma and Blood Cell Binding

A close relationship exists between the degree of binding to plasma proteins and to red blood cells. Because both binding processes are equilibrium reactions, the unbound concentration of the xenobiotic in both plasma and red cells must be equal at equilibrium. A simple relationship that relates the unbound fraction in blood and the unbound fraction in plasma to total concentrations in these tissues is based on the definition of the unbound fraction in blood, f_B (Eq. 9)[14]:

$$f_B = C_{P,u}/C_B \tag{9}$$

This is the ratio of the unbound concentration in plasma ($C_{P,u}$), and also equals that in red cells, divided by the total whole blood concentration, C_B. By using a similar definition for the unbound fraction in plasma, f_P (Eq. 10),

$$f_P = C_{P,u}/C_P \tag{10}$$

as the unbound concentration in plasma divided by the total plasma concentration, and by combining and rearranging Eqs. (9) and (10), the unbound fraction in blood, f_B (Eq. 11) is the unbound fraction in plasma divided by the blood/plasma ratio (C_B/C_P).[14]

$$f_B = f_P/(C_B/C_P) \tag{11}$$

When the ratio $C_B/C_P = 1.0$, $f_B \simeq f_P$ and indicates that the xenobiotic is evenly distributed in vascular components or that no binding results. When the ratio is greater than 1.0, preferential distribution into red cells occurs. Conversely, when $C_B/C_P < 1.0$, the xenobiotic is more concentrated in plasma.

Actually, the value of f_B is a complex function, being dependent on the concentration of the xenobiotic, on the affinity constants for binding to the respective vascular components, on the concentrations of the binding constituents in plasma and blood cells, and on the hematocrit.[15]

C. Tissue Binding

The interaction between a xenobiotic and blood components causes a transient immobilization of the xenobiotic. Because the driving force of transport between blood and other tissues is the concentration difference in the "mobile" or unbound form of the compound, binding limits distribution. Distribution between phases will continue as long as there is a difference in the concentration of the unbound compound. But in addition to xenobiotic binding to plasma proteins and blood cells, the degree of tissue binding is equally important in determining the extent of distribution.[16] Foreign compounds are known to bind to most tissues, intercalate with DNA, and bind even to the pigmented structures of the eye.[17] A group of proteins, the ligandins, which are actually glutathione S-transferases, are avid binding proteins within the cytosol of liver, with appreciable concentrations found in the renal tubular cells and mucosal cells of the small intestine. These proteins bind a vast number of lipophilic compounds that include such physiological metabolites as folate, bilirubin, and corticosteroids, but also the lipophilic xenobiotics that include azo dyes and bromosulfophthalein.[18]

Again, the physicochemical properties of the xenobiotic prevail in the distributive process. Because weak acids are more ionized at physiological pH (7.4), weak acids tend to be less distributed (Eq. 2) than weak bases, the latter being more un-ionized than weak acids. Consequently, weak acids are present at higher concentrations in blood and plasma than weak bases. Lipophilic compounds that readily gain access during the absorptive process also tend to be more extensively distributed and remain mainly in adipose tissue stores or in the lipid components of cells. Polar compounds do not enter the brain readily because of the endothelial lining that separates the brain from the circulation. In contrast to other organs and tissues, brain capillaries are devoid of pores that intersperse the membrane continuum. Rather, the capillary endothelial cells in brain are joined to each other by continuous, tight intercellular junctions, a topography that suggests that materials must pass through cells, rather than between them, to move from blood to brain and vice versa. This effectively poses a barrier for drug transport that is commonly referred to as the *blood–brain barrier*. In the choroid plexus, however, the choroidal cells themselves are joined to each other with continuous, tight junctions similar to those of the brain capillaries, and this device is regarded as an additional barrier, the *cerebrospinal fluid barrier*, in the transport of substances into the brain. Nevertheless, compounds of relatively large molecular weight (e.g., ferritin) move quite freely between the cells into the extracellular fluid of the brain. There exists an extracellular fluid space in brain, approximately similar to that of muscle and other body tissues, which must have pores greater than 50–100 Å to effect the transfer. Another active process has been demonstrated for the transport of weak acids (e.g., penicillin and salicylic acid) out of the brain and into blood against a concentration gradient. Generally, lipophilic and un-ionized molecules traverse membranes easily and enter the brain.[19]

When the body is viewed as the sum of two major components, blood and other tissues, a larger percentage of the same body burden of a xenobiotic in tissue, versus that in blood, implies that the xenobiotic is highly distributed; if a large proportion of the xenobiotic is in blood, the distribution is limited.

D. Substrate Specificity

The structure of a xenobiotic dictates the manner in which it is eliminated (metabolized and excreted) by an organism. Most metabolic reactions are highly specific, that is, the reaction is dependent on the type of substrate and the enzyme system in each metabolizing organ. Even when the same enzyme system is present in several organs, the availability of cosubstrate, the total amount of enzyme, and abundance of an isozymic form of enzyme that is primarily responsible for the metabolic reaction, and the enzymatic parameters (V_{max} and K_m) in each organ may differ.

Although excretory mechanisms are usually not considered as being substrate specific, a polar configuration and a molecular size requirement must exist before a molecule is excreted into bile. Moreover, specific, active processes exist in the kidneys for secretion, one for weak acids and another for weak bases.

III. HEMODYNAMICS

In considering the processes of absorption, distribution, and elimination, in addition to the physicochemical profile of the specific xenobiotic, the rate of delivery and removal of the xenobiotic is of paramount importance. Although the blood serves as the physiological medium of translocation and exchange for all tissues, most tissues have access only to a fraction of that supply. The lungs, however, receive the entire cardiac output. Knowledge of differential perfusion rates (blood flow per 100 g of tissue) has permitted a classification of tissues into those that are highly perfused (the adrenals, kidneys, liver) and those that are relatively poorly perfused (muscle, fat).[20] Because of constraint in blood flow to specific tissues, absorption and distribution of xenobiotics are also categorized as two types. For very lipid-soluble compounds that readily gain access to the circulation and subsequently to the tissues, absorption[21] and distribution may be limited by the rate of blood flow of those tissues that have the greatest tendency to accommodate them. For xenobiotics with poor lipid solubility, whose ability to penetrate membrane barriers is low, diffusion, and not blood flow, limits the absorptive and distributive processes.

The importance of *in vivo* organ perfusion rates on uptake has been demonstrated adequately with inhalation anesthetics. Increasing perfusion of pulmonary tissue by increasing cardiac output leads to an enhanced uptake of anesthetic into the systemic circulation; lowering the output has the opposite effect. Changes are most pronounced for gases exhibiting the greatest solubility in blood.[22] Because the role of perfusion in xenobiotic transport is similar at any blood–membrane interface, the dynamic interplay of hemodynamics with factors such as the physicochemical properties of a xenobiotic and tissue pH must also be considered in accounting for the rate of delivery from the site of administration to the target sites.

For illustrative purposes (Fig. 3), two weak acids of widely differing lipophilicity, thiopental (pK_a = 6 and 8)[23] and salicylic acid (pK_a = 3),[24] are profoundly different in their disposition. Because of good lipid solubility and limited dissociation, thiopental distributes rapidly to the brain and other highly perfused tissues after intravenous administration. However, because of rapid equilibration of the anesthetic with other tissues, the concentration of thiopental in blood or plasma drops rapidly. The brain relinquishes its contents with equal rapidity in equilibration with blood. The result is a sharp decline in brain levels, thereby contributing to the short duration of action of the anesthetic. Accumulation of this highly lipophilic compound in adipose tissue occurs only slowly, mainly because of the slow rates of perfusion into fatty depots and is *perfusion limited*.[19] By contrast, salicylic acid is highly ionized and poorly lipid soluble, accounting for its minimal accumulation in brain. An active process also transports the acid out of brain; the unbound concentration of salicylic acid in brain is less than that in blood.[25] Accumulation in adipose stores is poor as a result of poor lipid solubility, and the restricted distribution of this weak acid exemplifies *diffusion limitation*.

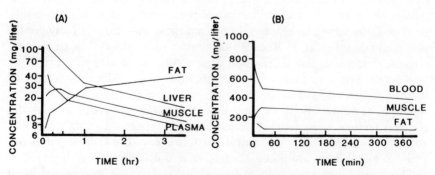

Fig. 3. The disposition of (A) thiopental and (B) salicylic acid in various organs (tissues) *in vivo*. From Refs. 23 and 24, respectively, with permission of the publishers.

IV. XENOBIOTIC METABOLISM

Elimination of xenobiotics is effected primarily by biotransformation and excretion mechanisms.[8] Some compounds, especially if lipophilic, are readily reabsorbed by the peritubular cells of the kidney. Unless the xenobiotic is metabolized to more polar metabolites that are ultimately excreted, it will remain, mostly in fatty tissues, for a long time. In general, xenobiotics will be metabolized by relatively nonspecific enzyme systems, collectively labeled the *enzymes of detoxication*.[26-28] Biotransformed products are usually, but not always, more polar than the parent molecular entity.

A. Concept of Organ Clearance and Total Body Clearance

A simplistic way of describing the elimination activities of an organ is by the measurement of xenobiotic disappearance in its passage through the organ. By Fick's principle (Eq. 12), the difference between the rate of entry of the xenobiotic to the organ (product of organ blood flow Q and the entering concentration of the xenobiotic C_{In}) and the rate of appearance of the xenobiotic in venous blood (product of organ blood flow and the outflow concentration of the xenobiotic C_{Out}) is the rate of loss. At any instant, the rate of loss of a compound across the organ is attributed to its rate of binding within tissues, the rate of metabolism, and the rate of excretion of unchanged xenobiotic. When the organ approaches a steady state in which the rate of change in the concentration of the xenobiotic is zero, the rate of loss (v_{ss}) is totally accounted for by the rates of metabolism and excretion.

$$v_{ss} = Q(C_{In_{ss}} - C_{Out_{ss}}) \qquad (12)$$

A physiological approach in presenting the rate of loss of a xenobiotic across an eliminating organ is modeled after clearance concepts that originated in the description of the renal excretion of urea.[29] Clearance of any substance is defined as the volume of biological fluid that is cleared of the substance per unit of time. This is not a real volume. As blood passes through the kidneys, no single milliliter of blood has all of its urea removed in one transit. Rather, a little urea is removed from each of the many milliliters of blood perfusing the kidneys. If all urea is added up and expressed as though it were derived by completely clearing a much smaller volume of all of the contained urea, the clearance of the kidney for urea may be calculated. Mathematically, clearance by an organ at any instant, Cl (Eq. 13), denotes the rate of loss of the xenobiotic relative to the input concentration.

$$Cl = v/C_{In} = Q(C_{In} - C_{Out})/C_{In} \tag{13}$$

The concept of clearance applies not only to the kidneys but to each of the eliminating organs irrespective of the process or mechanism involved. Strictly speaking, clearance should be defined with respect to the eliminating organ, the sampling site, and the eliminatory (metabolic/excretory) process involved. For example, the clearance of a xenobiotic by the liver as measured in blood is its hepatic blood metabolic clearance. Because blood is the perfusing medium to an organ, clearance values should be strictly correlated with the concentrations of the xenobiotic in blood that is flowing into the organ and to the site of organ blood flow. Unfortunately, plasma is usually the assayed medium, and a correction factor that denotes the blood/plasma ratio must be incorporated.

Because the body is the composite of tissues and organs, total body clearance has also been taken as an additive property, that is, the sum of individual organ clearances.[30] This concept will apply well if all eliminating organs receive their blood supply directly from the aorta in a parallel fashion, for example, the kidney, the brain, and the intestines. But the intestines, the liver, and the lung are three potential metabolizing organs that are arranged in sequence. The outflow from the intestines drains into the portal vein, accounting partially for the flow into liver; the remainder, 25% of the total liver blood flow, arises from the aorta and represents the hepatic arterial flow. The outflow from the liver, or hepatic venous flow, combines with all other venous returns to the right heart and perfuses the lung before the total cardiac output returns to the left heart. If individual organ clearances for the intestines, liver, and lung are summed, a value greater than the true value will be obtained.[31] The concept of total body clearance as the sum of all clearances does not apply to eliminating organs that are arranged in sequence. What is appropriate is an effective clearance for all organs in sequence; this aspect is discussed more fully in Section VI,D.

Total body clearance is also known as *systemic clearance*. It is often sought following intravenous administration and, more appropriately, by intraarterial (ia) administration of the xenobiotic if the lung is an eliminating organ. A proven and useful relationship that describes total body clearance (Cl_T), the intraarterial dose (Dose[ia]), and the area under the blood concentration–time curve, or AUC, is expressed in Eq. (14).[30]

$$Cl_T = \text{Dose}^{ia}/AUC = \text{Dose}^{ia}\bigg/\int_0^\infty C_B \, dt \tag{14}$$

8. Fate of Xenobiotics

The *AUC* is the plot of the concentration of xenobiotic in blood versus time, or the integral of the concentration with time from zero to infinity. The Cl_T obtained from Eq. (14) will provide a time-averaged value that is identical to summing all organ clearances during steady state. Actually, venous sampling from a noneliminating organ, such as an arm, is frequently used, and the venous arm concentration is taken to reflect both arterial arm concentration and all other arterial concentrations that enter eliminating organs. It should be emphasized that, if plasma rather than blood concentration of a xenobiotic is used for the *AUC*, Cl_T will be underestimated when the xenobiotic resides mainly in plasma (C_B/C_P or $\lambda \ll 1.0$) and overestimated when extensive distribution occurs into red cells ($\lambda \gg 1.0$). The total body clearance should be corrected by Eq. (15).[32]

$$Cl_T = \text{Dose}^{ia} \Big/ \int_0^\infty \lambda C_P \, dt \qquad (15)$$

B. Extraction Ratio

An alternative method for describing clearance measurements for the efficiency of an eliminating organ in removing a foreign compound is the extraction ratio, *E*, or that fraction removed when a xenobiotic is passed through an organ (Eq. 16).[32]

$$E = Q(C_{In} - C_{Out})/QC_{In} = (C_{In} - C_{Out})/C_{In} \qquad (16)$$

This fraction can be viewed as the rate of loss per rate of presentation of the compound (QC_{In}). Clearance by an eliminating organ, *Cl*, is defined by Eq. (17).[32]

$$Cl = QE \qquad (17)$$

The relationship among organ clearance, blood flow, and extraction ratio is more complex than it may appear. In liver, for example, increasing hepatic blood flow has essentially no effect on the hepatic clearance of antipyrine, whereas the hepatic clearances of lidocaine and propranolol increase almost proportionally with increasing blood flow rate; the hepatic clearance of chromic phosphate colloid also increases with increasing blood flow rate, but not proportionally.[15]

Classification of a xenobiotic on the basis of the removal capacity of an eliminating organ has greatly aided in the understanding of the underlying physiological processes influencing its clearance. A xenobiotic that is cleared rapidly by an eliminating organ is considered as possessing a high extraction ratio ($E \simeq 1.0$); the output concentration is a small fraction of the input concentration. Conversely, a xenobiotic that is cleared poorly by an eliminating organ is considered as having a low extraction ratio ($E \simeq 0$); outflow concentration nearly equals input concentration.

C. Intrinsic Clearance

An additional concept requires definition, that of organ intrinsic clearance, Cl_{int}.[33] This term has been developed in an attempt to measure the enzymatic activity of an eliminating organ in a manner that is independent of organ blood flow and binding within the vascular system. The intrinsic clearance relates the rate of elimination to the (unbound) concentration of xenobiotic (C_u) surrounding the eliminatory systems. Thus, *intrinsic clearance*[15,33] (Eq. 18) may be defined as the volume of cellular water of an eliminating organ that is effectively cleared of xenobiotic per unit of time.

$$Cl_{int} = v/C_u \qquad (18)$$

A feature distinguishing Cl_{int} from Cl is that intrinsic clearance describes the actual eliminatory capacity of the organ for metabolism/excretion of the xenobiotic, whereas organ clearance describes the overall efficiency of the organ in removing the xenobiotic. Notably, intrinsic clearance is the rate of loss of the xenobiotic relative to the substrate concentration, or unbound concentration in the organ C_u, whereas organ clearance is the rate of loss of the xenobiotic relative to the input concentration, both bound and unbound (cf. Eqs. 18 and 13). Organ clearance will be restrained by physiological variables such as organ blood flow, binding characteristics to vascular components, and the intrinsic clearance of the organ for xenobiotic removal.

As an approach to organ clearances, each eliminating organ is viewed as operationally uniform, that is, the enzymatic and excretory mechanisms responsible for xenobiotic removal are assumed to be evenly distributed within the elimination organ, so that mixing between substrate and the removal mechanism occurs evenly. Distribution equilibrium is achieved so rapidly in the organ that the xenobiotic in the emergent venous blood is in equilibrium with that in the organ. Assuming passive diffusion and assuming that a substrate diffuses rapidly through membranes to the site of elimination, that is, perfusion limitation, when the rate of diffusion is much greater than the rate of elimination, it follows that the concentration of unbound xenobiotic in venous blood, $C_{Out,u}$ and in the organ, C_u, are equal. Insofar as elimination obeys Michaelis–Menten kinetics for n competing pathways, the following mass balance relationship applies in the steady state.[15]

8. Fate of Xenobiotics

$$v_{ss} = Q(C_{In,ss} - C_{Out,ss}) = \sum_{i=1}^{n} \frac{V_{max,i} C_{u,ss}}{(K_{m,i} + C_{u,ss})} \tag{19}$$

By recalling the definition of Cl_{int} as in Eq. (18), the Cl_{int} in terms of the enzymatic parameters becomes Eq. (20),

$$Cl_{int} = \sum_{i=1}^{n} \frac{V_{max,i}}{(K_{m,i} + C_{u,ss})} \tag{20}$$

and is reduced to a constant under linear kinetic conditions when $K_{m,i} \gg C_{u,ss}$.

$$Cl_{int} = \sum_{i=1}^{n} \frac{V_{max,i}}{K_{m,i}} \tag{21}$$

Also, Cl_{ss} becomes

$$Cl_{ss} = \frac{v_{ss}}{C_{In,ss}} = \frac{1}{C_{In,ss}} \sum_{i=1}^{n} \frac{V_{max,i} C_{u,ss}}{(K_{m,i} + C_{u,ss})} \tag{22}$$

By appropriate rearrangement and acknowledging that $C_u = C_{Out,u} = f_B C_{Out}$, we have

$$Cl_{ss} = \sum_{i=1}^{n} \frac{Q f_B Cl_{int,i}}{(Q + f_B Cl_{int,i})} \tag{23}$$

and

$$E_{ss} = \sum_{i=1}^{n} \frac{f_B Cl_{int,i}}{Q + f_B Cl_{int,i}} \tag{24}$$

Equation (23) has been used to estimate the intrinsic clearance of eliminating organs such as the liver. The difficulty with this technique lies in the uncertainty of a value for the concentration of unbound substrate at the hepatocytes of an *in vivo* system. Despite the unknown value for the *in vivo* substrate concentration, investigators often attempt to equate *in vitro* with *in vivo* values. In some cases, a correlation is found fortuitously, resulting in the misconception that a correspondence between the two systems exists. Reasons for the lack of correlation are plentiful because the conditions of the *in vitro* system are generally highly manipulated. There is a lack of binding components, lack of blood flow, lack of spatial integrity of an eliminating organ, absence of barriers, absence of competing eliminating organs, an abundance of cosubstrates, and the presence of isolated enzyme systems that channel metabolism in an unphysiological direction, thereby distorting all estimates for an *in vivo* system. Nonetheless, both sets of data may be applicable and correct within the limitations of each system.

D. Role of Organ Blood Flow in Area Clearance

Returning to the classification of compounds on the basis of organ extraction ratios, it is apparent from Eq. (24) that, for values of E approaching 1.0, the product, $f_B Cl_{int}$, greatly exceeds the value of organ blood flow. In this sense, the organ possesses an immense ability to eliminate a xenobiotic, and that capacity exceeds the rate at which xenobiotic molecules are brought to the organ by way of blood flow. The rate-limiting step in organ clearance, therefore, is the delivery of substrates in blood to the site of elimination by means of flow, that is, $Cl \simeq Q$ (Eq. 23). E becomes independent of Q (Eq. 24), that is, the substance delivered to the organ by way of blood is almost completely removed because of high organ efficiency.[15]

By deduction, the organ intrinsic clearance of a poorly extracted compound ($E \simeq 0$) is very low, much less than organ blood flow. The Cl of this compound, however, will approach $f_B Cl_{int}$ (Eq. 23) and becomes virtually independent of changes in blood flow. This absence of a hemodynamic role in the clearances of poorly extracted xenobiotics exists because the rate-limiting step is the inability of the organ to eliminate the xenobiotic. Hence, perturbations in organ blood flow will affect clearances of poorly cleared compounds only minimally. But a contrasting relationship is evident for the extraction ratio, E. Values of E for poorly cleared xenobiotics will change inversely with organ blood flow (Eq. 24). Generally speaking, a slower blood flow rate to an eliminating organ will increase the extraction ratio, whereas an increase in flow will decrease E.[15] The effect may be explained on the basis of transit time of a xenobiotic within the organ; a faster flow brings about a shortening of the transit time of blood, resulting in a decrease in the fraction eliminated; the converse holds for faster flow rates.

Compounds that have intermediate extraction ratios will be affected by organ blood flow but in a nonlinear fashion. The clearances of such compounds change proportionally, as do compounds with a high E, but the increases are less than proportional.

E. Role of Xenobiotic Binding to Vascular Components in Organ Clearance

The binding of a xenobiotic to macromolecules such as plasma proteins and erythrocytes tends to decrease organ clearance. For xenobiotics that are highly cleared, however, the transient interaction between a xenobiotic and proteins may not decrease organ clearance appreciably. As soon as the unbound compound is removed by the eliminating organ, bound

8. Fate of Xenobiotics

xenobiotic rapidly dissociates and furnishes more of the unbound species for elimination. All xenobiotics, bound and unbound, will be effectively eliminated by the organ when the xenobiotic is highly cleared ($E \simeq 1.0$). An example is the highly bound (95%) yet highly cleared ($E = 0.7$–0.9) β-blocker, propranolol, during hepatic removal.[34] When the extraction ratio of a xenobiotic is small, great sensitivity of organ clearance to the unbound fraction in blood is seen. This relationship is evident from Eq. (23) under conditions in which $Q \gg f_B Cl_{int}$ and $Cl \simeq f_B Cl_{int}$. A better explanation of the sensitivity of clearance to binding is the poor ability of the organ in removing the xenobiotic (Cl_{int} is very small) that requires the xenobiotic to be in a form preferred for elimination. This sensitivity of organ clearance to the unbound fraction is seen with the anticoagulant, warfarin, when clearance changes proportionally with the unbound fraction.[10]

Two descriptive terms, *nonrestrictive* and *restrictive* clearances have been used in describing the absence and the presence, respectively, of a role of binding on clearance.[35] Nonrestrictive clearance describes the lack of a role of binding in organ clearance, and the ratio (rate of removal)/(total drug concentration) will accurately predict the efficiency in eliminating xenobiotics that are rapidly cleared. Restrictive clearance, however, describes a direct dependence on xenobiotic–protein binding; the usual description of clearance that is based on total concentration should be modified to include this influence for poorly cleared xenobiotics. A better description would be clearance based on the unbound fraction in blood (Eq. 25):

$$Cl_u = Cl/f_B \qquad (25)$$

This sensitivity will become evident only when the dissociation rate constant (k_2 from Eq. 3) or the $t_{1/2}$ for dissociation ($0.693/k_2$) of the xenobiotic–protein complex is much longer than the transit time of blood through the organ.[36]

F. Role of Organ Intrinsic Clearance in Organ Clearance

As organ intrinsic clearance describes the overall capacity of the organ to eliminate a xenobiotic, it is the sum total of the individual intrinsic clearances of all eliminatory pathways within that organ. For metabolic reactions mediated by enzyme systems that can be described by simple Michaelis–Menten kinetics, the *in vivo* V_{max} and K_m are the critical parameters in intrinsic clearance (Eq. 20). For those pathways in which poor cosubstrate availability prevails, the measured *in vivo* V_{max} and K_m may be those for the synthesis of the cosubstrate, that is, the rate-limiting step. For excretory pathways, the same principles for metabolic reactions may be applied, although V_{max} may be in the guise of the transport maximum, T_{max}, as in biliary excretion (see Chapter 10, this volume).

G. Role of the Diffusional Barrier in Organ Clearance

Despite the assumption that xenobiotics must possess some lipophilic characteristics to assure transport through lipoidal membrane barriers to the site of elimination or effectiveness, numerous compounds have difficulty entering and leaving cells. Quaternary ammonium compounds, for example, are charged and not readily absorbed, extensively distributed, or rapidly eliminated because of their inability to cross membranes.[37,38] A diffusional barrier may exist for polar compounds in gaining access to eliminating organs when the rate of diffusion of a compound constitutes the rate-limiting step in the overall rate of disappearance of a xenobiotic; the diffusional clearance, Cl_{dif}, measured as the flux or rate of transport divided by C_{in}, is much less than the intrinsic clearance, that is, less than the capacity of the organ to remove the compound. Organ clearance under this instance becomes[39]

$$Cl = Qf_B Cl_{dif}/(f_B Cl_{dif} + Q) \qquad (26)$$

But when both diffusional clearance and intrinsic clearance are of comparable value, organ clearance is influenced by both parameters, as shown by Eq. (27).[39]

$$Cl = Qf_B Cl_{int} Cl_{dif}/[Cl_{int}(f_B Cl_{dif} + Q) + QCl_{dif}] \qquad (27)$$

Some xenobiotics may not experience difficulty entering eliminatory sites, but exhibit difficulty in transferring from the eliminating organ into the blood that is sweeping through the organ. When the rate of elimination of the xenobiotic is much faster than the rate of back-diffusion from the eliminating organ, a constant extraction (E = constant) and a constant clearance may result. The composite data on harmol metabolism suggested such an example when the compound was used as a model for sulfation and glucuronidation. The paradox of a constant extraction ratio of harmol with input concentrations of 50–200 μM, which resulted in varying proportions of sulfate and glucuronide conjugates in perfused rat liver, differed from the expected first-order kinetics.[40] When 2,6-dichloro-4-nitrophenol (DCNP), an inhibitor of sulfation, was added to the perfusate entering the liver, hepatic extraction ratios remained constant over the range of input harmol concentrations (10–200 μM) and were similar to those obtained in the absence of DCNP. Sulfation, which was a dominant metabolic pathway, was suppressed to 10% of its control value in the presence of DCNP. The suppression of sulfation was compensated for by increased glucuronidation.[41] This aberrant kinetic behavior is explicable only by formulating a diffusional barrier retarding exit of harmol from liver.[42]

H. Role of Substrate Concentration in Organ Clearance

Most of the enzymes for detoxication usually accept more than one substrate, that is, cosubstrates, which are usually endogenous compounds.[27] As a result, it is not always appreciated that the observed *in vivo* K_m and V_{max} would necessarily reflect the K_m and V_{max} for the purified enzyme system; the estimates for the *in vitro* system are also dependent on the randomness or the order of attachment of both substrate and cosubstrates.[43]

I. Role of Spatial Organization of Organ and Enzyme Localization

The distribution of enzyme systems in a drug-metabolizing organ has been studied mostly in the liver, although other metabolizing organs such as the kidney, intestine, and lung are being investigated. *In vitro* techniques have permitted the intracellular separation of several particulate fractions from "soluble" enzymes. The sulfotransferases, for example, are present mostly in the cytosol,[44] whereas the cytochrome P-450 system is located in the microsomal fraction.[45]

It has been suggested that the distribution of enzymes is not uniform throughout a given organ. Early metabolic studies had suggested a preponderance of the cytochrome P-450 system in the centrilobular region of the liver,[46] from which hepatic venous blood drains. This was confirmed by direct immunohistochemical and staining techniques.[47] Indirect evidence on the distribution of aryl sulfotransferases suggests their greater abundance in the periportal region,[48,49] an area that receives the blood from both the hepatic artery and the portal vein. These observations confirm the heterogeneous distribution of drug-metabolizing enzymes, that is, to an intercellular localization of enzymes (see Chapter 4, this volume).

The intercellular enzymatic distributions in an organ are of paramount importance in determining the types and amounts of metabolites formed. For example, two enzyme systems, A and B, compete for the same sub-

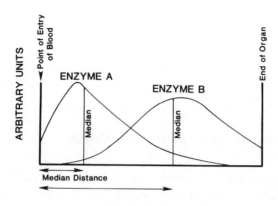

Fig. 4. The distribution of enzymatic activities for enzyme A and enzyme B, and their median values along the length of the liver. The distance between the median and the point of blood entry to the organ is the median distance. Designations of enzymes A and B for the metabolic reactions are explained in the text. Adapted from Ref. 50, with permission of the publisher.

strate, but the distributions of A and B are localized with respect to the flow path through the organ (Fig. 4).[50] The flow path may be expressed in terms of time elapsed after the entry of blood into the organ (Fig. 5). When the distribution of enzyme A is closer to the entry of blood and that of enzyme B is skewed toward the point of exit of blood for the organ, enzyme A will receive the entire concentration or amount of substrate entering the organ, whereas enzyme B will receive only the residual substrate, or the amount that remains after biotransformation by enzyme A. Given equal enzymatic activities for enzymes A and B, metabolites from enzyme A will be more abundant in the effluent blood than metabolites from enzyme B because of the anterior localization of A along the flow path.

Enzymatic distribution is also important in the formation of sequential metabolites. Following the introduction of a substrate, a primary metabolite formed in the organ may be metabolized immediately *in situ* before leaving the organ, an occurrence known as *sequential first-pass metabolism* of a primary metabolite.[51] The dependence of the extent of sequential metabolism on enzymic localization is illustrated by the metabolism of a compound D to a primary metabolite, M_I and subsequently to a terminal metabolite M_{II} by enzymes A and B, as presented in Fig. 4. Two are used. In the first, $D \rightarrow M_I \rightarrow M_{II}$ is mediated by enzymes A and B, respectively; in the second, $D \rightarrow M_I \rightarrow M_{II}$ is mediated by the same two enzymes but acting in reverse order. When D is introduced to the organ, it is met by

8. Fate of Xenobiotics

enzyme A undergoing rapid conversion to M_I before encountering enzyme B, in the first case. In the second example, D needs to travel farther in order to reach enzyme B for conversion to M_I, and M_I has to recycle to enzyme A in order to form M_{II}. Any diversion in the ability of M_I to reach enzyme A will result in a decreased conversion of M_I to M_{II} when compared to the first example or to a situation in which a preformed metabolite is delivered directly into the organ.[52]

This dependence of sequential metabolism on localization of enzymes is also explicable in terms of blood transit time, substrate "duration time," and metabolite "duration time."[50] Because blood flowing through an organ has a finite transit time (organ blood flow divided by the volume of the organ) and because the localization of an enzymatic system is described with respect to the flow path [translated as time elapsed after entry of substrate into the organ (Fig. 4)], the constraint of a finite blood transit time, along with the time required for a substrate to reach an enzyme system, as well as the rapidity with which conversion is carried out (intrinsic clearance for D), will determine the substrate duration time. The time required for a substrate to reach its enzyme site is dependent on the median distance (Fig. 4); the median, or center, is taken as the point on the scale of observations on each side of which there are equal areas of the enzyme system. If the median distance is short, less time will be required for the substrate D to traverse the flow path in reaching the enzyme, and a longer interval remains for the substrate to be metabolized. The converse will hold for longer median distances. Analogously, the metabolite duration time will be dependent on the blood transit time, the substrate duration time, the time required for M_I to reach the second enzyme system for further metabolism (the metabolite transfer time, or the time required for M_I to travel between the first and second enzyme systems), and the time required for M_I conversion to M_{II} (or the intrinsic clearance for formation of M_{II}).[50]

The dependence of metabolite duration time on blood transit time and the intrinsic clearances of formation of M_I from D is depicted by simulations of concentrations of a substrate D, its primary metabolite M_I, and the terminal metabolite M_{II} after the entry of a substrate into the organ (Fig. 5A). The intrinsic clearances for the conversion of D to M_I are varied from 0.4 to 0.1 and 0.025 ml/sec, but the intrinsic clearance for the terminal metabolic reaction of M_I to M_{II} is kept constant at 0.05 ml/sec. Varying decay profiles of concentrations of D result from the differences in intrinsic clearances for D to M_I, but the profiles of M_I and M_{II} are subsequently affected (Fig. 5A). The concentrations of D, M_I, and M_{II} within the organ become those concentrations in blood leaving the organ when the time elapsed equals the blood transit time. When this simulation is presented

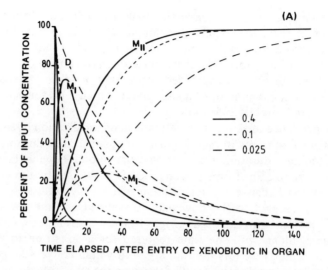
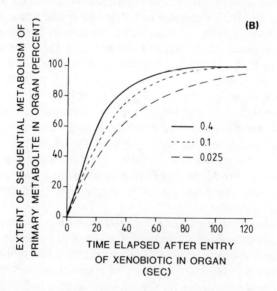

Fig. 5. (A) Simulations of the concentrations of a precursor (D) and its primary (M_I) and terminal (M_{II}) metabolites in an organ and (B) the extent of sequential metabolism by the organ. The hepatic intrinsic clearances for M_I formation from D were varied (0.4, 0.1, and 0.025 ml/sec), whereas the hepatic intrinsic clearance for M_I metabolism was kept constant (0.05 ml/sec). The precursor decayed after a period of time, whereas the primary metabolite followed a biphasic upswing and downswing curve, and terminal metabolite accumulated. Adapted from Ref. 50, with permission of the publisher.

alternatively as the extent of sequential metabolism $[M_{II}]/([M_I] + [M_{II}])$ at various times in the organ (Fig. 5B), a faster sequential metabolism of M_I accompanies a higher intrinsic clearance for its formation, and a lesser extent occurs as a consequence of a lower intrinsic clearance. Again, differences in the extent of metabolism for M_I from its various precursors will be dependent on the blood transit time in the organ.

In contrast, when the intrinsic clearance for the conversion of M_I to M_{II} is very high compared to that for the formation of M_I (ratio of enzyme activities = 100), the metabolite detected in the effluent blood from the organ will be mostly M_{II}, that is, the extent of sequential metabolism is mostly complete. Enzymatic distributions, as well as the intrinsic clearance for the formation of the primary metabolite, will share less influence on the extent of sequential metabolism of the primary metabolite (K. S. Pang and R. N. Stillwell, unpublished results).

An understanding of this concept of metabolite duration time, which is dependent on (1) blood transit time, (2) enzymatic distributions, (3) intrinsic clearance for primary metabolite formation, (4) intrinsic clearance for conversion of the primary metabolite, and (5) the manner in which predominance of one of these factors affects the extent of sequential metabolism, will provide the basis for an explanation of varying degrees of toxicity arising from metabolic bioactivation of putative toxicants.

V. XENOBIOTIC EXCRETION

A. The Kidneys

The kidneys receive the largest proportion of the cardiac output (20–25%) and are the major organs of excretion,[53] although they should also be recognized as organs of metabolism. Three mechanisms are involved in renal excretion: glomerular filtration, secretion, and reabsorption.[54] Glomerular filtration is a physical process whereby particulate matter such as red cells and colloidal material such as proteins are retained while the filtrate (plasma water) passes through pores (>70 Å) that exist in the glomerulus. The driving force is the hydrostatic pressure or the intracapillary blood pressure; opposing filtration is the osmotic pressure exerted by the plasma proteins and the hydrostatic pressure of the fluid in Bowman's capsule. It is pressure, and not the blood or plasma flow in the glomerular capillaries, that determines the glomerular filtration rate (GFR). Glomerular filtration rate (GFR) can be determined by the clearance of a substrate that is neither reabsorbed nor secreted. The GFR measured with such a substance (e.g., inulin) averages 125 ml/min in humans.

Organic acids and bases with diverse structures and properties are actively secreted in the proximal tubules. These secretory processes somehow lack general specificity, although weak acids compete for the same active process for secretion, and weak bases for another.

Reabsorption of xenobiotics occurs in the distal tubules, where water in the ultrafiltrate is reabsorbed and a concentration gradient is established between the lumen of the nephron and blood. As a result, passive diffusion of the undissociated forms of the compound, regardless of whether they are actively secreted or filtered, may take place along the tubules from the peritubular cells to the luminal fluids. Reabsorption by diffusion will thereby be affected by the urinary pH, because the concentration of unionized species may change depending on the pK_a. Reabsorption by diffusion is decreased by an increase in urine flow, the result of low concentration of the diffusible compound in the tubular fluid, and a decreased residence time in the tubules (see also Chapter 10, this volume).

1. Role of Vascular Binding and Blood Flow in Renal Excretion

Because plasma water is the ultrafiltrate resulting from glomerular filtration, a xenobiotic that is transiently bound is not filtered. The excretion rate due to glomerular filtration is therefore proportional to the concentration in plasma water, that is, the unbound concentration in plasma.

$$\begin{aligned} \text{Rate of excretion due to filtration} &= (\text{GFR}) \times C_{P,u} \\ &= \text{GFR} \times f_p \times C_P \end{aligned} \quad (28)$$

By contrast, the degree of binding to vascular components affects renal secretion only minimally. Again, the unbound form is the presumed moiety removed by secretion. Rapid dissociation of the bound form follows, so that effectively, all bound and unbound forms of the xenobiotic are removed by secretion. The limiting step for secretion is the rate of delivery of molecules by renal blood flow. This condition is analogous to that for a highly extracted compound that is eliminated by an organ with a high intrinsic clearance in which clearance is limited by blood flow.

2. Renal Clearance

For any compound excreted by filtration, secretion, and reabsorption, renal clearance (Cl_R) is a complex function of a number of determinants: protein binding, which determines the amount excreted by glomerular filtration; the structure of the compound, which determines the amount of secretion; the pK_a of the compound; and its lipid solubility, which determines the extent of nonionic reabsorption. In addition, physiological vari-

8. Fate of Xenobiotics

ables such as urinary pH, urine flow rate, and renal blood flow contribute to the determination of net renal clearance.

What is usually measured as urinary excretion rate is the product of the volume of urine (V_u) and the concentration of the xenobiotic in urine (C_u). Excretion rate is the amount excreted divided by the time interval for urine collection, Δt. Renal clearance, Cl_R, is defined in Eq. (29),

$$Cl_R = V_U C_U / \Delta t C_A$$
$$= (\text{GFR} \times f_p C_p + \text{secretion rate} - \text{reabsorption})/C_A \quad (29)$$

where C_P and C_A denote the plasma and blood concentrations, respectively, that are entering the kidneys.

B. The Liver

The hepatobiliary system is another important means for the excretion of xenobiotics by secretion in bile. The physicochemical characteristics of the xenobiotic or its metabolite, particularly its molecular weight and polarity, determine the extent of biliary excretion. A molecular weight greater than 300 seems to be required, but the optimum molecular weight for biliary excretion varies with the animal species.[55] This subject is considered in detail in Chapter 10, this volume.

VI. ORGANS OF ELIMINATION

A. The Liver

The liver (3-4% body weight) is unique in its anatomical position in relationship to circulation. The mesenteric vein and the splanchnic vein drain into the portal vein and constitute about 75% of total liver blood flow. The remainder is delivered by the hepatic artery that arises from the celiac artery. Despite the divergent blood supply to the liver, close association and compensatory mechanisms exist between both streams. When a reduction in portal venous flow occurs, hepatic arterial flow rises and partially compensates for the reduction.[56]

Blood flow in the liver differs from that to all other organs. The difference may be ascribed to a highly ramified network of cells that are interspersed with blood spaces, the sinusoids.[57] Functionally, liver tissue is organized into units known as *acini*[58] that are centered on the portal venules that drain the blood. The cells in the acini are organized as one-cell-thick sheets and beams that in an overall sense are parallel to the lines of flow. Blood from the terminal portal venules and hepatic arterioles (bathing Zone I cells) pours into the sinusoids, bathing each side of single-cell sheets (approximately 25 µm thick) and leaving by the hepatic venules (bathing Zone III). The region between Zones I and III is called Zone II of the liver acinus (see Chapter 4, this volume).

The sinusoids are lined by filmy endothelial cells perforated by open fenestra of varying size, so that solutes freely enter the underlying space, Disse's space, which serves as the extracellular space of the liver. Disse's space is also continuous with the intracellular cleft. Moreover, the surface of liver cells underlying the sinusoidal lining is very large because of the area provided by microvilli that form the first structural barrier between blood and the hepatocyte. Because the depth of the space of Disse is small, diffusional equilibration in the lateral direction will be rapid and virtually instantaneous. But in the axial direction, the length of the sinusoid is such that the process may take longer. The mechanism responsible for carrying material from input to output is, therefore, that of flow and not of diffusion.[59]

The liver is the major organ for biotransformation of most endogenous compounds. It regulates intermediary metabolism and is primarily responsible for the metabolism of exogenously administered compounds. It is also an organ for sequential metabolism of biotransformed products. Certain xenobiotics and some of their metabolites are excreted into bile.

B. The Intestine

The contribution of intestinal mucosal metabolism to the overall metabolism of a xenobiotic is firmly acknowledged. For example, salicylamide[60] and isoproterenol[61] are conjugated by intestinal mucosal enzymes. But quantitative assessment of the contribution by the intestines during passage of a xenobiotic "across" the mucosa remains obscure. The difficulty lies in evaluating the effect of the bacterial flora in the gut lumen that plays a role in metabolizing foreign compounds.[62,63] The preponderance of bacteria differs with diet, age, and the presence of drugs, particularly antibiotics. Moreover, a variety of factors make interpretation of the role of this organ difficult to quantitate. Consideration must be given to the cellular makeup, the thickness of the mucosa, the occurrence of microvilli with their large surface area, and the distribution of intestinal blood flow, which differs greatly for each of the segmental divisions.[64] The location and distribution of enzymes in each segment, and their proximity to the lumen (the site of orally administered compounds) and to the capillary network (delivery into the circulation), are generally unknown. There is considerable evidence that the cytochrome P-450 system[65] and the

8. Fate of Xenobiotics

glutathione transferases[66] are located mainly in the epithelial lining of the villous tip, the most mature cells, rather than the crypt cells, whereas thiol S-methyltransferase[67] is more evenly distributed. Some of these enzymes, the cytochrome P-450 system,[68] and the glutathione transferases[66] are found considerably higher in the upper segments of the small intestines, whereas thiol S-methyltransferase is highest in the cecum.[67] Differential rates of metabolism of xenobiotics may result when differences exist for the accessibility of the enzymes to the delivery site, namely, gut lumen or intestinal blood.

Interpretation of *in vivo* data on intestinal metabolism is further complicated when a xenobiotic and its metabolites participate in the enterohepatic circulation. This process has been used to describe the passage of bile salts from the liver into the bile duct, which pours its contents into the upper duodenum, allowing reabsorption into the portal circulation. The release of bile into the duodenum, however, is discontinuous, as bile is usually stored in the gall bladder, which concentrates the contents before release. Xenobiotic conjugates that are secreted into bile may undergo a similar process. They may be reabsorbed directly in the small intestine or may undergo hydrolysis either by bacterial action or with intestinal mucosal enzymes to yield the parent compound for reabsorption. Enterohepatic circulation of xenobiotics acts as a mechanism whereby the duration of the xenobiotic in the body is prolonged (see Chapter 10, this volume).

The intestine as a secretory tissue has been documented for chlordecone,[69] sulfanilic acid,[70] and quaternary ammonium compounds.[71] The ability of the intestine to secrete xenobiotics into its lumen tends to provide an overestimation of the overall metabolizing capacity of the intestinal mucosa.

C. The Lung

The lung is unique in that it receives the entire cardiac output; all venous return passes through the lungs before the circulation is redistributed as arterial blood. The drug-metabolizing activity of the lung has not been considered to be as important as that of the liver, despite the similar specific activities that are present for the mixed-function oxidases in the two organs[72]; the total enzymatic activity is only 5% that of the liver because of the smaller mass of the lung. However, the blood flow rate to the lung is about four times that of the liver. On the basis of these reasons, clearance by the lung may contribute significantly to the overall clearance of some xenobiotics. Some enzymes in lung appear to be as potent as those in liver, for example, in biphenyl metabolism[73] and sulfation of phenol.[74] Such localization of enzymatic activities differs significantly among animal species; rabbit lung is found to be among the most enzymatically active organs.

D. Sequential Organs: The Intestine, the Liver, and the Lung

Anatomically, the intestine, the liver and the lung are viewed as three eliminating organs arranged in a sequential fashion; the venous flow from the first becomes the input flow to the second organ, and the venous flow from the second becomes the input flow to the third. By contrast, other eliminating organs, for example, the kidney and brain, receive arterial blood from the heart and may be treated as if arranged in a parallel fashion.

Clearance for organs that are arranged in sequence should not be summed. Rather, an effective clearance (Cl_{eff}) should be used instead to denote the overall efficiencies of these sequential organs in the removal of a xenobiotic. As an illustration, consider three sequential organs, I, II, and III (Fig. 6A), with respective clearance values of 30, 20, and 10 ml/min. Organ blood flow is 100 ml/min, and this flow is preserved through each compartment. The input concentration of I is arbitrarily set at 1 μg/ml. The extraction ratios for I, II, and III that result in the noted clearance values are 0.3, 0.2, and 0.1, respectively. The output concentration of I is 0.7 μg/ml and becomes the input concentration of II. Analogously, the output concentration of II is 0.56 μg/ml and becomes the input concentration of III; the output concentration of III is 0.504 μg/ml. The effective clearance, calculated as the effective rate of elimination (100 $\times$ [1 $-$ 0.504] μg/min) divided by the input concentration (1 μg/ml), is about 50 ml/min. This effective clearance is less than the sum of the clearances, 60 ml/min, by the three organs.

The effective clearance across the intestine, liver, and lung is more complex because the rate of blood flow through these organs is not a constant. The intestinal blood flow (Q_I) multiplied by the arterial concentration (C_A) is the rate of presentation of substrate to the intestine; the rate of exit of substrate is $Q_I C_{V,I}$, where $C_{V,I}$ is the venous concentration from the intestine. The substrate leaving the intestine at the rate $Q_I C_{V,I}$ becomes the presentation rate to the liver by the portal flow and is joined by the rate of presentation, $(Q_H - Q_I)C_A$, from the hepatic artery (hepatic arterial flow is $Q_H - Q_I$; Q_H is the total liver blood flow). Analogously, the rate at which a substrate leaves the liver, $Q_H C_{V,H}$, is joined by the rate of substrate presentation by all other venous return, $(Q_T - Q_H)C_{V,\text{Others}}$ (blood flow from all other organs is [$Q_T - Q_H$]; Q_T is total cardiac output);

8. Fate of Xenobiotics

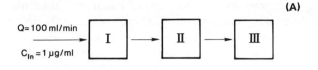

(A)

	I	II	III
CL(ml/min)	30	20	10
E	0.3	0.2	0.1
C_{Out} (µg/min)	0.7	0.56	0.504

$CL_T = (30+20+10)$ ml/min $= 60$ ml/min

$CL_{eff} = \dfrac{100(1-.504)}{1}$ ml/min $= 49.6$ ml/min

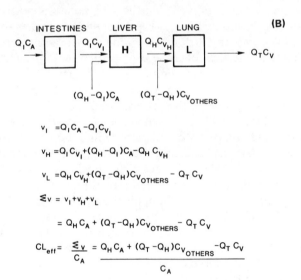

Fig. 6. Eliminating organs arranged in sequence. A simple illustration of three eliminating organs, I, II, and III, in sequence. Constant flow ($Q = 100$ ml/min) occurs through all organs; the input concentration to I is 1 g/ml. (A) The mathematical calculations demonstrate an overestimation of the effective clearance (Cl_{eff}, 49.6 ml/min) by the sum of the three organ clearances (60 ml/min). (B) The intestines, the liver, and the lung and the rates of presentation to the organ are described. The equations describe the rate of removal by the intestine (v_I), the liver (v_H), and the lung (v_L), and the effective removal rate (v) as well as the effective clearance by the organs in sequence.

the venous concentrations of substrate from liver, and from all organs other than liver, are denoted by $C_{V,H}$ and $C_{V,Others}$, respectively. The rate of removal by each organ, v_I, v_H, and v_L, and the effective rate of removal (Σv) are given in Fig. 6B. The effective clearance, in turn, can be calculated by the effective removal rate divided by the arterial concentration (Fig. 6B).

E. Effect of Route of Administration

The site of introduction of a foreign compound into the organism qualitatively governs the initial exposure of the xenobiotic to the several organs. One such direct mechanism of administration vascularly is by intravenous injection (Route A, Fig. 7); the entire amount is passed through the lungs before it is returned to the heart and distributed by arterial blood to the rest of the body. In contrast, an orally administered compound has to traverse the intestines, the liver, and the lung before it reaches the systemic circulation (Route B, Fig. 7). The amount that reaches the systemic circulation intact may be much less than the amount administered as it is dependent on the efficiency of these organs for xenobiotic removal. This presystemic removal of a compound that accompanies the extraarterial administration is known as the *first-pass* effect[75]; the phenomenon has been most commonly seen with oral administration. The intraperitoneal route (Route C, Fig. 7) is often chosen in animal studies on the assumption that the entire dose will be absorbed directly into the portal circulation to the liver. However, nonspecific absorption may occur from the capillaries lining the abdominal walls or directly by the intestinal wall.[76] Consequently, the entire dose is not absorbed directly into the liver; the extent of absorption will depend on the lipophilicity of the compound as well as on the exact site of injection into the abdomen.

Inhalation, intramuscular, and subcutaneous routes (Routes D–F, Fig. 7) will resemble those of intravenous administration, although the rate of delivery into the circulation may depend on the drainage of blood through the specific tissue; the circulation will bring the absorbed molecules into the lung before the compound is distributed into the arterial system. If the lung acts as an eliminating organ, first-pass elimination will occur.

By contrast, intraarterial (ia) administration delivers the entire dose to the site of sampling (Route G, Fig. 7). The concentration–time profile (or AUC^{ia}), therefore, serves as a reference point for the paraintraarterial (pia) routes of administration. When elimination remains linear for all organs, the systemic availability (F_{sys}) or the fraction that escapes the first-pass effect multiplied by the administered dose is the effective dose that reaches the systemic circulation intact. This amount equals the total

8. Fate of Xenobiotics

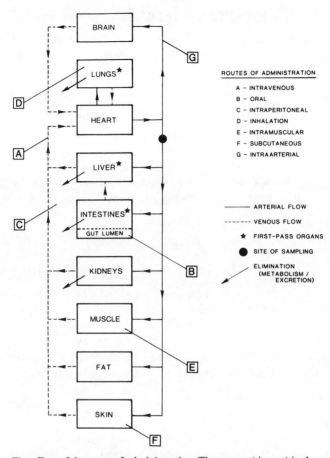

Fig. 7. The effect of the route of administration. The organs (tissues) in the organism are placed in anatomical perspective, interconnected by the circulation. The first-pass organ for intravenous, subcutaneous, intramuscular routes of administration is the lung; the first-pass organs by oral and intraperitoneal routes of administration are the intestine, the liver, and the lung.

body clearance multiplied by the area under the curve in the following relationship (Eq. 30):

$$F_{sys}Dose^{pia} = AUC^{pia}Cl_T \tag{30}$$

The relationship for ia administration is shown in Eq. (31),

$$Dose^{ia} = AUC^{ia}Cl_T \tag{31}$$

where F_{sys} is unity. Dividing Eq. (30) by Eq. (31) yields the following comparison.

$$F_{sys} = (AUC^{pia}/Dose^{pia})/(AUC^{ia}/Dose^{ia}) \qquad (32)$$

The systemic availability is a complex function of the availability of each first-pass organ. Organ availability F is related to the extraction ratio as in Eq. (33),

$$F = 1 - E \qquad (33)$$

and the overall availability (or systemic availability) is the product of the availability of all of the first-pass organs (Eq. 34).[77]

$$F_{sys} = F_{gut}F_{liver}F_{lung} \qquad (34)$$

The oral (po), intraportal (ipl), and intravenous (iv) administration of a compound will result in the following relationships.

$$F_{gut}F_{liver}F_{lung}Dose^{po} = AUC^{po}Cl_T \qquad (35a)$$

$$F_{liver}F_{lung}Dose^{ipl} = AUC^{ipl}Cl_T \qquad (35b)$$

$$F_{lung}Dose^{iv} = AUC^{iv}Cl_T \qquad (35c)$$

Under first-order conditions, a comparison of the AUC values of intravenous to intraarterial administration will provide the availability of the lung; intraportal to intravenous, the availability of the liver; and oral to intraportal, the availability of the gut. The availability of the gut is a complex function of the fraction metabolized by bacterial action, the fraction absorbed into the portal circulation, and the availability of the intestines.

VII. COMMENTS

The interactions between xenobiotics and body constituents, the physiological processes, and the sites of elimination have been outlined. This knowledge hopefully will aid in the understanding of biological events that occur *in vivo*. All of the physiological constraints and the dynamic forces interacting on a foreign compound must be assessed.

REFERENCES

1. Hansch, C., and Clayton, J. M. (1973). Lipophilic character and biological activity of drugs. II. The parabolic case. *J. Pharm. Sci.* **62,** 1–21.
2. Martin, A. N., Swarbrick J., Cammuratra, A., and Chun, A. H. C. (1969). "Physical Pharmacy," p. 237. Lea & Febiger, Philadelphia.
3. Brobeck, J. R. (1973). "Best and Taylor's Physiological Basis of Medical Practice," 9th ed. Williams & Wilkins, Baltimore, Maryland.

8. Fate of Xenobiotics

4. Meyer, M. C., and Guttman, D. E. (1968). The binding of drugs by plasma proteins. *J. Pharm. Sci.* **57,** 895–918.
5. Vallner, J. J., and Chen, L. (1977). Lipoproteins: Possible plasma transport proteins for basic drugs. *J. Pharm. Sci.* **66,** 420–423.
6. Scatchard, G. (1949). The attractions of proteins for small molecules and ions. *Ann. N.Y. Acad. Sci.* **51,** 660–672.
7. Goldstein, A. (1949). The interaction of drugs and plasma proteins. *Pharmacol. Rev.* **1,** 102–165.
8. Davison, C. (1972). Protein binding. *In* "Fundamentals of Drug Metabolism and Drug Disposition" (B. N. La Du, H. G. Mandel, and E. L. Ways, eds.), pp. 63–75. Williams & Wilkins, Baltimore, Maryland.
9. Martin, B. K. (1965). Potential effect of the plasma proteins on drug distribution. *Nature (London)* **207,** 274–276.
10. Levy, G., and Yacobi, A. (1974). Effect of protein binding on elimination of warfarin. *J. Pharm. Sci.* **63,** 805–806.
11. Gazzard, B. G., Ford-Hutchinson, A. W., Smith, M. I. H., and Williams, R. (1973). The binding of paracetamol to plasma proteins to man and pig. *J. Pharm. Pharmacol.* **25,** 964–967.
12. Hinderling, P. H., and Garrett, E. R. (1974). Protein binding and erythrocyte partitioning of disopyramide and its monodealkylated metabolite. *J. Pharm. Sci.* **63,** 1684–1690.
13. Wallace, S. M., and Riegelman, S. (1977). Uptake of acetazolamide by human erythrocytes *in vitro*. *J. Pharm. Sci.* **66,** 729–731.
14. Wilkinson, G. R., and Shand, D. G. (1975). Commentary. A physiological approach to hepatic drug clearance. *Clin. Pharmacol. Ther.* **18,** 377–390.
15. Pang, K. S., and Rowland, M. (1977). Hepatic clearance of drugs. I. Theoretical consideration of a "well-stirred" model and a "parallel tube" model. Influence of hepatic blood flow, plasma and blood cells binding, and the hepatocellular enzymatic activity on hepatic drug clearance. *J. Pharmacokinet. Biopharm.* **5,** 625–653.
16. McNamara, P. J., Levy, G., and Gibaldi, M. (1979). Effect of plasma protein and tissue binding on the time course of drug concentration in plasma. *J. Pharmacokinet. Biopharm.* **7,** 195–206.
17. Jusko, W. J., and Gretch, M. (1976). Plasma and tissue protein binding on drugs in pharmacokinetics. *Drug Metab. Rev.* **5,** 43–149.
18. Jakoby, W. B., and Habig, W. H. (1980). Guthione transferases. *In* "Enzymatic Basis of Detoxification" (W. B. Jakoby, ed.), Vol. 2, pp. 63–94. Academic Press, New York
19. Pall, D. P. (1972). Drug entry into brain and cerebrospinal fluid. *In* "Fundamentals of Drug Distribution" (B. N. La Du, H. G. Mandel, and E. L. Way, eds.), pp. 76–87. Williams & Wilkins, Baltimore, Maryland.
20. Price, H. L., Kovnat, P. J., Safer, J. N., Conner, E. H., and Price, M. L. (1960). The uptake of thiopental by body tissues and its relation to the duration of narcosis. *Clin. Pharmacol. Ther.* **1,** 16–22.
21. Winne, D. (1979). Influence of blood flow on intestinal absorption of drugs and nutrients. *Pharmac. Ther.* **6,** 333–393.
22. Wilkinson, G. R. (1975). Pharmacokinetics of drug disposition: Hemodynamic Considerations. *Pharmacol. Rev.* **27,** 11–27.
23. Brodie, B. B. (1952). Physiological disposition and chemical fate of thiobarbiturates in the body. *Fed. Proc., Fed. Am. Soc. Exp. Biol.* **11,** 632–639.
24. Chen, C. N., Coleman, D. L., Andrade, J. D., and Temple, A. R. (1978). Pharmacokinetic model for salicylates in cerebrospinal fluid, blood, organs and tissues. *J. Pharm. Sci.* **67,** 38–45.

25. Gonzalez, M. A., Tozer, T. N., and Chang, D. T. T. (1975). Non-linear tissue disposition of salicylic acid in rat brain. *J. Pharm. Sci.* **64**, 99–103.
26. Williams, R. T. (1959). "Detoxification Mechanism," 2nd ed. Chapman & Hall, London.
27. Jakoby, W. B., ed. (1980). "Enzymatic Basis of Detoxication." Academic Press, New York.
28. Jakoby, W. B., Bend, J., and Caldwell, J., eds. (1982). "Metabolic Basis of Detoxication." Academic Press, New York.
29. Grehánt, N. (1904). Physiologique des reins par le dosage de l'urée dans le sand et dans l'urine. *J. Physiol. Pathol. Gen.* **6**, 1–12.
30. Wagner, J. G., Northam, J. I., Always, C. D., and Carpenter, O. S. (1965). Blood levels of drug at the equilibrium state after multiple dosing. *Nature (London)* **201**, 1301–1302.
31. Gillette, J. R. (1982). Sequential organ first-pass effects: Simple methods for constructing compartmental pharmacokinetic models from physiological models of drug disposition by several organs. *J. Pharm. Sci.* **71**, 673–677.
32. Rowland, M. (1972). The influence of route of administration of drug availability. *J. Pharm. Sci.* **61**, 70–74.
33. Shand, D. G., Kornhauser, D. M., and Wilkinson, G. R. (1975). Effects of route of administration and blood flow on hepatic drug elimination. *J. Pharmacol. Exp. Ther.* **195**, 424–432.
34. Kornhauser, D. M., Wood, A. J. J., Vestal, R. E., and Wilkinson, G. R. (1978). Biological determinants of propranolol disposition in man. *Clin. Pharmacol. Ther.* **23**, 165–174.
35. Shand, D. G., Mitchell, J. R., and Oates, J. A. (1975). Pharmacokinetic drug interaction. *Handb. Exp. Pharmakol.* [N.S.] **28**, Part 3, 272–314.
36. Gillette, J. R. (1975). Overview of drug-protein binding. *Ann. N.Y. Acad. Sci.* **226**, 6–17.
37. Koelle, G. B. (1971). Neuromuscular blocking agents. *In* "The Pharmacological Basis of Therapeutics" (L. S. Goodman and A. Gilman, eds.), 4th ed., p. 614–616. Macmillan, New York.
38. Hallén, B., Sundwall, A., Elwin, C. E., and Nissman, (1979). Renal, kidney and intestinal clearance of a quaternary ammonium compound, emepronium in the dog. *Acta Pharmacol. Toxicol.* **44**, 43–59.
39. Gillette, J. R., and Pang, K. S. (1977). Theoretical aspects of pharmacokinetic drug interactions. *Clin. Pharmacol. Ther.* **22**, 623–639.
40. Pang, K. S., Koster, H., Halsema, I. C. M., Scholtens, E., and Mulder, G. J. (1981). Aberrant pharmacokinetics of harmol in the perfused rat liver: Sulfate and glucuronide conjugation. *J. Pharmacol. Exp. Ther.* **219**, 134–140.
41. Koster, H., Halsema, I. C. M., Scholtens, E., Pang, K. S., and Mulder, G. J. (1982). Kinetics of sulfation and glucuronidation of harmol in the perfused rat liver preparation. Disappearance of aberrancies in glucuronidation kinetics by inhibition of sulfation. *Biochem. Pharmacol.* **31**, 3025–3038.
42. Koster, H., and Mulder, G. J. (1982). Apparent aberrancy in the kinetics of intracellular metabolism of a substrate by two enzymes. An alternative explanation for anomalies in the kinetics of sulfation and glucuronidation. *Drug Metab. Dispos.* **10**, 330–335.
43. Segel, I. H. (1975). "Enzyme Kinetics." Wiley-Interscience, New York.
44. Roy, A. B. (1971). Sulfate conjugation enzymes. *Handb. Exp. Pharmakol.* [N.S.] **28**, Part 1, 526–563.
45. Gillette, J. R. (1968). Factors that affect the stimulation of the microsomal drug enzymes induced by foreign compounds. *Adv. Enzyme Regul.* **1**, 215–225.
46. Wattenberg, L. W., and Leong, J. L. (1962). Histochemical demonstration of reduced pyridine nucleotide dependent polycyclic hydrocarbon metabolizing systems. *J. Histochem. Cytochem.* **10**, 412–420.

47. Baron, J., Redick, J. A., and Guengerich, F. P. (1978). Immunohistochemical localization of cytochrome *P*-450 in the rat liver. *Life Sci.* **23**, 2627–2632.
48. De Baun, J. R., Smith, J. Y. R., Miller, E. C., and Miller, J. A. (1971). Reactivity *in vivo* of the carcinogen *N*-hydroxy-2-acetylaminofluorene: Increase by sulfate ion. *Science* **167**, 184–186.
49. Pang, K. S., and Terrell, J. A. (1981). Retrograde perfusion to probe the heterogeneous distribution of hepatic drug metabolizing enzymes in rats. *J. Pharmacol. Exp. Ther.* **216**, 339–346.
50. Pang, K. S., Waller, L., Horning, M. G., and Chan, K. K. (1982). Metabolite kinetics. Formation of acetaminophen from deuterated and non-deuterated phenacetin and acetanilide on acetaminophen sulfation kinetics in the perfused rat liver preparation. *J. Pharmacol. Exp. Ther.* **222**, 14–19.
51. Pang, K. S., and Gillette, J. R. (1979). Sequential first-pass elimination of a metabolite derived from a precursor. *J. Pharmacokinet. Biopharm.* **7**, 275–290.
52. Pang, K. S., and Gillette, J. R. (1978). Kinetics of metabolite formation and elimination in the perfused rat liver preparation. Differences between the elimination of preformed acetaminophen and acetaminophen formed from phenacetin. *J. Pharmacol. Exp. Ther.* **207**, 178–194.
53. Pritchard, J. B., and Janus, M. D. (1982). Metabolism and urinary excretion. *In* "Metabolic Basis of Detoxication" (W. B. Jakoby, ed.), pp. 339–357. Academic Press, New York.
54. Weiner, I. M., and Mudge, G. H. (1964). Renal tubular mechanisms for excretion of organic acids and bases. *Am. J. Med.* **36**, 743–762.
55. Smith, R. L. (1973). "The Excretory Function of Bile. The Elimination of Drugs and Toxic Substances in Bile." Chapman & Hall, London.
56. Lautt, W. W. (1980). Control of hepatic arterial blood flow: Independence from liver metabolic activity. *Am. J. Physiol.* **239**, H559–H564.
57. Brauer, R. (1963). Liver circulation and function. *Physiol. Rev.* **43**, 115–213.
58. Rappaport, A. M. (1979). Physioanatomical basis of toxic liver injury. *In* "Toxic Injury of the Liver" (E. Farber and M. M. Fisher, eds.), Part A, pp. 1–57. Dekker, New York.
59. Goresky, C. A. (1981). Tracer kinetics in the liver. *In* "Hepatic Circulation in Health and Disease" (W. W. Lautt, ed.), pp. 25–39. Raven Press, New York.
60. Barr, W. H., and Riegelman, S. (1970). Intestinal drug absorption and metabolism. Comparison of methods and models to study physiological factors *in vitro* and *in vivo* intestinal absorption. *J. Pharm. Sci.* **59**, 154–163.
61. Dollery, C. T., Davies, D. S., and Connally, M. B. (1971). Differences in the metabolism depend upon their routes of administration. *Ann. N.Y. Acad. Sci.* **179**, 108–112.
62. Scheline, R. R. (1973). Metabolism of foreign compounds by gastrointestinal organisms. *Pharmacol. Rev.* **25**, 451–523.
63. Goldman, P. (1982). The role of the microbial flora in detoxication. *In* "Metabolic Basis of Detoxication" (W. B. Jakoby, ed.), pp. 340–360. Academic Press, New York.
64. Svanik, J. (1973). Mucosal blood flow circulation and its influence on passive absorption in the small intestine. *Acta Physiol. Scand., Suppl.* **385**, 1–44.
65. Dawson, J. R., and Bridges, J. W. (1981). Intestinal microsomal drug metabolism. A comparison of rat and guinea pig enzymes, and of rat crypt and villous tip cell enzymes. *Biochem. Pharmacol.* **30**, 2415–2420.
66. Pinkus, L. M., Ketley, J. N., and Jakoby, W. B. (1977). The glutathione-*S*-transferase as a possible detoxication system of rat intestinal epithelium. *Biochem. Pharmacol.* **26**, 2359–2363.
67. Weisiger, R. A., Pinkus, L. M., and Jakoby, W. B. (1980). Thiol *S*-methyltransferase: Suggested role in detoxication of intestinal hydrogen sulfide. *Biochem. Pharmacol.* **29**, 2885–2887.

68. Hoensch, H., Woo, C. H., and Schmid, R. (1975). Cytochrome P-450 and drug metabolism in intestinal villous and crypt cells of rats: Effect of dietary iron. *Biochem. Biophys. Res. Commun.* **65**, 399–406.
69. Boylan, I. J., Cohn, W. J., Egle, J. L., Blanke, R. V., and Guzelain, P. S. (1978). Excretion of chlordecone by the gastrointestinal tract. Evidence of a nonbiliary mechanism. *Clin. Pharmacol. Ther.* **25**, 579–585.
70. Bredo, R., and Lauterbach, F. (1978). Intestinal secretion of sulfanilic acid by isolated mucosa of guinea pig ileum. *Acta Pharmacol. Toxicol.* **43**, 331–338.
71. Turnheim, K., and Lauterbach, F. (1980). Interaction between intestinal absorption and secretion of monoquaternary ammonium compounds in guinea pigs. A concept for the absorption kinetics of organic cations. *J. Pharmacol. Exp. Ther.* **212**, 418–424.
72. Litterest, C. L., Mamnaugh, E. C., Reagan, R. L., and Gram, T. E. (1975). Comparison of *in vitro* drug metabolism by lung, liver, and kidney in several common laboratory species. *Drug. Metab. Dispos.* **3**, 259–265.
73. Bend, J. R., Hook, G. E., Easterling, R. E., Gram, T. E., and Fouts, J. R. (1972). A comparative study of the hepatic and pulmonary mixed-function oxidase systems in the rabbit. *J. Pharmacol. Exp. Ther.* **183**, 206–217.
74. Cassidy, M. K., and Houston, J. B. (1980). *In vivo* assessment of extrahepatic conjugation metabolism in first-pass effects using the model compound phenol. *J. Pharm. Pharmac.* **32**, 57–59.
75. Gibaldi, M., Boyes, R. N., and Feldman, S. (1970). Influence of first-pass effect on availability of drugs. *J. Pharm. Sci.* **59**, 1338–1340.
76. Pang, K. S., and Gillette, J. R. (1979). Complications in the estimation of hepatic blood flow *in vivo* by pharmacokinetic parameters: The area under the curve after concomitant intravenous and intraperitoneal (intraportal) administration of acetaminophen in rat. *Drug Metab. Dispos.* **6**, 567–576.
77. Rowland, M. (1973). Effect of some physiologic factors on bioavailability of oral dosage form. *In* "Current Concepts in the Pharmaceutical Sciences. Dosage Form Design and Bioavailability" (J. Swarbrick, ed.), pp. 181–222. Lea & Febiger, Philadelphia.

CHAPTER 9

Excretion Mechanisms

Walter G. Levine

I.	Introduction	251
II.	Renal Handling of Organic Anions and Cations	252
III.	Biliary Excretion of Xenobiotics	254
	A. Physiology of Bile Formation	255
	B. Characteristics of Compounds Excreted in the Bile	257
	C. Hepatic Uptake versus Biliary Transport	259
	D. Intracellular Binding	260
	E. Metabolism	261
	F. Effects of Other Xenobiotics	263
	G. Effects of Bile Salts	266
	H. Miscellaneous Factors	267
	I. Enterohepatic Circulation	268
	J. Regeneration	269
IV.	Salivary Excretion	270
V.	Excretion into Milk	271
VI.	Excretion into Expired Air	272
VII.	Comments	273
	References	274

I. INTRODUCTION

The body is provided with several means to rid itself of the myriad foreign chemicals to which it is exposed. Interruption of any of the major routes of excretion appreciably increases the toxicity of most xenobiotics. Quantitatively, the kidney is the most important excretory organ, and the vast majority of xenobiotics and/or their metabolites appear in the urine to some extent. The bile ranks next in importance and, in addition to an excretory pathway, provides a means of retaining drugs through the enterohepatic circulation. Other routes of excretion include expired air, sweat, saliva, feces, and milk. Although these are generally of less quantitative importance than urine and bile, they become consequential for specific types of compounds and circumstances. Gases and vapors, including most anesthetic agents, are excreted by the lungs. The finding that a number of drugs and metabolites appear in expired air and in saliva has given rise to a means of assessing metabolism and plasma levels of drugs in humans by simple, noninvasive procedures. The secretion of drugs and potentially toxic chemicals into the milk is of great concern at a time when breast feeding of infants is again finding increasing acceptance. Limitations of space preclude extensive coverage of all routes and mechanisms of xenobiotic excretion. Biliary excretion will be the primary focus of this chapter, and renal mechanisms will be considered to a lesser extent. Other routes are of necessity relegated to a few brief paragraphs. The reader is cautioned to regard this chapter as a source of information and not as an evaluation of the relative pharmacological or physiological importance of the several routes of excretion.

II. RENAL HANDLING OF ORGANIC ANIONS AND CATIONS

Most xenobiotics or their metabolites are weak organic anions or cations. Mechanisms of renal excretion of xenobiotics have been reviewed by Pritchard and James[1] and by Irish and Grantham,[2] and a more pharmacokinetic analysis of renal processes has been presented by Garrett.[3] Renal excretion will be considered briefly here. The process of urine formation and drug excretion begins with the passage of blood through the glomerulus to yield a tubular fluid that is essentially an ultrafiltrate of plasma. Compounds with a molecular weight of approximately 5000 or less are readily filtered by the glomerulus. This would include most xenobiotics and their metabolites. The driving force for filtration is the hydrostatic pressure of the blood, and formation of tubular fluid is sensitive to changes in blood flow and pressure. Approximately 180 liters of filtrate is produced per day in the average human. All but 1.5 liters is reabsorbed by the tubules. Glomerular filtration of xenobiotics not bound to plasma proteins, although a passive process, is nearly 100% efficient. Protein binding is a major limitation in renal excretion of xenobiotics because it diminishes the free drug concentration gradient across the capillary membrane within the glomerulus. Consequently, drugs such as warfarin, which are extensively bound to plasma proteins, have long half-lives. The rate of blood flow to the kidney and the volume of distribution also influence the amount of xenobiotic that appears in the glomerular filtrate. Both of these factors may be appreciably altered in disease states and thus affect the elimination of drugs and other chemicals. Glomerular filtration rate is commonly measured by clearance of creatinine, an endogenously formed amino acid derived from creatine phosphate of muscles. Although a small amount of tubular secretion occurs, the method is suitable for obtaining a first approximation of glomerular filtration rate.

Xenobiotics also enter the tubular fluid by active transport across tubular cells. The mechanisms involved are identical to those that actively transport physiological organic anions and cations. Active tubular transport of organic anions and cations is relatively unhampered by plasma protein binding because the process is very rapid and the rate of dissociation from plasma protein binding is typically measured in milliseconds and, thus, is negligible. A case in point is the organic anion, penicillin G, 65% of which is bound to plasma proteins. However, its plasma half-life is only 30 min because of active tubular transport. The half-life of penicillin G can be significantly extended by administration of another organic anion, probenecid, which successfully competes for the tubular anion transport mechanism. Methotrexate excretion is inhibited by salicylates that compete for tubular transport. This important drug interaction may lead to toxic blood levels of the antitumor agent. Other inhibitors of active transport such as 2,4-dinitrophenol and ouabain decrease accumulation of organic anions in tubular cells. According to current theory, organic anions are actively transported into tubular cells and then move passively into the lumen down an electrochemical gradient. Facilitated diffusion may also be involved. Among the many substances secreted by the tubular organic anion system are bile salts, cyclic AMP, hippurate, hydroxyindoleacetic acid, prostaglandins, urate, acetazolamide, cephalothin, furosemide, penicillin G, probenecid, and saccharin.

An analogous tubular system is responsible for the transport of organic cations, mainly amines carrying a positive charge within the physiological range of pH. Although not as extensively studied as the anionic system, organic cationic transport appears to follow a similar mechanism. Competitive inhibition by transported cations has been demonstrated, although saturation of the putative carrier has been difficult to demonstrate in view of the toxicity of high concentrations of most organic cations. Metabolism of some organic cations by tubular cells adds an interesting aspect to their excretion. For example, morphine is taken up by the cells in cationic form but conjugated to sulfate and excreted as the anion. The anion inhibitor, probenecid, does not block renal excretion of basic drugs that are sulfated by the kidney but does block excretion of the preformed sulfate, supporting the concept of active transport of organic anions only in the uptake phase of tubular transport. The cationic and anionic systems are generally considered to be independent of each other, although exceptions may exist. Substances transported by the organic cation system include acetylcholine, creatinine, dopamine, epinephrine, 5-hydroxytryptamine, histamine, atropine, hexamethonium, morphine, neostigmine, paraquat, quinine, and tetraethylammonium.

Both anionic and cationic systems are deficient in the neonate. This is probably because of the efficiency of the mother's kidney and, consequently, the lack of exposure to organic anions and cations during the fetal state. Maturation of these systems begins shortly after birth as the kidney in the neonate assumes responsibility for dealing with xenobiotics.

Passage into the tubular fluid by glomerular filtration may be followed by passive transport through tubular membranes back into the blood. The degree of reabsorption is dependent on urine flow, is proportional to the lipid solubility of the drug, and is influenced by binding to plasma proteins that will again occur when the drug reenters the blood. Highly lipid-soluble compounds are readily reabsorbed, and there is little or no urinary excretion despite nearly complete glomerular filtration. For drugs with a pK near the pH of urine, the rate of excretion can be increased or decreased by manipulation of urinary pH. Phenobarbital, a weak organic anion, is slowly excreted unchanged in the urine. Alkalinizing the urine increases the charge on the molecular, resulting in decreased tubular reabsorption and increased total excretion. Diuretics can alter passive reabsorption of other drugs by increasing urinary flow, thus decreasing exposure time to luminal membranes, and by changing urinary pH.

Most xenobiotic metabolism is considered to be conducive to urinary excretion in view of the conversion of relatively lipophilic compounds to highly polar, charged metabolites. This reduces the likelihood of passive tubular reabsorption and enhances the possibility of active transport by the organic cationic and anionic systems. The relation between metabolism and urinary excretion has been reviewed in this series by Pritchard and James.[1] The initial metabolism of many, if not most, xenobiotics is catalyzed by enzyme systems containing cytochromes P-450 situated within the hepatic endoplasmic reticulum. Although the resulting metabolites are generally more polar than the parent compounds, conjugation, the second metabolic step, makes the major contribution to increased polarity and charge. In particular, glucuronide and sulfate conjugates are strong anions, and many drugs appear in the urine in this form.

III. BILIARY EXCRETION OF XENOBIOTICS

Although fewer drugs are excreted by humans in the bile than in the urine, it is an important route for many compounds, and its function in the enterohepatic circulation is responsible for the persistence of many xenobiotics within the body. Lower animals excrete drugs in the bile to a greater extent; compounds with molecular weights as low as 200 are excreted in rat bile. Our understanding of biliary excretion is somewhat less than that of urinary excretion because our knowledge of biliary physiology is relatively sketchy. Several articles[4-8] review current concepts of formation and flow of bile that will be considered briefly here.

A. Physiology of Bile Formation

1. Structural Considerations

Because of the central direction of blood flow within the lobule of the liver, peripheral (periportal) hepatocytes are exposed to higher concentrations of blood solutes than are central hepatocytes. Periportal bile cancliculi are larger, suggesting greater transport activity.[6] However, anastomosing networks of sinusoids and branches of the hepatic artery emptying intermittently along the sinusoids suggest that the rigid concept of portal to central gradient within the lobule should be more flexible (see Chapter 10, this volume).

The canaliculus, a morphologically and functionally unique area of the hepatocyte membrane, is characterized by many microvilli, indicative of an extensive transport system. Because of the tight junction between hepatocytes, it had been believed until recently that under normal conditions essentially all blood-to-bile transport is transcellular. Under conditions of continuous bile salt infusion, paracellular movement of water may occur in association with the choleretic response.[9] In addition, inulin and sucrose, which equilibrate slowly with cellular water, equilibrate quite rapidly with bile, suggesting paracellular transport. Uncharged solutes may undergo paracellular transport more readily than anionic solutes.[10]

Intracellular events associated with bile formation and biliary transport are not well defined. The Golgi area may participate in biliary secretion. Normally, the periportal hepatocytes contain considerably more Golgi membranes than do centrilobular cells, possibly reflecting their greater biliary activity. Bile salt-induced choleresis leads to an increase in the size of pericanalicular Golgi membranes but few other organelle changes, supporting a role in the secretion process.[5] Chronic treatment with phenobarbital leads to choleresis and a profound proliferation of the smooth endoplasmic reticulum. However, no role in biliary excretion other than metabolism can as yet be assigned to these membranes. Accumulation within the lysosomes is seen for some basic drugs, such as d-tubocurarine, and the bound drug appears not to be readily available for biliary excretion.[11]

Acid hydrolases appear in the bile,[12] probably derived from lysosomal vesicles, although the significance of this organelle in biliary excretion is not clear. Electron microscopy reveals a system of microfilaments in the pericanalicular area of the hepatocyte.[13] These may influence biliary transport by contributing to the contraction and relaxation of the canalicular microvilli and to intracytoplasmic migration of vesicles in the pericanalicular region. The conclusion is supported by the observation that phalloidin, an accelerator of actin polymerization into microfilaments, and norethandrolone, an inducer of marked alterations in the pericanalicular microfilaments, will each suppress bile flow.[14,15]

Biochemical mechanisms in bile formation have received relatively little attention. Cellular levels of cAMP may influence bile formation because theophylline, an inhibitor of phosphodiesterase, and glucagon, which stimulates adenyl cyclase, in turn stimulate bile salt–independent flow in dogs.[16] However, there is no parallel between bile flow and cellular levels of cAMP in dogs and rats, casting doubt on a causal relationship.[17] Similarly, no increased biliary cAMP is seen during taurocholate-induced choleresis in the dog.[18]

The contribution of the ductule system to bile formation is poorly defined. In species other than rabbit and rat, secretin is a potent choleretic and increases bicarbonate and chloride excretion, probably at the ductule level.[4] However, certain work suggests a canalicular origin as well for bicarbonate.[19] Reabsorption of water probably occurs in the ductule, as demonstrated directly by retrograde instillation of ^3H_2O into the rat biliary tree.[20] Several xenobiotics are also absorbed by the ductules, indicating bidirectional exchange of solutes within the biliary tree. Thus, the canaliculus–ductule system appears to be functionally analogous to the nephron in that primary bile is formed at the proximal end, whereas compositional change occurs as the bile flows through the ductular system. Molecules as large as inulin and albumin are apparently reabsorbed through the biliary tree,[21] giving rise to speculation that the protein normally found in the bile[22] may also undergo bidirectional flow within the biliary tree; however, direct evidence is lacking. Even a highly charged compound such as phenolphthalein glucuronide appears in the plasma after retrograde injection into the bile duct.[23] Thus, permeability to macromolecules and to highly charged metabolites of xenobiotics has been demonstrated, although it has yet to be proven whether the canaliculi or the ductules, or both, are involved.

2. Bile Salt–Dependent Bile Formation

Bile secretion, unlike that of the urine, is relatively independent of hydrostatic pressure. A primary step in bile formation is the transport of bile acids across the canalicular membrane, resulting in an osmotic

9. Excretion Mechanisms

movement of water into the canalicular lumen.[24] The relationship between bile flow and the secretion of bile salts is referred to as *bile salt–dependent bile flow*. Numerous other organic anions, such as sulfobromophthalein (BSP) and iodipamide exert a similar osmotic effect. Bile salts are normally present as micelles within the bile, and consequently their osmotic activity is somewhat lower than would be predicted on a molar basis. Therefore, bile salts that do not form micelles cause a proportionally greater increase in bile flow than do those that form micelles.[25] Counter ions transported with bile salts may make a major contribution to the osmotic gradient. Nonosmotic factors may also play a role in bile salt–dependent flow.[8]

3. Bile Salt–Independent Bile Formation

From analysis of bile flow at varying levels of bile salt infusion, it has been calculated that bile flow can occur in the absence of bile salts, that is, the *bile salt–independent flow*. Quantitation of this flow is difficult because it assumes a straight-line relationship between bile salt secretion and bile flow, even at a low concentration of bile salts, that is, below that critical for micelle formation. Certain studies[26,27] reveal that this relationship may be curvilinear at low concentrations of bile salts, casting doubt on the conventional procedure for determination of bile salt–independent flow. Nevertheless, nearly normal bile flow is maintained in the rat in the virtual absence of bile salts after prolonged bile drainage[28] and in the isolated perfused rat liver.[29] The mechanism of the bile salt–independent flow, although as yet unsettled, revolves around active transport of sodium associated with (Na^+, K^+)-ATPase activity. This theory presumes that (Na^+, K^+)-ATPase present in the canalicular membrane generates a gradient across the membrane by means of the sodium pump. Induction of synthesis of (Na^+, K^+)-ATPase and of bile salt–independent flow by phenobarbital are both blocked by inhibitors of protein synthesis.[30] However, ouabain and ethacrynic acid, each of which inhibits (Na^+, K^+)-ATPase and therefore should suppress bile salt–independent flow, actually increase flow under certain circumstances,[31,32] possibly because of the osmotic effect of their own excretion. Doubt has been cast on the existence of (Na^+, K^+)-ATPase in the canalicular membranes,[6] and highly purified membrane preparations show no enrichment of the enzyme compared to liver homogenates (M. Inouye, personal communication).

B. Characteristics of Compounds Excreted in the Bile

Over 200 xenobiotics or their metabolites have been detected in the bile.[33] These include nearly every class of pharmacological agent as well as numerous environmental, carcinogenic, and otherwise toxic chemicals. Many metals, toxic as well as physiological, are excreted in the bile. Excretion has been studied in rodents and other common laboratory animals, fish, several domesticated and wild animals, and humans, but there is considerable variation among species.

Brauer[34] originally classified compounds based on their bile/plasma concentration ratios. Class A substances have a ratio of approximately 1.0 and include sodium, potassium, chloride, and glucose. Class B substances have a ratio considerably greater than 1.0 and include bile salts, bilirubin, a number of dyes such as iodipamide and BSP, and many xenobiotics. Class C substances are relatively excluded from the bile, and the bile/plasma ratio is less than 1.0. This last group includes many macromolecules such as phospholipids and proteins. For xenobiotics readily excreted in the bile (class B), a distinction is made between organic anions, cations, and noncharged compounds. Each group is considered to be transported into bile by an independent mechanism, although a number of subgroups and exceptions to the theory are seen. Bile salts are transported by mechanisms independent of that for BSP and apparently other organic anions as well. Biliary excretion of organic cations, although proceeding by an apparently independent mechanism, may also show a multiplicity of pathways.[35] Partially independent pathways for mono- and diquaternary amines have been demonstrated for procainamide ethobromide and d-tubocurarine.[36,37] Even two bile acids, dehydrocholate and taurocholate, may not entirely share the same pathway.[38] Another intragroup distinction is seen in the biliary excretion of the anions morphine sulfate and morphine glucuronide; phenobarbital treatment enhances excretion of the former but not the latter.[39] In the rat, BSP may be transported by two pathways, only one of which is shared with bilirubin.[40] After treatment with nafenopin, the isolated perfused rat liver takes up BSP rather poorly, whereas bilirubin transport is unaffected.[41] The transport mechanism for uncharged compounds is typically studied using ouabain, a cardiac glycoside. It has a steroidlike structure, and a number of steroid hormones inhibit its uptake into liver slices as well as its secretion in the bile. A distinct pathway for noncharged compounds has been questioned in view of findings that the organic anions, BSP and phenolphthalein, inhibit the uptake of ouabain into liver slices and into the bile.[42,43] It is apparent that the question of independent hepatic mechanisms for transport of anions, cations, and noncharged compounds has yet to be resolved.

A molecular weight threshold exists for biliary organic anions that varies considerably among species. The minimum molecular weight is 325, 400, 475, and 500–700 for rats, guinea pigs, rabbits, and humans, respectively.[44] These figures are only approximate, and there is considerable latitude within each species. In general, lower molecular weight compounds are excreted in the urine. For organic cations, the threshold is 200 in the rat and appears to exhibit little or no species variation.[45] For compounds well below the mean threshold, renal ligation does not lead to a greater biliary excretion. Similarly, for compounds well above the threshold, bile duct ligation does not increase urinary excretion. In contrast, a group of compounds of intermediate molecular weight (325–465) are excreted in both urine and bile of rats. For this last group, occlusion of one route of excretion leads to compensatory excretion in the other.[46] The mechanism for such a distinction based on molecular weight remains unknown.

C. Hepatic Uptake versus Biliary Transport

Biliary excretion is commonly studied from the viewpoint of movement from plasma to bile, overlooking the distinction between hepatic uptake and canalicular transport. Plasma disappearance rates are often equated with hepatic uptake, neglecting extrahepatic distribution. The latter problem can be overcome using the isolated perfused liver. Studies on xenobiotic uptake involved the use of isolated hepatocytes.[47–50] In many cases, the data obtained compare well with those from perfused liver. However, the structural and functional polarity of the hepatocyte *in vivo* is lost during isolation, and exposure of both sinusoidal and canalicular faces to the medium occurs in suspensions. Therefore, it is impossible to determine if release of a compound into the medium is analogous to efflux into the plasma or secretion into the canaliculus *in vivo*. Consequently, uptake studies must be carried out over a short period to minimize release. The ratio of concentration within the cell to that in the medium is often greater than 1.0, implying active transport. Uptake of taurocholate depends on active (Na^+, K^+)-ATPase and a transmembranal Na^+ gradient, whereas transport of BSP, indocyanine green, and ouabain does not. This correlation suggests different mechanisms and confirms findings *in vivo* and in the perfused liver that indicate that bile salts are taken up by mechanisms differing from those of most xenobiotics.

Despite the limitation mentioned, taurocholate secretion from isolated hepatocytes has been studied by incubating cells with bile salts and subsequently measuring the efflux in fresh medium.[51] Transport is carrier and energy dependent. Secretion of intracellularly formed glutathione conjugate of BSP[49] and of 5-methyl tetrahydrofolate[48] has also been reported.

To distinguish between hepatic uptake and canalicular transport, use was made of probenecid, an inhibitor of renal transport of organic anions.

It was found to depress hepatic uptake of ouabain *in vivo* and in liver slices[51a] but did not effect biliary excretion. Nafenopin, in contrast, inhibits the biliary excretion of bilirubin *in vivo*,[52] but not its hepatic uptake in the isolated perfused liver.[41] Bucolome, an anti-inflammatory drug, enhances the biliary excretion of ouabain but inhibits its disappearance in plasma,[53] although clearance of BSP and bilirubin is enhanced. The selective response to these drugs undoubtedly reflects multiple transport mechanisms within the hepatocyte plasma membrane that are far more complex than the simple distinction between cations, anions, and noncharged molecules. The independence or interdependence of these systems requires additional study.

D. Intracellular Binding

For bile salts, bilirubin, and most xenobiotics, uptake into the liver is followed by intracellular binding prior to metabolism and excretion. A major cytosolic binding protein is ligandin.[54] It normally comprises about 5% of hepatic cytosolic protein and may increase to 10–15% after treatment with phenobarbital. It binds bilirubin, glutathione, and a number of xenobiotics[54] and was shown to be identical to an azo dye binding protein[55] and a cortisol metabolite binding protein.[56] Rat liver ligandin is identical to glutathione S-transferase B, one of several such transferases found within the cytosol.[57] Indeed, all of the glutathione transferases appear to have similar binding activity and constitute a family of ligandins.[58] Treatment of rats with phenobarbital, DDT, dieldrin, or pregnenolone 16α-carbonitrile, inducers of the hepatic mixed-function oxidase system, increases hepatic but not renal levels of ligandin. Treatment with tetrachlorodibenzo-p-dioxin (TCDD) induces both hepatic and renal ligandin.[59] Phenobarbital also enhances uptake of bilirubin and BSP through decreased efflux.[60] Such direct evidence is not available for other xenobiotics, although binding constants for a number of compounds are similar to that for bilirubin, suggesting a comparable mechanism.[61,62] Elasmobranchs (sharks and skates) are virtually devoid of hepatic ligandin, although these species are able to concentrate and excrete organic anions.[63] This suggests that alternative mechanisms for hepatic transport are operative in elasmobranchs.

A second cytosolic binding protein Z (aminoazo dye–binding protein A), also binds a diverse group of xenobiotics. There is little evidence regarding its role in hepatic transport. Treatment with nafenopin markedly increases hepatic concentration of Z[64] and leads to decreased uptake of BSP in the perfused liver, whereas bilirubin uptake is unaffected.[41] The Z protein also binds fatty acids,[65] although the significance of this is not clear. A third type of cytosolic binding proteins are those that nonenzymatically and covalently bind a large number of highly reactive electrophilic compounds. This group may simply represent a general pool of proteins, although those with hydrophobic binding sites appear to be the most susceptible. No function in hepatobiliary transport has been ascribed to this group.

Bile contains more than 12 proteins,[66] but their role, if any, in the binding and transport of xenobiotics has received little attention. Indocyanine green is bound approximately 50% to biliary proteins in guinea pigs,[67] but the contribution to dye excretion is unknown. Differential binding of metabolites of N,N-dimethyl-4-aminoazobenzene (DAB) to fractions of biliary proteins has been demonstrated (A. R. Samuels, M. M. Bhargava, and W. G. Levine, unpublished experiments). The specificity of this binding suggests a possible function for certain biliary proteins in the biliary excretion of DAB metabolites.

E. Metabolism

Most xenobiotics are converted to metabolites prior to excretion in bile or urine. The enzymatic and chemical aspects of biotransformation have been described in this series of publications.[68,69] To summarize briefly, metabolism generally may be divided into two major steps. In the first phase, a number of oxidative enzymes catalyze hydroxylation, dealkylation, sulfoxidation, aromatization, or nitro and azo reduction reactions. Following this are several synthetic reactions that result in the formation of one of a number of conjugates (e.g., glucuronides and sulfates) or glutathione thioethers, among others. The appropriate enzymes are found in the microsomal and cytosolic fractions of the liver. The latter group of reactions increases greatly the polarity of the compound. This is conducive to rapid excretion, and the significant increase in molecular weight facilitates biliary excretion.

The importance of biotransformation in biliary excretion is well illustrated by the Gunn rat, which is devoid of bilirubin-UDPglucuronyltransferase. Being unable to excrete bilirubin, the animal remains jaundiced throughout its lifetime. Induction of drug-metabolizing enzymes frequently leads to increased biliary excretion of xenobiotics. This has been demonstrated for a number of inducing agents and xenobiotics,[33] although the significance of such studies is sometimes difficult to assess. Phenobarbital induces both types of enzymes[70] but also increases liver size, hepatic blood flow,[71,72] and bile salt–independent bile flow[73,74] unrelated to its effects on cytochrome P-450. Because clearance of compounds with high hepatic extraction is dependent on blood flow, enhanced biliary excretion after phenobarbital might in some cases be attributable mainly to increased hepatic blood flow. In at least one instance, phenobarbital paradoxically induced glucuronidation of morphine *in vitro* but depressed excretion of the conjugate *in vivo*,[75] possibly by inhibition of biliary transport. Other inducing agents, spironolactone and pregnenolone 16α-carbonitrile, also increase bile flow.[76] In contrast, polycyclic hydrocarbons do not increase liver weight or bile flow in the short period of time required for significant enzyme induction, usually less than 24 hr.

In the absence of enhanced bile flow or increased liver size, an increase in the rate of excretion of a metabolized xenobiotic after pretreatment with these agents leads to the conclusion that metabolism is the rate-limiting step in the hepatobiliary fate of that particular xenobiotic. This is seen for methadone[77] and a number of chemical carcinogens.[78-82] For example, benzo[*a*]pyrene, which appears in the bile entirely in the form of its metabolites, is excreted more rapidly after induction with 3-methylcholanthrene and with phenobarbital even when increased liver size and bile flow are taken into account.[78] Conversely, inhibitors of the oxidase system depress the rate of biliary excretion of benzo[*a*]pyrene metabolites. Sex differences for benzopyrene metabolism in rats are also reflected in rates of biliary excretion. Similar observations have been made with the carcinogens, 3-methylcholanthrene, 7,12-dimethylbenzanthracene, 2-acetylaminofluorene, and *N,N*-dimethyl-4-aminoazobenzene.[78-82]

Phenobarbital has been shown to increase the rate of plasma disappearance and biliary excretion of two nonmetabolized compounds, phenol-3,6-dibromphthalein and indocyanine green.[83] The barbiturate also increases biliary excretion of bilirubin[84] and BSP,[85] both of which are conjugated prior to excretion. Both enhanced formation and transport of the conjugate are involved. In such studies, metabolism is considered the rate-limiting step in the excretion of the xenobiotic. However, an investigation of the biliary excretion of phenolphthalein, 3-methylumbelliferone, and 8-hydroxyquinoline, each of which appears in the bile as its glucuronide, revealed that canalicular transport and not metabolism was the rate-limiting process.[86] In general, conjugation reactions are carried out far more rapidly than are the initial oxidative steps. Therefore, it is less likely that compounds requiring only conjugation would be limited by metabolism in their hepatobiliary excretion. Generalizations concerning rate-limiting steps for xenobiotic excretion must take into account whether initial or conjugation phases, or both, are involved, as well as the nonmetabolic effects of inducing agents. A case in point is the effect of nafenopin, which has no effect on metabolism but inhibits biliary transport of xenobiotics. Oxidative metabolism is rate limiting in the excretion of

9. Excretion Mechanisms

benzo[a]pyrene, and nafenopin, predictably, does not alter the rate of biliary excretion. However, if metabolism is markedly induced with 3-methylcholanthrene or is entirely bypassed through injection of preformed metabolites, canalicular transport becomes rate limiting and inhibition by nafenopin is seen.[87]

It follows from this explanation that glutathione conjugation, as for example with BSP, would not be rate limiting in excretion of the dye. However, depletion of hepatic glutathione by diethylmaleate or methyliodide depresses dye excretion,[88,89] which is attributable to the loss of substrate (glutathione) for the transferase. The azo dye carcinogen N,N-dimethyl-4-aminoazobenzene undergoes N-demethylation and ring hydroxylation and is excreted in the bile as a mixture of metabolites, including conjugates of glutathione, glucuronic acid, and sulfate.[90,91] Depletion of hepatic glutathione specifically depresses N-demethylation, whereas ring hydroxylation is relatively unaffected.[92] N-demethylation probably is critical in the excretion of the azo dye, and depressed excretion is seen after glutathione depletion. Some of the inhibition of excretion is probably attributable to depressed formation of glutathione conjugate, although the total mechanism of this interaction has yet to be elucidated. A direct toxic effect of depleting agents has been ruled out because the effect is readily reversed upon restoration of hepatic glutathione and very high concentrations of these agents are required to inhibit N-demethylation *in vitro*. Specific inhibition of cytochrome P-450 may be a consequence of glutathione depletion, possibly mediated through heme oxygenase and lipid peroxidation. Both are increased after glutathione depletion,[93,94] and each could lead to loss of cytochrome P-450 activity. In particular, the cytochrome is sensitive to effects of endogenous and exogenous lipid peroxides. Whether the loss of N-demethylating activity after glutathione depletion is specifically attributable to this mechanism deserves further investigation.

F. Effects of Other Xenobiotics

Many xenobiotics directly or indirectly influence hepatobiliary excretion. Some of these compete for transport based on similarity of charge. Others induce or inhibit drug-metabolizing enzymes. Still others are hepatotrophic and increase the functional size of the liver, and some are choleretic. A xenobiotic may affect its own excretion. BSP is commonly used to assess hepatobiliary function but is toxic in high doses. During infusion for measurement of maximum transport, decreased bile flow and dye excretion have been seen.[95,96] Liver ATP falls precipitously,[97] possibly through an action of the dye on the mitochondrial membrane.[98] The glutathione conjugate of BSP does not show similar toxicity. In view of these effects, the use of BSP for determination of the transport capacity of the liver is open to question.[96]

Administration of the hypolipidemic compound, nafenopin, increases liver size and evokes a marked choleresis accompanied by inhibition of biliary excretion of organic anions and uncharged xenobiotics.[52,99,100] Transport of organic cations is not affected, indicating a selectivity in effect and not simply general hepatotoxicity. The effect is seen *in vivo* and in the isolated perfused liver. Unlike most hepatotrophic chemicals, nafenopin does not induce drug metabolism,[52] and there is no effect on excretion where metabolism is rate limiting. In contrast to other organic anions, chlorothiazide[101] and harmol sulfate[102] are excreted more rapidly after nafenopin pretreatment, probably in association with increased bile flow. This provides further evidence for the multiplicity of anion transport mechanisms in the bile. Studies show that nafenopin blocks both the uptake[41] and the biliary excretion[52] of BSP; bilirubin uptake is not affected, although its excretion is depressed. Apparently, nafenopin inhibits the transport of highly charged compounds. Consequently, there is no influence on the hepatic uptake of bilirubin, which is essentially uncharged at physiological pH but is excreted into the bile as the highly charged glucuronide. The choleretic effect of nafenopin is bile salt–independent but otherwise not entirely explained. The increased flow is of canalicular origin and requires pretreatment with the drug as well as the presence of the drug and/or its metabolites in the bile.[100] Thus, a change in the functional state of the liver and a possible osmotic effect may contribute to the choleretic response.

Many other xenobiotic and physiological compounds influence hepatobiliary function.[33] Most, although not all, of the effects are detrimental. The substances include carbon tetrachloride, the pesticide kepone, the toxic chemicals, TCDD and polychlorinated biphenyl (PCB), organic forms of certain metals, and others. Such commonly used drugs as diazepam and acetaminophen and such hormones as aldosterone, hydrocortisone, and estrone also affect biliary function, at least in experimental animals. Extrapolation of these results to humans is difficult, because relatively few biliary studies of this sort have been carried out in humans. Pregnancy and chronic administration of estrogens are accompanied by decreased bile flow and depressed biliary excretion of a number of substances.[103,104] Sensitive women may become jaundiced with the use of oral contraceptives or during the last trimester of pregnancy. Some of this response may be related to the known inhibitory effect of estrogens on drug-metabolizing enzymes. However, 17β-estradiolglucuronide, ordinarily considered a detoxication product, inhibits bile flow in rats.[105] Other

9. Excretion Mechanisms

estrogens and their conjugates are not inhibitory. Whether this is responsible for estrogen-induced depression of biliary function in women remains an open question.

Ethinyl estradiol suppresses bile salt–independent bile flow,[106] biliary excretion of bile acids,[107] bilirubin,[107] and BSP,[108] as well as hepatic (Na^+, K^+)-ATPase activity.[109] Administration of Triton WR-1339 restores bile flow and bile acid excretion, whereas pretreatment with phenobarbital corrects bile flow only.[110] The Triton effect is independent of protein synthesis. It has been inferred that ethinylestradiol primarily alters lipid structure and, consequently, fluidity of the cell membrane, which leads to changes in (Na^+, K^+)-ATPase and transport activities.

In addition to estrogens, many xenobiotics induce cholestasis (reviewed by Plaa and Priestly[106]). Of considerable clinical importance are the propionate and estolate forms of erythromycin, phenothiazines such as chlorpromazine, and tricyclic antidepressants. Less significant are the responses to novobiocin, rifampicin, and nitrofurantoin, each of which may elicit mild jaundice and retention of BSP. An extensively studied cholestatic agent is α-naphthoisothiocyanate, which affects rats and mice more than other species. Accompanying the rapid inhibition of biliary function is dilatation of bile ducts with a loss of microvilli. Hyperplasia of bile ducts is seen after chronic administration. Inhibition of the mixed-function oxidase system is also seen, although the relationship to cholestasis is unknown. Phenobarbital pretreatment potentiates and SKF 525-A reduces hepatotoxicity, implying metabolic activation of α-naphthoisothiocyanate. Inhibitors of protein synthesis block the effect of this agent as well as excretion of its metabolites; protein synthesis is apparently required for metabolic activation. The active metabolite(s) requires an intact enterohepatic circulation for effect.

Chelating agents, as might be predicted, often influence the fate of metals that ordinarily are excreted in the bile to a minor extent. Trisodium calcium diethylenetriaminepentaacetic acid increases the excretion of plutonium in bile,[111] and deferoxamine promotes the excretion of iron.[112] In contrast, neither citrate, cysteine, nor penicillamine alters biliary levels of cadmium, although each readily forms a complex with the metal.[113] Excretion of methylmercury is altered only slightly by penicillamine, dimercaprol, and diethyldithiocarbamate.[114] Spironolactone, although not a chelating agent, enhances biliary excretion of mercury,[115] decreases excretion of silver, and does not affect excretion of lead, manganese, and arsenic.[116]

Pretreatment of homozygous and heterozygous Gunn rats with PCBs accelerates biliary excretion and depresses plasma levels of thyroxine,[117] which is excreted in great part as its glucuronide. However, homozygous

Gunn rats are incapable of conjugating thyroxin with glucuronic acid, and it was found that PCBs act directly on the thyroid gland to lower plasma thyroxine levels.

The vehicle in which a xenobiotic is administered may also affect its disposition. When benzo[a]pyrene and methylcholanthrene are injected as solutions in dimethyl sulfoxide, biliary excretion of metabolites is more rapid than when an albumin suspension of the carcinogens is injected.[118] This is attributable to the formation of a carcinogen–solvent complex that binds tightly to the hepatic microsomes. This leads to more rapid metabolism and, consequently, faster biliary excretion.

G. Effects of Bile Salts

Uptake, conjugation, and secretion of bile salts are primary functions of the hepatobiliary system. Effects of bile salts on xenobiotic disposition have been studied extensively. Taurocholate infusion facilitates hepatic uptake of BSP and of its nonmetabolized analog DBSP in the dog,[119] although inhibition has also been reported.[120] The apparent discrepancy may be related to the dose. High doses of bile salts inhibit hepatic uptake of DPSP in the rat, whereas low doses stimulate it.[121] In the rat, taurocholate infusion increases excretion of BSP and its conjugate[122] and lowers hepatic levels of the dye, implying depressed uptake. Large doses of taurocholate also reduce biliary excretion of succinylsulfathiazole.[123] Enhanced excretion of the cation acetylprocainamide ethobromide is seen during taurocholate infusion,[124] although inhibited excretion during high rates of bile salt infusion (260 μmol/hr) has also been reported.[125] The variation in techniques (perfused liver versus *in vivo*), species, and doses precludes comparison among various laboratories and leaves the total picture of bile salt–xenobiotic interaction essentially unsettled. Taurocholate-stimulated BSP excretion[126,127] is not simply due to its choleretic effect because other choleretics, theophylline,[128] umbelliferone,[129] and SC2614[130] do not increase dye excretion. A direct interaction between the dye and the bile salts has been suggested, the latter acting as a carrier. The micellar sink theory[131] proposed that certain xenobiotics could be incorporated into mixed micelles, in which form bile salts and lipid appear in the bile. Several studies performed to test the theory[132,133] yielded conflicting information. Among the problems is the dependence of micellar formation on a critical concentration of bile salts, which differs from one bile salt to another. Dilution or concentration of the micellar constituents during analysis is difficult to avoid, impairing interpretation of results. Cerulein increases the transport maximum (Tm) of BSP and rose bengal in chickens, but only if bile salts are present,[134] again

9. Excretion Mechanisms

supporting a specific role for bile salts in the effect whatever the mechanism.

An association of quaternary amines with bile salt micelles[135,136] has been demonstrated as a possible means for intestinal absorption of quaternary amines; the mechanism may be applicable to biliary excretion as well.

H. Miscellaneous Factors

Sex differences in biliary function are rarely seen. The dye tartrazine is excreted unchanged in the bile far more rapidly in female than in male rats,[137] but the sex difference is not seen in guinea pigs or rabbits. The difference is hormonally controlled; estradiol enhances excretion in males and testosterone depresses excretion in females. The structural analog lissamine fast is excreted equally well by male and female rats. During liver regeneration, tartrazine excretion is depressed in both male and female rats, but the sex difference disappears on the second and third days.[138] The difference reappears with time and is equal to that of control animals upon restoration of original liver size. No sex differences are seen in the biliary excretion of DBSP, ouabain, amaranth, or indocyanin green in rats. Sex differences are evident where xenobiotic excretion is limited by metabolism. Thus, benzo[a]pyrene metabolites are excreted more rapidly in male rats than in females in parallel to their relative rates of metabolism *in vitro*.[78] In contrast, aldosterone, which appears in the bile mainly in the form of its metabolites, is excreted more rapidly in female than in male rats.[139] Male rats metabolize and excrete pentobarbital more rapidly than do females.[140] Sex-related variations in drug metabolism have been thoroughly reviewed[141] but are mainly restricted to studies with rats. Mice, guinea pigs, hamsters, rabbits, dogs, monkeys, and humans show no clear-cut sex differences. Sex differences in human drug responses are therefore likely to be related to factors other than hepatic metabolism. Metabolism of drugs exhibiting substantial sex differences in rats is less susceptible to induction with 3-methylcholanthrene than are those with smaller sex differences. This implies that the sex difference is related to specific isoenzymes of cytochrome P-450 because 3-methylcholanthrene induces a highly selective isoenzyme of cytochrome P-450.

The metabolic fate of xenobiotics varies considerably with age. Many aspects of drug disposition are poorly developed in the newborn and, after attainment of adult levels, begin to wane with the onset of old age. The toxicity of a number of xenobiotics is higher in immature compared to adult animals, although similar comparisons have not been made for older animals. Developmental aspects of oxidative[142] and conjugative metabolism[143] have been reviewed, as have the consequences of aging for pharmacokinetics.[144] Eosine, which is excreted in the bile unmetabolized, is excreted more efficiently in 30- to 60-day-old rats than in animals somewhat younger or older.[145] Older, but not younger, rats excrete the dye more rapidly after phenobarbital pretreatment, indicating an ability to respond to the inducing agent in older rats that is absent in younger rats. However, bile flow does increase in younger animals in the face of unchanged dye excretion. Increased hepatic transport and decreased toxicity of ouabain in newborn rats are observed after induction with phenobarbital, pregnenolone 16α-carbonitrile, and spironolactone.[146] Decreased ouabain excretion in older rats is returned to adult levels after pretreatment with spironolactone.[147] Development of the enterohepatic circulation of diethylstilbestrol in the rat proceeds with age in parallel to the rates of glucuronidation, conjugate transport, and conjugate hydrolysis by gut flora.[148] Carbon tetrachloride inhibits plasma-to-liver transport of ouabain in young rats (<35 days) and liver bile transport in older rats (>35 days).[149]

I. Enterohepatic Circulation

The recycling of biliary constituents by reabsorption from the gut, passage into the portal circulation, and transport again into bile is referred to as the *enterohepatic circulation* (EHC).[150,151] Any compound absorbed from the gut and present in the bile might be expected to undergo the EHC cycle. In actuality, convincing evidence is available only for a few compounds. Metabolism is a major consideration; hepatic pathways typically decrease lipophilicity of a xenobiotic, which should impede intestinal absorption by passive diffusion. However, further metabolism by bacteria of the large intestine[152–154] often reverses this process and is a key step in determining which compounds actually enter the EHC. During metabolism within the gut, the original compound may be regenerated, oxidative metabolites may be liberated because of conjugate hydrolysis, or the liberated metabolites may then be further metabolized. The reactions catalyzed by intestinal flora may be considered complementary and often antagonistic to those of the liver. For example, hepatic reactions include hydroxylation, oxidation, and conjugation, whereas the intestinal flora catalyze dehydroxylation, reduction, and conjugate hydrolysis. Hydrolysis of glucuronides is probably the most common reaction that bears on the EHC of the compounds studied. The best example of EHC is that for bile salts. The conjugated salts are carried with the bile into the intestinal tract, where hydrolysis occurs; the bile salts and their metabolites are reabsorbed into the liver by way of the portal vein. The effectiveness of the EHC is such that only 5% of the circulating bile salts are lost each day.

Evidence for EHC is obtained in a number of ways.[154] Where hydrolysis of glucuronides is believed to be crucial, glucuronidase from intestinal bacteria are the focus of study. Administration to rats of D-glucaro-1,4-lactone, an inhibitor of β-glucuronidase, leads to a shortened duration of action of phenobarbital and progesterone, both of which are excreted in the bile partially as glucuronides.[155] The response to zoxazolamine is unchanged by this inhibitor because it is excreted in the bile as the glucuronide of the pharmacologically inactive 6-hydroxy metabolite. Suppression of the flora by antibiotics is also used to assess EHC. Under such conditions, there are changes in the disposition of morphine,[156] diethylstilbestrol,[157] mestranol,[158] estradiol,[158] and etorphine.[159] Cautious interpretation is required because the antibiotics may cause diarrhea, altering transit time in the gut. Blood levels and urinary excretion are also measured with and without bile drainage. Considerable urinary excretion of those xenobiotics that readily appear in the bile is possible from extensive EHC. Each recycling will result in transport of some of the compound to the kidney, provided the first-pass effect of the liver is less than 100%. Also used is the technique of channeling bile directly into the intestine of a second animal and again measuring biliary excretion. Differences between biliary and fecal excretion are also interpreted as a reflection of EHC.

Enterohepatic circulation has been demonstrated for more than 30 compounds.[33] For the most part, they have been investigated in a single species, and extrapolation to other species is not advisable. Species variation is seen for indomethacin[160] and diazepam.[161,162] The intestinal lesions observed after feeding indomethacin are attributed to its extensive EHC[163] and show a parallel species variation. In contrast, although there is extensive EHC of acetylmethadol,[164] a drug that requires metabolic activation, the EHC makes little contribution to maintenance of the blood levels of active metabolites.

In an interesting clinical application, the ion-exchange resin cholestyramine is used to bind bile salts within the intestine, interrupting their EHC and consequently reducing plasma levels of cholesterol.[165] This has led to the use of cholestyramine in overdose of digitalis and poisoning with the pesticide kepone.[166]

J. Regeneration

Surgical removal of up to 80% of the liver is followed by a rapid regeneration of liver mass in the surviving lobes in every species studied.[167] Accompanying regeneration are a large number of functional and structural changes. Most of those functions that are directed toward replenishment of hepatic tissue receive high priority, whereas others, such as metabolism and transport of xenobiotics, often are diminished during regeneration. Biliary excretion of the glutathione conjugate of injected BSP is diminished during regeneration in proportion to the decrease in conjugation activity.[138] Immediately following partial hepatectomy, bile flow, biliary excretion of bile salts, and BSP per gram of residual tissue are actually increased, reflecting compensatory activity in the remaining lobes. If bile salts are depleted through biliary drainage, BSP excretion per gram of liver is no greater than that of control animals, supporting a role of bile salts in the enhancement of biliary excretion of BSP. The rate of biliary excretion of diethylstilbestrol, indocyanine green, and procainamide ethobromide per gram of residual liver is increased during regeneration.[168] Despite the relatively small decrease in total excretion of diethylstilbestrol, its toxicity is increased greatly as the result of partial hepatectomy. This raises the question of whether the toxicity of the drug under these circumstances is governed solely by its rate of biliary excretion or whether other factors, associated with partial hepatectomy, are also involved. Tartrazine excretion is suppressed during regeneration even on a liver weight basis.[138] Its biliary excretion shows sex dependency that is altered considerably during regeneration. Furthermore, its biliary excretion is not altered by nafenopin, which depresses excretion of BSP and indocyanine green.

IV. SALIVARY EXCRETION

Many xenobiotics can be found in the saliva,[169-177] and salivary concentrations generally reflect those in the plasma. For a comprehensive review, the reader is referred to that of Horning *et al.*[177] Impetus for much of the work in this area stems from the need to monitor plasma levels of drugs by noninvasive methods. Analytical techniques applied to plasma are usually suitable for saliva. Transport from plasma to saliva is mainly passive and is related to the plasma/saliva concentration gradient, lipid solubility, and pK of the xenobiotic. Only the nonbound fraction within the plasma is readily diffusible, and salivary concentrations should reflect this fraction and thus be a relatively easy means of determining the degree of binding to plasma proteins. Salivary protein concentration is less than 10% of that of plasma and should have little influence on passive diffusion of xenobiotics. It is apparent that salivary analysis is in many cases a satisfactory means of determining plasma levels and other pharmacokinetic parameters. Because salivary pH (6.7–6.9) is somewhat less than that of plasma, the saliva/plasma concentration ratios for xenobiotics with pK values in the range 5.5–8.5 may be somewhat less or more than 1.0. Organic bases in this group tend to be concentrated in the saliva, whereas

9. Excretion Mechanisms

organic acids diffuse into the saliva less readily. Examples of drugs within the latter group are the organic acids phenobarbital and salicylic acid, salivary concentrations of which are less than those predicted on the basis of plasma protein binding alone.[177] In contrast, lithium ion appears to be actively transported from plasma to saliva, and the latter has higher concentrations than the former. In a study of six antibiotics, almost no correlation was obtained between salivary and plasma concentrations.[178] Although secretion was predicted on the basis of pK, several antibiotics were not detectable in saliva.

V. EXCRETION INTO MILK

It is somewhat more practical importance to consider xenobiotic secretion into milk than into other body fluids because milk is a link in the nutritional chain rather than an excretion in the usual sense. The reader is referred to several reviews on this subject.[179-184] Very few preformed endogenous systemic substances are found in milk, suggesting in effect a blood–milk barrier. The mammary duct is permeable to water, but milk is isoosmotic with plasma. Estrogens and progesterone facilitate milk secretion, and the high levels of prolactin following delivery and suckling promote secretion. Multiple interactions of other hormones on milk secretion have been described.[183] These include growth hormones, ACTH, insulin, coritsol, thyroid, and parathyroid. The quantity of milk secreted also depends on blood flow, which is very high to the breast during lactation.

Xenobiotics diffuse across the capillary wall into the alveolar breast cells and then into the milk by passive diffusion or reverse pinocytosis. Concentrations within the milk are independent of milk volume. If these were the sole considerations for plasma-to-milk transport, then concentrations of nonbound xenobiotic in milk and plasma should be equal. However, the lipid solubility and pK of organic anions or cations influence the extent of passive diffusion, as described for saliva. The pH of milk is slightly below that of plasma and influences the secretion of these substances. The high fat content of milk certainly affects the secretion of highly lipid-soluble xenobiotics. Secretion into milk may occur through binding to milk protein or through binding on or within fat globules. Binding of xenobiotics to milk proteins is far less than to plasma proteins and probably has little influence on their diffusion from plasma to milk. Free diffusion of ionized xenobiotics is limited to substances with a molecular weight below 200. The diffusion of larger molecules is dependent on lipid solubility, which in turn is dependent upon pK. The diurnal variation in composition of milk may influence the concentration of drugs within the milk.

A considerable number of drugs appear in milk, but few have been studied systematically, and the significance to the nursing infant is not always clear. A study of antipyrine, a freely diffusible drug, showed similar milk and plasma concentration over 24 hr after a single oral dose.[184] Sulfasalazine, in contrast, is undetectable in milk, although it readily passes into urine.[184] A major metabolite, sulfapyridine, is found in plasma and mlik in approximately equal concentrations but does not appear to be hazardous to the infant. Studies and reviews[185-192] list more than 50 drugs, radioisotopes, and other xenobiotics that appear in milk. Some of the findings raise serious questions as to the advisability of nursing an infant while taking drugs. For example, the concentrations of codeine and morphine in human milk are more than twice those in plasma.[192] Several environmental pollutants, including DDT and PCB, are also found in breast milk[193]; the high but variable fat content of milk facilitates their secretion. Nevertheless, reports on overt toxicity from such compounds transmitted in the milk are rare. Obstructive jaundice due to chlorinated hydrocarbon in milk has been seen.[194] Methemoglobin was found in an infant whose mother was treated with phenytoin.[195] The toxicity of some xenobiotics for the infant is so great that they are proscribed during nursing; if they are required by the mother, nursing should cease. Suppression of infants' thyroid gland activity is possible after administration of radioactive iodine or propylthiouracil to nursing mothers.[196,197] Other proscribed drugs include lithium, which is concentrated in milk, methadone, isoniazid, chloramphenicol, and certain antineoplastic agents.[198] There are no reports that tetrahydrocannabinol passes into human milk, but it does appear in the milk of sheep[199] and squirrel monkeys.[200] Alcohol passes freely into milk, although there is little evidence that an occasional drink by the mother is harmful to a nursing infant. A woman who smokes 10–20 cigarettes per day will have 0.4–0.5 mg of nicotine per liter of milk.

Information on the excretion of xenobiotics in milk, although reported sporadically, is gradually increasing. Quantitative data and information on their relevance to toxicity in the infant are greatly needed, as are comparisons between xenobiotic levels in milk and plasma. In the absence of such information, a conservative approach would be to limit the use of drugs in nursing mothers.

VI. EXCRETION INTO EXPIRED AIR

Analysis of drugs in the expired air is a noninvasive technique developed for assessing plasma levels of drugs in humans and, by inference, their rate of metabolism.[201-208] The most commonly used drug is

9. Excretion Mechanisms

aminopyrine that is N-demethylated; the methyl groups are rapidly converted to carbon dioxide, which appears in the expired air. A tracer dose of [N-$dimethyl$-^{14}C]aminopyrine is administered, and the $^{14}CO_2$ in expired air is collected over a suitable period of time and counted (see Chapter 10, this volume). It must be appreciated that $^{14}CO_2$ analysis reflects the endpoint of a complex series of steps that begins with N-demethylation. Quantitation of the first step by this method is possible only if there is assurance that none of the succeeding reactions is rate limiting. The technique is also applicable to deacetylation and decarboxylation. CO_2 production is a normal physiological process and varies with physical activity, metabolic state, and body temperature. Therefore, these factors must be considered when utilizing this technique, particularly in humans. It has been suggested that the technique is suitable for assessment of liver function in cases of cirrhosis, hepatitis, and liver cancer,[203] although patients with cholestasis are not readily distinguished from controls. The technique is also useful in experimental animals, and changes in aminopyrine demethylation in response to phenobarbital, bile duct ligation, partial hepatectomy, and portacaval shunt reflect those predicted from *in vitro* drug metabolism studies.[204,206] Assessment of other aspects of liver function through breath analysis has been discussed,[201,205,207] and wider application of the technique can be expected.

One notable application has been in the noninvasive determination of lipid peroxidation, during which saturated short-chain hydrocarbons are split from unsaturated fatty acids such as linoleic acid and passed into the expired air.[207] The results do not always agree with *in vitro* measurement of lipid peroxidaton,[208] and further work is needed in this area.

VII. COMMENTS

Accurate assessment of excretion of xenobiotics is essential in many aspects of metabolism, pharmacokinetics, toxicology, and diagnosis of diseases affecting the organs of excretion, for example, liver and kidney. Until recently, this was accomplished principally through analyses of urinary products, reflecting the facts that urinary excretion is the major route for most xenobiotics and that the physiology of renal excretion is well understood. More current investigation into other excretory routes recognizes the need to assess xenobiotic metabolism, particularly in humans, and the practical necessity of noninvasive methods. Thus, the use of breath and salivary analyses will undoubtedly continue to develop in this direction. Much of this work has begun with the assumption that the principles governing the transport and excretion of xenobiotics in other areas are applicable here as well. However, inconsistent findings indicate that physiological mechanisms for these routes of excretion may not be well understood, and data on excretion in salivary and expired air must be interpreted with due caution.

Excretion of drugs in breast milk is greatly in need of expanded investigation. Because more than half of the infants in the world are breast-fed, and this number is increasing, possible exposure to maternal xenobiotics is a major consideration. There is a lack of pharmacokinetic studies in this area and of information on the effects on infants of long-term exposure to maternal xenobiotics through breast feeding. Along with such practical necessities is a need to enlarge our understanding of the physiological mechanisms for secretion of drugs and other chemicals into the milk. Another area that lacks sufficient quantitative information is the normal excretion of xenobiotics into human bile. Virtually all collection of human bile is associated with surgical procedures and some drug treatment. Because both of these factors could influence biliary excretion, the studies published must be interpreted with caution.

REFERENCES

1. Pritchard, J. B., and James, M. O. (1982). Metabolism and urinary excretion. *In* "Metabolic Basis of Detoxication" (W. Jakoby, J. R. Bend, and J. Caldwell, eds.), pp. 339–357, Academic Press, New York.
2. Irish, J. M., and Grantham, J. J. (1981). Renal handling of organic anions and cations. *In* "The Kidney" (B. M. Brenner and F. C. Rector, eds.), pp. 619–649. Saunders, Philadelphia.
3. Jones, R. S., and Meyers, W. C. (1979). Regulation of hepatic biliary secretion. *Annu. Rev. Physiol.* **41**, 67–82.
4. Chenderovitch, J. (1973). Bile secretion. *Clin. Gastroenterol.* **2**, 31–47.
5. Garrett, E. R. (1978). Pharmacokinetics and clearances related to renal processes. *Int. J. Clin. Pharmacol.* **16**, 155–172.
6. Boyer, J. L. (1980). New concepts of mechanisms of hepatocyte bile formation. *Physiol. Rev.* **60**, 303–326.
7. Erlinger, S. (1981). Hepatocyte bile secretion: Current views and controversies. *Hepatology* **1**, 352–359.
8. Blitzer, B. L., and Boyer, J. L. (1982). Cellular mechanisms of bile formation. *Gastroenterology* **82**, 346–357.
9. Layden, T. J., Elias, E., and Boyer, J. L. (1978). Bile formation in the rat—The role of the paracellular shunt pathway. *J. Clin. Invest.* **62**, 1375–1385.
10. Bradley, S. E., and Herz, R. (1978). Permselectivity of biliary canalicular membrane in rats: Clearance probe analysis. *Am. J. Physiol.* **235**, E570–E576.
11. Weitering, J. G., Lammers, W., Meijer, D. K. F., and Mulder, G. J. (1977). Localization of d-tubocurarine in rat liver lysozomes. Lysozomal uptake, biliary excretion and displacement by quinacrine *in vivo*. *Naunyn-Schmiedeberg's Arch. Pharmacol.* **299**, 277–281.

9. Excretion Mechanisms

12. LaRusso, N. F., and Fowler, S. (1979). Coordinate secretion of acid hydrolases in rat bile hepatocyte exocytosis of lysosomal protein. *J. Clin. Invest.* **64,** 948–954.
13. Oda, M., Price, V. M., Fisher, M. M., and Phillips, N. J. (1974). Ultrastructure of bile canaliculi with special reference to the surface coat and the pericanalicular web. *Lab. Invest.* **31,** 314–323.
14. Phillips, M. J., Oda, M., and Funatsu, K. (1978). Evidence for microfilament involvement in norethandrolone-induced intrahepatic cholestasis. *Am. J. Pathol.* **93,** 729–744.
15. Dubin, M., Maurice, M., Feldmann, G., and Erlinger, S. (1978). Phalloidin-induced cholestasis in the rat: Relation to changes in microfilaments. *Gastroenterology* **75,** 450–455.
16. Barnhart, J. L., and Combes, B. (1975). Characteristics common to choleretic increments of bile induced by theophylline, glucagon and SQ-20009 in the dog. *Proc. Soc. Exp. Biol. Med.* **150,** 591–596.
17. Poupon, R. E., Dol, M. L., Dumont, M., and Erlinger, S. (1978). Evidence against a physiological role of cAMP in choleresis in dogs and rats. *Biochem. Pharmacol.* **27,** 2413–2416.
18. Kaminski, D. L., Ruwart, M. J., and Deshpande, Y. G. (1979). The role of cyclic AMP in canine secretin-stimulated bile flow. *J. Surg. Res.* **57,** 57–61.
19. Barnhart, J. L., and Combes, B. (1978). Characterization of SC2644-induced choleresis in the dog. Evidence for canalicular bicarbonate secretion. *J. Pharmacol. Exp. Ther.* **206,** 190–197.
20. Peterson, R. E., and Fujimoto, J. M. (1973). Retrograde intrabiliary injection: Absorption of water and other compounds from the rat biliary tree. *J. Pharmacol. Exp. Ther.* **185,** 150–162.
21. Olson, J. R., and Fujimoto, J. M. (1980). Evaluation of hepatobiliary function in the rat by the segmented retrograde intrabiliary injection technique. *Biochem. Pharmacol.* **29,** 205–211.
22. Kakis, G., and Yousef, I. M. (1978). Protein composition of rat bile. *Can. J. Biochem.* **56,** 287–290.
23. Gustafson, J. H., and Benet, L. Z. (1974). Biliary excretion kinetics of phenolphthalein glucuronide after intravenous and retrograde biliary administration. *J. Pharm. Pharmacol.* **26,** 937–944.
24. Sperber, I. (1959). Secretion of organic anions in the formation of urine and bile. *Pharmacol. Rev.* **11,** 109–134.
25. Vonk, R. J., Jekel, P., and Meijer, D. K. F. (1975). Choleresis and hepatic transport mechanisms. II. Influence of bile salt choleresis and biliary micelle binding on biliary excretion of various organic anions. *Naunyn-Schmiedeberg's Arch. Pharmacol.* **290,** 375–387.
26. Balabaud, C., Kron, K. A., and Gumucio, J. J. (1977). The assessment of the bile salt nondependent fraction of canalicular bile in the rat. *J. Lab. Clin. Med.* **89,** 393–399.
27. Baker, A. L., Wood, R. A. B., Moosa, A. R., and Boyer, J. L. (1979). Sodium taurocholate modifies the bile acid independent fraction of canalicular bile flow in the Rhesus monkey. *J. Clin. Invest.* **64,** 312–320.
28. Klaassen, C. D. (1974). Bile flow and composition during bile acid depletion and administration. *Can. J. Physiol. Pharmacol.* **52,** 334–348.
29. Boyer, J. L., and Klatskin, G. (1970). Canalicular bile flow and bile secretory pressure. Evidence for a non-bile salt dependent fraction in the isolated perfused rat liver. *Gastroenterology* **59,** 853–859.
30. Simon, F., Sutherland, E., and Accatino, L. (1977). Stimulation of hepatic sodium and potassium-activated adenosine triphosphatase activity by phenobarbital. Its possible role in regulation of bile flow. *J. Clin. Invest.* **59,** 849–861.

31. Graf, J., Korn, P., and Peterlik, M. (1972). Choleretic effects of ouabain and ethacrynic acid in the isolated perfused rat liver. *Naunyn-Schmiedeberg's Arch. Pharmacol.* **272,** 230–233.
32. Shaw, H., Caple, I., and Heath, T. (1972). Effect of ethacrynic acid on bile formation in sheep, dogs, rats, guinea pigs and rabbits. *J. Pharmacol. Exp. Ther.* **182,** 27–33.
33. Levine, W. G. (1981). Biliary excretion of drugs and other xenobiotics. *Prog. Drug Res.* **25,** 361–420.
34. Brauer, D. S. (1959). Mechanisms of bile secretion. *JAMA, J. Am. Med. Assoc.* **169,** 1462–1466.
35. Meijer, D. K. F. (1977). The mechanisms for hepatic uptake and biliary excretion of organic cations. *In* "Intestinal Permeation" (M. Kramer and F. Lauterbach, eds.), pp. 196–207. Exerpta Medica, Amsterdam.
36. Cohen, E. N., Winstow, B. H., and Smith, D. (1967). The metabolism and elimination of d-tubocurarine-H^3. *Anesthesiology* **28,** 309–317.
37. Meijer, D. K. F., and Weitering, J. G. (1970). Curare-like agents: Relation between lipid solubility and transport into bile in perfused rat liver. *Eur. J. Pharmacol.* **10,** 283–289.
38. Meijer, D. K. F., Vonk, R. J., Scholtens, E. J., and Levine, W. G. (1976). The influence of dehydrocholate on hepatic uptake and biliary excretion of ^{3}H-taurocholate and ^{3}H-ouabain. *Drug Metab. Dispos.* **4,** 1–7.
39. Peterson, R. E., and Fujimoto, J. M. (1973). Biliary excretion of morphine-3-glucuronide and morphine-3-ethereal sulfate by different pathways in the rat. *J. Pharmacol. Exp. Ther.* **184,** 409–418.
40. Clarenburg, R., and Kao, C. (1973). Shared and separate pathways for biliary excretion of bilirubin and BSP in rats. *Am. J. Physiol.* **225,** 192–200.
41. Gartner, U., Stockert, T. J., Levine, W. G., and Wolkoff, A. (1982). Effect of nafenopin on the uptake of bilirubin and sulfobromophthalcin by isolated perfused rat liver. *Gastroenterology* **83,** 1163–1169.
42. Erttmann, R. R., and Damm, K.H. (1975). Influence of bile flow, theophylline and some organic anions on the bililary excretion of ^{3}H-ouabain in rats. *Arch. Int. Pharmacodyn. Therm.* **218,** 290–298.
43. Erttmann, R. R., and Damm, K. H. (1975). On the problem of a common hepatic transport process for steroids. Uptake of ^{3}H-ouabain, ^{3}H-taurocholic acid and ^{3}H-corticosterone into rat liver slices. *Arch. Int. Pharmacodyn. Ther.* **214,** 232–239.
44. Millburn, P. (1975). Excretion of xenobiotic compounds in bile. *In* "The Hepatobiliary System. Fundamental and Pathological Mechanisms" (W. Taylor, ed), pp. 109–129. Plenum, New York.
45. Hughes, R., Millburn, P., and Williams, R. T. (1973). Molecular weight as a factor in the excretion of monoquaternary ammonium cations in the bile of the rat, rabbit and guinea pig. *Biochem. J.* **136,** 967–978.
46. Hirom, P. C., Millburn, P., and Smith, R. L. (1976). Bile and urine as complementary pathways for the excretion of foreign organic compounds. *Xenobiotica* **6,** 55–64.
47. Vonk, R. J., Jekel, P. A., Meijer, D. K. F., and Hardonk, M. J. (1978). Transport of drugs in isolated hepatocytes, the influence of bile salts. *Biochem. Pharmacol.* **27,** 397–405.
48. Horne, D. W., Briggs, W. T., and Wagner, C. (1978). Transport of 5-methyltetrahydrofolate and folic acid in freshly isolated hepatocytes. *J. Biol. Chem.* **253,** 3529–3535.
49. Schwarz, L. R., Summer, K. H., and Schwenk, M. (1979). Transport and metabolism of bromosulfophthalein by isolated rat liver cells. *Eur. J. Biocem.* **94,** 617–622.

50. Schwenk, M. (1980). Transport systems of isolated hepatocytes. Studies on the transport of biliary compounds. *Arch. Toxicol.* **44,** 113–126.
51. Schwarz, L. R., Schwenk, M., and Greim, H. (1976). Excretion of taurocholate from isolated hepatocytes. *Eur. J. Biochem.* **71,** 369–373.
51a. Erttmann, R. R., and Damm, K. H. (1976). Probenecid-induced effects on bile flow and biliary excretion of ^{3}H-ouabain. *Arch. Int. Pharmacodyn. Ther.* **223,** 96–106.
52. Levine, W. G. (1974). Effect of the hypolipidemic drug nafenopin (2-methyl-2-[p-(1,2,3,4-tetrahydro-1-napthyl)phenoxy] propionic acid; TPIA: SU-13,437) on the hepatic disposition of foreign compounds in the rat. *Drug Metab. Dispos.* **2,** 178–186.
53. Kitani, K., Kanai, S., and Miura, R. (1978). Increased biliary excretion of ouabain induced by bucolome in the rat. *Clin. Exp. Pharmacol. Physiol.* **5,** 117–124.
54. Arias, I. M. (1979). Ligandin: Review and update of a multifunctional protein. *Med. Biol.* **57,** 328–334.
55. Ketterer, B. R., Ross-Mansell, P., and Whitehead, J. K. (1967). The isolation of carcinogen-binding protein from livers of rats given 4-dimethylaminoazobenzene. *Bichem. J.* **103,** 316–324.
56. Morey, K. S., and Litwack, G. (1969). Isolation and properties of cortisol metabolite binding proteins of rat liver cytosol. *Biochemistry* **8,** 4813–4821.
57. Habig, W. H., Pabst, M. J., Fleischner, G., Gatmaitan, Z., Arias, I. M., and Jakoby, W. B. (1974). The identity of glutathione S-transferase with ligandin, a major binding protein of liver. *Proc. Natl. Acad. Sci. U.S.A.* **71,** 3879–3882.
58. Ketley, J. N., Habig, W. H., and Jakoby, W. B. (1975). Binding of non-substrate ligands to glutathione S-transferases. *J. Biol. Chem.* **250,** 8670–8673.
59. Kirsch, R., Kamisaka, K., Fleischner, G., and Arias, I. M. (1975). Structural and functional studies of ligandin, a major renal organic anion binding protein. *J. Clin. Invest.* **55,** 1009–1019.
60. Meijer, D. K. F., Vonk, R. J., Keulemans, K., and Weitering, J. G. (1977). Hepatic uptake and biliary excretion of dibromosulphthalein. Albumin dependence, influence of phenobarbital and nafenopin pretreatment and the role of Y and Z protein. *J. Pharmacol. Exp. Ther.* **202,** 8–21.
61. Ketterer, B., Tipping, E., Meuwissen, J., and Beale, D. (1975). Ligandin. *Biochem. Soc. Trans.* **3,** 626–630.
62. Sugiyama, Y., Iga, T., Awazu, S., and Hanano, M. (1979). Multiplicity of sulfobromophthalein-binding proteins in Y-fraction from rat liver. *J. Pharmacobio-Dyn.* **2,** 193–204.
63. Boyer, J. L., Schwartz, J., and Smith, N. (1976). Biliary secretion in elasmobranchs. II. Hepatic uptake and biliary excretion of organic anions. *Am. J. Physiol.* **230,** 974–981.
64. Fleischner, G. M., Meijer, D. K. F., Levine, W. G., Gatamitan, Z., Gluck, R., and Arias, I. M. (1976). Effect of hypolipidemic drugs, nafenopin and clofibrate, on the concentration of ligandin and Z protein in rat liver. *Biochem. Biophys. Res. Commun.* **67,** 1401–1407.
65. Ockner, R. K., Manning, J. A., Poppenhauser, R. B., and Ho, W. K. L. (1972). A binding protein for fatty acids in cytosol of intestinal mucosa, liver, myocardium and other tissues. *Science* **177,** 56–58.
66. Kakis, G., and Yousef, I. M. (1978). Protein composition of rat bile. *Can. J. Biochem.* **56,** 287–290.
67. Hwang, S. W. (1975). Plasma and hepatic binding of indocyanine green in guinea pigs of different ages. *Am. J. Physiol.* **228,** 718–724.

68. Jakoby, W. B., ed. (1980). "Enzymatic Basis of Detoxication," Vols. 1 and 2. Academic Press, New York.
69. Jakoby, W. B., Bend, J. R., and Caldwell, J., eds. (1982). "Metabolic Basis of Detoxication." Academic Press, New York.
70. Estabrook, R. W., and Lindenlaub, E., eds. (1979) "The Induction of Drug Metabolism." Schattauer, Stuttgart.
71. Branch, R. A., Shand, D. G., Wilkinson, G. R., and Nies, A. S. (1974). Increased clearance of antipyrine and d-propranolol after phenobarbital treatment in the monkey. Relative contributions of enzyme induction and increased hepatic blood flow. *J. Clin. Invest.* **53,** 1101–1107.
72. Ohnhaus, E. E., and Tilvis, R. (1976). Liver blood flow, metabolic heat production and body temperature before, during and after phenobarbitone administration. *Acta Hepato-Gastroenterol.* **23,** 404–408.
73. Berthelot, P., Erlinger, S., Dhumeaux, D., and Preaux, A. (1970). Mechanism of phenobarbital-induced hypercholeresis in the rat. *Am. J. Physiol.* **219,** 809–813.
74. Maxwell, J. D., Hunter, J., Stewart, D. A., Carrella, M., and Williams, R. (1973). Effect of phenobarbitone on bile flow and bilirubin metabolism in man and the rat. *Digestion* **9,** 138–148.
75. Roerig, D., Hasegawa, A., Peterson, R., and Wang, R. (1974). Effect of chloroquine and phenobarbital on morphine glucuronidation and biliary excretion in the rat. *Biochem. Pharmacol.* **23,** 1331–1339.
76. Zsigmond, G., and Solymoss, B. (1972). Effect of spironolactone, pregnenolone-16β-carbonitrile and cortisol on the metabolism and biliary excretion of sulfobromophthalein and phenol-3,6-dibromophthalein disulfonate in rat. *J. Pharmacol. Exp. Ther.* **183,** 499–507.
77. Roerig, D. L., Hasegawa, A. T., and Wang, R. I. H. (1976). Role of metabolism in the biliary excretion of methadone metabolites. *J. Pharmacol. Exp. Ther.* **199,** 93–102.
78. Levine, W. G. (1970). The role of microsomal drug-metabolizing enzymes in the biliary excretion of 3,4-benzopyrene in the rat. *J. Pharmacol. Exp. Ther.* **175,** 301–310.
79. Levine, W. G. (1971). Metabolism and biliary excretion of N-2-fluorenylacetamide and N-hydroxy-2-fluorenylacetamide. *Life Sci.* **10,** 727–735.
80. Levine, W. G. (1972). Biliary excretion of 3-methylcholanthrene as controlled by its metabolism. *J. Pharmacol. Exp. Ther.* **183,** 420–426.
81. Levine, W. G. (1974). Hepatic uptake, metabolism, and biliary excretion of 7,12-dimethylbenzanthracene in the rat. *Drug Metab. Dispos.* **2,** 169–177.
82. Levine, W. G. (1980). Induction and inhibition of the metabolism and biliary excretion of the azo dye carcinogen, N,N-dimethyl-4-aminoazobenzene (DAB) in the rat. *Drug Metab. Dispos.* **8,** 212–217.
83. Klaassen, C. (1970). Effects of phenobarbital on the plasma disappearance and biliary excretion of drugs in rats. *J. Pharmacol. Exp. Ther.* **175,** 289–300.
84. Roberts, R. J., and Plaa, G. L. (1967). Effect of phenobarbital on the excretion of an exogenous bilirubin load. *Biochem. Pharmacol.* **16,** 827–835.
85. Whelan, G., and Combes, B. (1975). Phenobarbital-enhanced biliary excretion of administered unconjugated and conjugated sulfobromophthalein (BSP) in rat. *Biochem. Pharmacol.* **24,** 1283–1286.
86. Mulder, G. H. (1973). The rate-limiting step in the biliary elimination of some substrates of uridine diphosphate glucuronyltransferase in the rat. *Biochem. Pharmacol.* **22,** 1751–1763.
87. Levine, W. G., and Bognacki, J. (1976). Biliary excretion of 3,4-benzpyrene in nafenopin treated rats. *J. Pharmacol. Exp. Ther.* **196,** 486–492.

88. Varga, F., Fisher, E., and Szily, T. S. (1974). Biliary excretion of bromsulfphthalein and glutathione conjugate of bromsulphthalein in rats pretreated with diethyl maleate. *Biochem. Pharmacol.* **23,** 2617-2623.
89. Priestly, B. G., and Plaa, G. L. (1970). Sulfobromophthalein metabolism and excretion in rats with iodomethane-induced depletion of hepatic glutathione. *J. Pharmacol. Exp. Ther.* **174,** 221-231.
90. Levine, W. G., and Finkelstein, T. T. (1978). Biliary excretion of N,N-dimethyl-4-aminoazobenzene (DAB) in the rat. *Drug Metab. Dispos.* **6,** 265-272.
91. Ketterer, B., Kadlubar, F., Flammang, T., Carne, T., and Enderby, G. (1979). Glutathione adducts of N-methyl-4-aminoazobenzene formed *in vivo* and by reaction of N-benzoyl-N-methyl-4-aminoazobenzene with glutathione. *Chem.-Biol. Interact.* **25,** 7-21.
92. Levine, W. G., and Finkelstein, T. T. (1979). A role for liver glutathione in the hepatobiliary fate of N,N-dimethyl-4-aminoazobenzene. *J. Pharmacol. Exp. Ther.* **208,** 399-405.
93. Burk, R. F., and Correia, M. A. (1979). Stimulation of rat hepatic microsomal heme oxygenase by diethyl maleate. *Res. Commun. Chem. Pathol. Pharmacol.* **24,** 205-207.
94. Younes, M., and Siegers, C. P. (1980). Lipid peroxidation as a consequence of glutathione depletion in rat and mouse liver. *Res. Commun. Chem. Pathol. Pharmacol.* **27,** 119-128.
95. Schulze, P., and Czok, G. (1975). Reduced bile flow in rats during sulfobromophthalein infusion. *Toxicol. Appl. Pharmacol.* **32,** 213-224.
96. Dhumeaux, D., Berthelot, P., Preaux, A. M., Erlinger, S., and Fauvert, R. (1970). A critical study of the concept of maximal biliary transport of sulfobromophthalein (BSP) in the Wistar rat. *Rev. Eur. Etud. Clin. Biol.* **15,** 279-286.
97. Laperche, Y., and Oudea, P. (1974). Inhibition by sulfobromophthalein of mitochondrial translocation of anions and adenine nucleotides: Effects upon liver adenosine triphosphate and possible correlation with inhibition of bile flow in the rat. *J. Pharmacol. Expl Ther.* **197,** 235-244.
98. Schwenk, M., Burr, R., Baur, H., and Pfaff, E. Interaction of bromosulfophthalein with mitochondrial membranes: Effect on ion movements. *Biochem.Pharmacol.* **26,** 825-832.
99. Miejer, D. K. F., Bognacki, J., and Levine, W. G. (1975). Effect of nafenopin (SU-13,437) on liver function. Hepatic uptake and biliary excretion of ouabain in the rat. *Drug Metab. Dispos.* **3,** 220-225.
100. Levine, W. G., Braunstein, I. R., and Meijer, D. K. F. (1975). Effect of nafenopin (SU-13,437) on liver function. Mechanism of the choleretic effect. *Naunyn-Schmiedeberg's Arch. Pharmacol.* **290,** 221-234.
101. Uesugi, T., and Levine, W. G. (1976). Effect of nafenopin (SU-13,437) on liver function. Influence on the hepatic transport of phenolphthalein glucuronide and chlorothiazide. *Drug Metab. Dispos.* **4,** 107-111.
102. Jorritsma, J., Meerman, J. H. N., Vonk, R. J., and Mulder, G. J. (1979). Biliary and urinary excretion of drug conjugates: Effect of diuresis and choleresis on excretion of harmol sulfate and harmol glucuronide in the rat. *Xenobiotica* **9,** 247-252.
103. Mueller, M. N., and Kappas, A. (1964). Estrogen pharmacology. I. The influence of estradiol and estriol on hepatic disposal of sulfobromophthalein (BSP) in man. *J. Clin. Invest.* **43,** 1905-1914.
104. Reyes, H., and Kern, F. (1979). Effect of pregnancy on bile flow and biliary lipids in the hamster. *Gastroenterology* **76,** 144-150.
105. Meyers, M., Slikker, W., Pascoe, G., and Vore, M. (1980). Characterization of chole-

stasis induced by estradiol-17-β-glucuronide in the rat. *J. Pharmacol. Exp. Ther.* **214,** 87–93.
106. Plaa, G. L., and Priestly, B. G. (1976). Intrahepatic cholestasis induced by drugs and chemicals. *Pharmacol. Rev.* **28,** 207–273.
107. Gumucio, J. J., Accatino, L., Macho, A. M., and Contreras, A. (1973). Effect of phenobarbital on the ethynyl estradiol-induced cholestasis in the rat. *Gastroenterology* **65,** 651–657.
108. Forker, E. L. (1969). The effect of estrogen on bile formation in the rat. *J. Clin. Invest.* **48,** 654–663.
109. Davis, R. A., Kern, F., Showalter, R., Sutherland, E., Sinensky, M., and Simon, F. R. (1978). Alterations of hepatic (Na^+, K^+)-ATPase and bile flow by estrogen: Effects on live surface membrane lipid structure and function. *Proc. Natl. Acad. Sci. U.S.A.* **75,** 4130–4134.
110. Simon, F. R., Gonzalez, M., Sutherland, E., Accatino, L., and Davis, R. A. (1980). Reversal of ethinyl estradiol-induced bile secretory failure with Triton WR-1339. *J. Clin. Invest.* **65,** 851–860.
111. Bhattacharyga, M. H., and Peterson, D. P. (1979). Action of DTPA on hepatic plutonium. III. Evidence for a direct chelation mechanism for DTPA-induced excretion of monomeric plutonium into rat bile. *Radiat. Res.* **80,** 108–115.
112. Figueroa, W. G., and Thompson, J. H. (1968). Biliary iron excretion in normal and iron loaded rats after desferrioxamine and Ca DTPA. *Am. J. Physiol.* **215,** 807–810.
113. Kiyozuma, M., and Kojima, S. (1978). Studies on poisonous metals. Excretion of cadmium through bile and gastrointestinal mucosa and effect of chelating agents on its excretion in cadmium pretreated rats. *Chem. Pharm. Bull.* **26,** 3410–3415.
114. Norseth, T. (1974). The effect of diethyldithiocarbamate on biliary transport, excretion and organ distribution of mercury in the rat after exposure to methyl mercuric chloride. *Acta Phacmacol. Toxicol.* **34,** 76–87.
115. Kitani, K., Miura, R., Kanai, S., and Morita, Y. (1977). The effect of spironolactone pretreatment on the biliary excretion and renal accumulation of inorganic mercury in the rat. *Biochem. Pharmacol.* **26,** 1823–1824.
116. Klaassen, C. D. (1979). Effect of spironolactone on the biliary excretion and distribution of metals. *Toxicol. Appl. Pharmacol.* **50,** 41–48.
117. Collins, W. T., and Capen, C. C. (1980). Biliary excretion of ^{125}I-thyroxine and fine structure alterations in the thyroid glands of Gunn rats fed polychlorinated biphenyls. *Lab. Invest.* **43,** 158–164.
118. Levine, W. G. (1975). Effect of DMSO on the hepatic disposition of chemical carcinogens. *Ann. N.Y. Acad. Sci.* **243,** 185–193.
119. Marinovic, Y., Glasinovic, J., Semelle, B., Boivieux, J., and Erlinger, S. (1977). Facilitation of hepatic uptake of phenol-3,6-dibromophthalein disulfonate by taurocholate. *Am. J. Physiol.* **232,** E560–E564.
120. Delage, Y., Erlinger, S., Duval, M., and Benkamon, J. P. (1976). Influence of dehydrocholate and taurocholate on bromsulphthalein uptake, storage, and excretion in the dog. *Gut* **16,** 105–108.
121. Vonk, R. J., Danhof, M., Coenraads, T., Van Doorn, A. B. D., Keulemans, K., Scof, A. H. J., and Meijer, D. K. F. (1979). Influence of bile salts on hepatic transport of dibromosulphthalein. *Am. J. Physiol.* **237,** E524–E534.
122. Gregus, Z., Fischer, E., and Varga, F. (1979). Effect of sodium taurocholate on the hepatic transport of bromsulphthalein in rats. *Arch. Int. Pharmacodyn. Ther.* **238,** 124–133.
123. Abou-El-Makarem, M. M., Millburn, P., and Smith, R. L. (1967). Biliary excretion of [^{14}C]succinylsulphathioazole in the rat and rabbit. *Biochem. J.* **105,** 1295–1299.

124. Kuo, S. H., and Johnson, G. E. (1975). The influence of [^{3}H]taurocholate on the biliary excretion of [^{14}C]acetylprocaine amide ethobromide. *Can. J. Physiol. Pharmacol.* **53**, 888–894.
125. Vonk, R. J., Scholtens, E., Keulemans, G. T. P., and Meijer, D. K. F. (1978). Choleresis and hepatic transport mechanism. IV. Influence of bile salt choleresis on the hepatic transport of the organic cations, *d*-tubocurarine and N^4-acetyl procainamide ethobromide. *Naunyn-Schmiedeberg's Arch. Pharmacol.* **302**, 1–9.
126. O'Maille, E. R. L., Richards, T. G., and Short, A. H. (1966). Factors determining the maximal rate of organic anions secretion by the liver and further evidence on the hepatic site of action of the hormone secretin. *J. Physiol. (London)* **186**, 424–438.
127. Boyer, J., Schieg, R. L., and Klatskin, G. (1970). Effect of sodium taurocholate on the hepatic metabolism of sulfobromophthalein sodium (BSP). The role of bile flow. *J. Clin. Invest.* **49**, 206–215.
128. Barnhart, J., Ritt, D., Ware, A., and Combes, B. (1973). *In* "The Liver. Quantitative Aspects of Structure and Function" (G. Paumgartner and R. Preisig, eds.), pp. 315–325. Karger, Basel.
129. Erlinger, S., and Dumont, M. (1973). Influence of canalicular bile flow on sulfobromophthalein transport maximum in bile in the dog. *In* "The Liver. Quantitative Aspects of Structure and Function" (G. Paumgartner and R. Preisig, eds.), pp. 306–314. Karger, Basel.
130. Gibson, G. E., and Forker, E. L. (1974). Canalicular bile flow and bromosulfophthalein transport maximum: The effect of a bile salt independent choleretic, SC-2644. *Gastroenterology* **66**, 1046–1053.
131. Scharschmidt, B. F., and Schmid, R. (1978). The micellar sink. A quantitative assessment of the association of organic anions with mixed micelles and other macromolecular aggregates in rat bile. *J. Clin. Invest.* **62**, 1122–1133.
132. Vonk, R. J., Van Doorn, A. B. D., Scaf, A. H. J., and Meijer, D. K. F. (1977). Choleresis and hepatic transport mechanisms. III. Binding of ouabain and K-strophanthoside to biliary micelles and influence of choleresis on their biliary excretion. *Naunyn-Schmiedeberg's Arch. Pharmacol.* **300**, 173–177.
133. Binet, S., Delage, Y., and Erlinger, S. (1979). Influence of taurocholate, taurochenodeoxycholate and taurodehydrocholate on sulfobromophthalein transport into bile. *Am. J. Physiol.* **236**, E10–E14.
134. Angelucci, L., Linari, G., and Baldieri, M. (1975). Caerulein effect on the biliary excretion of colorants. *Pharmacol. Res. Commun.* **7**, 311–322.
135. Kellaway, I. W., and Marriott, C. (1977). The interaction of quaternary ammonium compounds with bile salts. I. The interaction of dodecyl-trimethylammonium bromide with sodium deoxycholate. *Can. J. Pharm. Sci.* **12**, 70–73.
136. Kellaway, I., and Marriott, C. (1977). The interaction of quaternary ammonium compound with bile salts. II. The influence of alkyl chain length on the interaction of alkylmethylammonium bromides with sodium deoxycholate. *Can. J. Pharm. Sci.* **12**, 74–76.
137. Bertagni, P., Hirom, P. C., Millburn, P., Osiyemi, F. O., Smith, R. L., Turbert, H. B., and Williams, R. T. (1972). Sex and species differences in the biliary excretion of tartrazine and lissamine fast yellow in the rat, guinea pig, and rabbit. The influence of sex hormones in tartrazine excretion in the rat. *J. Pharm. Pharmacol.* **24**, 620–624.
138. Uesugi, T., Bognacki, J., and Levine, W. G. (1976). Biliary excretion of drugs in the rat during liver regeneration. *Biochem. Pharmacol.* **25**, 1187–1193.
139. Morris, D. H., and Silverman, J. A. (1975). Biliary excretion of [^{3}H]aldosterone and its sex dependence in adrenalectomized rats. *Endocrinology* **96**, 1386–1391.

140. Buttar, H. S., Caldwell, B. B., and Thomas, B. H. (1974). Excretion of ^{14}C-pentobarbital and its metabolites into the bile and urine of male and female rats. *Arch. Int. Pharmacodyn. Ther.* **208**, 279–288.
141. Kato, R. (1974). Sex-related differences in drug metabolism. *Drug Metab. Rev.* **3**, 1–32.
142. Neims, A. H., Warner, M., Loughnan, P. M., and Aranda, J. V. (1976). Developmental aspects of the hepatic cytochrome *P*-450 monooxygenase system. *Annu. Rev. Pharmacol. Toxicol.* **16**, 427–445.
143. Dutton, G. J. (1978). Developmental aspects of drug conjugation with special reference to glucuronidation. *Annu. Rev. Pharmacol. Toxicol.* **18**, 17–35.
144. Richey, D. P., and Bender, A. D. (1977). Pharmacokinetic consequences of aging. *Annu. Rev. Pharmacol. Toxicol.* **17**, 49–65.
145. Fischer, E., Barth, A., Varga, F., and Klinger, W. (1979). Age dependence of hepatic transport in control and phenobarbital-pretreated rats. *Life Sci.* **24**, 557–562.
146. Klaassen, C. D. (1974). Effect of microsomal enzyme inducers on the biliary excretion of cardiac glycosides. *J. Pharmacol. Exp. Ther.* **191**, 201–211.
147. Kitani, K., Kanai, S., Miura, R., Morita, Y., and Kashara, M. (1978). The effect of aging on the biliary excretion of ouabain in the rat. *Exp. Gerontol.* **13**, 9–17.
148. Fischer, L. F., and Weissinger, J. L. (1972). Development in the newborn rat of the conjugation and de-conjugation processes involved in the enterohepatic circulation of diethylstilbestrol. *Xenobiotica* **2**, 399–412.
149. Cagen, S. Z., and Gibson, J. E. (1977). Liver damage following paraquat in selenium-deficient and diethyl maleate pretreated mice. *Toxicol. Appl. Pharmacol.* **40**, 193–200.
150. Duggan, D. E., and Kwan, K. C. (1979). Enterohepatic recirculation of drugs as a determinant of therapeutic ratio. *Drug Metab. Rev.* **9**, 21–41.
151. Plaa, G. L. (1975). The enterohepatic circulation. *Handb. Exp. Pharmakol.* [N.S.] **28**, Part 3, 130–149.
152. Scheline, R. R. (1973). Metabolism of foreign compounds by gastrointestinal microorganisms. *Pharmacol. Rev.* **25**, 451–532.
153. Goldman, P. (1978). Biochemical pharmacology of the intestinal flora. *Annu. Rev. Pharmacol. Toxicol.* **18**, 523–539.
154. Goldman, P. (1982). The role of intestinal microflora. *In* "Metabolic Basis of Detoxication" (W. B. Jakoby, J. R. Bend, and J. Caldwell, eds.), pp. 323–337. Academic Press, New York.
155. Marselos, M., Dutton, G., and Hanninen, O. (1975). Evidence that D-glucaro-1,4-lactone shortens the pharmacological action of drugs being disposed via the bile as glucuronides. *Biochem. Pharmacol.* **24**, 1855–1858.
156. Walsh, C. T., and Levine, R. R. (1975). Studies on the enterohepatic circulation of morphine in the rat. *J. Pharmacol. Exp. Ther.* **195**, 303–310.
157. Fischer, L. J., Kent, T. H., and Weissinger, J. L. (1973). Absorption of diethylstilbestrol and its glucuronide conjugate from the intestine of five- and twenty-five-day old rats. *J. Pharmacol. Exp. Ther.* **185**, 163–170.
158. Brewster, D., Jones, R. S., and Symons, A. M. (1977). Effect of neomycin on the biliary excretion and enterohepatic circulation of mestranol and 17-β-oestradiol. *Biochem. Pharmacol.* **26**, 943–946.
159. Dobbs, H. E., Hall, J. M., and Steiger, B. (1970). Enterohepatic circulation of etorphine, a potent analgesic, in the rat. *Proc. Eur. Soc. Study Drug Toxic.* **11**, 73–79.
160. Yesair, D. W., Callahan, M., Remington, L., and Kensler, C. J. (1970). Role of the entero-hepatic cycle of indomethacin on its metabolism, distribution in tissues and its excretion by rats, dogs and monkeys. *Biochem. Pharmacol.* **19**, 1579–1590.

161. Inaba, T., Tsutasumi, E., Mahon, W. A., and Kalow, W. (1974). Biliary excretion of diazepam in the rat. *Drug Metab. Dispos.* **2**, 429–432.
162. Sellman, R., Kanto, J., and Pekkarinen, J. (1975). Biliary excretion of diazepam and its metabolites in man. *Acta Pharmacol. Toxicol.* **37**, 242–249.
163. Duggan, D., Hooke, K., Noll, R., and Kwan, K. (1975). Enterohepatic circulation of indomethacin and its role in intestinal irritation. *Biochem. Pharmacol.* **25**, 1749–1754.
164. Roerig, D. L., Hasegawa, A. T., and Wang, R. I. H. (1980). Enterohepatic circulation of 1-α-acetylmethadol in the rat. *J. Pharmacol. Exp. Ther.* **213**, 284–288.
165. Levy, R. I., Fredrickson, D. S., Stone, N. J., Bilheimer, D. W., Brown, M. V., Glueck, C. J., Gotto, A. M., Herbert, P. N., Kwiterovich, P. O., Laager, T., and La Rosa, J., Lux, S. E., Rider, A. K., Shulmon, R. S., and Sloan, H. R. (1973). Cholestyramine in type II hyperlipoproteinemia—A double-blind trial. *Ann. Intern. Med.* **79**, 51–58.
166. Cohn, W. J., Boylan, J. J., Blanke, R. V., Fariss, M. W., Howell, J. R., and Guzelian, P. S. (1978). Treatment of chlordecone (Kepone) toxicity with cholestyramine. Results of a controlled clinical trial. *N. Engl. J. Med.* **298**, 243–248.
167. Bucher, N. L. R., and Malt, R. A. (1971). "Regeneration of Liver and Kidney." Little, Brown, Boston.
168. Klaassen, C. D. (1974). Comparison of the effects of two-thirds hepatectomy and bile duct ligation on hepatic excretory function. *J. Pharmacol. Exp. Ther.* **191**, 25–31.
169. Borzellica, J. F., and Cherrick, H. M. (1965). The excretion of drugs in saliva. *J. Oral Ther. Pharmacol.* **2**, 180–187.
170. Rasmussen, F. (1964). Salivary excretion of sulphonamides and barbiturates by cows and goats. *Acta Pharmacol. Toxicol.* **21**, 11–19.
171. Glynn, J. P., and Bastain, W. (1973). Salivary excretion of paracetamol in man. *J. Pharm. Pharmacol.* **25**, 420–421.
172. Koysooka, R., Ellis, E. F., and Levy, G. (1974). Relationship between theophylline concentration in plasma and saliva of man. *Clin. Pharmacol. Ther.* **15**, 454–460.
173. Matin, S. B., Wan, S. H., and Karam, J. H. (1974). Pharmacokinetics of tolbutamide: Prediction by concentration in saliva. *Clin. Pharmacol. Ther.* **16**, 1052–1058.
174. Huffman, D. H. (1975). Relationship between digoxin concentration in serum and saliva. *Clin. Pharmacol. Ther.* **17**, 310–312.
175. Cook, C. E., Amerson, E., and Poole, W. K. (1975). Phenytoin and phenobarbital concentrations in saliva and plasma measured by radioimmunoassay. *Clin. Pharmacol. Ther.* **18**, 742–747.
176. Dvorchik, B. H., and Vesell, E. S. (1976). Pharmacokinetic interpretation of data gathered during therapeutic drug monitoring. *Clin. Chem. (Winston-Salem, N.C.)* **22**, 868–878.
177. Horning, M. G., Brown, L., Nowlin, J., Lertratanagkoon, K., Kellaway, P., and Zion, T. E. (1977). Use of saliva in therapeutic drug monitoring. *Clin. Chem. (Winston-Salem, N.C.)* **23**, 157–164.
178. Stephen, K. W., McCrossan, J., Mackenzie, D., Macfarlane, C. B., and Speirs, C. F. (1980). Factors determining the passage of drugs from blood into saliva. *Br. J. Clin. Pharmacol.* **9**, 51–55.
179. O'Brien, T. E. (1974). Excretion of drugs in human milk. *Am. J. Hosp. Pharm.* **31**, 844–854.
180. Vorherr, H. (1974). Drug excretion in breast milk. *Postgrad. Med.* **56**, 97–104.
181. Giacoia, G. P., and Catz, C. S. (1979). Drugs and pollutants in breast milk. *Clin. Perinatol.* **6**, 181–196.
182. Arena, J. M. (1980). Drugs and chemicals excreted in breast milk. *Pediatr. Ann.* **9**, 452–457.

183. Wilson, J. T., Brown, R. D., Cherek, D. R., Dailey, J. W., Hilman, B., Jobe, P. C., Manna, B. R., Manna, J. E., Redetzki, H. M., and Stewart, J. J. (1980). Drug excretion in human breast milk. Principles, pharmacokinetics and projected consequences. *Clin. Pharmacokinet.* **5**, 1–66.
184. Berlin, C. M. (1980). The excretion of drugs in human milk. *Prog. Clin. Biol. Res.* **36**, 115–127.
185. Resman, B. H., Blumenthal, P., and Juska, W. J. (1977). Breast milk distribution of theobromine from chocolate. *J. Pediatr.* **91**, 477–480.
186. Varma, S. K., Collins, M., Row, A., Haller, W. S., and Varma, K. (1978). Thyroxine, tri-iodothyronine, and reverse tri-iodothyronine concentrations in human milk. *J. Pediatr.* **93**, 803–806.
187. Mischler, T. W., Corson, S. L., Larranaga, A., Bolognese, R. J., Neiss, E. S., and Vukovich, R. A. (1978). Cephradine and epicillin in body fluids of lactating and pregnant women. *J. Reprod. Med.* **21**, 130–136.
188. Koup, J. R., Rose, J. Q., and Cohen, M. E. (1978). Ethosuximide pharmacokinetics in a pregnant patient and her newborn. *Epilepsia* **19**, 535–539.
189. Gelenberg, A. J. (1979). Anoxapine, a new antidepressant appears in human milk. *J. Nerv. Ment. Dis.* **167**, 635–636.
190. Somogyi, A., and Gugler, R. (1979). Cimetidine excretion into breast milk. *Br. J. Clin. Pharmacol.* **7**, 627–629.
191. Berlin, C. M., Yaffe, S. J., and Ragni, M. (1980). Dispositin of acetaminophen in milk, saliva, and plasma of lactating women. *Pediatr. Pharmacol.* **1**, 135–141.
192. Findlay, J. W. A., De Angelis, R. L., Kearney, M. F., Welch, R. M., and Findlay, J. M. (1981). Analgesic drugs in breast milk and plasma. *Clin. Pharmacol. Ther.* **29**, 625–633.
193. Rogan, W. J., Bagniewska, M., and Damstra, T. (1980). Pollutants in breast milk. *N. Engl. J. Med.* **302**, 1450–1453.
194. Bagnell, P. C., and Ellenberger, H. A. (1977). Obstructive jaundice due to a chlorinated hydrocarbon in breast milk. *Can. Med. Assoc. J.* **117**, 1047–1048.
195. Mirkin, B. L. (1971). Diphenylhydantoin: Placental transport, fetal localization, neonatal metabolism, and possible teratogenic effects. *J. Pediatr.* **78**, 329–337.
196. Williams, R. H., Kay, G. A., and Jandorf, B. J. (1944). Thiouracil. Its absorption, distribution and excretion. *J. Clin. Invest.* **23**, 613–627.
197. Bland, E. P., Docker, M. J., Crawford, J. S., and Fan, R. F. (1969). Radioactive iodine uptake by thyroid of breast-fed infants after maternal blood-volume measurements. *Lancet* **2**, 1039–1040.
198. Amato, D., and Niblett, J. S. (1977). Neutropenia from cyclophosphamide in breast milk. *Med. J. Aust.* **1**, 383–384.
199. Jakubovic, A., Tair, R. M., and McGeer, P. L. (1974). Excretion of THC and its metabolites in ewe's milk. *Toxicol. Appl. Pharmacol.* **28**, 38–43.
200. Chao, F.-C., Green, D. E., Forrest, I. S., and Kaplan, J. N. (1976). The passage of ^{14}C-Δ-9-tetrahydrocannabinol into the milk of lactating squirrel monkeys. *Res. Commun. Chem. Pathol. Pharmacol.* **15**, 303–317.
201. Hepner, G. W. (1974). Breath analysis: Gastroenterological applications. *Gastroenterology* **67**, 1250–1256.
202. Hepner, G. W., and Vesell, E. S. (1974). Assessment of aminopyrene metabolism in man by breath analysis after oral administration of [^{14}C]aminopyrene. Effects of phenobarbital, disulfiram and portal cirrhosis. *N. Engl. J. Med.* **291**, 1384–1388.
203. Hepner, G. W., and Vesell, E. S. (1975). Quantitative assessment of hepatic function by breath analysis after oral administration of [^{14}C]aminopyrene. *Ann. Intern. Med.* **83**, 632–638.

204. Lauterburg, B. H., and Bircher, J. (1976). Expiratory measurement of maximal aminopyrine demethylation *in vivo:* Effect of phenobarbital, partial hepatectomy, portacaval shunt and bile duct ligation in the rat. *J. Pharmacol. Exp. Ther.* **196,** 501–509.
205. Bircher, J., and Preisig, R. (1981). Exhalation of isotopic CO_2. *In* "Methods in Enzymology" (W. B. Jakoby, ed.), Vol. 77, pp. 3–9. Academic Press, New York.
206. Houston, J. B., Lockwood, G. F., and Taylor, G. (1981). Use of metabolite exhalation rates as an index of enhanced mixed-function oxidase activity *in vivo*. *Drug Metab. Dispos.* **9,** 449–455.
207. Wendel, A., and Dumelin, E. E. (1981). Hydrocarbon exhalation. *In* "Methods in Enzymology" (W. B. Jakoby, ed.), Vol. 77, pp. 10–15. Academic Press, New York.
208. Wendel, A., Feuerstein, S., and Kong, K.-H. (1979). Acute paracetamol intoxication of starved mice leads to lipid peroxidation *in vivo*. *Biochem. Pharmacol.* **28,** 2051–2055.

CHAPTER 10

Impact of Nutrition on Detoxication

Juanell N. Boyd and
T. Colin Campbell

I.	Introduction	287
II.	Fasting/Starvation	289
III.	Protein	290
IV.	Carbohydrate	293
V.	Lipid	294
VI.	Trace Nutrients—Vitamins and Minerals	297
VII.	Comments	302
	References	302

I. INTRODUCTION

Nutritional status with respect to a specific nutrient or combination of nutrients can range from a severe deficiency disease to nutrient excess and toxicity. The pathological conditions that result from deficiencies, excesses, and imbalances can influence absorption, metabolism, and excretion of xenobiotic substances. Where pathological conditions exist, it is therefore important to consider if any change in xenobiotic toxicity is a primary effect of the nutrient on detoxication or is a secondary effect of the resultant nutrient-related pathology. The effects of pathological conditions on foreign compound metabolism have been reviewed by Kato.[1] In the absence of obvious pathology, nutrient imbalances may alter the extent or duration of a dose-dependent xenobiotic response. Changes in rate of disappearance of the parent compound from the blood and rate of appearance and patterns of excreted metabolites have been used as indicators of nutrient-induced changes in detoxication. Activities of individual components of xenobiotic-metabolizing systems may be altered and are usually measured by *ex vivo* experiments in which the treatment is applied *in vivo* but the effect is measured *in vitro*. Both direct and indirect activities of nutrients on such metabolizing systems may be measured with these procedures.

Not all nutrient-induced changes in xenobiotic activity are traced to the metabolizing enzyme systems. Nutritional modulation can alter rates of absorption, thereby influencing circulating levels even in the absence of any effect on metabolism. Tissue distribution of xenobiotics can be altered by nutritionally induced changes in body composition. Changes in renal function and in pH of body fluids can also result in changes in toxicity. Finally, target organ responsiveness may also be modified.

A change in any of the individual parameters measured *in vitro* may not necessarily result in an obvious alteration in biological effect. Apparent discrepancies in the interpretation of *in vitro* and *in vivo* effects can result if the activity measured *in vitro* is not rate limiting, if modified sequential enzyme reactions result in no net change in the concentration of the toxic metabolite at the target organ, or if there is a compensating change in target organ responsiveness. It is, therefore, important to weigh each type of experimental evidence in the context of related biochemical and physiological activities.

Although it is easiest to categorize nutritional effects by considering individual nutrients, it is important to consider the great difficulty of changing the intake or status of only one nutrient without influencing the activities of other nutrients. The macronutrients, protein, carbohydrate, and lipid, are particularly illustrative of this difficulty. Significant increases in one must be compensated for by significant decreases in another. The problem is confounded when lipid content is altered. Either caloric density will be changed or the intake of noncaloric dietary fiber must be modified, and that raises the additional question of the contribution of dietary fiber to the observed xenobiotic activities. It is possible to interchange protein and digestible carbohydrate without substantially changing caloric density, but the question still remains as to whether any observed effect is a "protein" effect or a "carbohydrate" effect. Different types of protein or carbohydrate may contribute different effects.

Another important consideration is that changes in diet composition will often lead to changes in total food intake. High-fat diets with their higher caloric densities will often lead to increased caloric intake when feeding is ad libitum, whereas diets deficient in essential fatty acid, total protein, or any of the essential amino acids will result in depressed intake.

10. Impact of Nutrition on Detoxication

Total food or caloric intake can be controlled by a pair-feeding protocol in which all groups are restricted to the intake level of the group that consumes the least. The animals on the restricted regimen, however, will be forced into a feeding/fasting cycle not exhibited by their ad libitum counterparts, thereby presenting yet another variable.

Nutrient–nutrient interactions are also important with the micronutrients, although it is usually possible to adjust their dietary content without significantly affecting the levels of other nutrients. Examples include the relationships among saturated and polyunsaturated fatty acids, vitamin E, and selenium as well as vitamin D, calcium, and phosphorus. Vitamin E and Se are both thought of as participating in the inhibition of peroxidative attack of membrane lipids; furthermore, the extent of protection depends on both intake and tissue levels of the polyunsaturated fatty acids. Their intake is, of course, related both to the total lipid and to the ratio of polyunsaturated to saturated fat in the diet, that is, a change in the intake of one type of fat often necessitates a change in the other. Quantitative assessment of the interdependence of polyunsaturated fatty acids, vitamin E, and Se, however, is not easily achieved and is likely to vary as a function of species and still other dietary conditions.

Because most studies of nutrient–toxicant interactions have been designed to investigate the effect of a single nutrient, this discussion will follow that convention. However, the reader must remain constantly aware of alternative interpretations due to nutrient interdependence. In evaluating past experiments, investigators should carefully consider caloric density, caloric intake, percentage of calories from each of the macronutrient sources, and nutrient–nutrient interactions. It may not be possible to state that an observed effect is due to a single nutrient, but the availability of all pertinent dietary information should make it possible to find nutrient combinations that either enhance or impair particular aspects of detoxication.

II. FASTING/STARVATION

The most severe form of nutritional modulation is starvation. The effect of starvation is generally decreased drug metabolism and clearance, with subsequent enhancement of pharmacological or toxic effects. These effects have been discussed in detail by Kato.[1] The effect of starvation on microsomal oxidase activity is species and sex dependent. Some reactions are decreased in male rats but increased in females, whereas other reactions are not affected at all. The sex-dependent effect is thought to be related to the ability of androgen to enhance binding of some substrates to cytochrome P-450, although the effect has not been observed in all species.

The effect of starvation on conjugation reactions has been studied less extensively than the effect on cytochrome P-450. A series of *ex vivo* studies indicate that glucuronide conjugation is decreased in starvation, although glucuronide transferase activity is not affected. Glucose administration prior to administration of a xenobiotic can diminish the depression of conjugate formation presumably due to an alteration in the availability of the cosubstrate, uridine diphosphoglucuronide.

Starvation can similarly influence relative rates of conjugation by different pathways. The ratio of glucuronide to sulfate conjugates in the urine is increased by starvation prior to administration of phenol to rats. This change in relative rates might also be explained by altered cosubstrate availabilities. Nonetheless, the overall effect of starvation is a reduction in the rate of conjugation.

Fasting can also influence the rate of excretion of xenobiotics by causing decreased urine flow and decreased urinary pH. The change in acidity results in more rapid nonionic diffusion of weak acids from the renal tubule back to blood.[2]

The effects of fasting and/or starvation on detoxication are complex. The ability to detoxify is generally impaired, but the effect on any one mechanism is not easily predicted. Although the effect on cytochrome P-450 system activities is species, sex, and substrate dependent, conjugation seems to be more consistently impaired. Influences on other factors such as absorption, plasma protein, and excretion are likely to depend on the severity of deprivation and on whether complicating pathological conditions exist.

III. PROTEIN

Nutritional status with respect to dietary protein is a function of both the quality and quantity consumed. Human protein intake varies from only a small fraction of that required for normal growth and maintenance to a severalfold excess. In humans, severely limited protein intake is usually accompanied by inadequate intake of all other nutrients, with a resultant difficulty in assigning specific pathological conditions to protein deficiency per se. Impaired hepatic function and hypoproteinemia are generally regarded, however, as consequences of protein deprivation, and similar effects can be produced in experimental animals by diets deficient in total protein or in individual essential amino acids.[3] It should be evident that protein intake will have a profound effect on toxicokinetic factors for xenobiotic disposition.

Dietary protein can influence the rate of absorption of orally administered xenobiotics. Alcohol intoxication is delayed by concomitant protein intake, and mildly acidic drugs such as aspirin and barbiturates are absorbed more slowly when gastric pH is increased by the buffering action of protein. Conversely, a high-protein meal results in enhanced absorption of the weak base theophylline.[4]

Hypoproteinemia induced by dietary protein deprivation can alter toxicity through decreased plasma binding of xenobiotics. This phenomenon has been demonstrated with the anticoagulant drug warfarin. Rats given decreased dietary protein (with an increased percentage of calories from carbohydrates and decreased total food consumption) had a lower concentration of albumin in plasma, an increased ratio of free to bound warfarin in the blood, and a more rapid increase in prothrombin time. Total plasma warfarin levels were also elevated, possibly as a consequence of impaired metabolism.[5]

The quality and quantity of dietary protein are known to affect the cytochrome P-450 system. For a more detailed discussion of this topic, the reader is referred to certain reviews.[6,7] A variety of *ex vivo* experiments have shown that protein deficiency (with increased dietary carbohydrate with or without differences in food consumption) can result in decreases in hepatic protein and DNA and in microsomal cytochrome P-450 and cytochrome P-450 reductase. Microsomal oxidation of a variety of substances is also impaired. Protein deficiency also influences microsomal enzyme induction by phenobarbital. The percentage increase is similar for sufficient and deficient animals, but the induced levels are considerably lower in deficient animals. In contrast, the induced levels in low-protein animals are similar to those in high-protein animals when expressed per kilogram of body weight. Similar effects have been observed with rainbow trout.[8] Decreased protein quality (lower biological value) also decreases microsomal oxidase activity.[9,10] In contrast to the effect of decreased dietary protein, supplementation of an adequate (18%) casein diet with tryptophan resulted in enhanced oxidase activity. The effect was similar to the inducing effect on these oxidases of other naturally occurring indoles.[11]

Conjugation is also influenced by dietary protein, but the effect is less consistent. Protein-deficient diets have been described as increasing, decreasing, or having no effect on conjugation reactions. The effect may vary with the species and substrate, but studies have been few, and the differences may simply reflect the study of different factors by different methods. For example, both decreased conjugation rates measurd *ex vivo* and decreased excretion of conjugated products have been observed, whereas microsomal transferase activities are either increased or not affected.[1,12,13] Except for glucuronidation, most of the other conjugation systems are primarily cytosolic, not microsomal. Some of these apparent discrepancies might be explained if decreased conjugation rates were due to reduced cosubstrate availability rather than reduced transferase activity.

It has long been recognized that alteration of xenobiotic metabolism by protein deprivation can result in either enhanced or decreased toxic response, depending on whether metabolites are more or less toxic than the parent compound. In a classic study, Kato and his colleagues[14] demonstrated that protein-deficient rats exhibited decreased metabolism and increased mortality with respect to strychnine, pentobarbital, and zoxazolamine. Mortality due to octamethylpyrophosphoramide, however, was decreased in protein-deficient animals; toxicity depended on the formation of an active metabolite by the cytochrome P-450 system. The results agreed with those of an earlier study[15] in which carbon tetrachloride hepatotoxicity was decreased in protein-deficient rats. A protein-deficient diet also protected against toxicity from heptachlor that is activated to the epoxide by the oxidases.[16] Interpretation of the influence of protein deficiency is confounded when rates of formation and removal of a reactive metabolite are both affected. Paracetamol is representative. The oxidases generate a reactive metabolite that binds to hepatic proteins, although such binding and subsequent hepatotoxicity occur only after glutathione has been depleted. Protein-deficient diets lead to increased toxicity despite decreased synthesis of the reactive metabolite because its subsequent removal is impaired by decreased hepatic glutathione levels.[17] Similarly, protein-deficient diets can either depress or enhance activation of procarcinogens to reactive metabolites. Aflatoxin B_1 and 7,12-dimethylbenzanthracene are examples of this phenomenon. A protein-deficient diet depresses binding of aflatoxin metabolites to DNA that is correlated with a decreased hepatic tumor response.[18] In contrast, a high-protein diet leads to less binding of 7,12-dimethylbenzanthracene metabolites to DNA and decreased mammary tumor response.[19] Altered carcinogen metabolism has been considered the primary mechanism whereby dietary protein can influence the carcinogenic response. Certain studies, however, suggest that postinitiation effects, for example, a depressed rate of cellular proliferation, may play at least as important a role as altered xenobiotic metabolism.[20] This complex issue is yet to be resolved.

In summary, protein deficiency in humans is virtually always accompanied by other deficiencies, and protein deficiency in experimental animals is generally accompanied by excess dietary carbohydrate and decreased caloric intake, making it difficult to attribute observed effects to protein alone. The overall effect is that protein deprivation and the accompanying nutritional perturbations are generally associated with a less rapid removal of xenobiotics from the body. This usually leads to increased toxicity, although there are certain exceptions. The effect of starvation on cytochrome P-450 system activity is species, sex, and substrate dependent, whereas the effect of protein deprivation is more consistently related to depressed activity. Starvation, in contrast, consistently depressed conjugation, whereas protein deprivation seems to have a varied effect.

IV. CARBOHYDRATE

The effect of dietary carbohydrate on xenobiotic metabolism has received relatively little attention. As noted, however, most studies employing decreased dietary protein were simultaneously employing increased levels of digestible carbohydrate. Generally, a high-carbohydrate (low-protein and/or lipid) diet results in a decreased rate of detoxication. Many reviews of nutrient–toxicant interactions have not considered carbohydrate effects, although Kato[1] and Campbell and Hayes[21] have discussed this nutrient. Microsomal oxidation is generally depressed when the carbohydrate/protein ratio is increased. Substitution of sucrose for a chow diet for 24–72 hr resulted in a marked decrease in oxidase activity.[1] These observations, however, cannot be considered solely a carbohydrate effect because all other nutrients, as well as anutrient inducers of the cytochrome P-450 system, were withheld. When the percentage of calories from carbohydrate was held constant, the nature of the carbohydrate source influenced oxidase activity. Sucrose or equal amounts of glucose and fructose resulted in the lowest values; glucose or fructose alone yielded intermediate values; and cornstarch gave the highest values.[22] Even nondigestible carbohydrates can influence hepatic microsomal oxidase activity. Substitution of pectin for cellulose in semipurified diets resulted in increases in both cytochrome P-450 and oxidative activity.[23]

As mentioned, glucuronide conjugation can be partially restored by administration of glucose to fasted rats. Clearance rates of antipyrine and theophylline were depressed when humans were switched from a high-protein diet to a high-carbohydrate diet.[24,25] Again, the effect cannot be attributed to carbohydrate alone. High-carbohydrate diets have been shown to increase barbiturate sleeping time in mice and benzylpenicillin mortality in rats.[21]

When a rat diet contained sucrose rather than equal parts of sucrose, glucose, and cornstarch, the lipogenic effect of phenobarbital and the emergence of aflatoxin-induced preneoplastic lesions were both enhanced.[26] Interestingly, when rats were fed isocaloric amounts of a semipurified diet containing either sucrose or glucose, the animals given sucrose had increased body weights and increased total body lipid but decreased total free fatty acid, phospholipid, and cholesterol. Liver lipids were not affected.[27] These rather profound biological effects of what appear to be relatively minor dietary modulations should merit consideration in the design and interpretation of studies evaluating toxicity and any effort to understand the mechanisms of nutrient–toxicant interactions.

V. LIPID

Lipid may be the most difficult of the macronutrients to study with respect to its influence on detoxication reactions because any change in the total lipid content of the diet involves a simultaneous change in caloric density. Furthermore, supplementation of a diet with additional lipid will result in depressed intake of all other nutrients by dilution whether or not caloric intake is held constant between supplemented and unsupplemented groups. All too often in the past, these factors were not given adequate consideration in the design and interpretation of experiments.

Studies of dietary lipid usually employ common food lipids such as lard, corn oil, coconut oil, safflower oil, hydrogenated vegetable oil (largely soy bean and cotton seed), herring oil, and menhadden oil, because clearly defined synthetic or purified lipids are prohibitively expensive for extensive feeding studies. Often, these sources of lipid have been used to elucidate the differential effects of saturated (S) and polyunsaturated (P) fatty acids. Those sources with a high P/S ratio are highly susceptible to autoxidation, especially in the presence of dietary iron, which serves as a catalyst. Autoxidation results in a decrease in the amount of P (and thus in a decreased P/S ratio) and an increase in the amount of lipid peroxides. Lipid peroxides further degrade to yield aldehydes and ketones that reduce the palatability of the diet. Organic peroxides are also toxic at high concentrations. Furthermore, conditions favoring autoxidation of fatty acids result in concurrent oxidation of fat-soluble vitamins. Antioxidants such as butylated hydroxyanisole (BHA), butylated hydroxytoluene (BHT), and ethoxyquin can be used to inhibit oxidation, but it is not

10. Impact of Nutrition on Detoxication

always clear to what extent this has been done in the past. The antioxidants may also directly influence detoxication.

The P/S ratio is not the only variable considered when these common food lipids are employed as experimental variables. For example, a high content of the C_{20} fatty acids can influence prostaglandin metabolism, cholesterol is a component of animal but not vegetable fats, and dietary lipid sources also vary in their content of fat-soluble vitamins. Finally, these lipid sources may contain such xenobiotics as secondary plant metabolites and environmental contaminants that may themselves be inducers of detoxifying enzymes.

Another important consideration is the duration of treatment. Brief treatment will result in differences in circulating lipid, whereas prolonged treatment can change the composition of membranes. Changes in microsomal membrane lipid may require as much as 8 to 10 weeks of dietary modification.[28] Lipid composition influences membrane fluidity, and the kinetics of membrane-bound enzyme systems will be altered by such changes in fluidity.[29,30] As the degree of unsaturation of membrane lipids increases, so does the potential for peroxidative damage. The effect of a dietary lipid modulation may depend, therefore, on whether membrane composition has been altered.

Despite the difficulty of controlling all the interdependent variables related to dietary lipid, numerous investigators have considered the effect of this nutrient on detoxication. Fat, like protein, can delay absorption when a drug or toxicant is administered orally during or soon after ingestion of food. Alcohol absorption and acute intoxication are influenced in this manner. In contrast, a high-fat meal can enhance absorption of highly lipophilic substances, as is the case with the antifungal agent griseofulvin.[4]

The influence of dietary lipid on detoxication has been investigated most extensively with respect to the cytochrome P-450 system. The hepatic endoplasmic reticulum contains 30–55% lipid, including cholesterol and cholesterol esters, free fatty acids, triglycerides, and predominantly phospholipid.[21] Phosphatidylcholine is an essential component of the cytochrome P-450 system, presumably playing a physicochemical role in maintaining architecture[13] and possibly influencing enzyme reaction rates.[31] Because changes in dietary lipid intake can result in changes in membrane composition, it is not surprising that such changes can influence the enzymatic activity of this membrane-bound system. Marshall and McLean[32] demonstrated that dietary lipids rich in polyunsaturated fatty acids permitted maximum phenobarbitone induction of the hepatic cytochrome P-450 system. Microsomal hydroxylase activity of rat kidney but not lung also increased with increased unsaturated lipid in the diet.[33] The increased oxidative activity corresponds to increased incorporation of linoleic and arachidonic acids into microsomal membranes.[34,35] Injection of unsaturated free fatty acids (intraperitoneally in dimethyl sulfoxide), however, caused depressed hydroxylase and demethylase activities, in some cases in the absence of any change in the content of polyunsaturated fatty acids in the microsomes.[35] It is not clear whether this apparent discrepancy is due to the route of administration, a nonmicrosomal membrane free fatty acid effect, or some other factor. Addition of unsaturated free fatty acids to microsomes *in vitro* has also been shown to inhibit metabolic activity.[36] It is quite clear from numerous investigations that a source of polyunsaturated fatty acids is required for optimal function of the cytochrome P-450 system.[37] Increasing the intake of highly unsaturated fat above 1–5% of the diet, however, has not been shown to alter oxidative activity significantly.[38]

Dietary cholesterol as well as the polyunsaturates have been shown to increase oxidative demethylation. The enhancing effort of 10% dietary herring oil was attributed primarily to the nonsaponifiable fraction rather than to the polyunsaturated fatty acids. The level of enhancement observed for this source, however, could not be explained on the basis of cholesterol content alone.[39] Microsomal enzyme-inducing environmental contaminants, if present, would occur in the nonsaponifiable fraction and might explain part of the difference between the nonsaponifiable fraction and the cholesterol in the herring oil. Hepatic aromatic hydrocarbon hydroxylase activity was not increased by cholesterol,[40] but this activity in the intestinal mucosa was increased by addition of cholesterol to the diet.[41] Cholesterol content was doubled in intestinal microsomes from cholesterol-supplemented animals. Because it has been demonstrated that polyunsaturated fatty acids enhance oxidase activity only when membrane composition is altered, the same may be true of cholesterol. In the work of Lambert and Wills,[39,40] cholesterol content of hepatic microsomes was not reported.

Although conjugation reactions have been studied less extensively, there is evidence that they also respond to changes in lipid intake. Administration of elaidic or linoleic acids intraperitoneally in dimethyl sulfoxide decreased UDPglucuronosyltransferase activity in hepatic microsomes, consistent with the decreased oxidase activity observed in the same experiment.[42] Activity was enhanced in intestinal microsomes by cholesterol supplementation.[41] The lipid influence on transferase activity, like the influence of the oxidases, is presumably a function of altered physicochemical characteristics of the endoplasmic reticulum.

Although biotransformation is generally enhanced and toxic and pharmacological effects are generally decreased by inclusion of polyunsaturated fat and cholesterol in the diet, extremely high polyunsaturated fatty acid and cholesterol intake can result in increased toxicity. Diets containing 17.5% of highly unsaturated sunflower seed oil resulted in depressed hepatic microsomal oxidative activity and increased phenobarbitone sleeping times, compared to the use of the less saturated 17.5% tallow diet.[43] Cardiotoxic effects of adriamycin were enhanced when rats were fed 53% chow with 39.7% butter and 5% cholesterol—an extreme nutritional modulation that resulted in numerous pathological changes prior to administration of the adriamycin.[44] It is well known that a high-fat diet, especially one high in polyunsaturates, results in increased carcinogenicity in a number of model systems. Epidemiological evidence suggests that a high-fat diet is also associated with increased cancer risk in humans.[45] Dietary fat, like dietary protein, can influence carcinogenicity by altering carcinogen metabolism. It appears, however, that the effect of dietary lipid is primarily on promotion rather than initiation. A high-fat diet has also been shown to increase excretion of bile acids, presumed promoters of colon carcinogenesis.[46] High-fat diets have enhanced carcinogenesis when given after but not before administration of the initiators, dimethylbenzanthracene or methylnitrosourea.[47,48] Thus, the effect of dietary fat on carcinogenesis may be more a function of target organ response than of altered detoxication.

In summary, modulation of dietary fat can influence absorption and biotransformation of xenobiotics. Activity of the cytochrome P-450 system is impaired when polyunsaturated fatty acid intake is inadequate. Some, but not all, oxidative activities, as well as UDPglucuronyltransferase, are enhanced by dietary cholesterol. Dietary fat appears to influence microsomal biotransformations by altering the physicochemical characteristic of the endoplasmic reticulum. Very high intake of polyunsaturated fatty acids may impair biotransformation and may result in an enhanced toxic response as a consequence of either altered pharmacokinetics or altered target organ responsiveness. Very high-fat diets may also result in altered toxicity secondary to diet-induced pathology. Some of the observed effects of diets high in unsaturates may be related to susceptibility both of diet components and of resultant cell membranes to autoxidation. The role of lipid peroxidation in detoxication will be discussed more fully in relation to vitamin E and Se.

VI. TRACE NUTRIENTS—VITAMINS AND MINERALS

Interactions are the rule rather than the exception in any consideration of effects of the trace nutrients on detoxication. Therefore, these nutrients will be discussed together rather than separately, although much of the research leading to current understanding of these interactions began as investigations related to evaluation of the role of individual nutrients. The New York Academy of Sciences, recognizing the importance of these interactions, in 1980 held a conference entitled Micronutrient Interactions: Vitamins, Minerals and Hazardous Elements. Many of the concepts discussed in this section were presented in the proceedings of that conference.[49]

The trace nutrients, like the macronutrients, can influence absorption of xenobiotics. It is recognized that minerals in the diet can influence the biological availability of each other, whether absorption is desirable of undesirable. Divalent cations can compete for chelation sites in intestinal contents as well as for binding sites on transport proteins. Zinc absorption is facilitated by formation of bidentate complexes with picolinic acid, a metabolite of tryptophan. The picolinic acid complexes of the toxic elements Cd and Pb are less stable than the Zn complex, thus providing a possible absorption-related explanation for the protective effect of excess Zn against Cd and Pb toxicities.[50] Competitive absorption of Pb and Ca is well documented and is probably due to competition for binding sites on intestinal mucosal proteins that could be vitamin D dependent. Iron deficiency can result in enhanced absorption of both divalent and trivalent cations, presumably because of increased synthesis of transferrin, which binds Fe preferentially but not exclusively.[51] Many other examples of competitive absorption can be found.[52]

It is important to realize that absorption of trace nutrients can be interdependent in a noncompetitive fashion as well. For example, ascorbic acid enhances absorption of nonheme iron both by chelation and by reducing Fe^{3+} to Fe^{2+}, the latter being more readily absorbed.[52,53] Vitamin A is required for normal differentiation of epithelial cells. In the absence of vitamin A, columnar cells become keratinized and may resemble squamous-type epithelium. It is likely that gastrointestinal absorption would be perturbed under such circumstances. Hypervitaminosis A causes, among other symptoms, drying and peeling of the skin, which would lead to enhanced dermal absorption. Absorption of inhaled xenobiotics might also be altered by changes in the respiratory epithelium induced by vitamin A deficiency. In chicks, Se deficiency causes pancreatic acinar cell atrophy with resultant inadequate secretion of pancreatic lipase. As a consequence, absorption of lipids and associated lipophilic substances is impaired. This means, of course, that absorption of fat-soluble vitamins is also impaired, resulting in secondary vitamin deficiencies. Gastrointestinal absorption, for example, could be perturbed directly as a result of lipid malabsorption or secondarily as a result of deficiencies of fat-soluble vitamins. With trace nutrients as well as with macronutrients, it is important to be aware of any pathological condition that might influence food intake, absorption of other nutrients, or absorption of xenobiotics.

The cytochrome P-450 system is influenced by a number of trace nutrients that are integral components of the system. Nicotinic acid is a necessary component for NADP and riboflavin for FMN or FAD in cytochrome P-450 reductase as well as other detoxifying enzymes.[54] Deficiency with respect to either of these vitamins has been associated with decreased oxidase activity.[1,21,55,56] Iron is an essential component of the heme moiety of cytochrome P-450, but surprisingly, iron-deficient diets have been associated with either no change or an increase in hepatic cytochrome P-450 activity without any difference in the specificativity of the cytochrome. Intestinal cytochrome P-450 content and oxidative activity, in contrast, are rapidly decreased when dietary iron is restricted. This discrepancy between the hepatic and intestinal effects may be due to the difficulty of depleting hepatic cells of iron by dietary restriction.[13,57,58] Iron excess can lead to depressed hepatic oxidative peroxidative damage to membrane lipids.[59]

Retinol (vitamin A) is not essential component of the reconstituted cytochrome P-450 system, but it may be a component of the endoplasmic reticulum and may influence activity by contributing to membrane stability.[1] Retinol-deficient diets have led to depressed hepatic oxidation, but activities in intestine and lung have been increased. *In vitro* addition of retinol to microsomes could not reverse the effect of deficient diets, thus suggesting an indirect effect of retinol, possibly involving its incorporation into the membrane.[60-62] Such studies have not always provided detailed information concerning the nature of the diet, duration of treatment, and tissue vitamin A levels. Vitamin A is susceptible to oxidation, and a high dietary concentration of vitamin A protects vitamin E, and vice versa. Because prolonged treatment is necessary to deplete tissues of these fat-soluble vitamins, it is possible that the supplemented diets may have influenced oxidase activity, not as a consequence of altered vitamin A status but as an effect on the oxidation of other dietary components.

There is substantial evidence that vitamin E (α-tocopherol) protects phospholipids of microsomal and mitochondrial membranes from peroxidative damage by trapping free radicals.[63] Because lipid peroxidation is associated with a decline in oxidase activities,[64,65] it is not surprising that activity can be influenced by dietary vitamin E. Maximum activity has been observed when diets contained both polyunsaturated fatty acids and vitamin E and when *ex vivo* membrane lipid peroxidation was minimal. The synthetic antioxidant BHT can protect dietary unsaturates and vitamin E from autoxidation but is not effective in preventing microsomal membrane peroxidation, presumably because it is not incorporated into the membrane. Dietary vitamin E has no effect on the cytochrome P-450 content of microsomes.

Selenium, like vitamin E, is thought to influence the cytochrome P-450 system primarily by protecting membrane lipids from peroxidative damage, but their mechanisms are entirely different. Whereas vitamin E probably functions as a membrane-bound antioxidant, Se, as does selenocystine, participates at the active site of glutathione peroxidase. This soluble enzyme[66] protects membrane lipids by destroying organic hydroperoxides and H_2O_2 before they can cause membrane disruption.[63,67] Because the enzyme is not membrane bound and occurs in the cytosol, *ex vivo* studies with isolated, resuspended microsomes may not demonstrate the protective effect of selenium. This may explain the lack of effect of selenium deficiency on microsomal oxidase activity in some studies.[68,69] Investigation of the relationship of microsomal oxidative activity to selenium and glutathione peroxidase activity are further complicated by the occurrence of a non-Se–dependent glutathione peroxidase activity due to glutathione S-transferase.[70] The ratio of non-Se to Se glutathione peroxidase activity has been shown to vary with both tissue and species.[71] When non-Se–dependent activity is high, total activity is not a sensitive indicator of Se status. The relative value of the non-Se and Se enzymes in protecting microsomal membranes is not clear.

Vitamin C (ascorbic acid) can also influence the cytochrome P-450 system. Decreased oxidative metabolism of xenobiotics has been observed in scorbutic guinea pigs.[13] As with any deficiency disease, however, it is difficult to establish whether altered metabolism is a direct or indirect effect of vitamin C. Chronic ascorbate deprivation without the complications of scurvy has been shown to result in depressed microsomal oxidase activity.[72] It has been suggested, but not proven, that vitamin C protects microsomal membranes from peroxidative damage by regenerating vitamin E when conditions favoring lipid autoxidation have resulted in the generation of vitamin E free radicals. The ascorbic acid free radical is then enzymatically reduced by an NADH-dependent system.[73] Vitamin C also enhances absorption of selenium and can thereby function indirectly by maintaining higher levels of Se-dependent glutathione perioxidase.[74]

In summary, trace nutrient deficiencies generally lead to depressed microsomal oxidase activity. Nicotinic acid, riboflavin, pantothenic acid, iron, and copper are directly involved in synthesis of components of the system. Vitamin E, selenium, and vitamin C are not necessary for activity in reconstituted microsomes, but appear to function by protecting membrane lipids from peroxidative damage.

10. Impact of Nutrition on Detoxication

A number of trace nutrients are required for conjugation reactions. Nicotinic acid is required for NAD, which is involved in the synthesis of UDPglucuronic acid. Coenzyme A is involved in hippuric acid synthesis, glutamine conjugation, and acetylation; pantothenic acid, in turn, is required for synthesis of coenzyme A. Folic acid is required for some methylations, and vitamin B_{12} is utilized in cyanide detoxication.[54] Because glucuronyltransferase is located in the endoplasmic reticulum and its activity is dependent on membrane structure,[75] the nutrients involved in membrane protection are likely to influence glucuronidation. Most investigations concerning these nutrients have involved only the oxidase systems, and little is known about the effect on conjugation reactions.

The influence of trace nutrients on xenobiotic clearance generally parallels their effect on metabolism. Thus, ascorbic acid deficiency has led to a decreased oxidative metabolism and an increased plasma half-life of acetanilide.[76] Magnesium deficiency has caused decreased metabolism of aniline both *ex vivo* and *in vivo*. Iron deprivation, in contrast, increased the oxidative metabolism and decreased the plasma half-life of the same substrate.[58]

The effects of trace nutrients on toxic response have often been related to their effects on xenobiotic metabolism. Increased duration of zoxazolamine paralysis in vitamin C–deficient guinea pigs is thought to be due to decreased oxidative metabolism.[76] Vitamin C may also decrease toxic response through its ability to trap superoxide and quinones, a suggestion offered as a mechanism for decreased hepatotoxicity of acetaminophen or cocaine in rats pretreated with ascorbate.[77] Sensitivity to acetaminophen hepatotoxicity is enhanced by the autoxidation-favoring conditions of increased dietary polyunsaturated fatty acids without vitamin E, suggesting that lipid peroxidation may be involved in the toxic mechanism.[78] Thus, the trace nutrients that protect against oxidative damage and maintain the integrity of the xenobiotic-metabolizing system also may be important in preventing cellular injury as a consequence of the peroxide-generating conditions that result directly from oxidative metabolism.

Any consideration of the influence of trace nutrients on detoxication must take into account the multitude of possibilities for interactions at all levels: absorption, metabolism, excretion, and target organ response. The dietary content of a trace nutrient, unlike that of a macronutrient, can be modified without substantially altering the content of other nutrients, but the required intake levels of many trace nutrients are interdependent, and mechanisms of influence on detoxication may overlap. Furthermore, deprivation or excess may lead to decreased food intake and/or pathological conditions, so that the status with respect to more than one nutrient is disturbed. The consequence of trace nutrient deprivation is generally impaired detoxication, but it is often difficult to establish the exact mechanism(s) of involvement of any given nutrient.

VII. COMMENTS

Nutritional modulations can have both profound and subtle effects on detoxification of xenobiotic substances. These effects can result from direct involvement of a nutrient in various detoxication reactions. Indirect effects on detoxication can also occur when nutritional manipulation results in pathological conditions and/or altered status with respect to other nutrients. Nutritional deprivation generally results in decreased rates of detoxication and increased toxic response, although it is somewhat hazardous to generalize. Exceptions occur when nutritional deprivation results in decreased synthesis of toxic metabolites or decreased responsiveness of the target organ, as in the influence of protein restriction on chemical carcinogenesis. The possibility of nutrient–nutrient and nutrient–toxicant interactions should be given serious consideration in the design and interpretation of any toxicological investigations.

REFERENCES

1. Kato, R. (1977). Drug metabolism under pathological and abnormal physiological states in animals and man. *Xenobiotica* **7**, 25–92.
2. Reidenberg, M. M. (1977). Obesity and fasting-effects on drug metabolism and drug action in man. *Clin. Pharmacol. Ther.* **22**, 729–734.
3. Goodhart, R. S., and Shils, W. E. (1978). "Modern Nutrition in Health and Disease," 5th ed., p. 80. Lea & Febiger, Philadelphia.
4. Roe, D. A. (1978). Diet-drug interactions and incompatibilities. *In* "Nutrition and Drug Interrelations" (J. N. Hathcock and J. Coon, eds.), pp. 319–346. Academic Press, New York.
5. Barber, D. L., and Colvin, N. W., Jr. (1980). Influence of dietary protein on the response of rats receiving toxic levels of warfarin. *Toxicol. Appl. Pharmacol.* **56**, 8–15.
6. Campbell, T. C., and Hayes, J. R. (1976). The effect of quantity and quality of dietary protein on drug metabolism. *Fed. Proc., Fed. Am. Soc. Exp. Biol.* **35**, 2470–2474.
7. Campbell, T. C. (1978). Effects of dietary protein on drug metabolism. *In* "Nutrition and Drug Interrelations" (J. N. Hatchcock and J. Coon, eds.), pp. 409–422. Academic Press, New York.
8. Stott, W. T., and Sinnhuber, R. O. (1978). Dietary protein levels and aflatoxin B_1 metabolism in Rainbow Trout (Salmo Gairdneri). *J. Environ. Pathol. Toxicol.* **2**, 379–388.
9. Miranda, C. L., Webb, R. E., and Ritchey, S. J. (1974). Effect of dietary amino acids on hepatic drug metabolism in the rat. *Bull. Environ. Contam. Toxicol.* **12**, 104–108.
10. Kato, N., Tani, T., and Yoshida, A. (1981). Effect of dietary quality of protein on liver

microsomal mixed function oxidase system, plasma cholesterol and urinary ascorbic acid in rats fed PCB. *J. Nutr.* **111**, 123–133.
11. Evarts, R. P., and Mostafa, M. H. (1981). Effects of indole and tryptophan in cytochrome *P*-450, dimethyl nitrosamine demethylase, and arylhydrocarbon hydroxylase activities. *Biochem. Pharmacol.* **30**, 517–522.
12. Wood, G. C., and Woodcock, B. G. (1970). Effects of dietary protein deficiency on the conjugation of foreign compounds in rat liver. *J. Pharm. Pharmacol.* **22**, Suppl., 60S–63S.
13. Parke, D. V., and Ioannides, C. (1981). The role of nutrition in toxicology. *Annu. Rev. Nutr.* **1**, 207–234.
14. Kato, R., Oshima, T., and Tomizawa, S. (1968). Toxicity and metabolism of drugs in relation to dietary protein. *Jpn. J. Pharmacol.* **18**, 356–366.
15. McLean, A. E. M., and McLean, E. K. (1966). The effect of diet and 1,1,1-trichloro-2,2-*bis*-(*p*-chlorophenyl) ethyl (DDT) on microsomal hydroxylating enzymes and on sensitivity of rats to carbon tetra chloride poisoning. *Biochem. J.* **100**, 564–571.
16. Weatherholtz, W. M., Campbell, T. C., and Webb, R. E. (1968). Effect of dietary protein levels on the toxicity and metabolism of heptachlor. *J. Nutr.* **98**, 90–94.
17. Mclean, A. E. M., and Day, P. (1975). The effect of diet on the toxicity of paracetamol and the safety of paracetamol-methionine mixtures. *Biochem. Pharmacol.* **24**, 37–42.
18. Preston, R. S., Hayes, J. R., and Campbell, T. C. (1976). The effect of protein deficiency on the *in vivo* binding of aflatoxin B_1 to rat liver macromolecules. *Life Sci.* **19**, 1191–1198.
19. Clinton, S. K., Truex, C. R., and Visek, W. (1979). Dietary protein, aryl hydrocarbon hydroxylase and chemical carcinogenesis in rats. *J. Nutr.* **109**, 55–62.
20. Appleton, B. S., and Campbell, T. C. (1982). The dietary protein level during the promotion phase of aflatoxin B_1 induced preneoplastic hepatic lesion development overwhelms the effect of protein on initiation. *Fed. Proc., Fed. Am. Soc. Exp. Biol.* **41**, 355.
21. Campbell, T. C., and Hayes, J. R. (1974). Role of nutrition in the drug-metabolizing enzyme system. *Pharmacol. Rev.* **26**, 171–197.
22. Basu, T. K., Dickerson, J. W. T., and Parke, D. V. (1975). Effect of dietary substitution of sucrose and its constituent monosaccharides on the activity of aromatic hydroxylase and the level of cytochrome *P*-450 in hepatic microsomes of growing rats. *Nutr. Metab.* **18**, 302–309.
23. Proia, A. D., McNamara, D. J., Edwards, K. D. G., and Anderson, K. E. (1981). Effects of dietary pectin and cellulose on hepatic and intestinal mixed-function oxidations and hepatic 3-hydroxy-3-methylglutaryl-coenzyme A reductase in the rat. *Biochem. Pharmacol.* **30**, 2553–2558.
24. Conney, A. H., Pantuck, E. J., Kuntzman, R., Kappas, A., Anderson, K. E., and Alvarez, A. P. (1977). Nutrition and chemical biotransformations in man. *Clin. Pharmacol. Ther.* **22**, 707–720.
25. Alvares, A. P., Anderson, K. E., Conney, A. H., and Kappas, A. (1976). Interactions between nutritional factors and drug biotransformations in man. *Proc. Natl. Acad. Sci. U.S.A.* **73**, 2501–2504.
26. Boyd, J. N., Misslbeck, N., Parker, R. S., and Campbell, T. C. (1982). Sucrose enhanced emergence of aflatoxin B_1-induced γ-glutamyl transpepeptidase-positive rat hepatic cell foci. *Fed. Proc., Fed. Am. Soc. Exp. Biol.* **41**, 356.
27. MacDonald, I., Grenby, T. H., Fisher, M. A., and Williams, C. (1981). Differences between sucrose and glucose diets in their effects on the rate of body weight change in rats. *J. Nutr.* **111**, 1543–1547.
28. Agradi, E., Spagnualo, C., and Galli, C. (1975). Dietary lipids and aniline and

benzpyrene hydroxylations in liver microsomes. *Pharmacol. Res. Commun.* **7,** 469–480.
29. Stier, A. (1976). Lipid structure and drug metabolizing enzymes. *Biochem. Pharmacol.* **25,** 109–113.
30. Haeffner, E. W., and Privett, O. S. (1975). Influence of dietary fatty acids on membrane properties and enzyme activities of liver mitochondria of normal and hypophysectomized rats. *Lipids* **10,** 75–81.
31. Hayes, J. R., and Campbell, T. C. (1974). Effect of protein deficiency on the inducibility of the hepatic microsomal drug-metabolizing enzyme system. IV. Effect of 3-methylcholanthrene induction on activity and binding kinetics. *Biochem. Pharmacol.* **23,** 1721–1731.
32. Marshall, W. J., and McLean, A. E. M. (1971). A requirement for dietary lipids for induction of cytochrome *P*-450 by phenobarbitone in rat liver microsomal fraction. *Biochem. J.* **122,** 569–573.
33. Paine, A. J., and McLean, A. E. M. (1973). The effect of dietary protein and fat on the activity of aryl hydrocarbon hydroxylase in rat liver, kidney and lung. *Biochem. Pharmacol.* **22,** 2875–2880.
34. Norred, W. P., and Wade, A. E. (1972). Dietary fatty acid induced alterations of hepatic microsomal drug metabolism. *Biochem. Pharmacol.* **21,** 2887–2891.
35. Norred, W. P., and Wade, A. E. (1973). Effect of dietary lipid ingestion on the induction of drug-metabolizing enzymes by phenobarbital. *Biochem. Pharmacol.* **22,** 432–436.
36. DiAugustine, R. P., and Fouts, J. R. (1969). The effects of unsaturated fatty acids on hepatic microsomal drug metabolism and cytochrome *P*-450. *Biochem. J.* **115,** 547–554.
37. Wade, A. E., and Norred, W. P. (1976). Effect of dietary lipid on drug-metabolizing enzymes. *Fed. Proc., Fed. Am. Soc. Exp. Biol.* **35,** 2475–2479.
38. Gaillard, D., Chamoiseau, G., and Derache, R. (1977). Dietary effects on inhibition of rat hepatic microsomal drug-metabolizing enzymes by a pesticide (Morestan). *Toxicology* **8,** 23–32.
39. Lambert, L., and Wills, E. D. (1977). The effect of dietary lipid peroxides, sterols, and oxidised sterols on cytochrome *P*-450 and oxidative demethylation in the endoplasmic reticulum. *Biochem. Pharmacol.* **26,** 1417–1421.
40. Lambert, L., and Wills, E. D. (1977). The effect of dietary lipids on 3,4-benzo[*a*]pyrene metabolism in the hepatic endoplasmic reticulum. *Biochem. Pharmacol.* **26,** 1423–1427.
41. Hietanen, E., and Laitinen, M. (1978). Dependence of intestinal biotransformation on dietary cholesterol. *Biochem. Pharmacol.* **27,** 1095–1097.
42. Lang, M. (1976). Depression of drug metabolism in liver microsomes after treating rats with unsaturated fatty acids. *Gen. Pharmacol.* **7,** 415–419.
43. Hopkins, G. J., and West, C. E. (1976). Effect of dietary fats on pentobarbitone-induced sleeping times and hepatic microsomal cytochrome *P*-450 in rats. *Lipids* **10,** 736–740.
44. Zbinden, G., Brandle, E., and Pfister, M. (1977). Modification of adriamycin toxicity in rats fed a high fat diet. *Agents Action* **7,** 163–170.
45. Armstrong, B. K., and Doll, R. (1975). Environmental factors and cancer incidence and mortality and mortality in different countries with special references to dietary practices. *Int. J. Cancer* **15,** 617–631.
46. Reddy, B. S., Cohen, L. C., McCoy, G. D., Hill, P., Weisburger, J. H., and Wynder, E. L. (1980). Nutrition and its relationship to cancer. *Adv. Cancer Res.* **32,** 237–345.
47. Hopkins, G. J., and Carroll, K. K. (1979). Relationship between amount and type of dietary fat in promotion of mammary carcinogenesis induced by 7,12-dimethyl benzo[*a*]anthrancene. *JNCI, J. Natl. Cancer Inst.* **62,** 1009–1012.
48. Cahn, P. C., Head, J. F., Cohen, L. A., and Wynder, E. L. (1977). Influence of dietary fat on the induction of mammary tumors by *N*-nitrosomethyl urea: Associated hormone

10. Impact of Nutrition on Detoxication

changes and differences between Sprague-Dawley and F-344 rats. *JNCI, J. Natl. Cancer Inst.* **59,** 1279–1283.
49. Levander, O. A., and Cheng, L. (1980). Micronutrient interactions: Vitamins, minerals and hazardous elements. *Ann. N.Y. Acad. Sci.* **355.**
50. Sandstead, H. H. (1980). Interactions of toxic elements with essential elements: Introduction. *Ann. N.Y. Acad. Sci.* **355,** 282–284.
51. Mahaffey, K. R., and Rader, J. I. (1978). Metabolic interactions: Lead, calcium, and iron. *Ann. N.Y. Acad. Sci.* **355,** 285–297.
52. Underwood, E. J. (1977). "Trace Elements in Human and Animal Nutrition." Academic Press, New York.
53. Lynch, S. R., and Cook, J. D. (1980). Interactions of vitamin C and iron. *Ann. N.Y. Acad. Sci.* **355,** 32–44.
54. Williams, R. T. (1978). Nutrients in drug detoxification reactions. *In* "Nutrition and Drug Interrelations" (J. N. Hathcock and J. Coon, eds.), pp. 303–318. Academic Press, New York.
55. Zannoni, V. G., and Sato, P. H. (1976). The effect of certain vitamin deficiencies on hepatic drug metabolism. *Fed. Proc., Fed. Am. Soc. Exp. Biol.* **35,** 2464–2469.
56. Miltenberger, R., and Oltersdorf, U. (1978). The B-vitamin group and the activity of hepatic microsomal mixed-function oxidases of the growing Wistar rat. *Br. J. Nutr.* **39,** 127–137.
57. Hoensch, H., Woo, C. H., Raffin, S. R., and Schmid, R. (1976). Oxidative metabolism of foreign compounds in rat small intestine: Cellular localization and dependence on dietary iron. *Gastroenterology* **70,** 1063–1070.
58. Becking, G. C. (1978). Dietary minerals and drug metabolism. *In* "Nutrition and Drug Interrelations" (J. N. Hatchcock and J. Coon, eds.), pp. 371–398. Academic Press, New York.
59. Maines, M. D., and Kappas, A. (1977). Regulation of cytochrome P-450–dependent microsomal drug-metabolizing enzymes by nickel, cobalt, and iron. *Clin. Pharmacol. Ther.* **22,** 780–790.
60. Colby, H. D., Kramer, R. E., Greiner, J. W., Robinson, D. A., Krause, R. F., and Canady, W. J. (1975). Hepatic drug metabolism in retinol-deficient rats. *Biochem. Pharmacol.* **24,** 1644–1646.
61. Hauswirth, J. W., and Brizuela, B. S. (1976). The differential effects of chemical carcinogens on vitamin A status and on microsomal drug metabolism in normal and vitamin A deficient rats. *Cancer Res.* **36,** 1941–1946.
62. Mirandaa, C. L., Mukhtar, H., Bend, J. R., and Chhabra, R. S. (1979). Effects of vitamin A deficiency on hepatic and extra hepatic mixed-function oxidase and epoxide-metabolizing enzymes in guinea pig and rabbit. *Biochem. Pharmacol.* **28,** 2713–2716.
63. Scott, M. L., Noguchi, T., and Combs, G. F. (1974). New evidence concerning mechanisms of action of vitamin E and selenium. *Vitam. Horm. (N.Y.)* **32,** 429–443.
64. Kamataki, T., and Kitagawa, H. (1973). Effects of lipid peroxidation on activities of drug-metabolizing enzymes in liver microsomes of rats. *Biochem. Pharmacol.* **22,** 3199–3207.
65. Jacobson, M., Levin, W., Lu, A. Y. H., Conney, A. H., and Kuntzman, R. (1973). The role of pentobarbital and acetanilide metabolism by liver microsomes: a function of lipid peroxidation and degradation of cytochrome P-450 heme. *Drug Metab. Dispos.* **1,** 766–774.
66. Wendel, A. (1980). Glutathione peroxidase. *In* "Enzymatyic Basis of Detoxication" (W. B. Jakoby, ed.), Vol. 1, p. 333–353. Academic Press, New York.

67. Tappel, A. L. (1980). Vitamin E and selenium protection from *in vivo* lipid peroxidation. *Ann. N.Y. Acad. Sci.* **355,** 18–31.
68. Shull, L. R., Buckmaster, G. W., and Cheeke, P. R. (1977). Dietary selenium status and pyrrolizidine alkaloid metabolism *in vitro* by rat liver microsomes. *Res. Commun. Chem. Pathol. Pharmacol.* **17,** 337–340.
69. Burk, R. F., Mackinnon, A. M., and Simon, F. R. (1973). Selenium and hepatic microsomal hemoproteins. *Biochem. Biophys. Res. Commun.* **56,** 431–436.
70. Jakoby, W. B. (1980). Glutathione transferases. *In* "Enzymatic Basis of Detoxification" (W. B. Jakoby, ed.), Vol. 2, pp. 63–94. Academic Press, New York.
71. Burk, R. F., Lane, J. M., Lawrence, R. A., and Gregory, P. E. (1981). Effect of selenium deficiency on liver and blood glutathione peroxidase activity in guinea pigs. *J. Nutr.* **111,** 690–693.
72. Peterson, F. J., Holloway, D. E., Duquette, P. H., and Rivers, J. M. (1982). Dietary ascorbic acid and hepatic mixed function oxidase activity. (Submitted for publication.)
73. Machlin, L. J., and Gabriel, E. (1980). Interactions of vitamin E, vitamin C, vitamin B_{12}, and zinc. *Ann. N.Y. Acad. Sci.* **355,** 98–108.
74. Combs, G. F., and Pesti, G. M. (1976). Influence of ascorbic acid on selenium nutrition in the chick. *J. Nutr.* **106,** 958–966.
75. Testa, B., and Jenner, P. (1976). "Drug Metabolism: Chemical and Biochemical Aspects," pp. 323–324. Dekker, New York.
76. Zannoni, V. G., and Lynch, M. M. (1973). The role of ascorbic acid in drug metabolism. *Drug Metab. Rev.* **2,** 57–69.
77. Peterson, F. J., and Lindemann, N. J. (1982). Protection against acetominophen- and cocaine-induced hepatic damage in mice by ascorbic acid. (Submitted, for publication.)
78. McLean, A. E. M., Witts, D. J., and Tame, D. (1980). The influence of nutrition and inducers on mechanisms of toxicity in humans and animals. *Ciba Found. Symp.* **76,** 275–288.

CHAPTER 11

Relationships between the Enzymes of Detoxication and Host Defense Mechanisms

Kenneth W. Renton

I. Introduction	307
II. Nonspecific Immunostimulants	308
III. Adjuvant-Induced Arthritis	311
IV. Reticuloendothelial System	312
V. Interferon	313
VI. Infection in Animals	316
VII. Viral Infection in Humans	318
References	321

I. INTRODUCTION

A large number of different factors are known to alter the normal rate of xenobiotic metabolism, which plays a major role in determining the intensity and duration of action of lipid-soluble drugs, chemicals, and carcinogens.[1] In practical terms, the therapy for disease with drugs is often complicated by changes in the rate of drug biotransformation and elimination. These alterations in elimination rates account for a sizeable proportion of all reported drug reactions and interactions. In addition to the many factors already described, it has been recognized that major changes in drug biotransformation occur during episodes of infection or following the administration of immunoactive agents. The interaction of host defense systems and drug elimination mechanisms might be an important consideration in the use of drugs and chemicals during infections.

The first indication that the elimination of drugs is impaired by a host defense mechanism was reported as early as 1953, when cytochrome P-450 was unknown and little was known about the regulation of xenobiotic metabolism. Samaras and Dietz[2] suggested that the blockade of the reticuloendothelial system following the administration of trypan blue hampered detoxication of pentobarbital, and it was reported that "narcosis in dye injected animals was deeper and longer, reflexes were absent and the animals resembled carcasses." Despite this very early recognition of such an interaction, it is only recently that the potential seriousness of altered drug elimination during infection or following immune system stimulation has been recognized. This chapter will review the relationship that exists between the immune and other host defense systems and the biotransformation and elimination of drugs in both animals and humans.

II. NONSPECIFIC IMMUNOSTIMULANTS

The administration of nonspecific immunostimulants usually results in a decrease in hepatic microsomal drug metabolism and a decrease in the capacity of the liver to eliminate drugs.[2-17] The effect of a large number of these agents on the pathways of xenobiotic metabolism is summarized in Table I. All of these agents are effective in depressing microsomal drug biotransformation when administered *in vivo,* but they have no effect when added *in vitro* to isolated preparations of hepatic microsomes. Although these immunostimulant agents share a number of properties, unfortunately they also have a wide variety of specific actions. It is therefore difficult to determine precisely which mechanisms are involved in their depressant effect on the enzymes of detoxication. The two most widely studied immunostimulants are *Corynebacterium parvum* and *Bordetella pertussis*.

Castro first observed that an abnormal sensitivity to pentobarbital occurred in animals treated with a high dose of *C. parvum,* which also produced hepatosplenomegaly and increased skin homograft survival in mice.[19] Later, this barbiturate sensitivity was shown to result from a decrease in the ability of the liver to metabolize pentobarbital. *Corynebacterium parvum* also depressed hepatic cytochrome P-450 levels and a number of associated oxidative activities, including those of aminopyrine N-demethylase, p-nitroanisole O-dealkylase, and aniline hydroxylase.[5]

11. Enzymes of Detoxication and Host Defense Mechanisms

TABLE I

Immunostimulants That Affect Hepatic Cytochrome P-450–Dependent Drug Biotransformation

Immunostimulant	Drug biotransformation affected[a]	Species	References
Bordetella pertussis vaccine	Ethylmorphine, aniline, pentobarbital Aminopyrine, cyt P-450[a]	Mouse	3, 4, 9
	Phenytoin, cyt P-450	Rat	
Corynebacterium parvum	p-Nitroanisole, hexobarbital, pentobarbital aminopyrine, aniline, cyt P-450, antipyrine	Mouse Human	5
M. butyricum	Pentobarbital, ketamine	Mouse	6
BCG	Phenytoin, aminopyrine, cyt P-450	Mouse	7
Freunds' adjuvant	Cyt P-450	Rat	8
Escherichia coli endotoxin	Ethylmorphine, aniline, cyt P-450	Rat	9
Zymosan	Hexobarbital	Mouse	10
Statolon	Aminopyrine, benzo[a]pyrene, cyt P-450	Rat	9
Colloidal carbon	Ethylmorphine, carbon tetrachloride, cyt P-450	Rat	11, 12
Dextrans	Aminopyrine, cyt P-450	Mouse	13
Latex beads	Aminopyrine, cyt P-450	Mouse	13
Trypan blue	Pentobarbital		2
Maleic anhydride divinyl ether copolymer	Aminopyrine, aniline, cyt P-450	Mouse	14
Tilorone and analogues	Ethylmorphine, aniline, benzo[a]pyrene, cyt P-450	Rat	9
Poly (rI·rC)	Cyt P-450, aniline, ethylmorphine	Rat, mouse	9, 15
Interferon	Aminopyrine, benzo[a]pyrene, phenytoin, cyt P-450	Mouse	16, 17
Quinacrine	Ethylmorphine, aniline, cyt P-450	Rat	9
N,N-Dioctadecyl-N^1,N^1-bis(2-hydroxyethyl)-propanediamine	Ethylmorphine, aniline, cyt P-450	Rat	9
OK 482	Aniline, aminopyrine, pentobarbital	Mouse	18

[a] Reference to changes in cyt P-450 refers to changes noted on spectral measurement of cytochrome P-450.

This effect was not universal to all components of hepatic microsomal membranes, as cytochrome b_5 and NADPH cytochrome c reductase were unaffected. This agent can also depress cytochrome P-450 and related biotransformation in the lung. Glucuronidation of o-aminophenol was also depressed in the liver. *Corynebacterium parvum* produces both a dose- and a time-dependent response on hepatic drug metabolism, with the maximum depression of cytochrome *P*-450 occurring within 24 hr of administration and lasting for at least 7 days. The first biochemical changes in drug metabolism occur well before the appearance of histological changes and signs of hepatic necrosis and inflammation. It is also apparent that the potency of *C. parvum* in its ability to depress drug metabolism is related to its potency as an antitumor agent, because strains of the organism (CN-5888) that have poor antitumor activity have little or no effect on drug metabolism in the liver.

In the rat, the administration of *Bordetella pertussis* causes depression in a number of the oxidative activities associated with the cytochrome *P*-450 system in the liver.[3] This agent also increases the half-life of a single dose of phenytoin approximately fourfold.[4] The decrease in phenytoin elimination correlates with a decrease in the ability of hepatic cytochrome *P*-450 to oxidase phenytoin to *p*-hydroxydiphenylhydantoin. *Bordetella pertussis* is a complex mixture, and Williams *et al.*[3] have suggested that at least two different components are involved in the depression of hepatic drug biotransformation. When *B. pertussis* is heated to 80°C, drug biotransformation is depressed significantly at 24 hr, although the levels return to normal by the fifth day. This contrasts to the effect of untreated *B. pertussis* on the organism when heated to only 56°C, which is not only effective at 24 hr but lasts for more than 5 days. Neither active component is likely to be the well-known histamine-sensitizing factor (HSF), which is also heat labile; *B. pertussis* is active in CDF_1 mice, which are resistant to the effects of HSF. Williams also demonstrated that partially purified HSF has no effect on the biotransformation of drugs in strains of mice that are responsive to HSF.

Although the mechanism by which *Bordetella pertussis* affects drug biotransformation is unknown, it appears to be quite different from the one involved in the effects of *Corynebacterium parvum*. Because factors affecting macrophage function (silica, radiation, or splenectomy) diminish the action of *C. parvum*, it is likely that this agent depresses drug metabolism by acting on macrophages. Unlike *C. parvum*, splenectomy does not impair the ability of *B. pertussis* to depress cytochrome *P*-450, nor does *B. pertussis* cause hepatosplenomegaly.[20] Williams *et al.*[20] demonstrated that *B. pertussis* decreased cytochrome *P*-450 for a short period of time in *nu/nu* athymic mice, but the effect quickly disappeared and the level of hemoprotein returned to normal. This pattern is identical to that observed with *B. pertussis* heated to 80°C, suggesting that the *nu/nu* mice are unresponsive only to the heat-sensitive component of *B. pertussis*. The lack of T cells in this strain of mice strongy suggests involvement of the T cell-mediated immune system in the action of the heat-sensitive component that inhibits xenobiotic metabolism. However, as Williams has pointed out, the possibility also exists that the levels of circulating organisms may be reduced by increased activation of macrophages, which do exist in this mouse strain, and that the reduced load of *B. pertussis* produces only a transient depression in cytochrome P-450.

Although the lower response to xenobiotics has been reported for the large number of agents listed in Table I, almost no information exists concerning their mode of action. Because they have such a wide variety of actions, and because at least two mechanisms appear to cause drug metabolism depression by *B. pertussis,* with a third involved in the depression caused by *C. parvum,* it is likely that these agents act by unrelated mechanisms. As will be discussed, many of these agents cause an activation of the reticuloendothelial system or induce the formation of interferon, both of which appear to be implicated in reducing hepatic drug metabolism. It is very unlikely that a single mode of action can be used to explain the activity of all of these agents.

III. ADJUVANT-INDUCED ARTHRITIS

The depression of drug metabolism caused by the agents listed in Table I generally occurs within 24 hr of administration. Such depression of hepatic drug metabolism is distinct from that associated with the development of adjuvant-induced arthritis.[21] This model of arthritis, which occurs after subcutaneous administration of immunoactive agents such as *Mycobacterium butyricum* or *Mycobacterium tuberculosis* suspended in paraffin oil, develops over a period of 15 days in rats. In arthritic rats, barbiturate sleeping time and zoxazolamine paralysis times are significantly increased and cytochrome P-450 and other indices of biotransformation in hepatic microsomes are depressed. The activation of cyclophosphamide is also depressed in arthritic rats. Anti-inflammatory agents such as phenylbutazone or indomethacin reverse both the arthritis and the lower rate of drug metabolism.[21] When the tubercle bacilli are injected via the lymph node, arthritis is not produced and depression in drug biotransformation does not occur. It therefore appears that the depression of drug biotransformation in this model of arthritis is totally dependent on the development of the arthritic lesions rather than on delayed hypersensitivity that would follow the administration of tuberculin.

Although the elimination of drugs in patients with rheumatoid arthritis is usually unchanged, several studies have reported increased rates of salicylate elimination.[21] This increase is probably due to lowered serum albumin levels and consequent diminished drug binding in arthritic patients, which then lead to a higher concentration of free drug for excretion. Such a mechanism has been confirmed in rats by Ferdinandi et al.,[23] who demonstrated that binding of etodolac and furobofen to albumin was diminished during adjuvant-induced arthritis. This resulted in lower drug concentrations in serum and higher concentrations in inflamed tissue.

IV. RETICULOENDOTHELIAL SYSTEM

Although most of the drug-metabolizing capacity of the liver occurs in the hepatocyte, it appears that the status of nonparenchymal cells exerts an influence on the steady-state levels of drug biotransformation in the hepatocyte. Such activity in the liver is decreased when the reticuloendothelial system (RES) is either activated or depressed.[10] When Kupffer cells, which are part of the RES of the liver, are loaded with colloidal carbon particles, metabolism and hepatotoxicity of carbon tetrachloride are markedly reduced.[12] It has been shown that the administration of carbon particles decreases the levels of cytochromes P-450 and b_5 and ethylmorphine N-demethylase activity in hepatic microsomes.[11] Previously, other workers had demonstrated that several different agents including methyl palmitate, thorium dioxide, and pyran copolymers, all of which are known to stimulate the RES, were capable of prolonging barbiturate anesthesia.[10] Heme oxygenase, which can degrade heme from cytochrome P-450, is induced in hepatic reticuloendothelial cells following treatment with zymosan or endotoxin.[24,25] Both agents are known to lower drug metabolism. Barnes and co-workers[14] have shown that the ability of maleic anhydride ether copolymers to decrease drug biotransformation correlated with an increase in their molecular weights, their antiviral and antitumor activity, and their ability to block phagocytosis. Because these copolymers are also immune modulators and weak interferon inducers, in addition to their potent ability to block the RES, it is reasonable that one or all of these effects may be involved in the depression of drug metabolism. Other work provides further evidence implicating RES function.[13] Agents such as high molecular weight dextrans, dextran sulfates, and preparations of latex beads (0.109 and 0.79 μm), which are predominantly phagocytosed by RES cells in the liver, also decrease cytochrome P-450 and related reactions. Although the possibility exists that these agents have a direct action on hepatic parenchymal cells, it is more likely that their activity is mediated solely through activation of the RES. Such a hypothesis would require some form of mediator, able to depress cytochrome P-450 in parenchymal cells, to be transferred from the activated cells of the hepatic RES, which themselves contain little or no cytochrome P-450.

The acetylation of drugs by N-acetyltransferase is another interesting aspect of biotransformation affected by the RES. This enzyme is located in the RES cells of the liver rather than in the parenchymal cells.[21] Several immunoactive agents, including Freunds' adjuvant[26] and zymosan,[27] increase the excretion of acetylated sulfonamides and isoniazid in the urine of animals, suggesting that the activity of this pathway is dependent on RES function. However, du Souich and Courteau[28] have disputed this hypothesis, suggesting that Freunds' adjuvant can induce the N-acetyltransferase directly and that the acetylation of drugs is independent of the RES.

V. INTERFERON

In simultaneous reports, two laboratories[29,30] demonstrated that the cytochrome P-450 monooxygenase system was depressed in rats following the administration of the interferon-inducing agent tilorone. This effect only occurred *in vivo*, was relatively specific for cytochrome P-450, and was not associated with any morphological damage to the endoplasmic reticulum of the liver. In an extension of these studies, Renton and Mannering[9] reported that all other interferon inducers tested had a similar effect on drug metabolism and proposed that the ability to depress cytochrome P-450 was a common property of all interferon inducers, possibly related to the production of interferon itself. Although this interferon hypothesis was attractive, interferon inducers also have a wide variety of other properties, and these experiments were not conclusive.

Other evidence, however, provided proof that interferon per se can directly affect the levels of cytochrome P-450. The time course of the loss of cytochrome P-450 corresponds exactly to the time course of the appearance of interferon in the serum following treatment of mice with the interferon-inducing agent poly(rI·rC) or during an infection with encephalomycarditis virus.[13] Evidence for a direct involvement of interferon is also suggested by experiments involving in inbred strains of mice carrying four distinct genetic loci that influence the levels of circulating interferon[31] produced by specific genetic loci, which influence the levels of circulating interferon produced by specific viruses.[32] For Newcastle disease virus (NDV), one autosomal locus (IF-1) determines a 10-fold difference in serum interferon levels. Strains of mice carrying the high (IF-1^h) or low (IF-1^l) production allele at the IF-1 locus was used to demonstrate that depression of hepatic cytochrome P-450 was correlated with circulating interferon levels. In C57BL/6J mice, containing the h allele at IF-1, cytochrome P-450 and aminopyrine N-demethylase were decreased by 35% and 48%, respectively, 24 hr after the injection of the virus; the mean circulating level of interferon was 2,440 PRD_{50} units/ml. In C3H/HeJ mice containing the l allele at IF-1, no significant change was observed in cytochrome P-450 concentration or activity following injection of the virus; the circulating level of interferon was below the lowest limits of detection.

In experiments in our laboratory, the administration of preparations of mouse fibroblast and leukocyte interferon to mice depressed cytochrome P-450.[16] Although this evidence appears to confirm the involvement of interferon, these preparations were relatively crude, and it could be argued that the contaminants that contribute the bulk of the protein in these preparations was responsible for the depression in drug biotransformation. Proof for the involvement of interferon in the depression of hepatic cytochrome P-450 has been obtained by using highly purified preparations of human interferon that were obtained by recombinant DNA techniques from *Escherichia coli*.[16] Although most human interferons have little effect in other species, one of the cloned hybrid α-types, interferon (LEIF-AD), has marked antiviral and antitumor effects in the mouse. This interferon also depressed cytochrome P-450, including aminopyrine N-demethylase and benzo[a]pyrene hydroxylase, in two different mouse strains. This experiment provides the first conclusive direct evidence to support the hypothesis that the production of interferon is a contributing factor in the depression of cytochrome P-450 and drug elimination that occurs during infection or following the administration of interferon-inducing agents.

This depressant effect on drug biotransformation is not confined to the α- and β-types of interferon. Sonnenfeld and his co-workers[17,33] have shown that cytochrome P-450 can also be depressed during the induction of γ-interferon or following the administration of crude preparations of γ-interferon. This type of interferon is more potent than α- or β-type interferon in the mouse and requires only 6000 units to depress levels of the hemoprotein significantly in the liver. This higher potency of γ-interferon might well be related to its potent immunomodulatory activity, a characteristic of this type of interferon; it is also subject to the caveat that preparations of γ-interferon are highly contaminated with other natural products.

Direct antiviral and antitumor effects of interferons appear to be mediated by biochemical events leading to protein synthesis.[34] Similar mechanisms could be the cause of the depression of cytochrome P-450 turnover, which would then be an inseparable side effect of interferon therapy. By utilizing recombinant DNA techniques, it may be possible, however, to provide interferons that lack such effects but retain their antiviral or antitumor properties. The effects of interferons on cytochrome P-450 metabolism are likely to be significant in relation to the practical clinical use of these agents.

Although the mechanism of any action of interferon is unknown, evidence suggests that it is the biosynthesis of cytochrome P-450 protein that is impaired in those processes in which interferon inhibits biotransformation of xenobiotics. Other hemoproteins in the liver, such as catalase and tryptophan 2,3-dioxygenase, are also depressed by interferon inducers, suggesting that the phenomenon involved may be confined to hemoproteins.[15] However, hemoproteins with a long turnover time, for example, cytochrome b_5 or cytochromes a, b, c, or c_1, were unaffected, suggesting that the effect is relatively short-lived and active only on hemoprotein having a fast turnover.

Several experiments have suggested that all species of cytochrome P-450 are not affected to the same degree. For example, the loss of N-demethylase activity does not always parallel the loss of other cytochrome P-450 activities following the administration of interferon inducers. The interferon inducers poly(rI · rC), tilorone, or Freunds' adjuvant depress different types of cytochrome P-450.[35] The separation of cytochrome P-450 species using SDS–polyacrylamide gel electrophoresis demonstrated clearly that these three agents specifically depressed hemoproteins of different molecular weights.

Because the loss of hepatic cytochrome P-450 caused by interferon or interferon inducers never exceeds 50% of control values, it appears that the hemoprotein attains a new steady-state level rather than being completely eliminated in a dose-dependent fashion, as is the case with cobaltous chloride or 2-allyl-2-isopropylacetamide. Changes in levels of cytochrome P-450 are regulated by the synthesis and degradation of the enzyme. Degradation of cytochrome P-450 occurs in a biphasic manner with half-lives of 7 and 50 hr, respectively. Using the incorporation of δ-aminolevulinic acid into hepatic microsomes, Mannering *et al.*[15] suggested that interferon-inducing agents lowered the concentration of cytochrome P-450 by increasing its degradation rather than affecting synthesis, and that only the fast phase turnover of cytochrome P-450 was affected. The magnitude of the degradation rates described in these experiments, however, could not account for the magnitude of the loss observed experimentally. However, Singh, in our own laboratory, has demonstrated that heme degradation, measured by the expiration of CO from the methene bridge carbon of the porphyrin ring, was increased threefold in mice treated with interferon inducers. The area under the curve for CO expiration was decreased in treated mice, suggesting that the synthesis of heme was also impaired in these experiments. The magnitude of the changes measured in this way is much more compatible with the observed decrease of cytochrome P-450.

In a series of experiments that examined the relative time course of the levels of heme oxygenase, δ-aminolevulinic acid synthetase, and holo- and apotryptophan 2,3-dioxygenase, interferon-inducing agents, caused an increase in the regulatory heme pool that controls the rate of synthesis of heme for hemoproteins such as cytochrome P-450.[15] This would occur if interferon inducers decreased the synthesis of the apoprotein of cytochrome P-450 or increased the dissociation of heme from cytochrome P-450. Decreased synthesis of the apoprotein following the administration of interferon inducers has been observed in our laboratory. Although the incorporation of ^{14}C-labeled amino acids into total microsomal proteins was increased, the incorporation of labeled amino acids specifically into isolated cytochrome P-450 was significantly depressed. Previously, interferon had been demonstrated as increasing the synthesis of some proteins while decreasing the synthesis of others.[34] From our experiments, it appears that interferon inhibits the synthesis of apocytochrome P-450. As suggested by Mannering,[15] this would lead to a relative increase in the regulatory heme pool that would result in disturbances of heme turnover. It now seems apparent that interferon or its inducers lower cytochrome P-450 concentrations in the liver by the combined mechanism of inhibiting synthesis of the apoprotein and increasing the rate of degradation of the hemoprotein. The magnitude of these effects is consistent with the extent of cytochrome P-450 loss that has been observed.

VI. INFECTION IN ANIMALS

Infections of several different types have been shown to alter drug biotransformation in animals. In the mouse, hexobarbital oxidation and cytochrome P-450 are depressed during infections with murine hepatitis virus.[15,36] In the duck, hepatitis[37,38] not only depresses several drug oxidation reactions but also enhances the induction of cytochrome P-450 by 1,1,1-trichloro-2,2-bis(p-chlorophenyl)ethane (DDT). In humans, hepatitis virus appears to have a variable effect; it has been reported as increasing, decreasing, or leaving unchanged the half-life of drugs (Table II).

Alteration of xenobiotic metabolism during infection with viruses is not, however, confined to episodes of hepatitis, which is known to cause both morphological and biochemical changes in the liver. Viral infections that do not cause primary pathological effects also depress the ability of the

TABLE II

Depression of Cytochrome P-450–Dependent Drug Biotransformation during Infections in Animals

Infection	Drug biotransformation[a]	Species	References
Viral hepatitis virus (MHV-3)	Aniline, cyt P-450[a]	Mouse	39
Murine hepatitis virus	Hexobarbital	Mouse	36
Duck hepatitis virus	Ethylmorphine	Duck	38
Mengo virus	Ethylmorphine, aniline, cyt P-450	Rat	9
Encephalomycarditis virus	Aminopyrine, cyt P-450	Mouse	40
Influenza virus (PR-8)	Benzo[a]pyrene	Mouse	41
Newcastle disease virus	Aminopyrine, cyt P-450	Mouse	32
Malaria (*Plasmodium berghei*)	Aniline, p-nitroanisole, hexobarbital, ethylmorphine	Rat	42
Fasciola hepatica	Aminopyrine, hexobarbital, aniline, zoxasolamine, cyt P-450	Rat	43

[a] Reference to changes in cyt P-450 refers to changes noted on spectral measurement of cytochrome P-450.

liver to metabolize drugs. Renton and Mannering[9] described changes in cytochrome P-450 and related drug biotransformations in the liver of rats infected with lethal doses of Mengo virus. Sublethal doses of encephalomyocarditis virus (EMC), antigenically indistinguishable from Mengo virus, have also been shown to depress cytochrome P-450 and aminopyrine N-demethylase activity in mice.[40] The time course for the depression of drug biotransformation during this infection correlated with the appearance of interferon in the serum. Newcastle disease virus (NDV) also produces a marked depression in cytochrome P-450 in strains of mice that provide large quantities of interferon.[32] These results with EMC and NDV are consistent with the idea that interferon produced during the course of a viral infection can depress drug biotransformation, as discussed earlier in this chapter. Depression of drug biotransformation in the lungs of mice has also been reported following infection with a mouse-adapted influenza virus[41]; benzo[a]pyrene hydroxylase activity in the lung was depressed to 10% of control levels, whereas the activity of this enzyme in the liver remained unchanged. Induction of benzo[a]pyrene hydroxylase by 3-methylcholanthrene was also impaired in the lung during the infective period. Identical results for the effect of a mouse-adapted influenza virus on the metabolism of benzo[a]pyrene in the lung and liver have been described by Renton,[13] and aminopyrine N-demethylase activity in the liver was also decreased by 30% on the fourth day following administration of the virus.

Several workers have reported that the induction of drug biotransformation in animals by certain chemical agents can be enhanced during a concomitant viral infection. Duck hepatitis enhances the induction of cytochrome P-450 by DDT.[38] Doshi et al.[44] demonstrated the occurrence of hyperplasia of the endoplasmic reticulum and an increase in pentobarbital hydroxylase activity during the regenerative phase of acute viral hepatitis in humans. A single case has been reported in which phenytoin elimination was increased in a patient with infectious mononucleosis.[45] Each of these reports concerning increased drug oxidation during a viral infection involves a situation in which a proliferation of the endoplasmic reticulum occurs. Drug biotransformation appears to be depressed in situations in which the endoplasmic reticulum is normal, as in most of the examples described to date.

In addition to the effect of viral infections, hepatic drug biotransformation is depressed during the course of infections with malaria (*Plasmodium berghei*)[42] and a helminth parasite (*Fasciola hepatica*).[43] In these experiments, it was not established if the effect of drug biotransformation resulted from a host defense response or if the parasites caused direct damage to the endoplasmic reticulum membrane. Although the effect of bacterial infections has not been studied, it is likely that the endotoxins released from organisms and the response of the immune system to them will produce changes in xenobiotic metabolism.

The effect of infections on drug metabolism raises some concerns about the use of experimental animals in general biological research. Investigators have now accepted that animals must be maintained in highly controlled facilities, with great care being taken in limiting their exposure to enzyme inducers in bedding and food. The work reviewed in this chapter also suggests that infections, which might not produce obviously sick animals, could also have effects on experiments in which drugs are used or evaluated. Our laboratory has received apparently healthy animals from reputable suppliers that had lower than normal cytochrome P-450 levels; these levels could not be lowered further by immunostimulants. This suggested that the drug-metabolizing system was already maximally depressed by the operation of the host defense system.

VII. VIRAL INFECTION IN HUMANS

Although the number of reports is limited, there is now no question that the metabolism of drugs is also impaired in humans during viral infections

11. Enzymes of Detoxication and Host Defense Mechanisms

TABLE III

Depression of Drug Biotransformation during Infections in Humans

Infection	Drug biotransformation	References
Influenza A	Theophylline	46, 47
Influenza B	Theophylline	48, 49
Influenza vaccine (trivalent)	Theophylline, aminopyrine, warfarin	50–52
Adenovirus	Theophylline	46
Haemophilus influenzae	Theophylline	46

(Table III). In 1978, Chang et al.[46] reported that theophylline elimination was altered during episodes of acute upper respiratory tract viral illness. The half-life of theophylline was increased in four young patients during infections with positively identified influenza A infection compared to the half-life determined one month after the viral illness in each patient. Acute theophylline toxicity and a serum level of 43.3 µg/ml were observed in one of these individuals. A similar change in theophylline elimination was noted in a patient with an adenovirus infection and in four patients with febrile illness in which serological evidence of viral disease could not be confirmed. Renton[47] subsequently suggested that these effects could be explained by a depression in the level of cytochrome P-450 in the liver that resulted from the infection or from interferon produced during the infection.

Several other reports have since confirmed the impaired elimination of theophylline during infection.[47] Although it was not recognized at the time, an earlier report described slower theophylline elimination in a patient with decompensated cor pulmonale during an episode of pneumonia. The infecting agents have not been identified in all cases, but a decrease in the ability of patients to eliminate theophylline appears to be a relatively common occurrence during upper respiratory tract virus infections. In individuals receiving chronic theophylline dosage, signs of theophylline toxicity and elevated drug levels are frequently observed. In two other reports, theophylline elimination was impaired during pneumonia, but these cases were complicated by other factors. However, one report is of particular interest because it is the only one that identifies a bacterial infection (*Haemophilus influenzae*) as affecting theophylline elimination.

The potential seriousness of changes in drug elimination during infections is dramatically illustrated in four children studied by Woo et al.[48] Each child was admitted with signs of theophylline toxicity, with serum theophylline levels ranging from 33 to 68 µg/ml. Toxicity was associated with the development of upper respiratory tract viral infection. The half-life of theophylline was greatly prolonged during the infective period compared to the half-life of the drug that was determined subsequent to the infection. During an epidemic of influenza B infection in Seattle in 1980, the occurrence of theophylline toxicity was noted in 11 children.[49] Each patient had clinical evidence of an upper respiratory tract viral infection and, in six of the children, positive serological evidence of an influenza B infection was found. Prior to the infection, no difficulties had been encountered by these patients being maintained regularly on a normal dosage of theophylline.

Because it is impossible for ethical reasons to study the effects of a well-controlled experimental viral infection in humans, the elimination of theophylline was examined in a study following the vaccination of patients and volunteers with a trivalent influenza vaccine.[50] In three patients with established, stable, steady-state levels of theophylline, the administration of influenza vaccine resulted in an increase in theophylline levels within 24 hr and produced obvious signs of theophylline toxicity. In four healthy volunteers, the elimination of a single dose of theophylline was also impaired 24 hr after the administration of influenza vaccine. The mean half-life increased by 122% and the mean clearance was decreased by 51% following vaccination compared to control values obtained in the same individuals prior to vaccination. These findings were confirmed in a report[51] on the dangerous increase in theophylline concentrations produced by the administration of influenza vaccine to a patient receiving chronic theophylline therapy for asthma. These studies clearly demonstrate that the elimination of theophylline can be impaired to such a degree following vaccination with influenza vaccine that dangerous levels of the drug can accumulate in a short time in patients receiving multiple theophylline dosages.

In addition to its effect on theophylline elimination, Kraemer and McClain[52] have reported that the administration of influenza vaccine impaired the metabolism and clearance of aminopyrine in human volunteers; the activity of aminopyrine N-demethylase was assessed directly by measuring the amount of $^{14}CO_2$ in expired air following the administration of [^{14}C]aminopyrine. The elimination of $^{14}CO_2$ was depressed in two patients within 2 days, and within 7 days the amount of $^{14}CO_2$ in expired air was depressed by 22–74% in all 12 subjects studied. Even after 21 days following vaccination, a significant reduction in aminopyrine metabolism remained, suggesting that this effect is relatively long-lasting. In the same report, a patient is described who had a massive upper gastrointestinal tract hemorrhage (prothrombin time, 48 sec) 10 days after the administration of influenza vaccine. This individual had received warfarin for 12 years, and had had a stable prothrombin time and no difficulty with the warfarin dosage. Because no other potential cause for the anticoagulation could be identified, it is likely that the influenza vaccine caused a decrease in the inactivation of warfarin by cytochrome P-450, resulting in higher concentrations of warfarin.

It appears that the ability of the human liver to metabolize drugs is impaired during episodes of naturally acquired influenza viral infection or following the administration of influenza vaccine. To date, only infections of the respiratory tract have been implicated, and only viruses of the influenza class have been positively identified as causing this effect in humans. Experiments in animals, however, suggest that other virus infections will produce the same result. It also appears that a wide variety of drugs may be affected by the interaction because both theophylline and aminopyrine, which are metabolized by different classes of cytochrome P-450, are degraded at lower rates. This suggests that all forms of cytochrome P-450 might be affected, causing an impairment in the elimination of a wide variety of drugs. However, the clinically important manifestations of this interaction will be confined to drugs with low therapeutic indices such as theophylline or warfarin.

REFERENCES

1. Peterson, F. J., and Holtzman, J. L. (1980). Drug metabolism in the liver—A perspective. *In* "Extrahepatic Metabolism of Drugs and Foreign Compounds" (T. R. Gram, ed.), pp. 1–121. Spectrum Publ., New York.
2. Samaras, S. C., and Dietz, N. (1953). Physiopathology of detoxification of pentobarbital sodium. *Fed. Proc., Fed. Am. Soc. Exp. Biol.* **12**, 400.
3. Williams, J. F., Lowitt, S., and Szentivanyi, A. (1980). Involvement of a heat-stable and heat-labile component of *Bordetalla pertussis* in the depression of the murine hepatic mixed-function oxidase system. *Biochem. Pharmacol.* **29**, 1483–1490.
4. Renton, K. W. (1979). The deleterious effect of *Bordetella pertussis* vaccine and poly(rI.rC) on the metabolism and disposition of phenytoin. *J. Pharmacol. Exp. Ther.* **208**, 267–270.
5. Soyka, L. F., Stephens, C., MacPherson, B. R., and Foster, R. S. (1979). Role of mononuclear phagocytes in decreased hepatic drug metabolism following administration of *Corynebacterium parvum*. *Int. J. Immunopharmacol.* **I**, 101–112.
6. Barbieri, E. J., and Ciacci, E. I. (1979). Depression of drug metabolism in the mouse by a combination of *Mycobacterium butyricum* and anaesthetics. *Br. J. Pharmacol.* **65**, 111–115.
7. Farquar, D., Loo, T. L., Gutterman, J. U., Hersh, E. M., and Luna, M. A. (1976). Inhibition of drug-metabolizing enzymes in the rat after *Bacillus calmette*–guerin treatment. *Biochem. Pharmacol.* **25**, 1529–1535.
8. Morton, D. M., and Chatfield, D. H. (1970). The effects of adjuvant-induced arthritis on the liver metabolism of drugs in rats. *Biochem. Pharmacol.* **19**, 473–481.

9. Renton, K. W., and Mannering, G. J. (1976). Depression of hepatic cytochrome P-450-dependent monooxygenase systems with administered interferon inducing agent. *Biochem. Biophys. Res. Commun.* **73**, 343-348.
10. Wooles, W. R., and Munson, A. E. (1971). The effect of stimulants and depressants of reticuloendothelial activity on drug metabolism. *RES-J. Reticuloendothel. Soc.* **9**, 108-119.
11. Letterrier, F., Reynier, M., and Mariaud, J. (1973). Effect of intracellular accumulation of inert carbon particles on the cytochrome P-450 and b_5-level in rat liver microsomes. *Biochem. Pharmacol.* **22**, 2206-2208.
12. Stenger, R. J., Petrelli, M., Segel, A., Williams, J. N., and Johnson, E. A. (1969). Modification of carbon tetrachloride hepatotoxicity by prior loading of the RES with carbon particles. *Am. J. Pathol.* **57**, 689-706.
13. Renton, K. W. (1981). Effects of interferon inducers and viral infections on the metabolism of drugs. *Adv. Immunopharmacol., Proc. Int. Conf. Immunopharmacol., 1st, 1980* pp. 17-24.
14. Barnes, D. W., Morahan, P. S., Loveless, S., and Munson, A. E. (1979). The effects of maleic anhydride-divinyl ether (MVE) copolymers on hepatic microsomal mixed function oxidases and other biologic activities. *J. Pharmacol. Exp. Ther.* **208**, 392-398.
15. Mannering, G. J., Renton, K. W., El Azhary, R., and DeLoria, L. B. (1980). Effects of interferon-inducing agents on hepatic cytochrome P-450 drug metabolizing systems. *Ann. N.Y. Acad. Sci.* **350**, 314-331.
16. Singh, G., Renton, K. W., and Stebbing, N. (1981). Effect of interferon on hepatic cytochrome P-450 drug metabolizing systems. *Pharmacologist* **23**, 283.
17. Sonnenfeld, G., Harned, C. L., Thaniyavarn, S., Huff, T., Mandel, A. D., and Nerland, D. E. (1980). Type II interferon induction and passive transfer depress the murine cytochrome P-450 drug metabolism system. *Antimicrob. Agents Chemother.* **17**, 969-972.
18. Matsubara, S. F., Suzuki, F., and Ishida, N. (1974). Induction of immune interferon in mice treated with a bacterial immunopotentiator OK-432. *Cancer Immunol. Immunother.* **6**, 41-45.
19. Castro, J. E. (1974). The effect of C. Parvum on structure and function of the lymphoid system in mice. *Eur. J. Cancer* **10**, 115-120.
20. Williams, J. F., Winters, A. L., Lowitt, S., and Szentivanji, A. (1981). Depression of hepatic mixed function oxidase activity by *B. pertussis* in splenectomized and athymic nude mice. *Immunopharmacology* **3**, 101-106.
21. Kato, R. (1977). Drug metabolism under pathological and abnormal physiological states in animals and man. *Xenobiotica* **7**, 25-92.
22. Sofia, R. D. (1977). Alteration of hepatic microsomal enzyme systems and the lethal action of non-steroidal anti-arthritic drugs in acute and chronic models of inflammation. *Agents Actions* **7**, 289-297.
23. Ferdinandi, E. S., Cayen, M. N., and Pace-Asciak, C. (1982). Disposition of etadolac, other anti-inflammatory pyranoindole-1-acetic acids and furobufen in normal and adjuvant arthritic rats. *J. Pharmacol. Exp. Ther.* **220**, 417-426.
24. Tenhunan, H., and Schmid, R. (1970). The enzymatic catabolism of hemoglobin: Stimulation of microsomal heme oxygenase by Lemin. *J. Lab. Clin. Med.* **75**, 410-421.
25. Gemsa, D., Woo, C. H., Fudenberg, H., and Schmid, R. (1974). Stimulation of heme oxygenase in macrophages and liver by endotoxiin. *J. Clin Invest.* **53**, 647-651.
26. Zidek, Z., Friebova, M., Janku, I., and Elis, J. (1977). Influence of sex and Freund's adjuvant on liver N-acetyltransferase activity and elimination of sulphadimidine in urine of rats. *Biochem. Pharmacol.* **26**, 69-70.

27. Notter, D., and Roland, E. (1978). Localisation des N-acetyltransferases dans les cellules sinusoidales hepatique. Influence du zymosan sur l'acetylation de la sulfamethazine et de Disoniazide chez le rat et dans le foie isole perfuse. C. R. Seances Soc. Biol. Ses Fil. **172**, 531–533.
28. du Souich, P., and Courteau, H. (1981). Induction of acetylating capacity with complete Fueund's adjuvant and hydrocortisone in the rabbit. Drug Metab. Dispos. **9**, 279–283.
29. Renton, K. W., and Mannering, G. J. (1976). Depression of the hepatic cytochrome P-450 mono-oxygenase system by tilorone (2,7-bis[2-diethylaminoethoxy]fluorene-9-one dihydrochloride). Drug Metab. Dispos. **4**, 223–231.
30. Leeson, G. A., Biedenback, S. A., Chan, K. Y., Gibson, J. P., and Wright, G. J. (1976). Decrease in the activity of the drug metabolizing enzymes of rat liver following the administration of tilorone hydrochloride. Drug Metab. Dipos. **4**, 232–238.
31. De Maeyer, E., and DeMaeyer-Guignard, J. (1979). Consideration on mouse genes influencing interferon production and action. In "Interferon 1 1979" (L. Gresser, ed.), pp. 75–100. Academic Press, New York.
32. Singh, G., and Renton, K. W. (1981). Interferon-mediated depression of cytochrome P-450-dependent drug biotransformation. Mol. Pharmacol. **20**, 681–684.
33. Sonnenfeld, G., Smith, P. K., and Nerland, D. E. (1982). Interferons and drug metabolism. In "Interferons: UCLA Symposia on Molecular and Cellular Biology" (T. C. Merigan, and R. M. Friedman, eds.), pp. 329–340, Academic Press, New York.
34. Friedman, R. M. (1981). "Interferons: A Primer." Academic Press, New York.
35. Zerkle, T. B., Wade, A. E., and Ragland, W. L. (1980). Selective depression of hepatic cytochrome P-450 hemoprotein by interferon inducers. Biochem. Biophys. Res. Commun. **96**, 121–127.
36. Kato, R., Nakamura, Y., and Chiesara, E. (1963). Enhanced phenobarbital induction of liver microsomal drug-metabolizing enzymes in mice infected with murine hepatitis virus. Biochem. Pharmacol. **12**, 365–370.
37. Ragland, W. L., and Buynitzky, S. J. (1975). Effects of viral replication on drug metabolism. Toxicol. Appl. Pharmacol. **33**, 187.
38. Ragland, W. L., Friend, M., Trainer, D. O., and Sladek, N. F. (1971). Interaction between duck hepatitis virus and DDT in ducks. Res. Commun. Chem. Pathol. Pharmacol. **2**, 236–244.
39. Budillon, G., Carella, M., and Coltori (1972). Phenobarbital liver microsomal induction in MHV-3 viral hepatitis of the mouse. Experientia **28**, 1011–1012.
40. Renton, K. W. (1981). The depression of hepatic cytochrome P-450-dependent mixed-function oxidase during infection with EMC virus. Biochem. Pharmacol. **16**, 2333–2336.
41. Corbett, T. H. (1973). Nettesheim, P.: Effect of PR-8 viral respiratory infection on benzo[a]pyrene hydroxylase activity in BALB/c mice. JNCI, J. Natl. Cancer Inst. **50**, 779–782.
42. McCarthy, J. S., Furner, R. L., Van Dyke, K., and Stitzel, R. E. (1970). Effects of malarial infection on host microsomal drug-metabolizing enzymes. Biochem. Pharmacol. **19**, 1341–1349.
43. Facino, R. M., Carini, M., Bertuletti, R., Genchi, C., and Malchiodi, A. (1981). Decrease of the in vitro drug metabolizing activity of the hepatic mixed function–oxidase system in rats infected experimentally with Fasciola hepatica. Pharmacol. Res. Commun. **13**, 731–742.
44. Doshi, J., Luisada-Opper, A., and Leevy, C. M. (1972). Microsomal pentobarbital hydroxylase activity in acute viral hepatitis. Proc. Soc. Exp. Biol. Med. **140**, 492–495.

45. Leppik, I. E., Ramani, V., Sawchuk, R. J., and Gumnit, R. J. (1979). Increased clearance of phenytoin during infectious mononucleosis. *N. Engl. J. Med.* **300,** 481–482.
46. Chang, K. C., Lauer, B. A., Bell, T. D., and Chai, H. (1978). Altered theophylline pharmacokinetics during acute respiratory viral illness. *Lancet* **1,** 1132–1133.
47. Renton, K. W. (1981). Effects of infection and immune stimulation on theophylline metabolism. *Semin. Perinatol.* **5,** 378–382.
48. Woo, F., Koup, J. R., Kraemer, M., and Robertson, W. (1980). Acute intoxication with theophylline while on chronic therapy. *Vet. Hum. Toxicol.* **22,** 48–51.
49. Kraemer, M. J., Furukawa, C. T., Koup, J. R., Shapiro, G. G., Pierson, W. E., and Fierman, C. W. (1982). Altered theophlline clearance during an influenza B outbreak. *Pediatrics* **69,** 476–480.
50. Renton, K. W., Gray, J. D., and Hall, R. I. (1981). Decreased elimination of theophylline after influenza vaccine. *Can. Med. Assoc. J.* **123,** 288–290.
51. Walker, S., Schreiber, L., and Middelkamp, J. N. (1981). Serum theophylline levels after influenza vaccination. *Can. Med. Assoc. J.* **125,** 243–244.
52. Kraemer, P., and McClain, C. J. (1981). Depression of aminopyrine metabolism by influenza vaccination. *N. Engl. J. Med.* **305,** 1262–1264.

CHAPTER 12

Metabolic Basis of Target Organ Toxicity

Gerald M. Cohen

I. Introduction	325
A. Factors Affecting Toxicity	328
II. Distribution of the Toxin as a Factor in Target Organ Toxicity	329
A. Renal Accumulation: Cephaloridine as an Example	329
B. Pulmonary Accumulation: Paraquat	331
C. Drug-Induced Phospholipidosis	332
III. Role of Metabolism in Determining Target Organ Toxicity	333
A. Toxicity Mediated by the Parent Compound	333
B. Toxicity Mediated by Reactive Metabolites	334
IV. Role of Specific Function of the Tissue	340
A. Organ Function	340
B. Other Functional Aspects	341
V. Comments	344
References	345

I. INTRODUCTION

Humans are exposed to a multitude of compounds of widely differing structures and biological activities in the air they breathe, the water they drink, the food they eat, the medicines they take, or the places in which they work. Many of these compounds will, under specific conditions, induce an encyclopedic list of different toxicities ranging from mild, reversible lesions to such fatal diseases as cancer. All tissues are susceptible to the toxic effects of different chemicals, but the majority of compounds that cause systemic toxicity do not affect all organs equally but generally exhibit toxicity in specific organs; these are known as the *target organs of toxicity*. Some examples of compounds and affected target organs are given in Table I. In addition, some of these compounds, depending on various factors including the dose, will affect other organs. For example, the widely used antimetabolite methotrexate can, in addition to producing toxic effects on the gastrointestinal tract, lead to bone marrow toxicity.

Toxicity may be due to the parent compound, a chemically reactive metabolite, or a chemically stable metabolite. Many xenobiotics are metabolized to reactive electrophiles that combine covalently with critical cellular macromolecules such as DNA, RNA or proteins (see Chapter 2, this volume). Many types of toxicity—carcinogenesis, mutagenesis, teratogenesis, cell necrosis, and hypersensitivity reactions—may be mediated by reactive metabolites[1] as shown in Table II. This subject is dealt with in greater detail in Chapter 13, this volume.

In order to exert their toxic effect, xenobiotics must first gain access to the body. Thus, portals of entry that include the lungs, intestine, and skin will be exposed to high concentrations of chemicals and, therefore, are themselves target organs of toxicity. High concentrations of drugs or metabolites are often excreted in the urine or feces via the kidney or intestinal tract, and these portals of exit are also common targets for toxicity. Many xenobiotics are lipid-soluble compounds that, if not metabolized to more polar, more readily excretable metabolites, would

TABLE I

Target Organs of Toxicity of a Number of Xenobiotics

Organ or tissue	Chemical	Toxicity
Nervous system	Acrylamide	Axonopathy
Liver	Paracetamol	Liver necrosis
Kidney	Cephaloridine	Nephrotoxicity
Respiratory system	Bleomycin	Pulmonary fibrosis
Eye	Chloroquine	Retinopathy
Ear	Streptomycin	Ototoxicity
Blood	Primaquine	Hemolytic anemia
Bone marrow	Chloramphenicol	Aplastic anemia
Reproductive system	1,2-Dibromo-3-chloropropane	Male sterility
Heart	Adriamycin	Cardiomyopathy
Skin	Phenylbutazone	Exfoliative dermatitis
Gastrointestinal tract	Methotrexate	Ulceration
Fetus	Phenytoin	Congenital abnormalities: cleft palate
Bone	Anticonvulsants	Osteomalacia

12. Metabolic Basis of Target Organ Toxicity

TABLE II

Toxicities That May Be Mediated by Reactive Metabolites

Toxicity	Parent compound	Reactive metabolite(s)
Carcinogenesis	Dimethylnitrosamine	CH_3^+
Mutagenesis	Benzo[a]pyrene	Several, including benzo[a]pyrene 7,8-diol 9,10-epoxide and benzo[a]pyrene 4,5-oxide
Cell necrosis	Bromobenzene	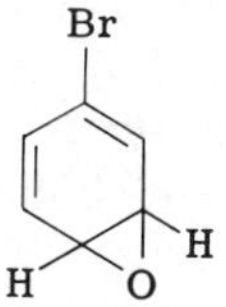 (bromobenzene 3,4-oxide structure)
Methemoglobinemia	Aniline	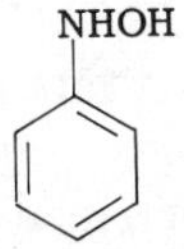 NHOH (phenylhydroxylamine structure)
Blood dyscrasias	Benzene	Not clearly established

remain in the body for long periods. Quantitatively, the major site of xenobiotic metabolism is the liver, and because many chemicals are converted to toxic metabolites, the liver is also a target organ for toxicity. The main target organs of toxicity are listed in Table III together with a major

TABLE III

Prime Factors Predisposing Specific Organs to Toxicity

Organ	Reason for susceptibility
Lung	Major portal of entry for all inhaled chemicals and pollutants
Skin	Major portal of entry for all chemicals encountered by contact
Gastrointestinal tract	Exposed to high concentrations of ingested compounds and those excreted in feces
Kidney	Major portal of exit for chemicals and their metabolites
Liver	Primary site of metabolism and of formation of reactive metabolites
Blood (and blood-forming elements)	Carrier of chemicals and reactive metabolites
Central nervous system	Vital site for controlling body functions

reason for their susceptibility. Obviously, this is a simplified picture, and many other factors influence the inherent susceptibility of any organ.

A. Factors Affecting Toxicity

The factors that affect the response of an organ may be divided into three broad categories:

1. Factors affecting the distribution or pharmacokinetics of the toxic compound to an organ
2. Metabolic fate of the chemical and its toxic metabolites within the organ and the body, for example, the generation and detoxication of reactive metabolites in the target tissue
3. The ability of the organ to respond to the effects of the chemical for example, its ability to repair chemically induced damage

These major factors may be affected to varying degrees by numerous other influences, some of which are outlined in Table IV; additional information is contained in this volume and in standard textbooks of pharmacology[2] and toxicology.[3]

Appropriate examples of the three major categories described above, that is, distribution and metabolism of the toxic chemical and the response of tissue, are used to illustrate each factor.

TABLE IV

Some Factors Affecting the Susceptibility of Organs to Toxicity

Factor	Characteristics
Pharmacokinetics	Physicochemical properties, pK_a, lipid solubility, absorption, distribution, plasma protein binding, and excretion of compound. Possibility of active transport or secretion. Dose-dependent kinetics.
Metabolism	Depends on whether the parent compound is active or a metabolite. If metabolism takes place in Liver and/or target organ(s). Qualitative and quantitative nature of enzymes present. Balance of activating and deactivating enzymes in liver and/or target organ(s). Possibility that different enzyme inducers or inhibitors may affect these tissues differently.
Specific biochemistry of tissue	Presence of a Particular Biochemical Pathway. Absence or low levels of key defensive mechanisms, e.g., Glutathione.
Host	Ability to repair a particular damage or lesion. Presence of impaired host function, e.g., liver or kidney disease.

II. DISTRIBUTION OF THE TOXIN AS A FACTOR IN TARGET ORGAN TOXICITY

The concentration of a xenobiotic or reactive metabolite in a target organ will generally determine the severity of the toxicological response (see also Chapter 8, this volume). However, high concentrations of a toxic compound in a tissue do not necessarily lead to toxicity in that tissue. For example, DDT and related insecticides accumulate in adipose tissue without exerting any apparently deleterious effect on it. In some cases, however, selective accumulation or concentration of a compound may be an important factor in determining the target organ. The nephrotoxicity of antibiotics such as cephaloridine, the pulmonary toxicity of paraquat, and drug-induced phospholipidosis are good examples.

A. Renal Accumulation: Cephaloridine as an Example

The kidney is susceptible to the toxicity of a large number of compounds.[4,5] One important predisposing factor to nephrotoxicity is the propensity of the organ to accumulate many chemicals. This is partly because of receipt of a particularly large blood flow (about 25% of cardiac output), but also because of other aspects of its normal physiological functions. Any potentially toxic chemical in the systemic circulation will necessarily be presented in significant quantities to the kidney, where it will be filtered at the glomerulus. Thus, toxic concentrations may be reached in the tubular fluid as salt and water are reabsorbed from the glomerular filtrate. Compounds may also become particularly concentrated in tubular cells because of active secretion or reabsorption. Any chemical that is actively secreted will first be concentrated in cells of the proximal tubule to levels greater than those found in plasma. These cells may therefore be exposed to potentially toxic concentrations. High concentrations of chemicals may also be attained in the renal medulla, although this tissue receives only about 10% of total renal blood flow; medullary cells will be exposed to high concentrations of chemicals in the tubular urine as they pass through the loop of Henle and the medullary collecting duct. The countercurrent effect in the medulla may also act to concentrate chemicals.[5]

Because of normal physiological functions, certain areas of the kidney may be exposed to excessively high concentrations of potential toxic compounds, resulting in nephrotoxicity. A number of different chemicals, including certain heavy metals, analgesics, anesthetics, and antibiotics, are capable of causing nephrotoxicity.

Nephrotoxicity due to high doses of certain antibiotics, for example, cephaloridine, streptomycin, neomycin, and gentamicin, may be a particular problem, especially in patients with impaired renal function. Cephaloridine is nephrotoxic in a number of species, including humans.[6] High doses of cephaloridine in the rabbit produce acute proximal tubular necrosis, and probenecid causes a dose-related decrease in the nephrotoxicity of the antibiotic that correlates with decreased cortical concentration of cephaloridine.[7] Previous work had shown that cephaloridine, in contrast to most cephalosporins and penicillins, does not undergo net secretion in the kidney.[8] These studies suggested that cephaloridine was dependent on the anion transport system of the kidney for uptake into cortical tubular cells but was not secreted by the anion transport system. Further support for the role of the anion transport system in the nephrotoxicity of cephaloridine was the observation that other organic anions such as benzylpenicillin and p-aminohippurate decreased both cortical uptake and nephrotoxicity of the antibiotic. The observation that newborn rabbits, with an immature transport system for organic anions, were less susceptible to cephaloridine nephrotoxicity provided additional support for the importance of the anion transport system in the toxicity. Finally, pretreatment of animals with penicillin G or p-aminohippurate stimulated the anionic transport system and increased the susceptibility of the animals to the nephrotoxicity of cephaloridine.[6]

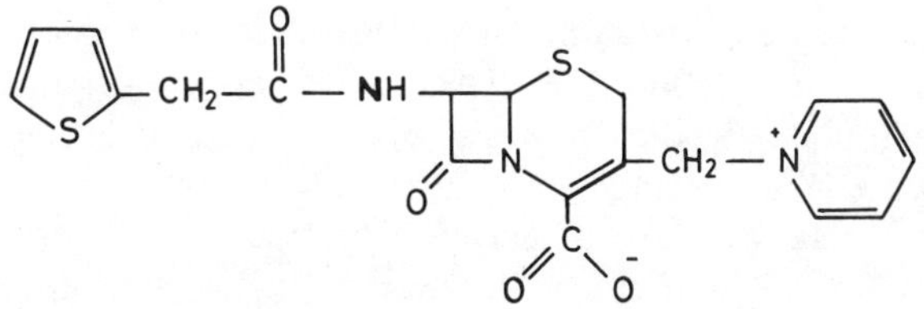

CEPHALORIDINE

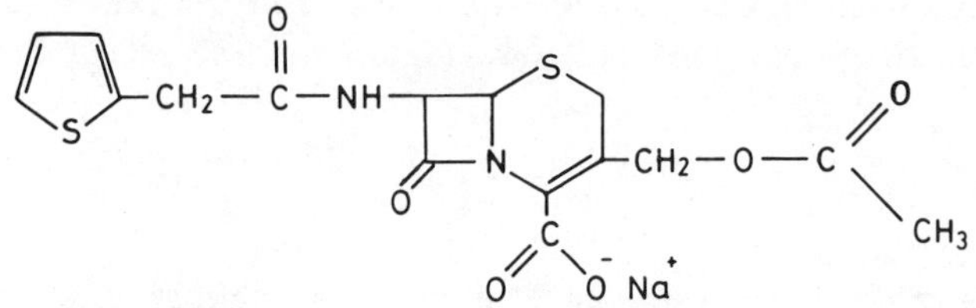

CEPHALOTHIN

Studies with a number of cephalosporins, including cephaloridine, cefazolin, cefamandole, and cephalothin, allow correlation of cephalosporin nephrotoxicity and cortical concentration of the antibiotic.[6] All of

12. Metabolic Basis of Target Organ Toxicity

these cephalosporins possess the carboxylic acid group of the β-lactam ring required for transport by this transport system. However, only cephaloridine also possesses a cationic functional group, that is, the quaternary nitrogen in the pyridinium ring. Wold[6] and collaborators have also noted the importance of this cationic grouping in the cortical accumulation of cephaloridine. Pretreatment of rabbits with either cyanine or mepiperphenidol, both inhibitors of renal cation transport, caused an increase in nephrotoxicity of cephaloridine but not of cefazolin. The cyanine appeared to act by slowing efflux of cephaloridine.

Thus, the target organ toxicity of cephaloridine for the kidney is partially determined by the organic anion transport system, which leads to high concentrations of the compound in the cortical tubular cells. The presence of the cationic group on cephaloridine prevents its secretion by the system responsible for the secretion of the other cephalosporins. The cationic transport system, being less efficient, allows a buildup of higher cortical concentrations of cephaloridine that are related to subsequent nephrotoxicity.[6] There is some disagreement as to whether cephaloridine toxicity is then mediated by an active epoxide formed on the thiophene ring, by inhibition of mitochondrial respiration, or by interaction with other cellular organelles.

B. Pulmonary Accumulation: Paraquat

Paraquat (1,1'-dimethyl-4,4'-bipyridylium) is a contact herbicide that has resulted in many human fatalities after deliberate ingestion. The most characteristic feature of paraquat toxicity to humans is pulmonary damage. The lung becomes progressively impaired, leading to death from pulmonary fibrosis. Similar pathology has been observed in rats and several other species exposed to paraquat.

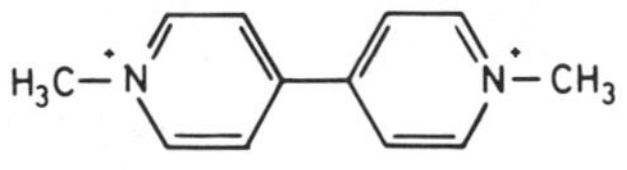

PARAQUAT

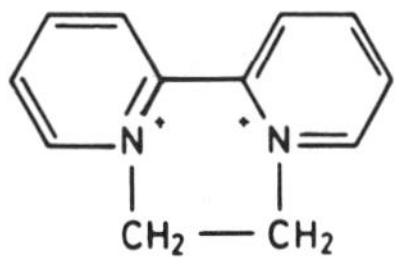

DIQUAT

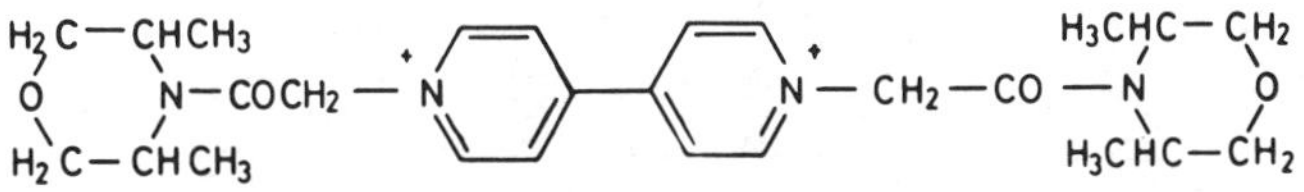

MORFAMQUAT

When paraquat is given to rats either intravenously or orally, the lungs are the organs found to have the highest concentration of paraquat, and they retain this compound selectively. Following oral adminstration of an approximate LD_{100} dose, the concentration of paraquat in the plasma is relatively constant, whereas the concentration in the lung shows a very significant time-dependent increase.[9] No other tissue studied shows a similar time-dependent increase in paraquat concentration. However the concentration in the kidney starts high and remains high, consistent with its role in the clearance of paraquat. Thus the lung, the organ selectively damaged by paraquat, accumulates the herbicide *in vivo*.[9]

Further studies with rat lung slices *in vitro* suggest that the pulmonary uptake of paraquat is due to an energy-dependent active transport process.[10] This uptake is apparently into alveolar type I and type II cells, which are the cells initially damaged by paraquat.[9] Paraquat in the lung undergoes a cyclical reduction and reoxidation with the concomitant production of superoxide anion, ultimately leading to cell death.[9] Although the mechanism of cell death is not clear, it is apparent that the selective accumulation of paraquat in the lung is a major factor in governing its target organ toxicity. Further evidence is obtained by consideration of the structurally related herbicide diquat. Diquat is neither toxic nor selectively accumulated by lung either *in vivo* or by lung slices *in vitro*.[11] Morfamquat, a structurally related herbicide, specifically affects the proximal convoluted tubules of the kidney. These examples illustrate not only the importance of distribution in determining target organ toxicity but also the difficulty in predicting from the chemical structure the likely target organ.

C. Drug-Induced Phospholipidosis

Phospholipidosis is a cellular disturbance produced by a wide variety of drugs, including anorectics (chlorphentermine and fenfluramine), coronary vasodilators (4,4'-diethylaminoethoxyhexoestrol), and antidepressants (imipramine).[12] The lesion, revealed ultrastructurally by the presence of cytoplasmic inclusion bodies of lamellated or crystalloid patterns, is observed most commonly in experimental animals such as the rat and the guinea pig but has also been described in more than 100 individuals treated with 4,4'-diethylaminoethoxyhexoestrol.[12] Biochemically, the affected tissues have a marked increase in phospholipid content. All compounds inducing phospholipidosis appear to be amphipathic in nature, possessing both a hydrophilic and a hydrophobic moiety in close proximity.

Chlorphentermine is accumulated in a number of tissues after a single injection.[12] Thus, 24 hr after injection, the lungs and kidneys have tissue/plasma ratios of between 6 and 12. However, after chronic treatment, an unusual pharmacokinetic distribution pattern has been observed. Tissue/plasma ratios increased with the duration of treatment, particularly in adrenal glands and the lungs. The increased accumulation was not due to simple partition or to the formation of metabolites but to the formation of new binding sites during treatment. It has been suggested that the chlorphentermine reaches the lysosomes, where it binds noncovalently with polar lipids and is then deposited as a lamellated or crystalloid body. Chlorphentermine-induced phospholipidosis affects many different cell types, including those in the lung and adrenal. It is apparent that rather high drug concentrations are required to induce phospholipidosis. Lowering the drug concentration within tissues by stopping administration of the drug leads eventually to complete dissociation of the drug–lipid complex and usually to reversibility of the lesions, provided the affected cells have not undergone a secondary irreversible change.[12]

III. ROLE OF METABOLISM IN DETERMINING TARGET ORGAN TOXICITY

Undoubtedly the metabolism of xenobiotics, in both hepatic and extrahepatic tissues, is of major importance in determining the relative susceptibility of these tissues (see Chapter 2, this volume). In analyzing the problem, it is necessary first to consider whether the toxicity is mediated by one of the following: (1) the parent compound, (2) a chemically reactive metabolite(s), or (3) a chemically stable metabolite(s). Second, in those cases in which toxicity is mediated by a metabolite, consideration must be given to whether (1) the formation of the toxic metabolites occurs in the liver, followed by transport to the target organs, (2) metabolic activation occurs in the target organ(s), or (3) the process requires a combination of both. These possibilities have been considered in detail by Boyd, and the schemes to be outlined, with minor modifications, are basically similar to those proposed by him.[13,14]

A. Toxicity Mediated by the Parent Compound

The effects of metabolism are self-evident when the parent compound is itself toxic (Fig. 1). Any change in the rate of metabolism of the compound, either in the liver or in any extrahepatic tissue, will alter its effective concentration in the target organ(s) and cause a corresponding effect on toxicity. It may also be readily appreciated from Fig. 1 that any other factors that serve to increase or decrease the concentration of the toxic compound in the target tissue will have a correspondingly predictable

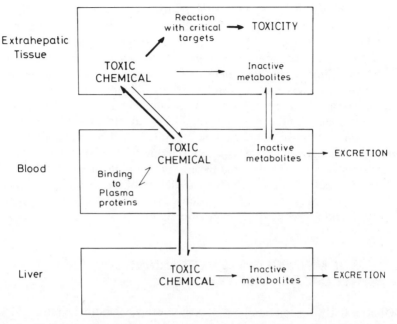

Fig. 1. Toxicity mediated by the parent chemical. Any factor that affects the concentration of the toxic chemical available to interact in the target organ will affect the toxicity of the chemical. Modified from Boyd.[14]

effect on toxicity provided, of course, that the toxicity is a dose-related phenomenon. Thus, interaction of a second compound with plasma proteins, leading to a displacement of the bound toxin and therefore an increase in its free concentration, would lead to an increase in toxicity. One compound whose toxicity or pharmacological activity follows the mechanism depicted in Fig. 1 is the muscle relaxant zoxazolamine. Induction of drug-metabolizing enzymes with inducing agents such as phenobarbitone or 3-methylcholanthrene results in increased metabolism of the active parent compound and therefore in decreased amounts available for pharmacological or toxicological activity.

B. Toxicity Mediated by Reactive Metabolites

The situation is more complex when toxicity is due to a metabolite. If the metabolite formed is chemically stable, then it may be formed in one organ and transported by the systemic circulation, as illustrated in Fig. 2. In contrast, some metabolites may be chemically reactive, that is, their half-lives are very short. Such reactive metabolites would most likely

12. Metabolic Basis of Target Organ Toxicity

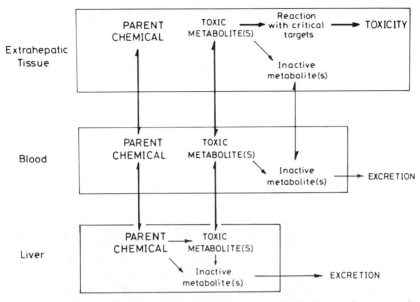

Fig. 2. Toxicity mediated by a toxic metabolite generated in the liver and transported to the target organ. The toxic metabolite is formed in the liver and is sufficiently stable to be then transported via the circulation to the target organ of toxicity. Modified from Boyd.[14]

react either intracellularly, within the cell in which they were formed, or, if they escaped from the cells, with constituents of blood. One must predict that the extrahepatic toxicity of these metabolites will be mediated by reactive intermediates generated *in situ* within their target organs, as illustrated in Fig. 3. In order to establish whether a compound is activated by either one or more of the previously discussed mechanisms, it is necessary to consider other factors, including the tissue distribution of the metabolizing enzymes, as illustrated in the following section.

1. Tissue Distribution of Drug-Metabolizing Enzymes

Implicit in the mechanism proposed in Fig. 3 is that the target organ should be able to activate the chemical metabolically, giving rise to a toxic metabolite. Although it is generally recognized that the major site of metabolism for most xenobiotics is the liver, a significant amount of extrahepatic metabolism also takes place. The ability of extrahepatic tissues to metabolize xenobiotics has been the subject of an increasing amount of study.[15] Although quantitatively the contribution of a particular extrahepatic tissue may represent only a small contribution to the overall

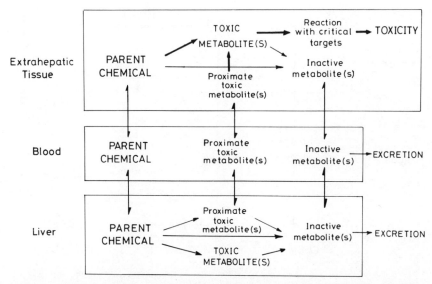

Fig. 3. Toxicity mediated by reactive metabolites formed in the target organ either from the parent chemical or from a proximate toxic metabolite. If the toxic metabolite is chemically very reactive, it may have to be generated *in situ* in the target organ in order to exert toxicity. Depending on the metabolic capability of the tissue, the metabolic activation by the parent chemical may be completely carried out in the target organ. Alternatively, part of the metabolic activation may be carried out in other tissues, such as the liver, but the ultimate toxic metabolites are formed *in situ* in the target organ.

metabolism, it may be of particular toxicological significance. The enzymes responsible for the activation of xenobiotics will often be the Phase I oxidative enzymes but may in some cases be the Phase II conjugating enzymes, such as the activation of N-hydroxyacetylaminofluorene following sulfate conjugation.[16–18] The tissue distributions of both Phase I and Phase II drug-metabolizing enzymes are therefore vital factors in determining the relative susceptibilities of different tissues to toxic chemicals.

It is readily apparent that a large number of tissues may, under appropriate conditions, metabolize foreign chemicals. Enzymes of the cytochrome P-450 system and the glutathione transferases, for example, are almost ubiquitous in distribution. Although metabolism may be demonstrated *in vitro,* such active tissues may not metabolize the same compound *in vivo,* because that will depend on many factors, such as blood flow, the nature of the substrate and its availability, and a supply of the necessary cofactors (see Chapter 9, this volume).

Important differences in the distribution of the drug-metabolizing enzymes in any particular tissue may also predispose particular cell types in

12. Metabolic Basis of Target Organ Toxicity

a tissue to toxicity. Even in the liver, considered as a relatively homogeneous tissue, large differences in tissue distributions of cytochrome P-450 and glutathione have been observed using quantitative cytochemistry[19] (see Chapter 4, this volume). Therefore, it is not surprising that in heterogeneous tissues such as the lung, with more than 40 cell types present, one observes an increased concentration of drug-metabolizing enzymes such as cytochrome P-450 in certain cell types.[20,21] In addition to quantitative differences in the distribution of these enzymes, important qualitative differences may also be observed that may contribute to susceptibility to target organ toxicity.

2. Importance of Balance of Activating and Deactivating Enzymes

In considering the generation of reactive metabolites in any tissue, it is vitally important that both activating and deactivating enzymes be taken into account. The balance of the activity of these enzymes and the availability of their respective cofactors ultimately determine how much of a reactive metabolite(s) is present in a particular tissue and therefore is subsequently available for interaction with critical cellular targets that ultimately lead to toxicity. The importance and role of many of the previously discussed concepts in determining the target organ toxicity is well illustrated by the work of Boyd[14,20] with 4-ipomeanol. In many species, this compound causes a specific pulmonary toxicity, and this toxic effect is apparently mediated by reactive metabolites generated in specific cells in the lung, that is, the mechanism shown in Fig. 3.

3. Reactive Metabolites Generated in the Target Organ—Illustrated by 4-Ipomeanol

The general role of extrahepatic metabolism in determining extrahepatic toxicity has been reviewed by Boyd.[13] In an excellent review, he has also illustrated the particular role of metabolic activation in chemical-induced lung damage.[14] It is apparent from this work that 4-ipomeanol, or 1-(3-furyl)-4 hydroxypentanone, a furan found on moldy sweet potatoes, produces a striking pulmonary toxicity in a number of species. The characteristic pulmonary toxicity observed is necrosis of the nonciliated bronchiolar epithelial or Clara cells.[14,20] Following *in vivo* administration of [^{14}C]ipomeanol to rats, significantly more covalently bound radioactivity was observed in lungs than any other tissues. Autoradiography revealed that the covalently bound radioactivity was predominantly associated with Clara cells, and *in vitro* studies suggested that a cytochrome P-450–dependent monooxygenase was required to activate 4-ipomeanol to a reactive metabolite(s) that bound covalently to macromolecules. Both rat lung and liver were capable of activating 4-ipomeanol to a binding species. However, when the results were expressed on the basis of covalent binding per molecule of cytochrome P-450, the lung appeared approximately eightfold more active than the liver. Boyd[20] and Serabjit-Singh[21] have independently shown that at least one type of cytochrome P-450 is specifically localized in the Clara cells. These observations demonstrate the importance of the qualitative nature of the cytochrome P-450, and its tissue distribution, in determining cells in a specific organ that may be predisposed to toxicity.

Further support for the key role of pulmonary metabolism of 4-ipomeanol in determining its pulmonary toxicity has come from a series of studies with enzyme inducers and inhibitors. Three inhibitors (pyrazole, piperonyl butoxide, and cobaltous chloride) decreased both hepatic and pulmonary binding of 4-ipomeanol both *in vitro* and *in vivo*, correlating well with the decreased pulmonary toxicity *in vivo*.[14] Pretreatment of rats with the inducing agents phenobarbitone and 3-methylcholanthrene caused a marked increase in the *in vitro* covalent binding of 4-ipomeanol to liver but not to lung microsomes.[14] These alterations were accompanied by decreased covalent binding to lung *in vivo* and decreased pulmonary toxicity. Treatment with 3-methylcholanthrene, however, caused an increase in *in vivo* covalent binding in the liver that correlated with centrilobular hepatic necrosis. Thus, differential alterations in the balance of the enzymes in the lung and liver caused by the 3-methylcholanthrene caused a shift in target organ toxicity. The observation that pretreatment with 3-methylcholanthrene increased both the toxicity and the covalent binding of 4-ipomeanol in the liver but decreased these actions in the lung supports the hypothesis that reactive metabolites of 4-ipomeanol will bind in those cells and tissues in which they are generated, that is, the mechanism in Fig. 3. Further support for this concept has come from the interspecies differences in the target organs of toxicity for 4-ipomeanol. In addition to causing bronchiolar necrosis, this compound leads to renal necrosis in adult male mice and hepatic necrosis in hamsters. In these two species, in addition to pulmonary covalent binding, high levels of binding to the kidney and liver were observed.

4. Reactive Metabolites Generated in Liver—Illustrated by the Pulmonary Toxicity of Pyrrolizidine Alkaloids

The pyrrolizidine alkaloids are highly toxic to liver and other tissues.[22] For example, monocrotaline, in addition to producing liver damage, causes severe lung injury. The data[14] strongly suggest that the pulmonary toxicity of monocrotaline is the result of toxic metabolites formed primar-

12. Metabolic Basis of Target Organ Toxicity

ily in the liver and subsequently transported in the circulation to the lung, as depicted in Fig. 2. This idea is supported by the observation that reactive metabolites may be formed by hepatic but not pulmonary enzymes, and that pulmonary endothelial toxicity produced by monocrotaline and dehydromonocrotaline are very similar. Work with enzyme inducers and inhibitors lends further support to this hypothesis. For example, pretreatment of rats with phenobarbitone increased liver microsomal enzyme activity *in vitro* and increased liver and lung concentrations of reactive pyrrole derivatives, which corresponds to the observed increase in liver and lung toxicity.

Investigations of the pulmonary toxicity of bromobenzene[23] have suggested that this toxicity may be due to a mixture of both of the mechanisms described in the preceding paragraph, that is, metabolites generated both *in situ* and in the liver and subsequently transported to the lung.[14]

5. The Possibility of Initial Metabolic Activation in the Liver and the Formation of the Ultimate Reactive Metabolite in the Target Tissue

The Millers noted the importance of proximate and ultimate carcinogens in their unifying concept of the role of metabolism in the generation of ultimate reactive electrophiles in chemical carcinogenesis (Chapter 2, this volume) (Fig. 4). It is realized that toxic chemicals, in addition to many carcinogens, require multistep metabolic activation before generating their ultimate toxic metabolites. Therefore, it is conceivable that a stable proximate toxic metabolite may be initially formed in one tissue and then be transported to the target organ where it is metabolized to the ultimate toxic metabolite, as illustrated in Fig. 3. The realization that many chemicals require such multistep activation suggests that this mechanism may be very important. Different enzymes could be involved at different stages of activation, and this may be of importance in target organ toxicity. For example, a target tissue may not be able to metabolize the parent compound to its proximate toxic metabolite but could convert the proximate toxic metabolite to the ultimate toxic metabolite. As an

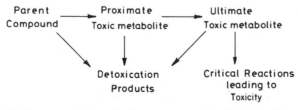

Fig. 4. Multistep metabolic activation to generate the ultimate toxic metabolite.

example, there are data indicating that prostaglandin endoperoxide synthetase may be involved in the metabolic activation of certain chemicals. The synthetase from guinea pig lung will metabolize the proximate carcinogen from benzo[a]pyrene,[24] that is, 7,8-dihydro-7,8-dihydroxybenzo-[a]pyrene, to the ultimate carcinogenic, a diolepoxide, but will not form the intermediate from benzo[a]pyrene.[25] It is conceivable that 7,8-dihydro-7,8-dihydroxybenzo[a]pyrene, formed in the liver, may be further metabolized in the lung of the guinea pig to its carcinogenic and mutagenic metabolite.

IV. ROLE OF SPECIFIC FUNCTION OF THE TISSUE

The presence of vital metabolic pathways and of specific receptors in specific tissues may predispose those tissues to toxicity. Undoubtedly, this is one of the least documented areas of target organ toxicity.[26] Much more is known about the factors governing both the disposition of toxic chemicals and their subsequent metabolic activation, if required, than about the interaction of the chemicals or their reactive metabolite(s) with cellular targets and the subsequent phenomena that are necessary before clinical signs of toxicity are apparent.[26]

Although the factors governing the inherent susceptibility of an organ are less well understood, it may be useful to subdivide them into those related either to (1) organ function or (2) biochemistry of the tissue.

A. Organ Function

The normal physiology of the tissue may predispose it to toxicity. An important example of this has been discussed in Section II,A in relation to the concentration of many chemicals in the kidney and their subsequent nephrotoxicity.

The normal function of a tissue may be impaired by disease, leading to impaired metabolism or excretion followed by an accumulation of toxic concentrations of the compound.[27] This may be a basis for the high incidence of adverse drug reactions observed in patients with impaired renal function. In addition to effects on the parent compound, impaired renal excretion of metabolites may also be observed in such patients, leading to increased concentrations of metabolites.[27,28] Metabolites may be toxicologically active per se, or they may interfere with the metabolism and distribution of the parent compound. This area and its clinical implications have been reviewed.[27,28]

12. Metabolic Basis of Target Organ Toxicity

The peripheral neuritis after nitrofurantoin therapy is thought to be due to accumulation of a toxic metabolite and is found almost exclusively in instances of impaired renal function.[27] Similarly, norpethidine, the N-demethylated metabolite of the narcotic analgesic pethidine, accumulates rapidly in patients with renal failure.[28] The severe irritability and twitching reported in some of these individuals has been associated with high levels of norpethidine, which has less analgesic but more convulsant activity than the parent compound.

Tissues with a high proportion of dividing cells (bone marrow, gastrointestinal tract, and hair follicles) are more sensitive to the toxic effects of alkylating agents used in the treatment of cancer than are tissues with a lower proportion of such cells.[29] Thus, treatment with antimetabolites or alkylating agents often results in damage to the bone marrow. Because the marrow contains stem cells, that is, the immature precursors of red cells, platelets, and white cells, the damage may lead to pancytopenia, a decrease in the circulating numbers of the three major groups of formed elements. If this condition is sufficiently severe, the marrow may no longer proliferate, resulting in aplastic anemia. Damage to platelets leads to thrombocytopenia and bleeding, whereas damage to leukocytes leads to an increased risk of infection in patients treated with many anticancer drugs.

A particular problem associated with the inhibition of rapidly dividing cells occurs with the fetus. Drugs inhibiting rapidly dividing fetal cells may produce developmental abnormalities, and some anticancer drugs have been shown to be teratogenic.[30]

B. Other Functional Aspects

1. Ability to Repair Lesions

Individuals suffering from xeroderma pigmentosum are very prone to develop skin cancer.[31] These individuals have a defect in their ability to repair DNA that is implicated as a major risk factor for them.[31] In animals, differences in DNA repair among organs have been correlated with the organ-specific carcinogenicity of nitrosamines and nitrosamides. Many nitrosamines and nitrosamides produce tumors in a wide variety of organs and species.[32] Certain N-nitroso compounds demonstrate very marked organ specificity in tumor induction, depending on the dose, route of administration, species, diet, and physiological status of the animal.[32] Some of the organ specificity may be clearly attributed to metabolism. The N-nitroso compounds are converted into alkylating agents, which react with DNA.[32,33] Whereas methylating agents such as dimethylnitrosamine can react with at least eight sites in DNA, a large body of evidence points to alkylation at a specific position, that is, O^6 of guanine, as being of special importance for biological activity. A good correlation has been obtained between alkylation of the O^6 of guanine and the incidence of kidney tumors in the rat and thymic lymphomas in the mouse.[32,33]

The importance of DNA repair in determining the organ specificity of nitrosamines and nitrosamides is well illustrated by the carcinogenicity of ethylnitrosourea.[34] *In vivo,* ethylnitrosourea undergoes nonenzymatic decomposition with a $t_{1/2}$ of less than 8 min; the ethyl carbonium ion formed reacts indiscriminately with whatever reactive molecule is nearby. However, despite the lack of tissue specificity in the reaction of the ultimate carcinogen, a single dose of the carcinogen given either to rats shortly after birth or to the fetus results in a high incidence of neuroectodermal neoplasms in the central and peripheral nervous systems.[34] A larger dose produces tumors in the kidney and in many other organs, but even the largest single dose does not induce liver tumors. The susceptibility of the organs parallels their relative ability to remove O^6-alkylguanine from their DNA, a measure of their presumed ability to repair DNA. The brain is unable to remove this lesion and the kidney removes it slowly, but liver removes it very rapidly.

It should be added that the sensitivity of the nervous system to ethylnitrosourea-induced carcinogenicity decreases drastically with increasing age of the animal and the increased maturity of the nervous system. This has been suggested as the result of fewer target cells in the appropriate proliferative or differentiated state.[34]

2. Presence of Specific Receptors

The differentiated or specialized functions of a cell or tissue may make it particularly susceptible to the action of compounds with an affinity for that function. The presence in a cell of specific receptors necessary for its specialized functions serves as an example.

a. Neurotoxic Esterase. The toxicity of certain organophosphorus esters is related to their ability to combine with a specific neurotoxic esterase in a reaction that is not related to the ability of these esterases to inhibit acetylcholinesterase.[35-37] Neurotoxic effects are not observed until 8 to 14 days after ingestion of the inhibitor, at which time ataxia and weakness develop in the lower limbs, progressing to paralysis. Adults are more severely affected than children or young animals. It is very difficult to produce the delayed neurotoxic effects in rodents, but hens, cats and other large species are affected by a single dose. Compounds such as tri-*o*-cresyl phosphate, diisopropylfluorophosphate (DFP), and the insec-

12. Metabolic Basis of Target Organ Toxicity

ticide mipafox are effective agents. Johnson[35,36] and his colleagues have provided evidence for the phosphorylation by these agents of a specific protein, now referred to as *neurotoxic esterase,* in the brain. Toxicity is dependent on the nature of the chemical group that is bound to the neurotoxic esterase. Other structurally related organophosphorus esters, tetraethyl pyrophosphate, for example, which do not phosphorylate neurotoxic esterase, are nontoxic.

b. The Ah Locus. Large differences in the susceptibility of different strains of mice to drug toxicity and chemical carcinogenesis, mutagenesis, and teratogenesis have been demonstrated.[38] Inducers of microsomal monooxygenase activity, including 2,3,7,8-tetrachlorodibenzo-*p*-dioxin, benzo[*a*]pyrene, 3-methylcholanthrene, and β-naphthoflavone, interact with cytosolic receptor proteins.[38,39] Genetically determined differences in the affinities for these receptor proteins in different strains of mice determine the response of the animal to the inducing agent. The induction receptor is a function of the *Ah* locus. The *Ah* locus and, thereby, the cytosolic induction receptor control the activities of a battery of different enzymes, including not only those due to monooxygenases but also microsomal UDPglucuronosyltransferase, cytosolic ornithine decarboxylase, and cytosolic reduced NAD(P):menadione oxidoreductase.[38] The presence or absence of the receptor, and defects in the receptor, have been related to a large number of different toxicities in mice, including that of 2,3,7,8-tetrachlorodibenzo-*p*-dioxin[39] and the induction of cataracts by paracetamol and naphthalene.[40]

3. Alterations in Enzyme Activities

Changes in the active concentration of specific enzymes in any given cell type or tissue may predispose that tissue to the effect of xenobiotics. The effect is well illustrated by the marked susceptibility to drug-induced hemolysis of erythrocytes in individuals with a deficiency of glucose-6-phosphate dehydrogenase.[2,3] A wide range of compounds including primaquine, phenylhydrazine, vitamin K, nitrofurantoin, sulfonamides, chloramphenicol, and fava beans are capable of inducing hemolysis in genetically susceptible individuals.[2,3]

The precise mechanism by which the glucose-6-phosphate dehydrogenase deficiency leads to hemolysis is not known. However, this deficiency does decrease the cell's ability to generate NADPH by the hexose monophosphate shunt. NADPH is required in part as a substrate for glutathione reductase that is necessary to maintain glutathione in the reduced state as GSH. GSH in turn is essential for the maintenance of protein sulfydryl groups in the reduced state, thereby preventing denaturation of enzymes and hemoglobin and possibly preserving the integrity of the erythrocyte membrane.[2,3]

4. Pharmacogenetic Differences in Humans

Atypical drug responses may be due to genetic variation in any of the processes—absorption, distribution, metabolism, and excretion—concerned with drug detoxication. This subject has been reviewed,[41] and the majority of documented examples relate to alterations in metabolism.[41,42] In addition to the genetic variations in glucose-6-phosphate dehydrogenase that have been described, a particularly significant example of such variation is the abnormal sensitivity of some individuals to succinylcholine, which is used to relax muscles during surgery.[2,41] Susceptible patients require artificial respiration because of paralysis of the respiratory muscles following normal doses. This response was associated with a mutant pseudocholinesterase that is less effective than the normal pseudocholinesterase in hydrolyzing succinylcholine.[2,41]

5. Additional Factors

Diet and age may be important modulators of target organ toxicity. Thus, dimethylnitrosamine given orally normally induces liver cancer, but if the animals are fed a low-protein diet, metabolic activation in the liver decreases and more of the carcinogen is available for activation in the kidney, resulting in kidney tumors.[43] The factor of age is underscored by the effect of calcium disodium ethylenediaminetetraacetic acid. In very young rats, the chelating agent causes liver necrosis, but in 12-week-old rats severe kidney lesions arise.[44]

Different animal species show varying responses. Induction of toxicity in one target organ in one species does not necessarily allow the conclusion that the same organ will be affected in another species.[45] In particular, when evaluating the safety of chemicals, the concern is with whether the responses of humans will be similar to those of test animals. During the next few years, greater use of human surgical tissues and cell culture systems may be expected in the evaluation of potential toxicity.

V. COMMENTS

A large number of factors, some shown in Table IV, may alter both the degree of toxicity of a xenobiotic and the organ affected, that is, the target organ of toxicity. Each of these factors has not received complete coverage, and some are discussed in other chapters. Particular emphasis has been placed here on the important role of the distribution and metabolism

12. Metabolic Basis of Target Organ Toxicity

of xenobiotics, as well as the specific biochemical and physiological functions of tissues in the determination of target organ toxicity.

REFERENCES

1. Gillette, J. R., Mitchell, J. R., and Brodie, B. B. (1974). Biochemical mechanisms of drug toxicity. *Annu. Rev. Pharmacol.* **14**, 271–288.
2. Goldstein, A., Aronow, L., and Kalman, S. M. (1974). "Principles of Drug Action: The Basis of Pharmacology," 2nd ed. Wiley, New York.
3. Doull, J., Klaassen, C. D., and Amdur, M. O. (1980). "Cassarett and Doull's Toxicology. The Basic Science of Poisons," 2nd ed. Macmillan, New York.
4. Hook, J. B., ed. (1981). "Toxicology of the Kidney." Raven Press, New York.
5. Hook, J. B., McCormack, K. M., and Kluwe, W. M. (1979). Biochemical mechanisms of nephrotoxicity. *Rev. Biochem. Toxicol.* **1**, 53–78.
6. Wold, J. S. (1981). Cephalosporin nephrotoxicity. *In* "Toxicology of the Kidney" (J. B. Hook, ed.), Raven Press, New York.
7. Tune, B. M., Wu, K. Y., and Kempson, R. L. (1977). Inhibition of transport and prevention of toxicity of cephaloridine in the kidney. Dose responsiveness of the rabbit and guinea-pig to probenecid. *J. Pharmacol. Exp. Ther.* **202**, 466–471.
8. Child, K. J., and Dodds, M. G. (1966). Mechanism of urinary excretion of cephaloridine and its effects on renal function in animals. *Br. J. Pharmacol. Chemother.* **26**, 108–119.
9. Smith, L. L., Rose, M. S., and Wyatt, I. (1979). The pathology and biochemistry of paraquat. *Ciba Found. Symp.* **65** (new ser.), 321–341.
10. Rose, M. S., Smith, L. L., and Wyatt, I. (1974). Evidence for energy-dependent accumulation of paraquat into rat lung. *Nature (London)* **252**, 314–315.
11. Rose, M. S., Lock, E. A., Smith, L. L., and Wyatt, I. (1976). Paraquat accumulation: Tissue and species specificity. *Biochem. Pharmacol.* **25**, 419–423.
12. Lüllmann, H., Lüllmann-Rauch, R., and Wassermann, O. (1975). Drug-induced phospholipidoses. *CRC Crit. Rev. Toxicol.* **4**, 185–218.
13. Boyd, M. R. (1980). Effects of inducers and inhibitors on drug-metabolizing enzymes and on drug toxicity in extrahepatic tissues. *Ciba Found. Symp.* **76** (new ser.), 43–76.
14. Boyd, M. R. (1980). Biochemical mechanisms in chemical-induced lung injury: Roles of metabolic activation. *Crit. Rev. Toxicol.* **7**, 103–176.
15. Gram, T. E., ed. (1980). "Extrahepatic Metabolism of Drugs and Other Foreign Compounds." MTP Press Ltd., Lancaster, England.
16. Jollow, D. J., Kocsis, J. J., Snyder, R., and Vainio, H., eds. (1977). "Biological Reactive Intermediates, Formation, Toxicity and Inactivation." Plenum, New York.
17. Ingelman-Sundberg, M. (1980). Bioactivation or inactivation of foreign compounds? *Trends Pharmacol. Sci.* **1**, 176–179.
18. Miller, E. C. (1978). Some current perspectives on chemical carcinogenesis in humans and experimental animals. *Cancer Res.* **38**, 1479–1496.
19. Smith, M. T., Loveridge, N., Wills, E. D., and Chayen, J. (1979). The distribution of glutathione in the rat liver lobule. *Biochem. J.* **182**, 103–108.
20. Boyd, M. R. (1977). Evidence for the Clara cell as a site of cytochrome *P*-450–dependent mixed-function oxidase activity in lung. *Nature (London)* **269**, 713–715.
21. Serabjit-Singh, C. J., Wolf, C. R., Philpot, R. M., and Plopper, C. G. (1979). Cytochrome *P*-450: Localization in rabbit lung. *Science* **207**, 1469–1470.
22. McLean, E. K. (1970). The toxic action of pyrrolizidine (Senecio) alkaloids. *Pharmacol. Rev.* **22**, 429–483.

23. Reid, W. D., Illett, K. F., Glick, J. M., and Krishna, G. (1973). Metabolism and binding of aromatic hydrocarbons in the lung. *Am. Rev. Dis.* **107,** 539–551.
24. Jerina, D. M., Lehr, R., Schaeffer-Ridder, M., Yagi, H., Karle, J. M., Thakker, D. R., Wood, A. W., Lu, A. Y. H., Ryan, D., West, S., Levin, W., and Conney, A. H. (1977). Bay region dihydrodiols: A concept explaining the mutagenic and carcinogenic activity of benzo(a)anthracene. *Cold Spring Harbor Conf. Cell Proliferation* **4,** Book B, 639–658.
25. Marnett, L. J. (1981). Polycyclic aromatic hydrocarbon oxidation during prostaglandin biosynthesis. *Life Sci.* **29,** 531–546.
26. Aldridge, W. N. (1981). Mechanisms of toxicity: New concepts are required in toxicology. *Trends Pharmacol. Sci.* **2,** 228–231.
27. Drayer, D. E. (1976). Pharmacologically active drug metabolites: Therapeutic and toxic activities, plasma and urine data in man, accumulation in renal failure. *Clin. Pharmacokinet.* **1,** 426–433.
28. Verbeeck, R. K., Branch, R. A., and Wilkinson, G. R. (1981). Drug metabolites in renal failure: Pharmacokinetic and clinical implications. *Clin. Pharmacokinet.* **6,** 329–345.
29. Pratt, W. B., and Ruddon, R. W. (1979). "The Anticancer Drugs." Oxford Univ. Press, London/New York.
30. Sieber, S. M., and Adamson, R. H. (1975). The clastogenic, mutagenic, teratogenic and carcinogenic effects of various antineoplastic agents. *In* "Pharmacological Basis of Cancer Chemotherapy," pp. 401–468. Williams & Wilkins, Baltimore, Maryland.
31. Setlow, R. B., Ahmed, F. E., and Grist, E. (1977). Xeroderma Pigmentosum: Damage to DNA is involved in carcinogenesis. *Cold Spring Harbor Conf Cell Proliferation* **4,** Book B, 889–902.
32. Pegg, A. E. (1977). Formation and metabolism of alkylated nucleosides: Possible role in carcinogenesis by nitroso compounds and alkylating agents. *Adv. Cancer Res.* **25,** 195–269.
33. Swann, P. F., and Kyrtopoulos, S. A. (1980). Reaction with DNA and DNA-repair in chemical carcinogenesis. *In* "Mechanisms of Toxicity and Hazard Evaluation" (B. Holmstedt, R. Lauwerys, M. Mercier, and M. Roberfroid, eds.), Elsevier/North-Holland, Amsterdam.
34. Rajewsky, M. F., Augenlicht, L. H., Biessmann, H., Goth, R., Hülser, D. F., Laerum, O. D., and Lomakina, L. Ya. (1977). Nervous-system–specific carcinogenesis by ethylnitrosourea in the rat: Molecular and cellular aspects. *Cold Spring Harbor Conf. Cell Proliferation* **4** Book B, 709–725.
35. Johnson, M. K. (1975). The delayed neuropathy caused by some organophosphorous esters: Mechanism and challenge. *CRC, Crit. Rev. Toxicol.* **3,** 289–316.
36. Johnson, M. K. (1980). Delayed neurotoxicity induced by organophosphorous compounds—Areas of understanding and ignorance. *In* "Mechanisms of Toxicity and Hazard Evaluation" (B. Holmstedt, R. Lauwerys, M. Mercier, and M. Roberfroid, eds.), pp. 27–38. Elsevier/North-Holland, Amsterdam.
37. Richardson, R. J. (1981). Toxicology of the nervous system. *In* "Toxicology: Principles and Practice" (A. L. Reeves, ed.), Vol. 1, pp. 107–143.
38. Nebert, D. W., and Jensen, N. M. (1979). The Ah locus: Genetic regulation of the metabolism of carcinogens, drugs, and other environmental chemicals by cytochrome P-450–mediated monooxygenases. *CRC Crit. Rev. Biochem.* **6,** 401–437.
39. Poland, A., and Glover, E. (1980). 2,3,7,8-Tetrachlorodibenzo-p-dioxin: Segregation of toxicity with the Ah locus. *Mol. Pharmacol.* **17,** 86–94.
40. Shichi, H., Tanaka, M., Jensen, N. M., and Nebert, D. W. (1980). Genetic differences in cataract and other ocular abnormalities induced by paracetamol and naphthalene. *Pharmacology* **20,** 229–241.

12. Metabolic Basis of Target Organ Toxicity

41. Nebert, D. W. (1980). Human genetic variation in the enzymes of detoxication. *In* "Enzymatic Basis of Detoxication" (W. B. Jakoby, ed.), Vol. 1, pp. 25–68. Academic Press, New York.
42. Nebert, D. W. (1981). Possible clinical importance of genetic differences in drug metabolism. *Br. Med. J.* **283,** 537–542.
43. McLean, A. E. M., and Magee, P. (1970). Increased renal carcinogenesis by dimethylnitrosamine in protein deficient rats. *Br. J. Exp. Pathol.* **51,** 587–590.
44. Reuber, M. D. (1967). Hepatic lesions in young rats given calcium disodium edetate. *Toxicol. Appl. Pharmacol.* **11,** 321–326.
45. Heywood, R. (1981). Target organ toxicity. *Toxicol. Lett.* **8,** 349–358.

CHAPTER 13

Enzymes in Selective Toxicity

C. H. Walker and F. Oesch

I.	Introduction	349
II.	The Toxicological Significance of Enzymatic Conversion	350
III.	The Enzymatic Factor in Selective Toxicity	352
	A. Organophosphate Insecticides	352
	B. Organochlorine Insecticides	355
	C. Carbamate Insecticides	356
	D. Carcinogens and Mutagens	358
IV.	The Enzymatic Factor in Resistance	361
	A. Organophosphate Insecticides	361
	B. Organochlorine Insecticides	362
	C. Carbamate Insecticides	363
V.	Discussion	363
	References	365

I. INTRODUCTION

The biochemical and physiological basis of selective toxicity is a matter of both scientific and practical interest. The identification of those differences between species and groups that are responsible for differential susceptibility to toxic compounds can give useful insight into the biochemical and physiological processes that determine toxic response. Such an approach can aid the definition of mechanisms of toxicity of carcinogens, pesticides, drugs, and other toxic substances to which humans are exposed.

Knowledge of these differences is also relevant to an understanding of the environmental effects of pesticides and other pollutants. The design of selective pesticides, the overcoming of resistance to pesticides, and the avoidance of bioaccumulation of pollutants are all practical problems whose resolution is facilitated by a better understanding of the mechanisms of selective toxicity. Finally, knowledge of species differences provides a useful guideline in the selection of species as models for humans in toxicity evaluation.

The toxicity of any compound depends largely on two factors: (1) the intrinsic activity of the active form(s) of the compound at the site(s) of action and (2) the efficiency with which the active form(s) reach the site(s) of action. The efficiency of transfer to the site of action depends upon the processes of absorption, distribution, metabolism, storage, and excretion, the relative importance of which depends very much upon the substance in question (see Chapter 8, this volume). Selective toxicity may be seen as an expression of variation in one or more of processes between different species or groups of animals.

Work on the selective toxicity of, and resistance to, pesticides and other toxic substances strongly indicates that metabolism is the most important single factor determining differences in susceptibility. In humans, individual differences in response to drugs are sometimes the consequence of metabolic differences. Thus, it is worth considering the role of enzymes in determining differences in susceptibility to toxic substances between species and groups and also between individuals (see Chapter 14, this volume). The enzymes themselves are discussed in Chapters 1 and 2 (this volume), and a major review of the enzymology of the enzymes of detoxication has been presented.[1]

II. THE TOXICOLOGICAL SIGNIFICANCE OF ENZYMATIC CONVERSION

To a large extent, the metabolism of lipophilic xenobiotics has a detoxifying function. Lipophilic compounds that cannot be excreted unchanged to any significant extent are converted to water-soluble, readily excreted metabolites and conjugates. It would appear that during the course of evolution vertebrates and insects have developed mechanisms of metabolism and excretion that have protected them against the naturally occurring xenobiotics to which they have been exposed. Nevertheless, metabolism may yield reactive products that have toxic effects. Thus, many organophosphate insecticides and industrial chemicals that include azo dyes, amines, and chlorinated compounds are all activated by oxidative processes.[2,3] When oxygen is introduced into structures that are otherwise unreactive, it can cause reactivity by withdrawal of electrons to form electron-deficient carbon atoms or carbonium ions. The production of reactive metabolites raises technical problems for the investigator. Because of their instability, such compounds are difficult or impossible to isolate, and their identification may depend upon the characterization of the adducts that they form with cellular structures (see Chapter 2, this volume).

The extent to which active compounds can cause cellular damage depends on several factors, one of which is how efficiently they are metabolized to inactive forms. Thus, the active oxons of organophosphates can be degraded by A esterases, and most epoxides of aromatic or olefinic structural elements can be degraded by epoxide hydrolase or a glutathione transferase; usually these conversions are protective. In some instances, those derivatives that are toxicologically the most active are not readily metabolized by deactivating enzymes.

Apart from the definition of the activating or deactivating role of enzymes, there is the problem of their toxicokinetic significance. Are differences in enzyme levels between species, strains, sexes, age groups, or tissues of any significance with regard to the levels of active metabolites that actually reach sites of action? Where metabolic capacity is high in relation to the rate at which a xenobiotic reaches an enzyme, then differences between species with respect to enzyme activity are unlikely to influence the differential rates of metabolism *in vivo*. In an experiment with [^{14}C]dieldrin and two of its analogs, the rate of excretion of radioactive metabolites was monitored in the bile of male rats following intraperitoneal injection. The rate of metabolism of the three substrates was studied in liver microsomes, where the analogs were found to be metabolized much more rapidly than dieldrin itself.[4] The excretion rate for dieldrin metabolites was very low, in keeping with the very slow rate of metabolism *in vitro*, and was increased threefold after induction of liver enzymes. By contrast, the rates of excretion of the metabolites of the two dieldrin analogs were much more rapid, although far short of the metabolic capacity indicated by *in vitro* studies, and the rate was not increased by induction of liver enzymes. The evidence clearly indicated that the enzymatic processes, mainly due to monooxygenases, were rate limiting for dieldrin but not for its more labile analogs.

In general, enzymatic processes will become rate limiting as they approach saturation. This may happen (1) when the metabolic capacity is low for a specific substrate and (2) when the rate of arrival of the substrate is relatively rapid. Under these circumstances, species, sex, strain, and age differences in enzyme activity may be reflected in differences in half-lives, clearances, and excretion rates for xenobiotics[5] (see Chapters 8 and 9, this volume).

When enzymatic activities do not limit rates of metabolism, species differences may still be important if more than one enzyme is involved in the initial metabolism of a xenobiotic. Different patterns of metabolites may be produced due to species differences in the relative amounts of enzymes involved in metabolic conversion.

In either of the two situations described, enzymatic differences between species may lead to corresponding differences in susceptibility to toxicants. In attempting to identify the role of enzymes in the regulation of toxicity, it is clearly important to take account of the realistic tissue levels of the toxicant. *In vitro* studies should include a proper measurement of the concentration dependence of enzymatic processes, so that metabolic rates at tissue concentrations may be estimated.

A useful technique for identifying the role of enzymatic processes in the regulation of toxicity involves the selective inhibition of enzymes. Differences in susceptibility shown between groups or species can sometimes be reduced or even eliminated by blocking metabolic detoxication. This approach has often been used for the identification of metabolic mechanisms of resistance to insecticides by insects.

III. THE ENZYMATIC FACTOR IN SELECTIVE TOXICITY

A. Organophosphate Insecticides

One of the best studied examples of enzymatically related selectivity concerns the insecticide malathion. Malathion has a generally low mammalian toxicity in spite of its strong insecticidal properties. The acute oral LD_{50} values for malathion in mammals are about 500–5500 mg/kg. Like many other organophosphates, malathion is activated by monooxygenase attack to produce the potent anticholinesterase malaoxon (Fig. 1). Malathion itself has little or no cholinesterase activity. It is rapidly detoxified in mammals (but not in insects) by carboxyesterase attack to produce a monoacid, thus preventing the generation of substantial quantities of malaoxon.[6] Carboxyesterase activity has the character of a B esterase, that is, it can be inhibited by organophosphates such as malaoxon. Thus, organophosphates such as EPN (*O*-ethyl *O*-*p*-nitrophenyl phenylphosphonothioic acid) can act as synergists for malathion in mammals. By inhibiting this detoxication pathway, mammals may be made almost as sensitive as insects to malathion,[6] providing good evidence for the impor-

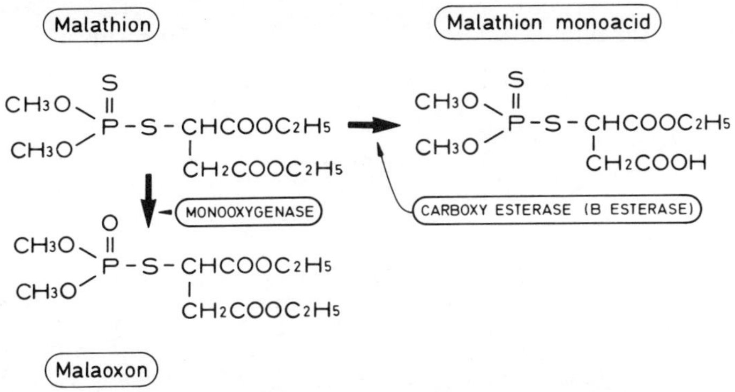

13. Enzymes in Selective Toxicity

Fig. 1. The metabolism of malathion.

tance of carboxyesterase in this striking case of selective toxicity. The carboxy ester bonds to which most of this selectivity is attributed are examples of what have been termed *selectrophores,* or groups that confer selectivity upon a toxic molecule.

Another example of selective toxicity among the organophosphates is shown by the related compounds diazinon and pirimiphos methyl, both of which have leaving groups derived from pyrimidine (Fig. 2). Both compounds are considerably more toxic to birds than to mammals. They are converted to their active forms, diazoxon and pirimiphos methyloxon, by microsomal monooxygenase attack. With oral intake, much of this activation occurs in the liver, and the subsequent toxic action depends upon the efficiency with which these active oxons can be transported by the blood to the brain. In the case of diazinon, studies with liver preparations show that birds and mammals produce similar quantities of the oxon under the same experimental conditions. In contrast, plasma from five mammalian species rapidly hydrolyzed the oxons, whereas virtually no hydrolysis occurred in the serum of the duck, chicken, or turkey.[7] This strongly suggests that mammals are protected against the toxic action of diazinon by rapid hydrolysis of its active oxon in the blood. With pirimiphos methyl oxon, rapid hydrolysis was observed in the plasma of 5 species of mammals, whereas little or no metabolism occurred in 14 species of birds belonging to 6 different orders.[8] Once again, the mammals were apparently protected by A esterase activity in the blood, that is, those that hydrolyze organic phosphates, in contrast to birds. The same difference between birds and mammals was found when measuring A esterase activity with paraoxon, although this substrate was hydrolyzed much more slowly than either pirimiphos methyl oxon or diazinon. It is of interest that paraoxon does not show the same selectivity between birds and mammals

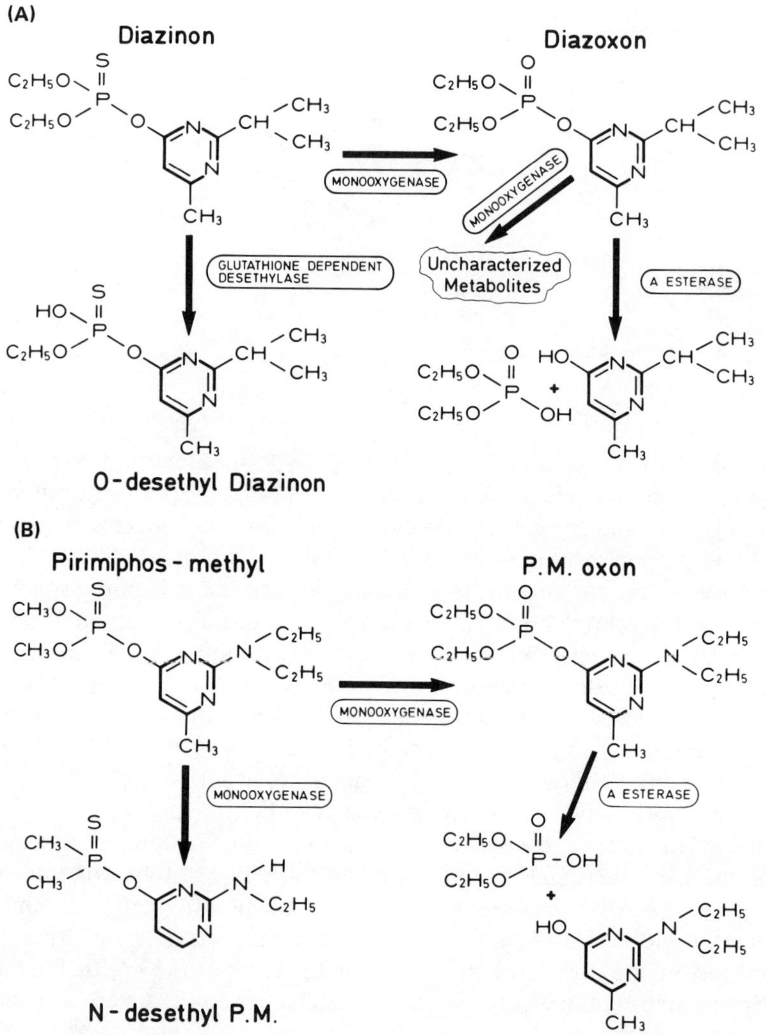

Fig. 2. The metabolism of (A) diazinon and (B) pirimiphos methyl.

found with the other two compounds, suggesting that paraoxon is not hydrolyzed sufficiently rapidly in mammalian blood to provide an effective defense mechanism.

In vivo work on pirimiphos methyl in the male Japanese quail (*Coturnix coturnix japonica*) and the male rat supports the above conclusions.[9] When the two species were given equitoxic doses of pirimiphos methyl (140 mg/kg in bird, 800 mg/kg in rat), the inhibition of cholinesterase 15 min after administration was 90% in the blood and 40% in the brain of the bird, but the rat enzymes were not inhibited. Studies with liver microsomes showed that although both species can hydrolyze pirimiphos methyl oxon, A esterase activity was much higher in the rat than in the Japanese quail.

Enzymatic differences, therefore, can be critical in providing the desired selectivity of organophosphate insecticides between vertebrates and insects. Such differences are also important in determining susceptibility between different vertebrate groups. Mammals detoxify the active oxons of diazinon and pirimiphos methyl by A esterase hydrolysis in liver and blood, but birds are deficient in this type of enzyme. A esterase activity is found in liver microsomes, blood serum, and plasma. Apart from pirimiphos methyl and diazinon, pirimiphos ethyl and coumaphos also show a marked selectivity between birds and mammals that may be related to rates of hydrolysis of oxons by A esterase.[7]

B. Organochlorine Insecticides

The organochlorine insecticides are a diverse group of highly lipid-soluble compounds, with water solubilities frequently below 1 ppm. The common ones are highly chlorinated, which frequently results in very slow rates of metabolism[10] leading to bioaccumulation. Before discussing these compounds, however, brief consideration will be given to some biodegradable members of the group.

Hexachlorocyclohexane (γHCH or γBHC) is readily biodegradable to chlorophenol conjugates and chlorophenylmercapturic acids[11]; the metabolic processes are complex. Dehydrochlorination appears to be the consequence of both glutathione conjugation and monooxygenase attack upon the ring. Although widely used as a seed dressing, this compound has not given rise to problems of bioaccumulation and secondary poisoning. Its limited persistence is a reflection of its high water solubility relative to other chlorinated insecticides (~7 ppm) and its relatively rapid metabolism.

A number of readily biodegradable analogs of dieldrin have been synthesized.[10,12] The analog 1,8,9,10,11,11-hexachloro-4,5-exoepoxy-2,3,7,6-endotricyclo[6,2,11,0]undec-9-ene (HEOM) and the isomeric 3,4-exoepoxy- and 3,6-endoepoxy compounds (HCE and ODA, respectively) are all considerably more toxic to tsetse flies (*Glossina* spp.) than to houseflies or certain species of mosquitoes.[12] Interestingly, the toxicity of these compounds is enhanced by microsomal monooxygenase inhibitors in the case of houseflies and mosquitoes but not to any significant degree in the case of tsetse flies. It appears that tsetse flies are vulnerable to these biodegradable insecticides because they are deficient in appropriate detoxifying enzymes, especially monooxygenases. This may be related to the fact that they feed exclusively on blood, a point that will be discussed in more detail.

Organochlorine insecticides such as DDT, aldrin, dieldrin, and heptachlor have given rise to problems of persistence. The metabolism of some of these compounds is shown in Fig. 3. Some of these metabolites are stable and persistent, as in the cases of p,p'-DDE, a metabolite of p,p-DDT, and dieldrin, a metabolite of aldrin; the metabolites are more persistent than the original insecticides.

The biological half-lives of these compounds differ markedly between species. Dieldrin half-lives have been estimated as follows: male rat 7.6–10.0 days, female rat 10.3–15.1 days, pigeons ~47 days, steers 74 days, sheep 97 days, and humans 270 days.[13] In a study of species differences in monooxygenase activity, an inverse relationship was found between enzyme activity and the half-lives of dieldrin and three other xenobiotics in a range of species.[14] It is noteworthy that the monooxygenases are responsible for most of the initial metabolism of these compounds, which tends to support the view that differences in enzyme activity are important in determining differences in half-lives.

Marked species differences also exist in the capacity for the bioaccumulation of compounds such as PCBs, DDE, and dieldrin when present at low concentrations in food. Gannon[15] gave food containing 0.4 ppm of dieldrin to a range of species over a period of 84 days and found residue levels in body fat ranging from 0.4 ppm in lambs to 10 ppm in hens; a much greater tendency for bioaccumulation was observed in the hens than in rats.[16] Field and laboratory studies have shown that fish-eating birds such as cormorants (*Phalacrocorax carbo*), shags (*P. aristotelis*), double-crested cormorants (*P. auritus*), and white pelicans (*Pelecanus erythrorhynchos*) have a marked tendency to bioaccumulate compounds of this type. Estimated bioaccumulation factors (conc. in tissues of bird/conc. in tissues of fish) ranged from 30- to 280-fold.[17] The bioaccumulation factors are inversely related to the hepatic microsomal monooxygenase activities of the species studied. Monooxygenase activities are highest in rats and lowest in fish-eating birds, with the hen occupying an intermediate position.[18]

Long half-lives and high bioaccumulation factors, therefore, appear to be correlated with low concentrations of hepatic microsomal monooxygenases. Other evidence strongly suggests that this is a causal relationship.[5]

C. Carbamate Insecticides

Most carbamate insecticides are rapidly detoxified by oxidative attack in both vertebrates and insects, as indicated by the strong synergistic

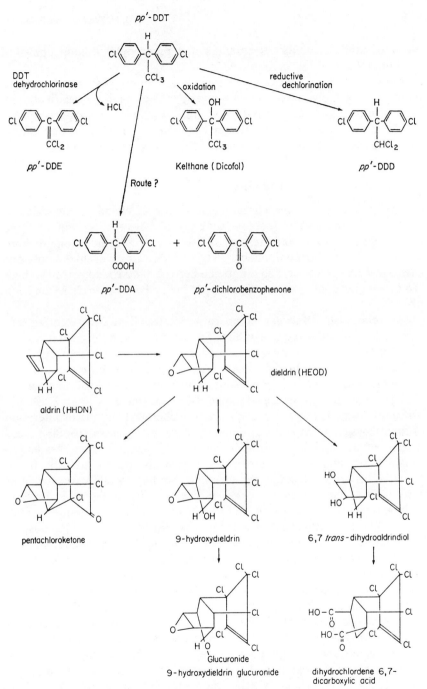

Fig. 3. The metabolism of DDT, aldrin, and dieldrin. From Walker, C. H. (1975).

effect of inhibitors such as sesamex and piperonyl butoxide upon the insecticidal activity of carbamates.[19]

It has been shown that bees are considerably more susceptible to propoxur than are the cockroach and housefly,[19] bees having a relatively low level of monooxygenase activity. On present evidence, birds are considerably more susceptible to carbamates than are mammals, and this is correlated with a relatively low level of monooxygenase in birds.[4]

D. Carcinogens and Mutagens

A recent review of the carcinogenic action of chemicals on different species[20] points to considerable differences between the rat and the mouse. Of 250 compounds tested, 147 were carcinogenic to one or both species. Of these 147 compounds, 38 (26%) were carcinogenic to only one species. With such a marked difference between two rodent species, there are inevitably doubts about the comparability between rodents and humans.

The relationship between comparative metabolism and selective toxicity was relatively easy to establish for the compounds discussed earlier because there was, in most cases at least, an acute lethal effect as the consequence of a well-established biochemical lesion. With carcinogens, however, much uncertainty remains concerning the critical biochemical changes that lead to the development of neoplasia, and there is often a substantial time lag between exposure to a chemical and the proliferation of tumor tissue. Furthermore, tumors generally appear in specific tissues when specific carcinogens are administered by defined routes, leading to considerable interest in metabolic activation and deactivation by target tissues. If a critical change, for example, binding to DNA, occurs after exposure to a carcinogen, tumors do not necessarily develop. The damage may be repaired, the cell may die or be destroyed by the immune system, or the cell carrying the DNA damage may remain "latent." Because of this complexity, the connection between species or strain differences in enzyme activity and differences in susceptibility to carcinogens is difficult to establish. Some of the best evidence for species differences in activation of carcinogens has been obtained with *in vitro* systems, which are valuable monitors of reactive metabolites but cannot take fully into account possible differences in the events that occur after the primary interaction of reactive metabolites with target tissue macromolecules.

In such systems, compounds are tested for their mutagenic activity toward bacteria, for example, by the Ames test, or they are added to cell preparations to determine whether they alter the rate of transformation. Activation by liver microsomes of rats or mice pretreated with enzyme inducers is frequently necessary to produce a response. In the Ames test, the effect is evidently due to the binding of material to DNA, and this may also be the main mechanism in tests of cell transformation.[21] It has been clearly demonstrated that metabolites of carcinogens such as benzo[a]pyrene and aflatoxin B are bound to DNA, when the original compounds are given both to living animals and to cell systems *in vitro*.[22]

Considerable interest therefore centers upon the influence of species differences in metabolism upon the production of active metabolites in *in vitro* systems and their importance in determining the outcome of tests. When liver microsomes from mouse and rat were used for the metabolic activation of benzo[a]pyrene in a test system for mutagenicity employing *Salmonella typhimurium,* only slight mutagenicity was observed with rat microsomes, whereas very high mutagenicity was found with mouse microsomes.[23] The mouse strain used had a higher monooxygenase activity (about twofold for most substrates, including benzo[a]pyrene) and a much lower epoxide hydrolase activity (six- to sevenfold for styrene oxide) than was found with the rat. Inhibition of microsomal epoxide hydrolase with 1,1,1-trichloropropene oxide[24] increased dramatically[25] the mutagenicity that was mediated by liver microsomes from rats, indicating that the relatively low mutagenicity mediated by microsomes from this species was due, at least in part, to the action of epoxide hydrolase. Addition of purified epoxide hydrolase[26] to this test system containing liver microsomes from untreated mice almost completely abolished the mutagenicity of benzo[a]pyrene.[27] This provided further evidence for the importance of epoxide hydrolase in determining the selective toxicity of benzo[a]pyrene in the test system.

Dog liver microsomes were less able than rat liver microsomes to activate benzo[a]pyrene in the Ames test.[28] In order to observe significant mutagenicity, the postmitochondrial supernatant, rather than microsomes, had to be used for metabolic activation of benzo[a]pyrene, probably due to higher stability of the monooxygenases in the former fraction. The lower activation by the dog, compared to the rat preparation, was due to several factors, an especially important and interesting one being the different sites of preferential oxidative attack in the benzo[a]pyrene molecule by liver microsomes of the two species. The microsomes of the dog showed a greater tendency to attack benzo[a]pyrene in the 3, 4, and 5 positions, whereas rat microsomes metabolized the molecule more in the 7, 8, 9, and 10 positions; the dog had higher epoxide hydrolase activity than the rat. This resulted in the formation of a lower proportion of benzo[a]pyrene 7,8-dihydrodiol 9,10-oxide in the dog. This last bay region derivative is especially chemically reactive and mutagenic, and represents an ultimate carcinogen of benzo[a]pyrene.[29] Induction of microsomal epoxide hydrolase[30] and shift of the monooxygenase-catalyzed oxidative attack from the 7,8,9,10-region of the angular ring, which carries the dihydrodiol bay region epoxide, to the 3,4,5-position was associated with a drastic reduction of mutagenicity.[31] Thus, differences between rat and dog with regard to the activities of microsomal epoxide hydrolase and of contrasting monooxygenase forms are responsible for the selective toxicity pattern observed.[25]

Another case of species differences in carcinogen metabolism and activity concerns the hepatocarcinogen 2-N-acetylaminofluorene (AAF). This compound is N-hydroxylated to the proximal carcinogen N-hydroxyacetylaminofluorene, a potent hepatocarcinogen in rat, hamster, and mouse but not in guinea pig or rhesus monkey.[32-35] It is presumed that the first step of the required metabolic activation of the precarcinogen AAF is its transformation to the N-hydroxy derivative as the proximate carcinogen.[36] Initially, it was believed that the resistance of the guinea pig liver is caused by its inability to perform the N-hydroxylation reaction. More recent results show that guinea pig liver possesses N-hydroxylase activity that transforms AAF to the N-hydroxy-AAF with efficiency similar to that of susceptible animal species such as the rat.[35] The liver of the monkey can also N-hydroxylate AAF.[37] However, both guinea pig and rhesus monkey liver metabolize N-hydroxy-AAF very rapidly to the inactive 7-hydroxy-AAF,[38] so that the proximate carcinogen, N-hydroxy-AAF, does not accumulate. Thus, one enzymatic mechanism responsible for the selective toxicity of AAF may be the metabolic detoxication of the proximate carcinogen, N-hydroxy-AAF. A second important enzymatic difference occurs in the conjugation of the proximate carcinogen, N-hydroxy-AAF, to an ultimate heptocarcinogen, the N,O-sulfate.[39] Although AAF does not produce tumors in the guinea pig at any site, N-hydroxy-AAF induces the development of adenocarcinoma of the small intestine and sarcoma at the site of injection.[35] In guinea pig liver there is relatively little accumulation of the proximate carcinogen, N-hydroxy-AAF, and only slight conversion to an ultimate carcinogen by aryl sulfotransferase.[40] Livers of male rats are much more susceptible to the carcinogenic activity of N-hydroxy-AAF than those of females or of male mice and hamsters; guinea pig liver is resistant. The male rat liver is consistently found to have much higher N-hydroxy-AAF sulfotransferase activity than have the livers of female rats, male mice, and hamsters; in guinea pig liver this activity is not detectable. That this correlation is not fortuitous but causally linked is supported by the fact that both the susceptibility of the male rat liver to the carcinogenic activity of AAF and its derivatives and the sulfotransferase activity for N-hydroxy-AAF can be modified by hormonal manipulation. Thyroidectomy, hypophysectomy, or castration followed by administration of estrogen markedly reduces the susceptibility of the male rat liver to the carcinogenic activity of AAF and N-hydroxy-AAF. These hormonal manipulations also lead to a decrease of N-hydroxy-AAF sulfotransferase activity, as well as a decrease of covalent binding to macromolecules in the liver of rats after administration of N-hydroxy-AAF. Both covalent binding to tissue macromolecules[41] and the carcinogenic activity[42] of N-hydroxy-AAF for the rat liver are increased by administration of sodium sulfate.

These examples indicate that species differences in enzymes can be important in determining susceptibility to carcinogens. The point should be considered given the often large differences in these enzymes between humans and certain experimental animals that are commonly used in testing carcinogens.

IV. THE ENZYMATIC FACTOR IN RESISTANCE

Resistance is a specific type of selective toxicity. Many examples are known of the development of resistant strains of insect in response to insecticides, both in the laboratory and in the field.[43,44] In general, this resistance is preadaptive, representing a selection of genes already present in the population. Metabolism is frequently the dominant physiological mechanism of resistance. In the following account, examples will be considered that provide illustrations of the importance of the enzymatic factor in resistance.

A. Organophosphate Insecticides

One of the best documented examples concerns malathion. Resistant strains of housefly, mosquito, and hide beetle (*Dermestes* spp.) show high concentrations of "carboxyesterase" (Fig. 1). This type of resistance can be overcome by synergists that inhibit the esterase, for example, EPN and triphenyl phosphate.[44] Thus, the resistant strains have a detoxication mechanism similar to the one that is well developed in mammals.

In recent years, strains of aphids with pronounced cross-resistance to a range of organophosphates have been studied. These strains had unusually large quantities of a protein with the character of a B esterase.[45] This esterase sequesters the active (oxon) forms of organophosphates that phosphorylate it. It is of interest that this resistance mechanism relies upon an esterase that is strongly inhibited by the organophosphate. A more efficient mechanism might have been provided by an A esterase that could rapidly break down the toxicant, which raises the possibility that aphids have no A esterase. Activity of this sort could not be found in three strains of *Tribolium castaneum* (rust-red flour beetle), two of which are resistant to organophosphates.[46]

The development of resistance by houseflies to diazinon appears to involve two distinct enzymatic mechanisms (Fig. 2). Diazinon is activated by conversion to diazoxon by microsomal monooxygenases. Paradoxically, one of the resistance mechanisms is based upon the further conversion of diazoxon by monooxygenases to as yet uncharacterized inactive metabolites.[47] The other mechanism involves O-deethylation of diazinon itself by the action of a glutathione S-transferase. When monooxygenases were inhibited with sesamex and the glutathione S-transferase with S,S,S-tributyl phosphorothioate, each inhibitor caused some loss of resistance. When they were used in combination, both resistant and susceptible strains were entirely similar in sensitivity to diazinon. There was no evidence that A esterases were attacking diazinon and thereby making a contribution to resistance, in contrast to the protective role of this enzyme in mammals.[8] These observations cast further doubt upon the existence of this type of esterase in certain species of insect.

B. Organochlorine Insecticides

Many cases of resistance have been reported for persistent organochlorine insecticides such as DDT and dieldrin, and it is widely believed that the marked persistence of these compounds tends to encourage development of resistance. Selection for resistance may occur over a relatively long period of time and, with movement of compounds through the food chain, may take place in areas some distance from the original place of application.

Increased enzyme activity has often been implicated in the development of resistance to DDT.[43] In most instances, this has been associated with the conversion of p,p'-DDT to its highly persistant but relatively nontoxic metabolite, p,p'-DDE (Fig. 3) by an enzyme entitled DDT-dehydrochlorinase. DDT-dehydrochlorinase was isolated from houseflies[48] and has a requirement for reduced glutathione, although the formation of glutathione conjugates of DDT has never been shown.

A contrasting mechanism in certain strains of DDT-resistant houseflies involves enhancement of monooxygenase activity.[44] Polar metabolites were formed in the resistant strains that have not yet been properly characterized. The resistance was overcome by using an inhibitor of monooxygenase. Not surprisingly, these strains showed cross-resistance to other insecticides. DDT resistance, based upon enhanced oxidation, has also been reported in strains of mosquito.[49] Here, the metabolite formed by the resistant strains was kelthane, an oxidation product of DDT (Fig. 3).

Resistance to organochlorine insecticides is not confined to insects. A strain of pine mouse (*Microtus pinetorum*) was reported in the United States as resistant to endrin.[50] Endrin is a stereoisomer of dieldrin that can be metabolized fairly rapidly by monooxygenase attack. The resistance factor was seven- to eightfold, and the resistant strain excreted hydrophilic metabolites in the feces twice as rapidly as did the susceptible strain.

C. Carbamate Insecticides

Resistance to carbamate insecticides appears to be due at least in part to enhanced monooxygenase activity.[19] Such resistance has been reported in houseflies and mosquito larvae, where it is partially overcome by the use of piperonyl butoxide, an inhibitor of monooxygenase.[19]

V. DISCUSSION

Certain clear phylogenetic trends have been reported in the activity of enzymes that participate in xenobiotic metabolism. The *in vitro* activity of the hepatic microsomal monooxygenases and epoxide hydrolase tends to be much higher in mammals than in fish, with birds occupying an intermediate position.[17,18] Hepatic microsomal glucuronyltransferase is more active in mammals than in fish. Serum A esterase activity is much higher in mammals than in birds.[8]

Sometimes trends are associated with diet. Thus, five out of six species of fish-eating sea birds had low monooxygenase activity compared to omnivorous or herbivorous birds.[17,18] Three carnivorous mammals had a low capacity for the p-hydroxylation of aniline compared to noncarnivores.[18] The blood-sucking tsetse flies appear to be very low in detoxifying enzymes in comparison to houseflies.[12]

It has been suggested that herbivorous and omnivorous species have evolved a greater capacity for the metabolism of lipid-soluble xenobiotics than have carnivores because of the relatively wide range of substances that are present in their food. Fish and amphibia represent a special case. Brodie and Maickel[51] suggested that fish have not needed to evolve enzymes of this type for the metabolism of xenobiotics because the excretion of lipid-soluble compounds can be achieved by passive diffusion across gills or skin into the surrounding water. This excretion route is effective for those lipid-soluble substances that have a favorable oil:water partition coefficient but is ineffective for very lipid-soluble compounds such as dieldrin, DDE, and highly chlorinated polychlorobiphenyls that have highly unfavorable coefficients. Fish can bioaccumulate compounds such as these, concentrating them by a factor of several thousand from the surrounding water. Whatever the truth of the evolutionary argument, excretion by passive diffusion clearly does not provide fish with effective protection against persistent lipophilic chlorinated compounds. There is, however, evidence suggesting that fish have developed improved detoxication systems in certain polluted areas, although it is not yet clear whether these are cases of resistance or induction.[52]

The idea that enhanced capacity for xenobiotic metabolism may have evolved in response to the selective action of xenobiotics gains support from studies of resistance to insecticides. There are many examples of insects developing resistance by increasing their capacity for metabolic detoxication.[43,44,53] This often occurs within a few generations. When enzymes of wide substrate specificity are involved, there may be cross-resistance extending over a range of insecticides, all of which are detoxified by the same enzyme. A specific instance is the appearance of large quantities of B-type esterase (carboxyesterase) in resistant aphids, which gives cross-resistance to many different organophosphates.[45] It has been suggested that the enhanced enzymatic activity is a consequence of gene duplication.[54]

Major trends in enzyme activity related to phylogeny and diet may be reflected in corresponding trends in susceptibility to toxicants. The low monooxygenase activity of fish and fish-eating birds is related to a marked tendency to bioaccumulate persistent organochlorine compounds; the low plasma A esterase activity of birds is related to a high susceptibility to organophosphates such as pirimiphos methyl and diazinon; and the relatively low monooxygenase of birds is correlated with a relatively high susceptibility to carbamate insecticides. As phylogenetic trends in enzymology become better defined and their physiological significance better understood, other relationships to patterns of selective toxicity may be expected.

The microsomal monooxygenases, epoxide hydrolase, glutathione transferases, and glucuronyl transferases are inducible by various exogenous and endogenous compounds, and this can have a profound effect upon the action of toxic substances. Induction generally reduces the half-life and toxic effect of a xenobiotic. In some cases, when the enzyme catalyzed reaction generates a toxic metabolite (e.g., in the oxidation of carcinogens or organophosphorothioates), increased toxicity will ensue.

Sometimes inducing agents alter the balance between enzymes: in rodents, 3-methylcholanthrene can increase the specific activity for substrates of cytochrome P_{448}-dependent monooxygenase by more than 10-fold, but the increase in microsomal epoxide hydrolase activity is only marginal.[55] Such changes in the balance among enzymes may well cause changes in the toxic response.

The foregoing discussion of species differences in enzymes was concerned with animals in the noninduced state. To what extent, then, can induction alter these differences? The microsomal monooxygenases of mammals, birds, and fish are inducible, although there are certain differences between fish and the other groups. The hepatic microsomal monooxygenase of fish can be induced by agents such as phenobarbitone and certain polycyclic aromatic hydrocarbons, but the induced cytochromes have spectral characteristics different from those of rats treated with the same inducers.[56] DDT and analogs and nonplanar PCBs, which are inducers in mammals, do not appear to induce readily in fish. Because mammalian and avian monooxygenases appear to be more sensitive to inducing agents than are those of fish, the presence of inducing agents in the environment is likely to emphasize rather than decrease the difference in oxidative capacity between fish and the other groups.

REFERENCES

1. Jakoby, W. B., ed. (1980). "Enzymatic Basis of Detoxication." Academic Press, New York.
2. Eto, M. (1974). "Organophosphorus Pesticides. Organic and Inorganic Chemistry." CRC Press, Boca Raton, Florida.
3. Searle, C. E. (1976). "Chemical Carcinogens," ACS Monogr. 173. Am. Chem. Soc., New York.
4. Chipman, J. K., and Walker, C. H. (1979). The metabolism of dieldrin and two of its analogues. *Biochem. Pharmacol.* **28**, 337–1345.
5. Walker, C. H. (1981). The correlation between *in vivo* and *in vitro* metabolism of pesticides in vertebrates. *Prog. Pestic. Biochem.* **1**, 247–285.
6. O'Brien, R. D. (1967). "Insecticides: Action and Metabolism." Academic Press, New York.
7. Machin, A. F., Anderson, P. H., Quick, M. P., Waddell, D. R., Skibniewska, J. K., and Howells, L. C. (1978). The metabolism of diazinon in the liver and blood of species of varying susceptibility to diazinon poisoning. *Xenobiotica* **7**, 104.
8. Brealey, C. J., Walker, C. H., and Baldwin, B. C. (1980). A-esterase activities in relation to the differential toxicity to birds and mammals. *Pestic. Sci.* **11**, 546–554.
9. Brealey, C. J. (1981). Comparative metabolism of pirimiphos-methyl in rat and Japanese quail. Ph.D. Thesis, Reading University.
10. Brooks, G. T. (1974). "The Chlorinated Insecticides." CRC Press, Boca Raton, Florida.
11. Stein, K., Portig, J., and Koransky, W. (1977). Oxidative transformation of

hexachlorocyclohexane in rats and with rat liver microsomes. *Naunyn-Schmiedeberg's Arch. Pharmacol.* **298**, 115–128.
12. Brooks, G. T., Barlow, F., Hadaway, A. B., and Harris, E. G. (1981). The toxicities of some analogues of dieldrin endosulfan and isobenzan to blood sucking diptera, especially tsetse flies. *Pestic. Sci.* **12**, 475–484.
13. Moriarty, F., ed. (1975). "Organochlorine Insecticides: Persistent Organic Pollutants," Chapter 2. Academic Press, New York.
14. Walker, C. H. (1978). Species differences in microsomal monooxygenase activities and their relationship to biological half-lives. *Drug Metab. Rev.* **7**, 295–323.
15. Gannon, N., Link, R. P., and Decker, G. C. (1959). Storage of dieldrin in tissues of steers, hogs, lambs and poultry fed dieldrin in their diets. *J. Agric. Food Chem.* **7**, 826–828.
16. Davison, K. L. (1973). Dieldrin ^{14}C balance in rats, sheep and chickens. *Bull. Environ. Contam. Toxicol.* **10**, 16–24.
17. Knight, G. C., Walker, C. H., Harris, M., and Cabot, D. C. (1980). The activities in sea birds of two hepatic microsomal enzymes which metabolize liposoluble xenobiotics. *Comp. Biochem. Physiol. C.* **68C**, 127–132.
18. Walker, C. H. (1980). Species variations in some hepatic microsomal enzymes that metabolize xenobiotics. *Prog. Drug Metab.* **5**, 113–116.
19. Wilkinson, C. F., ed. (1976). "Insecticide Biochemistry and Physiology." Heyden, London.
20. Purchase, I. F. H. (1980). Inter-species comparisons of carcinogenicity. *Br. J. Cancer* **41**, 454–468.
21. Clegg, J. C. S., Glatt, H. R., and Oesch, F. (1981). Coordinate mutation and transformation of mouse fibroblasts: Induction by nitroquinoline oxide and modulation by caffeine. *Carcinogenesis* **12**, 1255–1259.
22. Garner, R. C. (1975). The role of epoxides in bioactivation and carcinogenesis. *Prog. Drug Metab.* **1**, 77–128.
23. Oesch, F., and Glatt, H. R. (1976). Evaluation of the importance of enzymes involved in the control of mutagenic metabolites. *IARC Sci. Publ.* **1232**, 255–274.
24. Oesch, F., Kaubisch, N., Jerina, D. M., and Daly, J. W. (1971). Hepatic epoxide hydrase. Structure–activity relationships for substances and inhibitors. *Biochemistry* **10**, 4858–4866.
25. Oesch, F. (1980). Species differences in activating and inactivating enzymes related to *in vitro* mutagenicity mediated by tissue preparations from these species. *Arch. Toxicol., Suppl.* **3**, 179–194.
26. Bentley, P., and Oesch, F. (1975). Purification of rat liver epoxide hydratase to apparent homogeneity. *FEBS Lett.* **59**, 291–295.
27. Oesch, F., Bentley, P., and Glatt, H. R. (1970). Prevention of benzo[a]pyrene induced mutagenicity by homogeneous epoxide hydratase. *Int. J. Cancer* **18**, 448–452.
28. Golan, M. D., Bucker, M., Schmassmann, H. U., Raphael, D., Jung, R., Bindel, U., Brase, H. D., Tegtmeyer, F., Friedberg, T., Lorenz, J., Stasiecki, P., and Oesch, F. (1980). Characterization of dog hepatic drug-metabolizing enzymes and resultant effects on benzo[a]pyrene metabolite pattern and mutagenicity. *Drug Metab. Dispos.* **8**, 121–126.
29. Jerina, D. M., Lehr, R., Schaefer-Ridder, M., Yagi, H., Karle, J. M., Thakker, D. R., Wood, A. W., Lu, A. Y. H., Ryan, D., West, S., Levin, W., and Conney, A. H. (1971). Bay-region epoxides of dihydrodiols: A concept explaining the mutagenic and carcinogenic activity of benzo[a]pyrene and benzo[a]anthrocene. *Cold Spring Harbor Conf. Cell Proliferation* **4**, 639–658.

30. Schmassmann, H., and Oesch, F. (1978). *Trans*-stilbene oxide: A selective inducer of rat liver epoxide hydratase. *Mol. Pharmacol.* **14**, 834–847.
31. Bücker, M., Golan, M., Schmassmann, H. U., Glatt, H. R., Stasiecki, P., and Oesch, F. (1979). The epoxide hydratase inducer *trans*-stilbene shifts the metabolic epoxidation of benzo[*a*]pyrene from the bay- to the K-region and reduces its mutagenicity. *Mol. Pharmacol.* **16**, 656–666.
32. Breidenbach, A. W., and Argus, M. F. (1956). Attempted tumor induction in guinea-pigs. *Q. J. Fla. Acad. Sci.* **19**, 68–70.
33. Weisburger, J. H., and Weisburger, E. K. (1958). Chemistry, carcinogenicity and metabolism of 2-fluorenamine and related compounds. *Adv. Cancer Res.* **5**, 331–431.
34. Morris, H. P., Velat, L. A., Wagner, B. P., Dahlgard, N., and Roy, F. E. (1960). Studies of carcinogenicity in the rat of derivatives of aromatic amines related to *N*-2-fluorenylacetamide. *JNCI, J. Natl. Cancer Inst.* **24**, 149–180.
35. Miller, E. C., Miller, J. A., and Enomoto, M. (1964). The comparative carcinogenicities of 2-acetylaminofluorene and its *N*-hydroxy metabolite in mice, hamsters and guinea-pigs. *Cancer Res.* **24**, 2018–2031.
36. Miller, J. A., and Miller, E. C. (1976). In Screening Tests in Chemical Carcinogenesis. The metabolic activation of chemical carcinogens—Recent results with aromatic amines, safrole and aflatoxins. *IARC Sci. Publ.* **1232**, 153–176.
37. Razzouk, C., Mercier, M., and Roberfroid, M. (1980). Biochemical basis for the resistance of guinea-pig and monkey to the carcinogenic effects of aryl amines and aryl amides. *Xenobiotica* **10**, 565–571.
38. Razzouk, C., Batardy-Goegoine, M., and Roberfroid, M. (1982). Guinea-pig liver microsomal metabolism of *N*-hydroxy-2-acetylaminofluorene and *N*-hydroxy-2-aminofluorene. (Submitted for publication.)
39. Miller, J. A., and Miller, E. C. (1976). The metabolic activation of chemical carcinogens to reactive electrophiles. *In* "Biology of Radiation Carcinogenesis" (J. M. Yuhas, R. W. Tennant, and J. D. Regan, eds.), pp. 147–164. Raven Press, New York.
40. Miller, J. A. (1970). Carcinogenesis by chemicals: An overview. *Cancer Res.* **30**, 558–576.
41. DeBaun, J. R., Smith, J. Y. R., Miller, E. C., and Miller, J. A. (1970). Reactivity *in vivo* of the carcinogen *N*-hydroxy-2-acetylaminofluorene: Increase by sulfate ion. *Science* **167**, 184–186.
42. Weisburger, J. H., Yamamoto, R. S., Williams, G. H., Grantham, P. H., Matsushima, T., and Weisburger, E. K. (1972). On the sulfate ester of *N*-hydroxy-2-fluorenylacetamide as a key ultimate hepatocarcinogen in the rat. *Cancer Res.* **32**, 491–500.
43. Brown, A. W. A. (1971). Pest resistance to pesticides. *In* "Pesticides in the Environment" (R. White-Stevens, ed.), Vol. 1, pp. 437–551. Dekker, New York.
44. Oppenoorth, F. J., and Welling, W. (1976). Biochemistry and physiology of resistance. *In* "Pesticide Biochemistry and Physiology" (C. F. Wilkinson, ed.), pp. 507–554. Heyden, London.
45. Devonshire, A. C. (1977). The properties of a carboxylesterase from the peach-potato aphid *Myzus persicae* and its role in conferring insecticide resistance. *Biochem. J.* **167**, 675–683.
46. Mackness, M. I., Walker, C. H., Rowlands, D. G., and Price, N. G. (1982). Esterase activity in homogenates of three strains of the rust red flour beetle. *Comp. Biochem. Physiol.* (in press).
47. Lewis, J. B., and Lord, K. A. (1969). Metabolism of some organophosphorus insecticides by strains of housefly. *Proc. Br. Insectic. Fungic. Conf., 5th, 1969* pp. 465–471.

48. Lipke, H., and Kearns, C. W. (1960). DDT dehydrochlorinase. *Adv. Pest Control Res.* **3,** 253–288.
49. Busvine, J. R. (1971). The biochemical and genetic bases of resistance. *Pestic. Artic. News Serv.* **17,** 135–146.
50. Webb, R. E., Randolph, W. C., and Horsfall, (1972). Hepatic benzo[a]pyrene hydroxylase activity in endrin susceptible and resistant pine mice. *Life Sci.* **11,** Part 2, 477–484.
51. Brodie, B. B., and Maickel, R. P. (1962). Comparative biochemistry of drug metabolism. *Proc. Int. Pharm. Meet., 1st,* 6, pp. 299–324.
52. Ahokas, J. T., Pelkonen, O., and Karki, N. T. (1975). Metabolism of polycyclic hydrocarbons by a highly active aryl hydrocarbon hydroxylase. *Biochem. Biophys. Res. Commun.* **63,** 635.
53. Champ, B. R., and Dyte, C. E. (1976). "Report of FAO Global Survey of Pesticide Susceptibility of Stored Grain Pests." FAO, Rome.
54. Devonshire, A. L., and Sawicki, R. M. (1979). Insecticide resistant *Myzus persicae* as an example of evolution by gene duplication. *Nature (London)* **280,** 140–141.
55. Oesch, F. (1976). Differential control of rat microsomal aryl hydrocarbon monooxygenase and epoxide hydratase. *J. Biol. Chem.* **251,** 79–87.
56. Elcombe, C. R., Franklin, R. B., and Lech, J. J. (1981). Induction of hepatic microsomal enzymes in rainbow trout. *ACS Symp.* **99,** 319–337.

CHAPTER 14

Intraindividual and Interindividual Variations*

Elliot S. Vesell and
M. B. Penno

I. Introduction 369
   A. Attaining Near Basal Conditions in Subjects of Kinetic
      Studies 376
   B. Antipyrine as a Model Drug in Studies on Human
      Drug-Metabolizing Capacity 383
II. Sources of Intraindividual Variability 384
III. Sources of Interindividual Variability 391
   A. Evidence for Genetic Control of Large Interindividual
      Variations in Xenobiotic Metabolism 393
   B. Evidence for Environmental Control of Large Inter-
      individual Variations in Drug Metabolism 403
   References 406

I. INTRODUCTION

The topic of sources of variation in rates of hepatic metabolism of exogenous chemicals has become an independent, new area of scientific inquiry. A systematic investigation of causes of such variations has been launched. One facet of the subject, pharmacogenetics, concerns genetically determined interindividual variations in response to a specific envi-

* This work was supported in part by NIGMS 26,027.

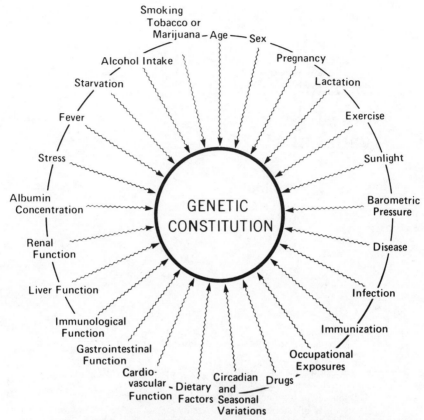

Fig. 1. This circular design suggests the multiplicity of either well-established or suspected host factors that may influence drug response in humans. A line joins all such factors in the outer circle to indicate their close interrelationship. Arrows from each factor in the outer circle are wavy to indicate that effects of each host factor on drug response may occur at multiple sites and through different processes that include drug absorption, distribution, metabolism, excretion, receptor action, and combinations thereof. Taken from Vesell, E. S. (1982). On the significance of host factors that affect drug disposition. *Clin. Pharmacol. Ther.* **31**, 1–7.

ronmental agent: drugs.[1-3] Other sources of such variations have been investigated and identified in numerous laboratories. Interactions among all currently recognized causes of interindividual variations are indicated in Figs. 1 and 2. Additional factors remain to be discovered, and the field itself is in the midst of rapid growth.

Because the methods used to identify sources of intraindividual and interindividual variations in drug metabolism profoundly influence the results obtained, a detailed examination of these methods is justified. This

14. Intraindividual and Interindividual Variations

examination includes a comparison of the advantages and limitations of the two principal techniques currently in use. Each of these methods uses subjects under distinctly different environmental conditions; therefore, this chapter emphasizes the critical role played by such conditions and selection criteria in determining experimental results.

Because so many factors affect the activity of hepatic "detoxication" enzymes, it is crucial to design experiments that permit, as much as possi-

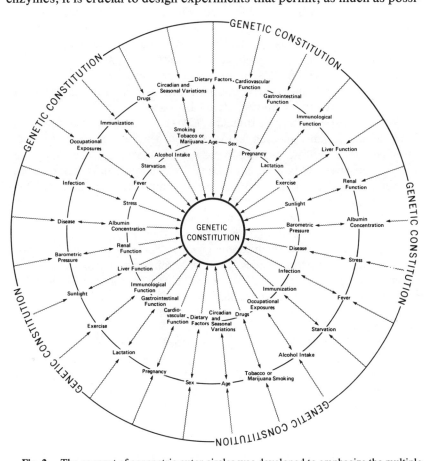

Fig. 2. The concept of concentric outer circles was developed to emphasize the multiple possibilities that exist for interaction among host factors and to suggest that the magnitude of the impact of host factors on drug response may be modulated by genetic constitution. Because in most cases these specific interactions and modulations have not yet been investigated, much less firmly established, this design is largely speculative and intended to stimulate future research rather than to depict the current state of knowledge in the field. Taken from Vesell, E. S. (1982). On the significance of host factors that affect drug disposition. *Clin. Pharmacol. Ther.* **31**, 1–7.

ble, isolation and investigation of each factor independently of the influence of multiple other factors. If these other factors are allowed to play a role, the results may be difficult to interpret because they will be confounded by effects from an indeterminate number of interacting variables. To facilitate identification of sources of interindividual variations, we developed the "model drug" approach in which the kinetics of a model drug are measured in each subject *in vivo* under near basal conditions and then again after imposition of a single environmental change.[5,6] This approach forms the basis of the antipyrine test[7]; antipyrine was selected because it possesses certain pharmacological properties that make it more suitable than any other currently recognized agent for use as a model drug. Differences in antipyrine kinetics between the first measurements, taken in normal volunteers under near basal conditions, and the last measurements, made under conditions altered by introduction of only a single environmental perturbation, indicate how that particular environmental factor affects hepatic metabolism of antipyrine.[8] In fact, repeated measurements of antipyrine kinetics in the same subject have been suggested as a conservative method for estimating the combined influence of all the environmental factors that simultaneously impinge on normal subjects living unrestricted life styles.[9] This method is considered conservative because it reduces or eliminates contributions from additional sources of interindividual variations that play a role when antipyrine kinetics are studied and contrasted in different groups. For example, each subject serves as his or her own control,[7] thereby eliminating all genetic factors as a source of variation. By contrast, genetic differences among subjects can confound results from the older, and still commonly used but much less sensitive, technique of comparing kinetic values of control and experimental groups of subjects.

Establishment of near basal conditions for subjects of pharmacokinetic studies helps to achieve reliability and reproducibility of results, regardless of the drug under investigation. If subjects are not carefully selected and controlled with respect to maintaining uniform, near basal conditions, the changing character of the subject's environment could by itself produce different results each time measurements are made, not only in the same subject but also in different subjects. Under such circumstances, no single kinetic value is truly representative of the subject or subjects, but only of some poorly defined, transient state that is difficult, if not impossible, to duplicate. In contrast, by confining subjects to near basal conditions, the antipyrine test yields results that may apply only under these restricted conditions. A subtle distinction exists between the terms *near basal* conditions and *stable environmental* conditions. Stable environmental conditions do not exclude the possibility that the subjects are exten-

14. Intraindividual and Interindividual Variations

sively induced or inhibited with respect to their hepatic detoxication enzyme activities, only that such inducing or inhibiting factors are constant and do not fluctuate in the environment. By contrast, in near basal conditions these activities are neither markedly induced nor inhibited.

Near basal does not imply that subjects are entirely unexposed to known environmental inducing or inhibiting agents; that would be virtually impossible given today's life-styles. Rather, this term suggests that subjects are minimally, rather than moderately or heavily, exposed to such agents or factors. In other words, subjects are not taking any drug regularly, including oral antifertility agents, are nonsmokers, and imbibe ethanol only rarely. In addition, their endocrinological, hepatic, renal, and cardiovascular functions are normal at the time of study. Moreover, they have no history of major organ system disease or drug allergy, thereby reducing the risk of an adverse drug reaction. The latter would be undesirable for many reasons, including the possibility that certain adverse reactions may themselves alter near basal rates of elimination of the drug under study.

Under the conditions described in the preceding paragraph, antipyrine presently appears to be a safe drug in normal volunteers. Assembling an adequate number of normal volunteers who fulfill these strict criteria is difficult. Although it may appear much simpler to admit for study any seemingly normal subject, in the long run application of strict selection criteria avoids the ambiguous and confusing results that have emerged from studies performed with inadequately screened subjects.

A principal advantage of the antipyrine test is its extreme sensitivity in registering environmental perturbations in normal volunteers selected according to strict criteria. In the antipyrine test, normal volunteers under near basal conditions drink antipyrine freshly dissolved in water; the usual dose of antipyrine (18 mg/kg) is administered before, during, and after imposition of a single environmental change.[6,10,11] Published a dozen years ago, Table I shows the results of an antipyrine test performed to quantitate the effects of a single environmental change, administration of the drug allopurinol, given for two different periods of time, on the antipyrine half-life of normal volunteers.[10] From such studies, dose–response curves can be constructed that indicate how increasing quantities of the single environmental agent or factor under study affect antipyrine kinetics.

Like all clinical tests, the antipyrine test has certain limitations. These have been enumerated previously.[7] Misconceptions still abound concerning potential applications of the antipyrine test. For example, it has been emphasized that antipyrine pharmacokinetics should not be expected to predict those of many other drugs[7] because antipyrine kinetics provide a heterogeneous measurement that reflects activities from at least three

TABLE I

Effect of Allopurinol (5 mg/kg by Mouth for 14 and 28 Days) on Plasma Antipyrine Half-Life[a]

Subject	Half-life (hr)			
	Before allopurinol	After two weeks	After four weeks	Two weeks after discontinuation
T. G.	6.0	19.0	35.0	13.5
D. H.	11.0	15.0	21.0	9.0
R. L.	12.0	27.5	33.0	10.0
A. M.	22.0	19.0	28.0	15.5
J. M.	9.0	17.3	29.0	9.5
R. S.	9.5	17.0	40.0	9.5
Mean ± SD	11.6 ± 5.5	19.1 ± 4.4 $p < 0.05$	31.0 ± 6.5 $p = 0.003$	11.2 ± 2.7 $p > 0.8$

[a] Reprinted, by permission of the *New England Journal of Medicine*, **283**, 1484–1488 (1970).

discrete detoxication enzymes.[12] Because environmental perturbations can differentially affect each of these enzymes, the need to maintain normal subjects under near basal conditions while the study is in progress cannot be overemphasized. Maintaining such a near basal state in ambulatory subjects is neither easy nor certain. Much cooperation from the volunteers is necessary.

A disadvantage of the antipyrine test, one that possibly could be turned to advantage in the identification of presently unsuspected environmental factors, is precisely the intrusion of such extraneous factors to alter the near basal state of the subjects under study. For example, Fig. 3 illustrates an occasionally large intraindividual variation in antipyrine kinetics that occurred seemingly spontaneously in a study of eight normal young male subjects who received antipyrine on six separate occasions over a 17-day period. This previously unpublished study was performed in 1977 under what we then considered to be near basal conditions. However, the subjects were not under continuous observation in a metabolic ward. Thus, significant changes such as diet, upper respiratory infections, sleep–wake cycle, time of eating, and environmental exposure to chemicals that can induce or inhibit the drug-metabolizing enzymes could have occurred in some subjects to produce their as yet unsatisfactorily explained occasional, but large, departures from mean values (Fig. 3). Consequently, results from an uncontrolled study could be difficult to reproduce or could be in conflict with the literature, or both. Occasionally, such anomalous results have been reported using the antipyrine test

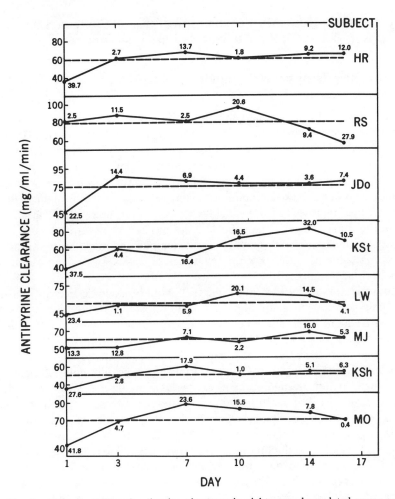

Fig. 3. Reproducibility of antipyrine clearance in eight normal unrelated young male subjects who received antipyrine in aqueous solution in a dose of 18 mg/kg on six separate occasions over a 17-day period. The subjects were not hospitalized. The dashed line indicates the mean value for each subject, and the number above each point is the percentage deviation of that particular value from the mean. The second dose of antipyrine, given on day 3, exhibited kinetic properties indicative of an inductive response from the antipyrine administered on day 1. Note the occasional large departure of a subject from the mean value shown.

in subjects who appeared to be, but may not have been, under stable, near basal conditions.[7,13–15]

Several investigators have expected the antipyrine test to accomplish a wide range of other tasks. Some of these expectations have not been met, occasionally because they were unrealistic.[7] For its failure to reach such unrealistic objectives, the antipyrine test has been criticized.

A. Attaining Near Basal Conditions in Subjects of Kinetic Studies

Only by admitting normal subjects, as defined previously, to a hospital metabolic ward where their environments, including diets, are rigidly controlled, can the investigator establish with certainty a near basal state. This state can be documented by obtaining antipyrine kinetics in each subject at regular intervals. Values should be highly reproducible within a subject. Under near basal conditions, intrasubject variability is less than 10% of the mean value for each subject. Larger intraindividual variations from the mean suggest that a basal state has not been attained and that environmental conditions need to be reexamined and made more uniform. In several studies, such extreme regulation of each subject's environment has been accomplished.[16-19]

Investigators using the antipyrine test should recognize that antipyrine can enhance its own metabolism if it is administered in too high doses or at too frequent intervals. Accordingly, antipyrine in a dose of 18 mg/kg should be given to a normal subject at intervals longer than every 4 days.[20] Now that more sensitive techniques are available for detecting antipyrine, the danger of inducing antipyrine metabolism can be eliminated if the dose is reduced below 18 mg/kg; doses as low as 1 mg/kg may now be used.[21]

The investigator can closely approximate near basal conditions in subjects even without hospitalization by establishing appropriate selection criteria. Carefully obtained histories together with full cooperation are necessary. Under such conditions, highly reproducible antipyrine kinetics can be demonstrated on repeat antipyrine administration to individuals who are not confined to metabolic wards (Fig. 4). Nevertheless, substantial intraindividual variations can and do occur sporadically in such nonhospitalized normal volunteers. What appeared to be near basal conditions proved not to be so because unsuspected environmental changes must have intruded episodically to cause the large shifts from baseline values illustrated in Fig. 3. The 10 subjects shown in Fig. 4 must have been maintained under more uniform conditions than those in Fig. 3; yet, even in them, a departure of more than 10% from a subject's mean value occurs on occasion. The subjects shown in Fig. 4 received antipyrine at 2-week intervals to avoid the inducing effect of antipyrine readministration. Such autoinduction of antipyrine metabolism is suggested by the enhanced antipyrine clearance in seven of the eight subjects (Fig. 3) on day 3 when the second dose of antipyrine was administered.

14. Intraindividual and Interindividual Variations

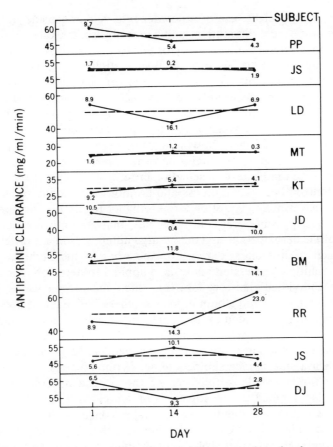

Fig. 4. Reproducibility of antipyrine clearance in ten normal unrelated young male subjects who received antipyrine in aqueous solution in a dose of 18 mg/kg on three separate occasions separated by 2 weeks. The symbols are the same as those in Fig. 3. No indication of autoinduction is apparent and reproducibility is high, suggesting that the subjects were maintained closer to near basal conditions than the subjects shown in Fig. 3. These 10 subjects were used to measure rate constants for formation of antipyrine metabolites.[22]

At the opposite extreme from this tedious, time-consuming approach, which requires repeat kinetic measurements on several different occasions in carefully selected and controlled subjects, is the method that employs everyday people.[23–26] Allowed to pursue a normal, almost unrestricted, life-style, the latter group of subjects is purposely recruited to encompass a broad spectrum of environmental differences that are assumed to be resolvable into components by a model based on multiple regression analysis. In some studies, antipyrine is even given as a single oral dose uncorrected for body weight; use of a powder in a gelatin capsule introduces problems of different dissolution, and hence absorption, rates in different subjects in whom the bioavailability of this form of antipyrine was shown to be incomplete.[26] Furthermore, in all such studies antipyrine is always administered on only one occasion. Kinetic values obtained once are assumed to represent accurately each subject's antipyrine-metabolizing capacity. However, this crucial assumption of high reproducibility of antipyrine kinetics is questionable in subjects under such nonbasal conditions; as suggested in the following discussion, reproducibility probably declines the more environmentally perturbed each subject becomes.

This approach imposes few selection criteria. Usually the subjects are not hospitalized because of illness and are not taking many other drugs, but some may smoke cigarettes, drink ethanol regularly, and even take certain drugs, possibly oral contraceptives, that are known to affect the activity of the detoxication enzymes. Large differences in age often occur, and subjects of different sex are used. According to this method, a model based on multiple regression analysis is applied to items obtained either from the individual's history or from such measurements as height, weight, hemoglobin, and albumin, among others.[23-26] Each of these independent variables is related to the dependent variable of antipyrine kinetics measured once in each subject. This model is relied on to quantitate the precise contribution made by each independent variable examined to the total variability among subjects in antipyrine kinetics.

Several fundamental assumptions of the particular model used are dubious when applied to investigations of sources of interindividual differences in antipyrine kinetics. Four major concerns arise. First, the number of subjects studied in most investigations has been small and does not represent a formal probability sample from a known population. Thus, the true representative character of the results is questionable, and the robustness of this model remains to be established through a demonstration that the results obtained with it can be replicated in another sample group drawn from the same larger population. The seriousness of this reservation arises from the multiple environmental factors that could exert some effect (Fig. 1) and the difficulty in obtaining a sufficiently large group of subjects to represent adequately most environmental factors. In fact, as indicated under the fourth concern, replicability of results has not been obtained with this model. By contrast, the antipyrine test reduces this difficulty by selecting only subjects who closely approximate a single condition: a near basal state similar for all subjects. Accordingly, all subjects might be expected to respond in a similar manner to an environmental perturbation because almost all sources of environmental inequity have been removed. However, even when all subjects are close to a near basal state, they do not all respond similarly to a single environmental change. Rather, considerable interindividual variation is observed, thereby raising a serious objection to a fundamental assumption of the model based on multiple regression analysis (see the third point).

Second, this method fails to account for the majority of the intersubject variability in antipyrine disposition that exists among even those few subjects in whom the method has been applied. For example, when smoking and oral contraceptive use were considered alone,[26] only 9% of the total variance in antipyrine kinetics among 207 normal volunteers was accounted for; none of the other factors examined by multiple regression analysis could provide a clue as to what was responsible for the other 91% of the variance.

The third and possibly most serious flaw in the particular multiple regression model used lies in the assumption that it accurately accounts for the large intersubject variability in antipyrine kinetics that arises in response not only to any single environmental factor acting alone but also to several such factors acting concomitantly. The multiple regression analysis model used to assess sources of phenotypic variation in antipyrine kinetics is sensitive only to the linear component of the relationship.[23-26] Distribution curves of antipyrine clearances themselves, as well as of environmental effects on antipyrine clearance, are not entirely linear. Nonlinear portions of this variability are insensitive to resolution by the model that was selected.[23-26] Other, more sensitive models based on multiple regression analysis might have been, and still can be, employed to resolve nonlinear systems or nonlinear portions of linear systems, providing these independent variables are properly represented, as, for example, by forming products or powers. However, with respect to phenotypic variations in antipyrine kinetics, our present knowledge of the correct distribution properties produced by each environmental factor is too incomplete to permit its appropriate representation as a linear or nonlinear function. The following considerations suggest that this fundamental property of linearity of the model based on multiple regression analysis renders it inadequate.

1. Individual responses to many environmental factors are unpredictable and highly variable. For example, induction of hepatic detoxication enzymes has been shown to be highly variable even though apparently normal subjects are exposed to similar doses of an inducing agent given by the same route. This variability in the magnitude of induction has been documented *in vivo* for phenobarbital[6] and *in vitro* for 3-methylcholanthrene.[27-30] In both systems, genetic factors that generate nonlinear distribution curves appear to be responsible for large interindi-

vidual differences in inducibility. These populations are generally composed of subjects under nonbasal conditions with highly diverse phenotypes for inducibility. Knowledge of such phenotypes would be necessary to predict satisfactorily how a given subject would respond to a particular inducing agent. The computer model selected for use cannot extract such phenotypes, much less the underlying genotypes, from the single kinetic value provided.

Hence, large interindividual variations exist in the extent to which hepatic detoxication enzyme activity of different subjects is enhanced by the same dose of certain inducing agents administered by the same route. Similarly, Table I shows large interindividual variations in the magnitude of enzyme inhibition produced by a drug administered in the same dose by the same route to normal subjects under near basal conditions. The computer model cannot decipher from the single antipyrine kinetic measurement sources for these large interindividual differences beyond the few specific environmental factors that the investigators have considered. Accordingly, it appears that such a model, by applying an average value for any given factor to all subjects, corrects inaccurately for variable responses to induction or inhibition. This model overcorrects certain subjects for the influence exerted by an environmental factor on antipyrine clearance while undercorrecting other subjects, thereby yielding contradictory conclusions (Table II).

2. When several concomitantly acting environmental factors influence antipyrine metabolism in nonbasal subjects, the extent of the variability produced probably exceeds that expected from the results of carefully controlled experiments in subjects under near basal conditions when only one factor is changed at a time. Although the antipyrine test can identify such departures from expectations due to interactions among factors, the model used based on multiple regression analysis, by contrast, assumes that the net effect of all factors can be accounted for simply by adding the effects of each factor taken alone. Thus, the particular model based on multiple regression analysis that was selected for use cannot detect synergism from interacting factors. It must be emphasized that the potential of multiple regression analysis to resolve sources of pharmacokinetic variation is much greater than has been realized by the particular "canned" model used previously. The technique is both sensitive and powerful. However, for multiple regression analysis to be used appropriately, a model must be developed that encompasses nonlinear as well as linear relationships. Error terms especially need to be appropriately modeled, rather than treated simply in an additive manner, as in previous applications of this method.[23-26]

Potential enhancement of variability among subjects from simultaneous

TABLE II

Effects of Age and Genetic Constitution on Antipyrine Kinetics as Determined by Different Methods and Investigators

Factor	Subjects (number and condition)	Effect on antipyrine kinetics	References
Age	19 elderly and 61 young subjects	Increased $t_{1/2}$ with increased age	31
	307 subjects, ages 18–92	Decreased clearance with increased age	32
	140 young subjects, ages 16–33; 16 elderly subjects, ages 34–80	Increased $t_{1/2}$ and decreased clearance with increased age	8
	51 subjects, ages 22–84	Increased $t_{1/2}$ and decreased clearance with increased age	33
	208 environmentally perturbed subjects, ages 15–72	No change with age	26
Genetic constitution	18 sets of MZ and DZ twins, all under similar environmental conditions	MZ more concordant in antipyrine $t_{1/2}$ than DZ twins, indicating genetic control of variation	5
	41 males and 16 females unrelated and in a near basal state	Trimodal distribution curve and high correlation with AHH inducibility suggested monogenic control over variations in plasma antipyrine $t_{1/2}$	34
	20 sets of MZ and DZ twins, all under similar environmental conditions	MZ more concordant than DZ twins in rate constant for formation of three main antipyrine metabolites, indicating genetic control of phenotypic variation	22
	12 two-generation families, all under similar environmental conditions	Pedigree analysis compatible with monogenic control of variations in rate constants for formation of three main metabolites of antipyrine	35
	208 volunteers from 78 families, ages 15–72, environmentally perturbed	After "correction," midparent–offspring, father–offspring, sib–sib, and parent–parent correlations all similar; no firm conclusions could be reached from these data concerning genetic factors	26

impingement of several environmental factors suggests that the particular model used previously based on multiple regression analysis might provide correlations appreciably lower than those that exist among relatives, were genetic mechanisms operating to produce intersubject variation. This expectation of a lower, rather than higher, correlation among relatives was indeed obtained with this model.[26] Accordingly, interpretations alternative to those favored by these authors should be considered because, as indicated, their results are consistent with a genetic explanation as well as with insensitivity of their model applied to environmentally heterogeneous subjects. Independent evidence based on pedigree analysis of carefully selected and environmentally controlled subjects supports these conclusions and suggests that in such subjects, twofold variability in antipyrine kinetics appears to be explained entirely by genetic factors. This evidence, described in the section on interindividual variations, is compatible with simple Mendelian transmission of alleles at a single locus that controls interindividual differences in the rate constant for formation of each major metabolite of antipyrine.

Fourth, in view of the preceding three major weaknesses in the model, it is not surprising that the results obtained with it differ markedly from much previous work on age, sex, and genetic constitution (Table II). For example, on the three occasions on which regression analysis was applied, three divergent conclusions were reached on the effect of sex on antipyrine metabolism. Despite the use of different populations in each study, ethnic differences alone probably do not explain the discrepancies. In one study, men metabolized antipyrine more rapidly than women,[26] whereas earlier work claimed, respectively, that men had longer plasma antipyrine half-lives and hence less rapid metabolism than women[23,24] and that no sex differences occurred.[25] In view of the discrepancies, it is difficult to understand the claim[26] that the results on sex differences confirmed those of earlier studies, including that of O'Malley et al.,[31] who reported that the plasma antipyrine half-life of young males was 30% greater than that of young females. Apparently, age can modulate the effect of sex on antipyrine metabolism, because O'Malley et al. observed a sex difference among young but not old subjects, a conclusion confirmed by others.[33] Age mediation of the sex effect probably does not explain the results of Blain et al.[26] because they also reached aberrant conclusions with respect to age, being unable to demonstrate the increased antipyrine half-life and decreased antipyrine clearance observed by all other groups that investigated the problem (Table II). Because they provided no information on antipyrine half-life or volume of distribution in their subjects, the source of their unusual results cannot be resolved kinetically. Collectively, these discrepancies suggest that the method and design used are insensitive and deficient.

An earlier example of application of the model based on multiple regression analysis concludes curiously that the following factors are statistically significantly correlated to antipyrine half-life: sex, cola nut consumption, hemoglobin in women, and height in men.[23] Despite the statistical significance, however, all correlations were low and of questionable biological significance. The highest r value was 0.40 for the relationship between cola nut consumption and antipyrine half-life.[23] As previously emphasized, such r values have low predictability because the level of predictability is reflected directly in r^2 rather than in r.[28]

The correlation between cola nut consumption and antipyrine kinetics[23] was reexamined under the conditions of a controlled prospective study performed to evaluate how chewing different amounts of cola nuts for different periods affected antipyrine kinetics. The results disclosed no effect of cola nut chewing on antipyrine disposition at any dose of cola nut chewing for any time period selected.[36] There is, therefore, the possibility that too much reliance may be placed on the statistical significance of P values derived from the model based on multiple regression analyses, particularly when the correlations themselves are low and the subjects are under highly perturbed environmental conditions.[4,37] Whereas the results of such studies on causes of variability in antipyrine kinetics may provide valuable clues, the conclusions and clues drawn from the model that uses multiple regression analyses should be validated by a carefully controlled, prospective antipyrine test before safe acceptance.[4,7]

B. Antipyrine as a Model Drug in Studies on Human Drug-Metabolizing Capacity

Depending on the objective of the investigation, it may or may not be appropriate to employ antipyrine as a "model drug." If the objective is to determine the disposition of other xenobiotics under diverse environmental or pathological conditions, it may be more useful to investigate directly the disposition of these other compounds. Thus, hazardous extrapolation could be avoided.[7] Alternatively, if direct measurement of the desired compound is not feasible, it might be advisable to include, besides antipyrine, several additional model compounds whose dispositions differ from that of antipyrine, but that, like antipyrine, are eliminated mainly by action of hepatic detoxication enzymes. Model drugs that have been used for this purpose include aminopyrine, amobarbital, phenylbutazone, phenytoin, theophylline, and warfarin. Investigators have used these compounds to determine whether an environmental factor, such as a dietary change, cigarette smoking, a disease state, and the like affects, in a manner similar to that produced with antipyrine, the disposition of drugs metabolized differently from antipyrine. For example, individuals who are slow hepatic 4-hydroxylators of debrisoquine do not appear to exhibit a corresponding retardation in any of the three main hepatic oxidation reactions involved in antipyrine metabolism.[38] The observation suggests that several types of cytochrome P-450 involved in hepatic 4-hydroxylation of debrisoquine differ in substrate specificity from the three other genetically distinct types of cytochrome P-450 that biotransform antipyrine.[39,40] However, individuals who metabolize debrisoquine slowly also metabolize sparteine, nortriptyline, phenacetin, phenformin, and phenytoin slowly, suggesting a common genetic factor in the biotransformation of the five drugs.[41–46]

Although debrisoquine, nortriptyline, phenformin, phenytoin, and sparteine can all produce undesirable pharmacological effects in normal volunteers, and hence are unsatisfactory as model drugs, antipyrine is without such side effects. Furthermore, antipyrine can be measured noninvasively in saliva. In addition to the safety advantages of antipyrine as a test drug, it is rapidly and completely absorbed from the gastrointestinal tract after administration in aqueous solution, distributed evenly in total body water, negligibly bound to tissue or plasma proteins, and almost completely metabolized by the liver detoxication enzymes. The hepatic extraction ratio of antipyrine is low. Thus, changes in liver blood flow do not markedly affect antipyrine disposition. Moreover, antipyrine exhibits negligible renal elimination and is easily, rapidly, and accurately measured, along with its urinary metabolites, by high-performance liquid chromatography.[7,12,22] For these reasons, changes in antipyrine concentration in plasma or saliva generally reflect changes in hepatic antipyrine metabolism.[8] No other test compound possesses so many desirable properties. Even other pyrazolon derivatives suffer from drawbacks: phenylbutazone is highly bound to albumin, and although aminopyrine is bound to only about 30%, it exhibits a considerable first-pass effect.[47]

II. SOURCES OF INTRAINDIVIDUAL VARIABILITY

The preceding introduction placed into broad perspective the topic of variability among subjects in the activity of the hepatic detoxication enzymes. Studies using antipyrine, the most commonly employed test substrate in humans, served as the main example from which underlying general principles were developed. The merits and limitations of two dif-

14. Intraindividual and Interindividual Variations

ferent methods employed to identify sources of interindividual variations in pharmacokinetic measurements were compared, mainly because the appropriateness of the model for this purpose based on multiple regression analysis has not been discussed by those using it.[23-26] Accordingly, a critical assessment seemed long overdue. Despite the lack of discussion on the suitability of this model, it has been used repeatedly over the past six years to identify sources of interindividual variations in antipyrine kinetics. By contrast, in an application of this approach to assessing factors that cause large interindividual variations in the clearance of theophylline,[48] the authors clearly recognized its limitations: "The factors identified as important in theophylline body clearances are associations found by retrospective statistical analysis which need not imply a cause-and-effect relationship, especially where a pathophysiologic or drug interaction rationale does not exist. Often these factors need further confirmation by prospective examination of cohorts of subjects with the disease or history in question" (p. 1364).

Several studies that measured intraindividual variations in the metabolism of the model compound, antipyrine, can now be discussed. In addition to the other reasons offered for use of antipyrine as a model, it is the most intensively studied drug from the point of view of intraindividual variability. Here, studies on intraindividual variations with antipyrine will be considered and conclusions compared to results with other xenobiotics wherever the investigation has been designed in a similar manner.

Drawn from studies performed more than 15 years ago,[5,49] Fig. 5 illustrates that the half-lives of antipyrine and dicoumarol are closely similar within sets of monozygotic (MZ) twins but vary markedly in most sets of dizygotic (DZ) twins. Similar results were obtained for phenylbutazone.[50] DZ twins share on average 50% of their genes; only very few sets of DZ twins have antipyrine and dicoumarol half-lives as similar as those of all MZ twins (Fig. 5). Although obviously different from kinetic studies performed several times in a single subject in estimating intraindividual variability, these studies in MZ twins present certain analogies. If the environments of the genetically identical MZ twins even approach the similarity of their genomes, then results obtained with them could be considered comparable to those obtained in a single subject studies on two separate occasions under near basal conditions. Figure 5 shows that such theoretical expectations were realized: no member of an MZ twinship deviated far in antipyrine or dicoumarol half-life from the value in his sib. These results in MZ twins conform to the highly reproducible values shown in Fig. 4 for antipyrine kinetics measured on three separate occasions in 10 normal unrelated volunteers maintained under near basal conditions.[22]

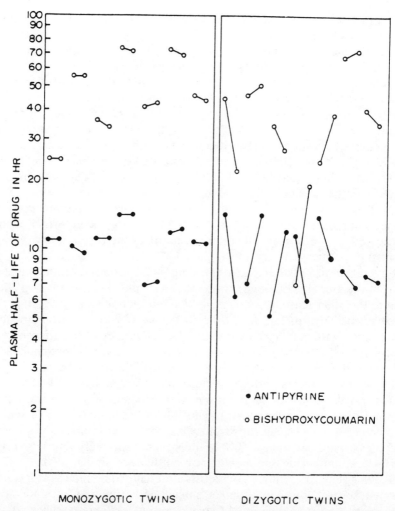

Fig. 5. Plasma half-lives of dicoumarol (bishydroxycoumarin) and antipyrine were measured separately at an interval of more than 6 months in healthy MZ and DZ twins. A solid line joins the values for each set of twins for each drug. Note that intratwin differences in the plasma half-life of both dicoumarol and antipyrine are smaller in MZ than in DZ twins. Data are from Vesell and Page.[5,49]

Figure 6 shows the results of another study on intraindividual variations in antipyrine kinetics.[9] In that study the subjects lived unrestricted lives, and Fig. 6 reveals an occasional large departure from mean values. Such deviations occurred even more frequently for the subjects depicted in Fig. 3. Although we hoped, when we performed the study, that the subjects in

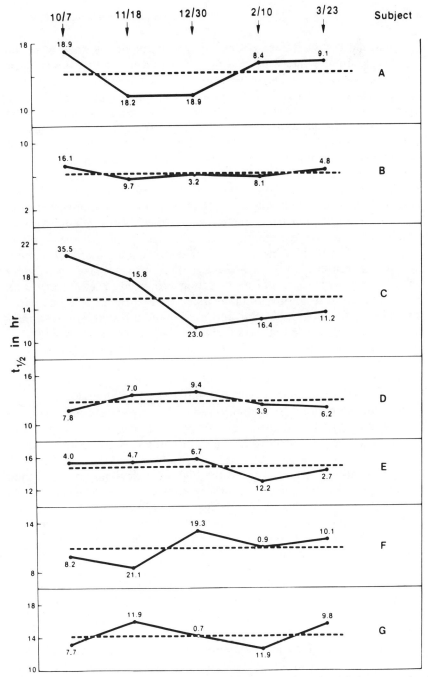

Fig. 6. Reproducibility of antipyrine clearance in seven subjects living unrestricted life-styles after five separate doses of antipyrine (18 mg/kg in aqueous solution). Reproduced by permission from Alvares et al.[51]

Fig. 3 would be maintained under near basal conditions, they were not confined to a metabolic ward and the results disclosed that the subjects were exposed episodically to environmental perturbations during the investigation. By contrast, the subjects shown in Fig. 4 appear to have been under more nearly basal conditions because they exhibited closely similar values on readministration of antipyrine. Even in Fig. 6, reproducibility of values for certain subjects (B, D, and E) was excellent. However, a few large deviations occurred, such as the first and second antipyrine half-lives of Subject C, which differed appreciably from the last three antipyrine half-lives. These large deviations can be accounted for by several special characteristics of Subject C. An oriental with a family history of intolerance to salicylates, he developed after the first two administrations of antipyrine an adverse reaction accompanied by mild diarrhea. However, on the last three occasions no adverse reactions occurred; accordingly, antipyrine half-lives on these last three measurements were very similar. Thus, environmental perturbation produced in a subject by an adverse reaction to antipyrine appears to be associated with increased variability. After Subject C, Subject A exhibited the next largest intraindividual variability. Like Subject C, Subject A had the most variability on the first few administrations of antipyrine, the last two half-lives being virtually identical. Although no historical details are available, one is tempted to speculate that Subject A, like Subject C, experienced an environmental perturbation after only the initial doses of antipyrine.

Phenylbutazone was also administered on five separate occasions to the same seven subjects.[9] If, during the antipyrine portion of the study, Subjects A and C are excluded from consideration, the extent of intraindividual variability for the phenylbutazone half-life is similar to that for the antipyrine half-life. By contrast to the high reproducibility of antipyrine and phenylbutazone half-lives in these subjects, phenacetin exhibited much greater intraindividual variability, possibly because it undergoes first-pass elimination through the liver. Antipyrine and phenylbutazone have low hepatic extraction, their dispositions being almost entirely unaffected by change in hepatic blood flow.

Because antipyrine is metabolized by three discrete types of hepatic cytochrome P-450,[39,40] intraindividual variations need to be assessed from the point of view of the reproducibility of rate constants for the production of each metabolite when antipyrine is administered on separate occasions to the same subject under near basal conditions. Table III shows highly reproducible values for each metabolite in 10 normal subjects, each studied in our laboratory on three separate occasions.

Although normal subjects maintained under near basal conditions exhibit highly reproducible antipyrine decay and rates of production of individual

14. Intraindividual and Interindividual Variations

TABLE III

High Reproducibility of Rate Constants for Metabolite Formation in 10 Unrelated Normal Male Subjects Who Received Antipyrene on Three Separate Occasions (I, II, and III)[a]

Subject	40HA[b] × 10^{-4}			30HMA[b] × 10^{-4}			NDA[b] × 10^{-4}		
	I	II	III	I	II	III	I	II	III
P. P.	2.2	3.1	2.6	1.4	1.6	1.4	1.2	1.2	1.0
J. S.	2.9	2.9	2.2	1.9	2.1	1.1	0.6	0.6	0.6
L. D.	1.7	1.9	1.9	0.8	1.9	1.2	0.9	1.1	1.4
M. T.	0.8	0.9	0.9	0.4	0.6	0.7	0.6	0.5	0.6
K. T.	1.5	1.3	1.3	0.8	0.8	1.3	0.8	0.6	0.8
J. D.	1.7	1.8	1.9	0.8	0.9	0.9	1.0	0.9	1.0
B. M.	1.8	1.5	1.8	0.9	0.7	0.8	0.7	0.6	0.6
R. R.	1.7	1.5	1.6	1.5	1.4	1.6	1.4	1.0	1.0
K. G.	3.5	3.8	3.5	2.1	2.4	2.4	1.1	1.5	0.8
J. S.	3.6	2.3	2.5	1.4	1.0	1.2	1.2	0.7	0.6

[a] Analysis of variance reveals significant interindividual variation ($p < 0.01$) but much lower intraindividual variation.

[b] The following abbreviations are used: 40HA, 4-hydroxyantipyrine; 30HMA, 3-hydroxymethylantipyrine; NDA, N-demethylantipyrine.

antipyrine metabolites, it must be reemphasized that highly reproducible antipyrine kinetics may not occur in subjects under perturbed environmental conditions. For example, Subject C in Fig. 6 showed lower reproducibility than all other subjects in that study, probably because of the mild adverse reaction encountered after initial doses of antipyrine. Unlike the data in Tables I and III and Figs. 4 and 6, there is no information on intraindividual variability for subjects in states of extensive environmental perturbation such as those used in typical regression analysis studies.[23-26] Such studies, characterized by reliance on only a single antipyrine measurement, assume that the single measurement is reproducible in these same subjects. Currently available evidence suggests an opposite conclusion: reproducibility of antipyrine kinetics may be inversely related to the number of environmental factors that impinge on an individual under study.

The influence of age on intraindividual and interindividual variations in the activity of detoxication enzymes should be recognized. For obvious reasons, no single study in humans has been able to measure such age-related changes in the same subjects throughout their lives. Nevertheless, the composite data clearly indicate that complex, marked changes in the activities of these enzymes occur from the fetal, to the neonatal, to the adolescent, to the adult, and finally to the geriatric period. Despite notable exceptions, a general pattern of developmental change for these enzymes can be offered. With respect to antipyrine and certain other substrates, enzyme activity appears low in the fetal and neonatal periods; the low activities, together with immaturity of renal function, explain why the fetus and neonate are particularly sensitive to toxicity from drugs that tend to accumulate more readily at these stages. The highest hepatic detoxication enzyme activity ever attained during life probably occurs shortly thereafter, at ages 1–3, when antipyrine and theophylline elimination rates are approximately twice as rapid as in normal adults.[51]

The developmental pattern from age 3 to adulthood needs to be defined to establish when the decline in the activity of these hepatic detoxication enzymes starts. Several studies noted a slight decline in hepatic antipyrine-metabolizing capacity from adulthood to old age.[31-33] As yet, we have no explanation for these complex age-related changes in the activity of hepatic drug-metabolizing enzymes. Highest activity at ages 1–3, with progressive declines thereafter, argues against attributing the changes to the extent of exposure to environmental inducing agents. If that theory were correct, a pattern opposite to that observed would be expected and the highest activities should appear in the geriatric period, when they seem to be least apparent. Interpretation is complicated by the knowledge that the response to induction declines with age, the precise temporal pattern of inducibility being as yet undetermined in humans.

From the point of view of a discussion of intraindividual and interindividual variability, these age-related changes require emphasis because they place into perspective the highly reproducible kinetic values obtained in a single subject for various test drugs. Such high reproducibility occurs only for subjects within a specific age range.

The relationship between intraindividual and interindividual variation also warrants discussion because some approaches have either ignored the former or assumed that it was negligible compared to the latter.[23-26] As indicated, the magnitude of intraindividual variation is not always small. The more subjects are environmentally perturbed, the larger is the magnitude of intraindividual variability relative to interindividual variability, although the former can never exceed the latter. Therefore, the magnitude of intraindividual variability should be measured before attempts are made to identify other causes of interindividual variability. A similar conclusion was reached earlier[52] in a paper that stressed large intraindividual variations in subjects in pharmacokinetic studies with clindamycin, ephedrine, ethosuximide, lincomycin, and warfarin. Interpretation of the values for intraindividual variation is rendered difficult because details concerning the environmental conditions of the subjects were not provided.

III. SOURCES OF INTERINDIVIDUAL VARIABILITY

In the late 1940s, methods were developed for differential extraction of lipid-soluble drugs and their more polar metabolites into organic solvents, thereby permitting accurate measurement of rates of disappearance of many parent drugs in human biological fluids. Such studies disclosed large differences among apparently normal human subjects in the kinetics of commonly used drugs.[53-60] Intraindividual variations appeared small compared to interindividual variations.

Table IV shows ranges of plasma half-lives of a number of drugs among normal subjects. The magnitude of interindividual variation depends on the drug selected for study, the number of subjects studied, the particular population from which the subjects are drawn, and the condition of the subjects, including their present and past health, genetic constitution, age, sex, diet, and exposure to environmental chemicals and drugs that induce or inhibit hepatic detoxication enzymes (Figs. 1 and 2).

Sources of interindividual variations have been investigated by different methods. Not unexpectedly, the results obtained are influenced by the specific method used and by the degree to which the fundamental assumptions underlying each method are fulfilled. Accordingly, this chapter

TABLE IV

Interindividual Variations in Plasma Half-Lives of Drugs Metabolized by Hepatic Detoxifying Enzymes

Drug	Plasma half-life (hr)[a]	Fold variation	Number of individuals investigated
Aminopyrine	1.1–4.5	4	12
Amobarbital	1.4–6.4	5	14 pairs of twins
Antipyrine	5–35	7	33
Carbamazepine	18–55	3	6
Dicoumarol	7–74	11	14 pairs of twins
Diazepam	9–53	6	22
Phenytoin	10–42	4	—
Indomethacin	4–12	3	15
Nortriptyline	15–90	6	25
Phenylbutazone	1.2–7.3	6	14 pairs of twins
Primidone	3.3–12.5[a]	4	—
Theophylline	4–18[a]	5	45
Tolbutamide	3–27[a]	9	50
Warfarin	15–70[a]	5	40

[a] Last four plasma half-lives given in days.

stresses the characteristics of each method and the assumptions upon which it is based.

Because the therapeutic consequences of the acknowledged large interindividual variations in rates of drug elimination are enormous, the sources of such interindividual variations are of critical interest. For example, individuals are known to vary markedly in the dose of a drug needed to achieve the desired therapeutic effect, thereby avoiding the extremes of undertreatment and toxicity. If each patient showed highly reproducible kinetic values from one test to another, each patient could be typed for his or her rate of drug elimination and all future drug doses in that individual could be guided by this value. By contrast, if each subject's kinetic behavior fluctuated extensively with time, such a typing for drug clearance would be useless. The true picture probably lies between the two alternatives and is most likely more complex. Whereas some subjects appear extremely stable and others extremely unstable, still others may be in the process of changing from stability to instability or vice versa.

Reproducibility depends on both the drug and the subject. With respect to the drug, those with low hepatic extraction (antipyrine and phenylbutazone) appear to yield more reproducible values in a given normal subject than those with high hepatic extraction (phenacetin).[9] With respect to the individual, those kept under near basal conditions, best approximated by confinement on a metabolic ward, most often show highly reproducible kinetics. By contrast, in normal subjects with unrestricted life-styles, sporadic fluctuations can occur frequently with antipyrine or phenylbutazone. Hence, a critical concept in attaining reproducible kinetic values requires close maintenance of near basal conditions. Accordingly, patients suffering from cardiovascular, hepatic, renal, or endocrinological disorders and receiving therapy would be expected to exhibit activities varying markedly from those measured in good health under near basal conditions. As specific diseases run their course, with consequent functional alterations in organ systems, the appropriate dose of a drug may change for that patient, that is, during the course of a disease process, enzyme activity can fluctuate. Because normal subjects exhibit large interindividual variations in clearance of most drugs that are metabolized, the kinetics of such xenobiotics in patients should be considered not just in comparison to those obtained in normal subjects with whom they may overlap, but rather with the values obtained in that same patient either before the onset of the disease or when clinical improvement is achieved.[61] Knowledge of these enzyme activities can help in evaluating patient management. Disease exemplifies one specific source of intraindividual pharmacokinetic variation with significant therapeutic implications.

A. Evidence for Genetic Control of Large Interindividual Variations in Xenobiotic Metabolism

The large interindividual variation in rates of drug elimination for normal subjects under near basal conditions, ranges from 3- to 11-fold (Table IV). One approach in comparing the relative contributions of genetic and environmental factors to such phenotypic variation is to examine the trait in twins.[62]

The twin method is suited for application to certain interindividual pharmacokinetic variations because so many diverse factors appear to be involved in their genesis (Figs. 1 and 2), and twin studies are traditionally useful in elucidating the relative contributions of nature and nurture to such polygenically controlled traits. Twin studies on the kinetics of almost a dozen drugs were performed in several different laboratories located in four countries to compare the relative contributions of genetic and environmental factors to the large interindividual variations. Results from all studies were remarkably similar: variation virtually disappeared within MZ twins but was preserved within most DZ twins. It was concluded that genetic factors primarily controlled large interindividual variations in the metabolism of amobarbital,[63] antipyrine,[5,22] dicoumarol,[49] ethanol,[11] halothane,[64] nortriptyline,[65] phenylbutazone,[50] phenytoin,[66] sodium salicylate,[67] and tolbutamide.[68] In most of these studies only the elimination rate of the parent drug was measured, whereas we now recognize that for drugs metabolized by multiple reactions, rate constants for the formation of each metabolite must be ascertained because such rate constants represent the product closest to the gene under study.

Most studies with twins employed normal adults who did not smoke, imbibe ethanol regularly, or take drugs, including antifertility agents, that is, individuals under near basal conditions. Furthermore, most twins lived in different households. Thus, the results, which showed greater similarity in kinetic values between MZ than between DZ twins, appeared to be independent of enhanced environmental similarities.

A fundamental assumption in the use of the twin method in estimating the relative contributions of genetic and environmental factors to phenotypic variation is that the environments of all twins are uniform with respect to critical factors. Only one independent variable, in this case genetic constitution, can be altered in a valid experiment that attempts to contrast phenotypic variation within MZ and DZ twins; the other independent variable, environment, must remain constant.

Although MZ twins living in separate households tend to create a more similar environment for themselves than do DZ twins living apart, the remarkable reproducibility of these diverse twin studies on drug disposition performed in different countries cannot be attributed to any currently identified environmental factor or to any combination of such factors more concordant among the MZ than among the DZ twins. The reasons for this conclusion are as follows.

First, a careful history taken from each twin disclosed no such environmental factor or factors that could be implicated in any of the twin studies. Second, if environmental inequality did arise, despite the care of the investigator, this imbalance could on occasion affect an MZ twin. Because of the close agreement in kinetic values within almost every MZ twinship examined, such a rare divergence among MZ twins would attract attention, and the causes would be pursued. For example, in a study of 78 twins in whom steady-state plasma concentrations of nortriptyline were measured, each of the 19 sets of MZ twins not treated concomitantly with other drugs exhibited concordant steady-state plasma nortriptyline concentrations; by contrast, DZ twins exhibited considerable intratwin discordance.[65] However, intrapair similarity of nortriptyline kinetics observed in otherwise unmedicated MZ twins did not occur if one member of an MZ twinship was simultaneously receiving other drugs.[65]

Third, Table V shows that when intratwin correlation coefficients are compared for twins, regardless of zygosity, living in the same household (0.42) and twins living in different households (0.42), the values were identical. This result suggests that environmental inequities among normal twins living together as opposed to those living apart, all under near basal conditions, play a negligible role in maintaining the large interindividual variations observed among all subjects.[22]

Fourth, when each environmental factor shown in Figs. 1 and 2 is considered from the point of view of those that could apply to normal twins of similar age and condition, the total combined contribution made by these factors to the marked interindividual variations exhibited by these healthy twins was relatively small. Hence, in normal subjects living under near basal conditions, the magnitude of the 3- to 11-fold interindividual variations exhibited for the metabolism of many drugs (Table IV) cannot be accounted for by any single currently identified environmental factor.

On the basis of the twin studies, rough estimates of the relative contributions of genetic and environmental factors to interindividual variations in drug metabolism are available. Critical assessment of these methods and a comparison of results derived from them are available.[3,69]

14. Intraindividual and Interindividual Variations

TABLE V

Intratwin Correlations for Antipyrine Half-Lives of Twins Living Together Compared to Twins Living Apart[a]

	Living together				Living apart		
		Half-life (hr)				Half-life (hr)	
Name	Zygosity	Sib 1	Sib 2	Name	Zygosity	Sib 1	Sib 2
H	MZ	13.3	13.8	Co	MZ	9.6	9.0
Ba	MZ	11.9	11.2	M	MZ	13.5	13.8
C	MZ	11.5	10.8	E	MZ	14.7	12.4
L	MZ	11.3	11.2	L	MZ	11.0	12.3
Bu	MZ	9.2	11.2	H	DZ	10.5	11.1
T	MZ	13.5	11.2	M	DZ	15.7	13.5
R	DZ	9.7	17.1	T_1	DZ	14.7	12.2
B	DZ	9.1	8.3	T_2	DZ	13.3	14.3
F	DZ	11.0	11.3	E	DZ	13.3	10.6
S	DZ	14.4	16.7	Ca	DZ	9.0	14.2

[a] $r_i = 0.42$; $n = 10$; $r_i = \Sigma(X_i - a)(Y_i - a)/ns^2$

Suffice it to say that some of these methods probably overestimate the genetic contribution. The different methods shown in Table VI also yield different results because the terms in each expression differ. The meaning of each term and the different equations containing them are described

TABLE VI

Heritability of Variations in Drug Metabolism of Twins Utilizing Different Methods of Data Analysis[a]

Data analysis	Antipyrine	Phenyl-butazone	Bishydroxy-coumarin	Ethanol	Halothane
$\dfrac{V_D - V_M}{V_D}$	0.98	0.99	0.97	0.98	0.88
r_M	0.85	0.83	0.85	0.82	0.52
r_D	0.47	0.33	0.66	0.38	0.36
$(r_M - r_D)/(1 - r_D)$	0.72	0.75	0.56	0.71	0.25
$2(r_M - r_D)$	0.76	1.00	0.38	0.88	0.32

[a] r is the intraclass correlation coefficient. This table is reproduced from Vesell.[3] Subscripts M and D refer to monozygotic and dizygotic twins, respectively.

elsewhere.[3,69] It is clear, however, that each expression places more or less weight on different aspects of variance.

If twin studies provide an initial clue that genetic factors play a major role in maintaining large phenotypic variations, pedigree analysis is the next step in defining the mode of genetic transmission. Family studies were performed to identify the Mendelian mode of transmission of interindividual variations in the metabolism of dicoumarol,[70] nortriptyline,[71] and phenylbutazone.[72] The results of pedigree analysis suggested polygenic modes of transmission for large interindividual variations in the kinetics of the three drugs. A family study based on rate constants for formation of the two main metabolites of amobarbital disclosed autosomal recessive inheritance of the deficiency of N-hydroxylation of amorbarbital,[73] clearly revealing the need to phenotype subjects as closely as possible to the gene product. This defect might have been missed if the kinetics had been performed only on amobaritol, rather than on its principal metabolites. Thus, the rates of production of each major product should be measured, rather than simply the rate of decay of the parent drug.

This approach permitted identification of a genetically controlled polymorphism in the hepatic detoxication of debrisoquine: approximately 7–9% of British subjects were deficient in the 4-hydroxylation of this drug.[74] The same genetic polymorphism is apparently responsible for the interindividual variability observed in the metabolism of sparteine, nortriptyline, phenacetin, phenformin, phenytoin, propranolol, and some other β-adrenergic blocking drugs. Nevertheless, several discrete genetic factors may play a role in controlling variations of the same drug in different subjects. Whereas deficient 4-hydroxylation of debrisoquine is transmitted as an autosomal recessive trait, apparently associated with slow hydroxylation of phenytoin, phenytoin metabolism is also affected by another mutation transmitted in a different way. Pedigrees from two laboratories suggest that toxicity to phenytoin given in the usual doses develops in individuals who inherit a defective capacity for detoxication of phenytoin in an autosomal dominant manner.[75,76] Furthermore, whereas debrisoquine phenotypes appear to be relatively resistant to environmental perturbation, this is not the case for rates of metabolism of phenytoin or propranolol. Thus, the debrisoquine phenotype of a patient may not be a reliable index of how that patient can eliminate phenytoin or propranolol under diverse environmental conditions.

Twin and family data suggested that a ninefold interindividual variation in rates of tolbutamide elimination was controlled predominantly by genetic factors.[68] A trimodal distribution of tolbutamide kinetic values in these subjects suggested regulation of interindividual variations by two alleles at a single genetic locus.

Another monogenically transmitted defect in the detoxication enzymes involves O-deethylation of phenacetin.[77] In a 17-year-old girl severe methemoglobinemia and hemolysis occurred after she received a small dose of phenacetin, presumably because of a defect in deethylation. Phenacetin toxicity produced severe illness. The severity of this illness probably distinguishes it from defective debrisoquine hydroxylation, which also involves reduced deethylation of phenacetin. The patient's 38-year-old sister resembled her in exhibiting an abnormal response to phenacetin, but two other sibs and both parents revealed the normal response to phenacetin, which consists in the appearance of more than 70% of the dose in urine as acetaminophen. Defective deethylation of phenacetin to produce acetaminophen in this family appeared to be transmitted as an autosomal recessive trait.[77]

The point has been made that large interindividual differences in antipyrine metabolism among carefully selected normal subjects maintained under near basal conditions appeared to be controlled predominantly by genetic factors. This conclusion was based on results of three independent studies, two in twins[5,22] and one in 57 unrelated subjects whose distribution curve of kinetic values was trimodal, thereby suggesting regulation by alleles at a single locus.[34] Two of these studies had measured only plasma antipyrine decay, which is a relatively distant and combined reflection of three separate gene products rather than a direct index of the three different genes involved in antipyrine metabolism.[39,40] Therefore, we assessed sources of interindividual variation in rate constants for formation of each of the main metabolites of antipyrine: N-demethylantipyrine, 4-hydroxyantipyrine, and 3-hydroxymethylantipyrine. These rate constants provide more direct information about each separate primary gene product. The three gene products investigated consist of the three discrete forms of cytochrome P-450 involved in antipyrine metabolism.

A twin study disclosed that genetic factors were mainly reponsible for twofold interindividual differences in rate constants for formation of each of these principal metabolites.[22] A subsequent study was intended to assess the Mendelian mode of transmission of these genetic factors in 12 two-generation families; the results were compatible with monogenic control of interindividual variations in rate constants for formation of each metabolite.[35] Family studies of this sort (Fig. 7) demonstrate the need to select only subjects under near basal conditions. The mother (Fig. 7a) seemed to be homozygous recessive for the trait, that is, the rate constant for production of each antipyrine metabolite. Her genotype was assigned according to her position on the trimodal distribution curve generated from all subjects in the study for each antipyrine metabolite. Because the father also was homozygous recessive by virtue of a value that placed him

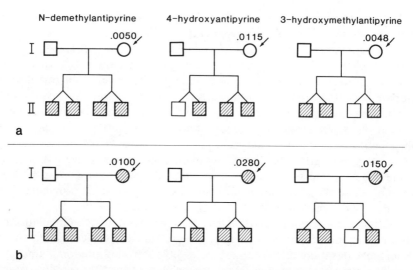

Fig. 7. Pedigree of a two-generation family consistent with monogenic transmission of interindividual variations in rate constants for formation of each main antipyrine metabolite (b) but not in (a), where the mother is in a perturbed environmental state because of drug ingestion. After cessation of this drug for 3 weeks, another dose of antipyrine was administered and the rate constants shown in (b) were obtained.

in the same portion of the distribution curve as the mother, all children of this union should be homozygous recessive for the trait. However, Fig. 7a shows that in only two of the 12 opportunities offered was a value observed among the children in the homozygous recessive range.

If valid and reproducible, this pedigree alone would disprove a monogenic hypothesis. But on closer questioning, the mother admitted taking cimetidine, a drug known to inhibit hepatic detoxication enzymes. Three weeks after discontinuation of cimetidine, she received antipyrine again and all rate constants for production of each antipyrine metabolite were redetermined (Fig. 7b). The more basal values suggest that antipyrine metabolite formation had been inhibited previously, because each rate constant on the second trial was at least doubled. Of particular significance is the fact that the pedigree in Fig. 7b, unlike that in Fig. 7a, is compatible with monogenic control of, and a trimodal distribution for, interindividual variations in the rate constant for formation of each antipyrine metabolite.

Figure 7 illustrates how maintenance of near basal conditions is essential for an adequate test of a genetic hypothesis and how investigators may be misled into believing that near basal conditions exist. This instance is not the first of this kind we have encountered, for example, in subject C

14. Intraindividual and Interindividual Variations

(Fig. 6), as well as in a nortriptyline study in which MZ twins were taking inducing or inhibiting drugs.

In humans, a group of drugs is detoxified polymorphically by a hepatic N-acetyltransferase (N-acetylase) that differs in several ways from the cytochrome P-450 oxidases described previously. Whereas the oxidases are located in smooth endoplasmic reticulum and require cytochrome P-450, molecular oxygen, and NADPH, the hepatic N-acetylases that show genetically determined differences in their activity in humans are located in the cytosol and transfer an acetyl group from acetyl coenzyme A to such drugs as dapsone, hydralazine, isoniazid, phenelzine, procainamide, sulfamaprine, and sulfamethazine. In contrast to the mixed-function oxidases, whose activities are sensitive to perturbation by many factors (Fig. 1), N-acetylases are relatively resistant to such changes and to alteration with age or after exposure to many drugs that induce the mixed-function oxidases. Ethanol, probably through its metabolic product, acetate, does accelerate N-acetylase activity.

Genetically controlled differences exist in N-acetylase activity. These differences have certain toxicological consequences. Slow acetylation of the drugs enumerated previously is a trait associated with a double dose of a recessive gene that causes reduced amounts of the hepatic cytoplasmic N-acetylase. On chronic administration of isoniazid, slow acetylators tend to develop higher isoniazid concentrations for longer periods of time than do rapid inactivators. Slow acetylators are also more sensitive than rapid inactivators to a principal form of toxicity from isoniazid: polyneuritis due to a deficiency of pyridoxal phosphate. Isoniazid interferes with utilization of this coenzyme in reacting with its carbonyl group *in situ*. Several drugs, including p-aminosalicylic acid and sulfanilamide, are acetylated monomorphically, probably by an N-acetylase different from the one that N-acetylates isoniazid, hydralazine, and procainamide.

From the point of view of this chapter, polymorphic acetylation of certain drugs illustrates not only the application of different methods to the elucidation of mechanisms underlying interindividual variations in rates of drug metabolism but also the consistency of results derived from diverse methods. Genetically determined variations in rates of isoniazid metabolism were first suggested by a twin study; after a single oral dose, the amount of isoniazid found in 24-hr urine collections was more similar in MZ than in DZ twins.[78] Among unrelated subjects, a distribution curve revealed bimodality in the percentage of a single oral dose of isoniazid excreted unchanged in urine.[79,80] Then, family studies disclosed the mode of inheritance of the rapid and slow acetylation phenotypes.[81] A dosage effect for this trait was established by the discovery that heterozygotes exhibited higher mean concentrations of isoniazid in plasma than did rapid acetylators who were homozygous for the dominant gene.[82] Finally, population studies showed that different races exhibited markedly different gene frequencies for this polymorphism, only 10% of Japanese being slow acetylators compared to approximately 50% of American whites and blacks.[83] Each study provided results consistent with a monogenic hypothesis.

Although consistency of conclusions derived from application of different methods might be anticipated for monogenically transmitted conditions in which the phenotype is relatively free from perturbation by environmental changes, as with the N-acetylase genetic polymorphism, similar consistency is found with most other drugs discussed in this section. Consistent conclusions were obtained with different genetic methods applied to the following three drugs, metabolized primarily by hepatic mixed-function oxidases: (1) dicoumarol, (2) phenylbutazone, and (3) antipyrine.

Before the influence of multiple environmental factors on drug disposition was recognized, a family study was performed to identify sources of interindividual variations in plasma dicoumarol half-life.[70] Subjects were probably not under near basal conditions, but rather were perturbed with respect to the factors shown in Fig. 1. This perturbation may explain why the results of this family study did not fit a single gene mechanism. To assess the potential role of multiple genetic factors in controlling interindividual variations in dicoumarol metabolism, correlation coefficients (r) between relatives were computed. The sib–sib correlation for dicoumarol half-life was 0.347 ± 0.091, but no parent–child or midparent–midoffspring correlations could be demonstrated.[70] Although Motulsky recognized that "the finding of sib-sib correlations could mean that a common environmental factor affects drug breakdown in sibs" (p. 62), he offered two reasons against this interpretation: (1) such an environmental factor operative in sibs actually did not exert, but would have been expected to exert, a role on parents living in the same household to produce a high parent–parent correlation; (2) reproducible dicoumarol half-lives in a given subject suggested the operation of genetic regulation, rather than episodic environmental perturbation. Therefore, Motulsky concluded that sib–sib correlation in the absence of parent–offspring correlation indicated the operation of recessive genes in dicoumarol disposition. A twin study also concluded that genetic factors primarily controlled large interindividual variations in dicoumarol half-life[49]; the twins were studied under near basal environmental conditions.

Another twin study showed that genetic factors were mainly responsible for large interindividual variations in phenylbutazone half-life.[50] An independent study of 43 unrelated random subjects and 28 two-generation families identified genetic factors as responsible for two-thirds of the ob-

14. Intraindividual and Interindividual Variations

served phenotypic variance in phenylbutazone half-life; a significant regression of mean offspring value on midparent value was observed.[72] The results of both studies, using different subjects and methods, indicated that control of interindividual differences in phenylbutazone metabolism was primarily genetic.

With respect to antipyrine and its principal metabolites, two independent twin studies on subjects under near basal conditions concluded that genetic factors were primary in controlling interindividual variations in the rate of antipyrine metabolism.[5,22] In another investigation, carefully selected, unrelated subjects exhibited a trimodal distribution of antipyrine half-life, suggesting monogenic control of interindividual variations; antipyrine half-life in each subject was highly correlated to inducibility of arylhydrocarbon hydroxylase activity measured in cultured lymphocytes.[34]

In confirmation of these observations, we administered antipyrine on separate occasions to 12 carefully selected two-generation families. The results were compatible with monogenic control over interindividual variations in the rate constant for formation of each antipyrine metabolite.[35] In performing such family studies, careful selection is necessary to ensure near basal states; Fig. 7 demonstrates clearly how the differential operation of environmental factors can perturb drug-metabolizing capacity and thereby conceal the transmission of genetic factors.

A family study on antipyrine that employed methods discussed earlier attempted to compensate for such differentially operating environmental factors in subjects under nonbasal conditions by assuming that the contribution of each factor could be accurately measured.[26] A model based on multiple regression analysis was relied on to correct the drug-metabolizing capacity of these environmentally perturbed subjects, with the expectation of restoring them to a near basal state in which the operation of genetic factors could be assessed. Sib–sib and parent–parent correlations, without significant parent–sib correlations, were offered as evidence of a primary role played by environmental factors and a negligible role for genetic factors in maintaining interindividual variations in antipyrine clearance.[26] However, when these correlations were corrected for weight, sex, contraceptive use, and smoking, little difference occurred between the midparent–offspring correlation, the sib–sib correlation, and the parent–parent correlation. The sib–sib correlation of 0.376 is only slightly low for a monogenic hypothesis, in which the expected value would be 0.500; so too is the midparent–midoffspring value of 0.227, which ideally should also be 0.500 according to a monogenic hypothesis. Thus, on the basis of a monogenic hypothesis that is supported by another family study[35] (Fig. 7), a midparent–offspring and sib–sib correlation of 0.500 would be expected, not far from the values obtained by Blain et al.,[26] who recognized that their data do not exclude a genetic hypothesis. They argued against such an effect because the form of the distribution curve of antipyrine clearance in their population appeared to be unimodal. However, their curve was skewed even after log transformation.

In concluding this section on genetically determined interindividual variations in rates of drug detoxication, attention is turned away from environmental conditions affecting subjects to the assets and limitations of each method used to test a genetic hypothesis in pharmacogenetics: distribution curves, twin studies, and pedigree analyses. Collectively, in subjects under near basal conditions, the previously described results derived from all these methods present a convincing picture, showing that genetic factors play a major role in the regulation of interindividual variations in drug metabolism. However, results of each method taken alone could be inconclusive or misleading. For example, distribution curves are unreliable as bases for deducing pharmacogenetic sources of variation. A unimodal curve may become bimodal or trimodal when more accurate and sensitive methods are used in establishing phenotypes. Nevertheless, a polymodal distribution curve does not necessarily indicate monogenic control over interindividual variation. Environmental differences among the subjects examined may also generate separate modes.

Twins have been useful in pharmacogenetics as initial screening procedures to assess whether genetic factors play a role in maintaining interindividual variation. The twin approach has inherent limitations that have been described earlier.[3,69] If twin studies suggest that genetic factors do not play a major role, family studies are obviated. However, if twin studies suggest that genetic factors may play a role, family studies should be performed as a second step in a tier system to identify the mode of transmission of these genetic factors.[3,69] If phenotyping is attempted by a pharmacokinetic measurement that depends on hepatic mixed-function oxidase activity, subjects of both twin and family studies should be normal and in a near basal condition with respect to most of the environmental factors shown in Fig. 1. In twin studies, estimates of heritability tend to be slightly exaggerated, probably because the underlying assumption of the twin method—identicality of environments of MZ and DZ twins during the study—is never completely met. Thus, a small portion of the increased phenotypic variation between DZ as compared to MZ twins that is attributed to heredity may arise instead from environmental factors.

The conclusion that genetic factors control interindividual variations in the pharmacokinetics of subjects under near basal conditions seemed then and now to be surprising because the drug response in humans, composed as it is of the discrete[5,49,50] processes of drug absorption, distribution, metabolism, excretion, and interaction with receptor sites, is exceedingly "plastic," subject to large perturbations by hundreds of environmental conditions, including other drugs. Yet in normal individuals under near basal conditions, the twin studies that have been described suggest that genetic constitution regulates extensive interindividual variations in the metabolism of many commonly used drugs. To apply the twin and family approaches in the evaluation of pharmacogenetic hypotheses, critical environmental factors affecting drug disposition should be kept balanced and equal among all subjects. Inequitable, perturbed environments will disguise the genetic contribution to pharmacokinetic variation (Fig. 7) and lead to the erroneous conclusion that environmental factors alone cause phenotypic variation, whereas they may only intercede to conceal the permanent underlying source of such variation.

It can be argued that the purpose of the twin or family study is not to disclose genetic, but rather environmental, contributions to phenotypic variation. However, for such a purpose twin and family studies are inefficient and inappropriate, because DZ twins and family members still differ genetically among themselves. It is possible to eliminate all genetic variation as a contribution to phenotypic variation by adopting another experimental design: the use of MZ twins reared apart or the use of each volunteer as his or her own control, as in the antipyrine test. Not only are genetic factors eliminated by this means, but many other environmental differences among subjects can be successfully excluded, permitting effects of a single factor to be examined in relative isolation. Thus, the antipyrine test offers a much more sensitive method for assessing the capacity of environmental factors to perturb a subject's genetically controlled, near basal state of drug metabolism.

Family studies have been considered the final step or "last word" in establishing a pharmacogenetic hypothesis.[3,69,84] Although definitive in excluding a genetic hypothesis, providing subjects are under near basal environmental conditions, the results of pedigree analysis in humans cannot prove a genetic hypothesis, only suggest its likelihood if a sufficient number of informative pedigrees are compatible with the hypothesis. By contrast, appropriate backcross matings to parent strains can be performed with animals or plants, thereby providing much stronger evidence for a genetic hypothesis than in humans.

B. Evidence for Environmental Control of Large Interindividual Variations in Drug Metabolism

All the material presented up to this point has illustrated the susceptibility of hepatic detoxication enzyme activity to alteration by a large number of environmental factors (Figs. 1 and 2). No doubts should exist concerning the profound effects that environmental factors can exert on drug metabolism and disposition, producing not only intraindividual but also interindividual variations. The extent of the influence of each environmental factor, in whom, under what circumstances, for how long, and with what degree of fluctuation in different subjects during their period of action, are questions that require answers. In addition, there is the extent of interaction among environmental factors to produce more or less effect than simple additivity would indicate. What methods are most appropriate and sensitive in measuring these effects?

Several of these questions have been addressed in the preceding sections. The model drug approach, in which a test compound such as antipyrine is administered on several occasions to normal subjects who serve as their own controls, permitted discovery of most of the factors illustrated in Fig. 1. Further studies using this method need to be performed to explore the possible synergism between such factors. This approach could be extended from subjects under near basal conditions to subjects under highly perturbed environmental conditions. The effect of several impinging environmental factors (Fig. 2) could be assessed in a controlled manner when serial measurements of the kinetics of a test compound are obtained. This approach permits measurement of the influence of only one impinging factor at a time, because only one would be manipulated independently. Thus, the simultaneous advantages of a controlled experiment and of studying subjects under perturbed environmental conditions that more closely reflect those of the real world are achievable.

The answer to the question of whether genetic or environmental factors contribute more to large interindividual variations in the disposition of antipyrine is deceptively simple: it depends on the condition of the particular subjects selected for study. The next question seems obvious but is difficult to answer: what proportion of subjects in a given population are under near basal environmental conditions, and what proportion are under perturbed environmental conditions?

Debates over the relative contributions of genetic and environmental factors to phenotypic variations of these enzymes tend to polarize the issues, distort the facts, and oversimplify the situation. The true picture is that the factors involved are complex and dynamic; both environmental and genetic factors are critical and interact dynamically. Genetic factors should not be ignored in view of the current emphasis on the role of the environment because genetic factors have been demonstrated in both twin and family studies to control most of the phenotypic variation in normal subjects under near basal conditions. Even under such basal conditions, the magnitude of interindividual variations still reaches 3- to 11-fold. This large phenotypic variation is independent of any identifiable environmental difference among the subjects studied. Nevertheless, many individuals in a population who smoke cigarettes, imbibe ethanol chronically, consume unusual diets, or are on antifertility agents and other drugs clearly have altered their genetically controlled basal xenobiotic-metabolizing activities. They are in a perturbed state that can be further perturbed by disease and the aging process. Some of these habits and conditions may change, thereby returning the individual to a more nearly basal state. Genetically controlled levels of metabolism exist for each subject, but they may be concealed in some who are in varying degrees of departure from or returning to these activities. The relative number of subjects in each group will vary with the population selected for study. Because most studies sample only a few subjects from a much larger population, the relative distribution of perturbing factors may not adequately represent their distribution in the large population from which they have been drawn, thereby contributing to erroneous results and conclusions.

To assess the causes of phenotypic variation in drug disposition, the presently used model based on multiple regression analysis appears to enjoy certain advantages, including rapidity, simplicity and ease, because only a single rate measurement has previously been performed in each subject. Thus, the method is more readily applied than the antipyrine test to large numbers of subjects who can be environmentally perturbed, that is, the method seems suitable under real world conditions. Nevertheless, this method has disadvantages as used. No controls have been included. Reproducibility of results has not been firmly established in the same population from which the subjects have been drawn, and because very few subjects are usually involved, they may be unrepresentative and the fit of the model overestimated. Also, the genetic constitution of these perturbed individuals may be concealed and unanalyzable. Relying on correlations with historical data of a qualitative nature, previous applications of this method have been retrospective rather than prospective, having all the attendant disadvantages of retrospective studies. Therefore, it is not surprising that several such correlations could not be confirmed in normal subjects under the conditions of a controlled prospective experiment. Furthermore, any given environmental factor is considered to produce similar effects in all subjects, whereas extensive interindividual variability occurs for which the model as used does not appear to compensate. Recent applications of the model could account for only a small portion of the total observed phenotypic variation in drug kinetics. Thus, the model appears unable to answer the main question it has been used to address: what causes the major part of phenotypic pharmacokinetic variation? For these several reasons, it is not surprising that the model has provided information that conflicts with well-established results (Table II), mainly because of a fundamental misapplication of multiple regression analysis, which has tried to *isolate* and *explain* causal agents, rather than to *predict* a criterion from another set of variables.

Models based on multiple regression analysis have been successfully applied in many other fields.[85] The potential sensitivity and power of the technique are much greater[85] than those that have been achieved in its past limited applications in unraveling sources of pharmacokinetic variability. Enzyme activities are affected by numerous complex interacting factors that have eluded sensitive resolution by the currently used canned models based on multiple regression analysis, probably because critical relationships among some of the factors involved are nonlinear. As emphasized,[85] once these linear and nonlinear relationships are identified, appropriately designed models based on multiple regression analysis can be developed. However, with respect to factors causing interindividual variations in antipyrine kinetics, information on linearity is not available.

ACKNOWLEDGMENTS

The authors express their gratitude to Dr. Marshall B. Jones and Dr. John R. Nesselroade of this institution for their help and advice.

REFERENCES

1. Kalow, W. (1962). "Pharmacogenetics: Heredity and the Response to Drugs." Saunders, Philadelphia.
2. LaDu, B. N. (1972). Pharmacogenetics: Defective enzymes in relation to reactions to drugs. *Annu. Rev. Med.* **23,** 453–468.
3. Vesell, E. S. (1973). Advances in pharmacogenetics. *Prog. Med. Genet.* **9,** 291–367.
4. Vesell, E. S. (1982). On the significance of host factors that affect drug disposition. *Clin. Pharmacol. Ther.* **31,** 1–7.
5. Vesell, E. S., and Page, J. G. (1968). Genetic control of drug levels in man: Antipyrine. *Science* **161,** 72–73.
6. Vesell, E. S., and Page, J. G. (1969). Genetic control of the phenobarbital-induced shortening of plasma antipyrine half-lives in man. *J. Clin. Invest.* **48,** 2202–2209.
7. Vesell, E. S. (1979). The antipyrine test in clinical pharmacology: Conceptions and misconceptions. *Clin. Pharmacol. Ther.* **26,** 275–286.
8. Sultatos, L. G., Dvorchik, B. H., Vesell, E. S., Shand, D. G., and Branch, R. (1980). Further observations on relationships between antipyrine half-life, clearance and volume of distribution: An appraisal of alternative kinetic parameters used to assess the elimination of antipyrine. *Clin. Pharmacokinet.* **5,** 263–273.
9. Alvares, A. P., Kappas, A., Eiseman, J. L., Anderson, K. E., Pantuck, C. B., Pantuck,

E. J., Hsiao, K.-C., Garland, W. A., and Conney, A. H. (1979). Intraindividual variation in drug disposition. *Clin. Pharmacol. Ther.* **26**, 407–419.
10. Vesell, E. S., Passananti, G. T., and Greene, F. E. (1970). Inhibition of drug metabolism in man by allopurinol and nortriptyline. *N. Engl. J. Med.* **283**, 1484–1488.
11. Vesell, E. S., Page, J. G., and Passananti, G. T. (1971). Genetic and environmental factors affecting ethanol metabolism in man. *Clin. Pharmacol. Ther.* **12**, 192–201.
12. Danhof, M., deGroat-vander Vis, E., and Breimer, D. D. (1979). Assay of antipyrine and its primary metabolites in plasma, saliva, and urine by high-performance liquid chromatography and some preliminary results in man. *Pharmacology* **18**, 210–223.
13. Vesell, E. S., Passananti, G. T., and Aurori, K. C. (1975). Anomalous results of studies on drug interactions in man. 1. Nortriptyline and antipyrine. *Pharmacology* **13**, 101–111.
14. Vesell, E. S., Passananti, G. T., and Glenwright, P. A. (1975). Anomalous results of studies on drug interaction in man. III. Disulfiram and antipyrine. *Pharmacology* **13**, 481–491.
15. Vesell, E. S., and Passananti, G. T. (1975). Anomalous results of studies on drug interaction in man. II. Halofenate (MK-185) and antipyrine, bishydroxycoumarin, and warfarin. *Pharmacology* **13**, 112–127.
16. Kappas, A., Alvares, A. P., Anderson, K. E., Pantuck, E. J., Pantuck, C. B., Chang, M. S., and Conney, A. H. (1978). Effect of charcoal-broiled beef on antipyrine and theophylline metabolism. *Clin. Pharmacol. Ther.* **23**, 445–450.
17. Kappas, A., Anderson, K. E., Conney, A. H., and Alvares, A. P. (1976). Influence of dietary protein and carbohydrate on antipyrine and theophylline metabolism in man. *Clin. Pharmacol. Ther.* **20**, 643–653.
18. Conney, A. H., Pantuck, E. J., Hsiao, K.-C., Garland, W. A., Anderson, K. E., Alvares, A. P., and Kappas, A. (1976). Enhanced phenacetin metabolism in humans fed charcoal-broiled beef. *Clin. Pharmacol. Ther.* **20**, 633–642.
19. Pantuck, E. J., Pantuck, C. B., Garland, W. A., Min, B. H., Wattenberg, L. W., Andersson, K. E., Kappas, A., and Conney, A. H. (1979). Stimulatory effect of brussels sprouts and cabbage on human drug metabolism. *Clin. Pharmacol. Ther.* **1**, 88–95.
20. Riester, E. F., Pantuck, E. J., Pantuck, C. B., Passananti, G. T., Vesell, E. S., and Conney, A. H. (1980). Antipyrine metabolism during the menstrual cycle. *Clin. Pharmacol. Ther.* **31**, 38–44.
21. Shively, C. A., Simons, R. J., and Vesell, E. S. (1979). A sensitive gaschromatographic assay using a nitrogen-phosphorus detector for determination of antipyrine and aminopyrine in biological fluids. *Pharmacology* **19**, 228–236.
22. Penno, M. B., Dvorchik, B. H., and Vesell, E. S. (1981). Genetic variation in rates of antipyrine metabolite formation: A study in uninduced twins. *Proc. Natl. Acad. Sci. U.S.A.* **78**, 5193–5196.
23. Fraser, H. S., Bulpitt, C. J., Kahn, C., Mould, G., Mucklow, J. C., and Dollery, C. T. (1976). Factors affecting antipyrine metabolism in West African villagers. *Clin. Pharmacol. Ther.* **20**, 369–376.
24. Fraser, H. S., Mucklow, J. C., Bulpitt, C. J., Kahn, C., Mould, G., and Dollery, C. T. (1977). Environmental effects on antipyrine half-life in man. *Clin. Pharmacol. Ther.* **22**, 799–808.
25. Fraser, H. S., Mucklow, J. C., Bulpitt, C. J., Kahn, C., Mould, G., and Dollery, C. T. (1979). Environmental factors affecting antipyrine metabolism in London factory and office workers. *Br. J. Clin. Pharmacol.* **7**, 237–243.
26. Blain, P. G., Mucklow, J. C., Wood, P., Roberts, D. F., and Rawlins, M. D. (1982). Family study of antipyrine clearance. *Br. Med. J.* **284**, 150–152.

27. Kellermann, G., Luyten-Kellermann, M., and Shaw, C. R. (1973). Genetic variation of aryl hydrocarbon hydroxylase in human lymphocytes. *Am. J. Hum. Genet.* **25**, 327–331.
28. Atlas, S. A., Vesell, E. S., and Nebert, D. W. (1976). Genetic control of interindividual variations in the inducibility of aryl hydrocarbon hydroxylase in cultured human lymphocytes. *Cancer Res.* **36**, 4619–4630.
29. Okuda, T., Vesell, E. S., Plotkin, E., Tarone, R., Bast, R. C., and Gelboin, H. V. (1977). Interindividual and intraindividual variation in aryl hydrocarbon hydroxylase in monocytes from monozygotic and dizygotic twins. *Cancer Res.* **37**, 3904–3911.
30. Paigen, B., Gurtoo, H. L., Minoada, J., Houten, L., Vincent, R., Paigen, K., Bejba Parker, N., Ward, E., and Thompson Hayner, N. (1977). Questionable relation of aryl hydrocarbon hydroxylase to lung-cancer risk. *N. Engl. J. Med.* **297**, 346–350.
31. O'Malley, K., Crooks, J., Duke, E., and Stevenson, I. H. (1971). Effect of age and sex on human drug metabolism. *Br. Med. J.* **3**, 607–609.
32. Vestal, R. E., Norris, A. H., Tobin, J. D., Cohen, B. H., Shock, N. W., and Andres, R. (1975). Antipyrine metabolism in man: Influence of age, alcohol, caffeine and smoking. *Clin. Pharmacol. Ther.* **18**, 425–432.
33. Greenblatt, D. J., Divoll, M., Abernethy, D. R., Harmatz, J. S., and Shader, R. I. (1982). Antipyrine kinetics in the elderly: Prediction of age-related changes in benzodiazepine oxidizing capacity. *J. Pharmacol. Exp. Ther.* **220**, 120–126.
34. Kellermann, G., Kellermann, M. L., Horning, M. G., and Stafford, M. (1976). Elimination of antipyrine and benzo[a]pyrene metabolism in cultured human lymphocytes. *Clin. Pharmacol. Ther.* **20**, 72–80.
35. Penno, M. B., and Vesell, E. S. (1982). Genetic control of interindividual variations in rate constants for formation of antipyrine metabolites in uninduced families. (Submitted for publication.)
36. Vesell, E. S., Shively, C. A., and Passananti, G. T. (1979). Failure of cola nut chewing to alter antipyrine disposition in normal male subjects in a small town in south central Pennsylvania. *Clin. Pharmacol. Ther.* **26**, 287–293.
37. Modell, W. (1981). On the significance of significant. *Clin. Pharmacol. Ther.* **30**, 1–2.
38. Danhof, M., Idle, J. R., Teunissen, M. W. E., Sloan, T. P., Breimer, D. D., and Smith, R. L. (1981). Influence of genetically controlled deficiency in debrisoquine hydroxylation on antipyrine metabolite formation. *Pharmacology* **22**, 349–358.
39. Danhof, M., Krom, D. P., and Breimer, D. D. (1979). Studies on the different metabolic pathways of antipyrine in rats: Influence of phenobarbital and 3-methylcholanthrene treatment. *Xenobiotica* **9**, 695–702.
40. Inaba, T., Lucassen, M., and Kalow, W. (1980). Antipyrine metabolism in the rat by three hepatic monooxygenases. *Life Sci.* **26**, 1977–1983.
41. Idle, J. R., Sloan, T. P., Smith, R. L., and Wakile, L. A. (1979). Application of the phenotyped panel approach to the detection of polymorphism of drug oxidation in man. *Br. J. Pharmacol.* **66**, 430–431P.
42. Inaba, T., Otton, S. V., and Kalow, W. (1980). Deficient metabolism of debrisoquine and sparteine. *Clin. Pharmacol. Ther.* **27**, 547–549.
43. Bertilsson, L., Mellström, B., Sjöqvist, F., Mårtensson, B., and Åsberg, M. (1981). Slow hydroxylation of nortriptyline and concomitant poor debrisoquine hydroxylation: Clinical implications. *Lancet* **1**, 560–561.
44. Sloan, T. P., Idle, J. R., and Smith, R. L. (1981). Influence of D^H/D^L alleles regulating debrisoquine oxidation on phenytoin hydroxylation. *Clin. Pharmacol. Ther.* **29**, 493–497.
45. Eichelbaum, M., Bertilsson, L., Säwe, J., and Zekorn, C. (1982). Polymorphic oxidation of sparteine and debrisoquine: Related pharmacogenetic entities. *Clin. Pharmacol. Ther.* **31**, 184–186.

46. Eichelbaum, M., Spannbrucker, N., Steincke, B., and Dengler, H. J. (1979). Defective N-oxidation of sparteine in man: A new pharmacogenetic defect. *Eur. J. Clin. Pharmacol.* **16,** 183–187.
47. Shively, C. A., Simons, R. J., Passananti, G. T., Dvorchik, B. H., and Vesell, E. S. (1981). Dietary patterns and diurnal variations in aminopyrine disposition. *Clin. Pharmacol. Ther.* **29,** 65–73.
48. Jusko, W. J., Gardner, M. J., Mangione, A., Schentag, J. J., Koup, J. R., and Vance, J. W. (1979). Factors affecting theophylline clearances: Age, tobacco, marijuana, cirrhosis, congestive heart failure, obesity, oral contraceptives, benzodiazepines, barbiturates, and ethanol. *J. Pharm. Sci.* **68,** 1358–1366.
49. Vesell, E. S., and Page, J. G. (1968). Genetic control of dicoumarol levels in man. *J. Clin. Invest.* **47,** 2657–2663.
50. Vesell, E. S., and Page, J. G. (1968). Genetic control of drug levels in man: Phenylbutazone. *Science* **159,** 1479–1480.
51. Alvares, A. P., Kapelner, S., Sassa, S., and Kappas, A. (1975). Drug metabolism in normal children, lead-poisoned children, and normal adults. *Clin. Pharmacol. Ther.* **17,** 179–183.
52. Wagner, J. G. (1973). Intrasubject variation in elimination half-lives of drugs which are appreciably metabolized. *J. Pharmacokinet. Biopharm.* **1,** 165–173.
53. Brodie, B. B., Lief, P., and Poet, R. (1948). The fate of procaine in man following its intravenous administration and methods for the estimation of procaine and diethylaminoethanol. *J. Pharmacol. Exp. Ther.* **95,** 359–366.
54. Brodie, B. B., and Axelrod, J. J. (1949). The fate of acetophenetidin (phenacetin) in man and method for the estimation of acetophenetidin and its metabolites in biological material. *J. Pharmacol. Exp. Ther.* **97,** 58–67.
55. Brodie, B. B., and Axelrod, J. (1950). The fate of antipyrine in man. *J. Pharmacol. Exp. Ther.* **98,** 97–104.
56. Brodie, B. B., and Axelrod, J. (1950). The fate of aminopyrine (pyamidon) in man and methods for the estimation of aminopyrine and its metabolites in man. *J. Pharmacol. Exp. Ther.* **99,** 171–184.
57. Weiner, M., Shapiro, B., Axelrod, J., Cooper, J. R., and Brodie, B. B. (1950). The physiological disposition of dicoumarol in man. *J. Pharmacol. Exp. Ther.* **99,** 409–420.
58. Burns, J. J., Rose, K. R., Chenkin, T., Goldman, A., Schulert, A., and Brodie, B. B. (1953). The physiological disposition of phenylbutazone (butazolidin) in man and a method for its estimation in biological material. *J. Pharmacol. Exp. Ther.* **109,** 346–357.
59. Burns, J. J., Weiner, M., Simson, G., and Brodie, B. B. (1953). The biotransformation of ethyl biscoumacetate (Tromexan) in man, rabbit and dog. *J. Pharmacol. Exp. Ther.* **108,** 33–41.
60. Brodie, B. B., Burns, J. J., Mark, L. C., Lief, P. A., Bernstein, E., and Papper, E. M. (1953). The fate of pentobarbital in man and dog and a method for its estimation in biological material. *J. Pharmacol. Exp. Ther.* **109,** 26–34.
61. Andreasen, P. B., and Ranek, L. (1975). Liver failure and drug metabolism. *Scand. J. Gastroenterol.* **10,** 293–297.
62. Galton, F. (1875). The history of twins as a criterion of the relative powers of nature and nurture. *J. Br. Anthropol. Inst.* **5,** 391–406.
63. Endrenyi, L., Inaba, T., and Kalow, W. (1976). Genetic study of amobarbital elimination based on its kinetics in twins. *Clin. Pharmacol. Ther.* **20,** 701–714.
64. Cascorbi, H. F., Vesell, E. S., Blake, D. A., and Helrich, M. (1971). Genetic and environmental influence on halothane metabolism in twins. *Clin. Pharmacol. Ther.* **12,** 50–55.
65. Alexanderson, B., Price Evans, D. A., and Sjöqvist, F. (1969). Steady-state plasma

levels of nortriptyline in twins. Influence of genetic factors and drug therapy. *Br. Med. J.* **4,** 764–768.
66. Andreasen, P. B., Froland, A., Skovsted, L., Andersen, S. A., and Hauge, M. (1973). Diphenylhydantoin half-life in man and its inhibition by phenylbutazone: The role of genetic factors. *Acta Med. Scand.* **193,** 561–564.
67. Furst, D. E., Gupta, N., and Paulus, H. E. (1977). Salicylate metabolism in twins: Evidence suggesting a genetic influence and induction of salicylurate formation. *J. Clin. Invest.* **60,** 32–42.
68. Scott, J., and Poffenbarger, P. L. (1979). Pharmacogenetics of tolbutamide metabolism in humans. *Diabetes* **28,** 41–51.
69. Vesell, E. S. (1978). Twin studies in pharmacogenetics. *Hum. Genet., Suppl.* **1,** 19–30.
70. Motulsky, A. (1964). Pharmacogenetics. *Prog. Med. Genet.* **3,** 49–74.
71. Åsberg, M., Price Evans, D. A., and Sjöqvist, F. (1971). Genetic control of nortriptyline in man: A study of relatives of propositi with high plasma concentrations. *J. Med. Genet.* **8,** 129–135.
72. Whittaker, J. A., and Price Evans, D. A. (1970). Genetic control of phenylbutazone metabolism in man. *Br. Med. J.* **4,** 323–328.
73. Kalow, W., Kadar, D., Inaba, T., and Tang, B. K. (1977). A case of deficiency of N-hydroxylation of amobarbital. *Clin. Pharmacol. Ther.* **21,** 530–535.
74. Mahgoub, A., Idle, J. R., Dring, L. D., Lancaster, R., and Smith, R. L. (1977). The polymorphic hydroxylation of debrisoquine in man. *Lancet* **2,** 584–586.
75. Kutt, H., Wolk, M., Scherman, R., and McDowell, F. (1964). Insufficient parahydroxylation as a cause of diphenylhydantoin toxicity. *Neurology* **14,** 542–548.
76. Vasko, M. R., Bell, R. D., Daly, D. D., and Pippenger, C. E. (1980). Inheritance of phenytoin hypometabolism: A kinetic study of one family. *Clin. Pharmacol. Ther.* **27,** 96–103.
77. Shahidi, N. T. (1967). Acetophenetidin sensitivity. *Am. J. Dis. Child.* **113,** 81–82.
78. Bönicke, R., and Lisboa, B. P. (1957). Über die Erbbedingtheit der intraindividuellen Konstanzder Isoniazidquasscheidung. *Naturwissenschaften* **44,** 314.
79. Biehl, J. P. (1956). The role of the dose and the metabolic fate of isoniazid in the emergence of isoniazid resistance. *Trans. Conf. Chemother. Tuberc.* **15,** 729–782.
80. Biehl, J. P. (1957). Emergence of drug resistance as related to the dosage and metabolism of isoniazid. *Trans. Conf. Chemother. Tuberc.* **16,** 108–113.
81. Knight, R. A., Selin, J., and Harris, H. W. (1959). Genetic factors influencing isoniazid blood levels in humans. *Trans. Conf. Chemother. Tuberc.* **18,** 52–58.
82. Price Evans, D. A. P., Manley, K., and McKusick, V. A. (1960). Genetic control of isoniazid metabolism in man. *Br. Med. J.* **2,** 485–491.
83. Harris, H. W., Knight, R. A., and Selin, M. J. (1958). Comparison of isoniazid concentrations in the blood of people of Japanese and European descent–therapeutic and genetic implications. *Am. Rev. Tuberc.* **78,** 944–948.
84. Cohen, S. N., and Weber, W. W. (1972). Pharmacogenetics. *Pediatr. Clin. North Am.* **19**(1), 21–36.
85. Cohen, J., and Cohen, P. (1975) "Applied Multiple Regression/Correlation Analysis for the Behavioral Sciences." Erlbaum, Hillsdale, New Jersey.

Index

A

Acetaminophen, 58, 59, 106, 119, 146
  biliary function, 264
  glucuronidation, 63
  metabolism, 35
  toxic effects, 106, 119
Acetazolamide, excretion, 253
2-Acetoxy-2-acetylaminofluorene, 41
2-Acetylaminofluorene, 16, 38, 40, 42, 46, 106, 116
N-Acetyl-2-aminofluorene, 32, 35, 36
  activation, 360
N-Acetylation, 16
Acetylcholine, excretion, 254
Acetylcholinesterase, 342
Acetylprocainamide ethobromide, 266
Acetylsalicylic acid, 146
N-Acetyltransferase
  genetic factors, 399
  ontogeny, 90
  social distribution, 400
Acesulfame, 163
Acid, unmetabolized, 159–165
Acid chloride, 51
Acrolein, 111
Acrylamide, 326
Activation, see also specific reaction, substance
  enzyme, tissue distribution, 105–127
  pathway, 37–39
  xenobiotic, 351, 358–361
Acyltransferase, 39
Adenosinetriphosphatase, 257, 259, 265
Adenosine triphosphate, 55
Adenovirus, 319
Adipic acid, 159, 161
Administration routes, 244–246
Adrenalin, in ontogeny, 83
Adriamycin, 195, 326
Aflatoxin, 294
Aflatoxin B_1, 32, 36, 38, 46, 117, 193
  regioselectivity, 48
Aflatoxin B_1 2,3-epoxide, 38
  hydrolysis, 40

Ah locus, 343
Ah-receptor agonist, 62
Albumin, 218
  esterase activity, 140
  hydrolytic activity, 140
Alcohol
  intoxication, 291
  sulfation, 60
Alcohol dehydrogenase, 111
Aldehyde, 144
Aldosterone, biliary function, 264
Aldrin, 356, 357
Aliflurane, 168
Alkaline phosphatase, localization, in liver, 111
Alkenal, 185
Alkoxy radical, 186, 187
  formation, 186, 187
Alkyldiazohydroxide, formation, 15
Alkyldiazonium compounds, 15
Allopurinol, 374
Allyl alcohol, 106
Allyl formate, 106
Ames test, 358
Amide, prodrug, 146
Amiloride, 167, 173
Amine
  N-hydroxymethylated, 146
  oxidation, 5, 9
  prodrug, 146
  quaternary, 165, 267
    biliary excretion, 258
  unmetabolized, 165, 166
Aminobenz[d]isothiazole-1,1-dioxide, 163
p-Aminohippurate, 330
δ-Aminolevulinic acid, 315
δ-Aminolevulinic acid synthetase, 316
o-Aminophenol, 309
p-Aminophenol, 120
Aminopyrine, 56, 57, 59, 85, 86, 309, 317, 319, 320, 383
  excretion, 273
  first-pass effect, 384
  half-life, plasma, 391
p-Aminosalicylic acid, genetic factors, 399

Amobarbital, 383
　genetic factors, 393, 396
　half-life, plasma, 391
Amphetamine, 50
Amygdalin, 157
Androgen, 92, 93
Anemia
　aplastic, 326
　hemolytic, 326
Anesthetic, 144, 223, 224
　unmetabolized, 168, 169
Aniline, 309, 317, 327
Aniline hydroxylase, activity, 122
Anion transport, 264
Anthracycline, 195
Anticonvulsant, 326
Antidepressant, tricyclic, effect on cholestasis, 265
Antioxidant, 189, 294
Antipyrine, 293, 309
　as assay, 372–406
　clearance, 375
　　model, 380
　　reproducibility, 377, 387
　cytochrome P-450, 388
　dosage, 376
　effect of allopurinol, 374
　genetic factors, 393, 398
　half-life, 374, 386
　　plasma, 388, 391
　　in twins, 395
　mean normal values, 374
　metabolites, 389, 397
　in milk, 272
　as model drug, 372
　pedigree, 398
　test, 372–402
　　assumptions, 378, 379
　　basal conditions, 376–383
　　effect of age, 381
　　environment, 374
　　genetic constitution, 381
　　as model drug, 383, 384
　　volunteers, 373
　in twins, 385
Antitumor agent, 195–197, see also specific substance
Arachidonic acid, 49, 296
Arachidonic hydroperoxide, 186
Arene oxide, 7
　formation, 13

Arthritis, model, 311, 312
Arylamide
　activation, 16–18
　formation, 16
Arylamine, 50, 142
　activation, 16–18
Aryl hydrocarbon hydroxylase, activity, 127
Arylhydroxylamine, 142
Aryl hydroxylamine O-glucuronide, stability, 40
Arylnitroso compounds, 142, 143
Aryl sulfotransferase, 40–43, 52, 60–62, 233, see also Sulfotransferase
　localization
　　liver, 119
　　skin, 125
Ascorbic acid, 298, 300
　deficiency, 301
ATP-sulfurylase, 58
Atropine, excretion, 254
Aurin tricarboxylic acid, 159–161
Autoxidation, 183, 187
Axonopathy, 326
Azo dye, GSH and, 59
Azo reduction, 4
Azoxy intermediate, 9

B

Barbiturate, sleeping time, 311
Bay-region theory, 14, 15
BCG, 309
Bentazon, 163
Benz[a]anthracene, 47
Benzene, 327
Benzenesulfonic acid, 161, 162
Benzil, 52
Benz[d]isothiazoline 1,1-dioxide, 163, 171
Benzodiazepine, 146
Benzo[a]pyrene, 32, 38, 41, 44, 46, 47, 49, 51, 52, 54, 148, 309, 317, 327, 340
　activation, 359
　as carcinogen, 145
　glucuronidation, 61, 64
　metabolism, 13
　nonenzymatic reactions, 139
　stereochemistry, 45
　sulfation, 61, 64
Benzo[a]pyrene 7,8-trans-dihydro-7,8-diol, 14
Benzo[a]pyrene dihydrodiol epoxide, 34

Index

Benzo[a]pyrene 7,8-dihydrodiol 9,10-oxide, 40, 49, 51, 359
Benzo[a]pyrene hydroxylase
  activity, 121
    localization, 123
  liver, 120
Benzo[a]pyrenequinone, 51
1,4-Benzoquinone, 148
Benzothiadiazines, unmetabolized, 164, 165
Benzoylecgonine, 145
Benzylpenicillin, 330
Bethanidine, 167
BHT, 299, *see also* Hydroxytoluene, butylated
Bile, 251
  flow, 110, 256, 257, 262
    bile salt-dependent, 257
    effect
      of age, 268
      of bile salts, 264, 266
      of starvation, 290
    liver regeneration, 270
    pregnancy, 264
  formation, physiology, 255–257
  proteins, 261
  secretion, effect of bile salts, 256
Bile acid, sulfation, 60
Bile duct hyperplasia, 265
Bile salt
  effects, 267
  enterohepatic circulation, 268
  excretion, 253, 258
  independent flow
    ATPase, 257
    quantitation, 257
  micelles, 267
  formation, 256
Biliary clearance, 153, *see also* Clearance
Biliary excretion, 254–270
  anatomic considerations, 255
  animal differences, 255
  ductule system, 256
  effect
    of bile salts, 255
    of carcinogens, 262
    of chelators, 265
    of dimethyl sulfoxide, 266
    of metals, 265
    of phenobarbital, 262
    of phenols, 262
  size of compounds, 258
  types of compounds, 258

Biliary function
  effect of age, 267, 268
  sex differences, 267
  species differences, 267, 268
Biliary transport, 259, 260
Bilirubin, 88
  transport, 258
Bilirubin-UDPglucuronyltransferase, 261
Binding
  constants, 219, 220
  equilibrium, 218
  intracellular, 260, 261
  xenobiotic, clearance, 230, 231
Binding protein, 252
  in bile, 261
  hepatic, 260
Binding protein Z, 260
Bioaccumulation, 350, 355, 356
Biotransformation, 240, *see also* Detoxication
  depressants, 308–311
  effect on rate, 307
  hepatitis, 316, 317
  host defense mechanism, 307
  immunostimulants, 308–311
  in infection, 308, 316–319
    viral, 318–321
  in influenza, cytochrome P-450, 321
  nonenzymatic, 137–149
Biphenyl, polychlorinated, biliary function, 264
1,4-Bis[2-(3,5-dichloropyridyloxy)]benzene, 45
Bladder cancer, 2, 64
Bladder tumor, 32, 36
Bleomycin, 196, 326
Blood
  cell, 220
    binding, plasma proteins, 221
  dyscrasias, 327
  duration time, 236
  flow
    in clearance, 230
    kidney, 252
    liver, 239, 240
    renal excretion, 238
  toxicity target, 326, 327
Blood-to-bile transport, 255
Blood-brain barrier, 222
Bone, toxicity target, 326
Bone marrow, toxicity target, 326, 327
Bordetella pertussis, 308–311
  splenectomy, 310

Bretylium, 165, 166
Bromobenzene, 106, 119, 327, 339
  activation, 66
  GSH depletion, 65
  toxicity, 66
Bromobenzene 3,4-epoxide, 66
Bronchial mucosa, 121
Bronchiolar epithelial cell, see Clara cell
Bucolome, 260
Butter yellow, see N-Methyl-4-aminoazobenzene
tert-Butyl hydroperoxide, 186
Butylhydroxyanisole, 62

C

Cadmium, toxicity, 298
Caerulein, 266
Calcium, dietary, 289
Cancer, see specific type
Carbamate insecticides, 356, 358, see also specific substance
  resistance, 363
Carbamazepine, half-life, plasma, 391
Carbamylation, 144
Carbamyl phosphate, 144
Carbene, 49, 51
Carbohydrate
  dietary, 293, 294
  effect, 288
  high-
    diet, 293
    effect on detoxication, 293
Carbon, colloidal, 309
Carbon tetrachloride, 21, 106, 122, 309
  biliary function, 264
  hepatotoxicity, 312
  in lipid peroxidation, 194, 195
Carboxyesterase, 351–353, 361
Carboxylic acid, unmetabolized, 159–161
Carcinogen, 8, 358–361, see also specific substance
  activation, 43, 50
  metabolism, 46–48
  proximate, 17, 339
  ultimate, 339
Carcinogenesis, 327, 339
  chemical, 44
  steps, 33
  high-fat diets, 297

induction, 32
metabolism, 31–67
  by 2-naphthylamine, 39
  role of coenzymes, 54
  sulfation, 64
Cardiomyopathy, 326
β-Carotene, 190
Carotenoid, 190
Catalase, 197, 201
  in cytotoxicity, 188
  interferon, 315
Cation excretion, 253
Cell necrosis, 327
Cephaloridine, 326, 329–331
  renal accumulation, 329–331
Cephalosporin, 330, 331
Cephalothin, excretion, 253
Cerebrospinal fluid barrier, 222
Ceruloplasmin, 187, 200
Chagas' disease, 197
Chelating agent, in bile, 265
Chemiluminescence, 200
Chemotactic factor, 198
Chemotherapy, 195–197
Chloramphenicol, 326, 343
Chlordecone, 241
2-Chloro-1,1-difluoroethylene, 50
Chloroethylene oxide, 38
Chloroform, 38, 49
Chloroquine, 326
Chlorothiazide, 163, 164
  excretion, 264
2-Chloro-1,1,1-trifluoroethane, 50
Chlorphentermine, 332, 333
Chlorpromazine, effect on cholestasis, 265
Cholestasis, 265
Choleresis
  effects, 266
    of nafenopin, 264
    of phenobarbital, 255
Cholesterol
  dietary, 296, 297
  effect on UDPglucuronyltransferase, 297
Cholinesterase, inhibition, 354, 355
Chromic phosphate, clearance, 227
Chrysene, 47
Cimetidine, 167
Clara cell, 106, 122, 123, 337
Clearance, 153, see also specific type
  diffusional barrier, 232, 233
  extraction ratio, 227, 228

intrinsic, 228, 229, 231, 232, 236, 237
nonrestrictive, 231
organ, 225–227
  blood flow, 230
  sequential, 242–244
  restrictive, 231
  substrate concentration, 233
  systemic, 226
  total body, 225–227
  xenobiotic binding, 230, 231
Cleft palate, 326
Cocaine, 145, 171
Codeine, in milk, 272
Coenzyme, carcinogenesis, 54
Coenzyme A, 193, 301
Congenital abnormality, 326
Conjugation reactions, 6
Copper, 301
Corticosteroid, 199, 200
Corticosterone, 91
  as inducer, 83
Corynebacterium parvum, 308–311
Creatinine
  clearance, 253
  excretion, 254
Cromoglycate, 159–161, 173
Cyclamate, 157, 161, 162, 171
Cyclic AMP, excretion, 253
Cyclohexylamine, 171
Cyclooxygenase, 198
Cyclopropane, 168
Cytochrome b_5, 49, 50, 57
  localization
    liver, 116
    lung, 121
  ontogeny, 84
Cytochrome c reductase, 84, 85
Cytochrome oxidase, localization, liver, 111
Cytochrome P-450, 2–4, 17, 19, 22, 37, 54, 62, 233, 240, 241, 261, 291, 317, 321
  antibodies, 44
  antipyrine, 388
  complexes, 50, 53
  dietary lipid, 295, 296
  distribution, 45
  effect of glucagon, 94
    of glucocorticoids, 92
    of trace nutrients, 299
  immunostimulants, 309
  induction, 44, 45, 94, 95, 113
  interferon effects, 313–316

intracellular location, 113
intralobular distributions, 113
localization
  liver, 110, 112–116, 120
  lung, 121, 122
  skin, 125, 127
macrophage, 124
metabolism, 42
multiplicity, 44–46
ontogenesis, 83–86
redox potential, 56
regioselectivity, 44–52
in starvation, 290
stereoselectivity, 44–52
Cytotoxicity, 198

D

Dapsone, genetic factors, 399
Daunorubicin, 195
DDE, 356, 357, 364
DDT, 260, 329, 356, 357, 362
  in milk, 272
DDT-dehydrochlorinase, 362
Dealkylation, 4
N-Dealkylation, 3, 4
Debrisoquine, 384
Decamethonium, 165
O-Deethylase, activity, 125
O-Deethylation
  activity, 123
  localization, 121
  ontogeny, 84, 85
Deferoxamine, 265
Dehydrochlorination, 355
Dehydrocholate, biliary excretion, 258
N-Demethylantipyrine, 389
  genetic factors, 397
Demethylase, activity, 320
N-Demethylase
  activity, 125
    localization, lung, 122, 123
O-Demethylase, activity, 125
N-Demethylation
  effect of glucocorticoids, 92
  inhibition, 263
  interferon, 315
  ontogeny, 85, 86
N-Demethylpirimiphos methyl, 354
Deoxyguanine, modification, 33, 34

Dermatitis, exfoliative, 326
Detoxication, *see also* Biotransformation
 diversionary, 41, 43
 effect of carbohydrate, 293
  of lipid, 295
  of trace nutrients, 297–302
 glutathione, 40–43
 host defense mechanisms, 307–321
 physiology, 214
 primary, 41, 42
Detoxication enzyme
 antipyrine test, 379
 effect of androgen, 92, 93
  of estrogen, 92, 93
  of glucocorticoids, 91, 92
  of growth hormone, 93
  of progesterone, 92, 93
 induction, 364, 379, 380
 species distribution, 363
 tissue distribution, 105–127
Dextran, 309
N,O-Diacetyl-p-aminophenol, 146
Diazene, 9, 19
 as alkylating agents, 20
 covalent adducts, 20
Diazepam, 269
 biliary function, 264
 half-life, plasma, 391
Diazinon, 353–355, 362
Diazoalkane, 19
Diazomethane, 8
 formation, 15
Dioxide, 163, 164
 metabolism, 156
 routes of administration, 156
 species specificity, 156
Dibenz[a,h]anthracene, 47
1,2-Dibromo-3-chloropropane, 326
p,p'-Dichlorobenzophenone, 357
2,3-Dichlorobiphenyl, 53
2,6-Dichloro-4-nitrophenol, 232
Dicoumarol, 400
 genetic factors, 393, 396
 half-life, 386
  plasma, 391
 in twins, 385
Dieldrin, 260, 351, 355, 356, 363, 364
 bioaccumulation, 356
 species differences, 356
4,4'-Diethylaminoethoxyhexoestrol, 332
Diethylenetriaminepentaacetic acid, 265

Diethyl maleate, 64, 65
Diffusion
 barrier, clearance, 232, 233
 facilitated, 253
 passive, 364
6,7-*trans*-Dihydroaldrindiol, 357
Dihydrochlordene 6,7-dicarboxylic acid, 357
7,8-Dihydro-7,8-dihydroxybenzo[a]pyrene, 340
Dihydrodiol, 8, 52
 formation, 53
trans-Dihydrodiol, 13, 14
Dihydrodiol epoxide, 66
Dihydrodiol oxide, hydrolysis, 40
N,N-Dimethyl-4-aminoazobenzene, 261
 as carcinogen, 16
Dimethylazoaminobenzene, GSH and, 59
Dimethylbenzanthracene, 297
7,12-Dimethylbenz[a]anthracene, 47, 125
 metabolites, 292
1,1'-Dimethyl-4,4'-bipyridylium, *see* Paraquat
Dimethylnitrosamine, 36, 38, 327, 344
 activation, 14, 15
 as carcinogen, 15
 metabolism, 48
2,4-Dinitrophenol, 58
 effect on organic anions, 253
N,N-Dioctadecyl-N^1,N^1-bis(2-hydroxyethyl)-
  propanediamine, 309
Diol epoxide, 23
Diquat, 331
Disulfamoylaniline, 171
Disulfide, oxidation, 11
Diuretics, 254
DNA
 adducts, 32, 35, 39, 51, 66
 alkylation, 54
 in carcinogenesis, 43
 modification, 33–35, 37
 repair, 341, 342
 stereoselectivity, 42
 as trap for epoxide, 54
Dopamine, excretion, 254
Drug metabolism, *see* specific substance
Ductule system, 256
Duration time, 236

E

Ear, toxicity target, 326
Ecgonine, 171

Index

Ecgonine methyl ester, 145
Electrophile
  as carcinogen, 33–36
  with DNA, 33–36
  with glutathione, 59
  nonenzymatic reaction, 144
Elimination, *see also* Excretion
  organs, 239–246
  sequential, 242–244
Ellipticine, 52
Emepronium, 165, 166
Encephalomyocarditis virus, 317
Endoplasmic reticulum, 109, 110
Endrin, 363
Enflurane, 168
Enterohepatic circulation, 268, 269
  of bile salts, 268, 269
  effect of age, 268
  inhibition, 269
  species differences, 269
Environment
  basal, 373
  factors, 404
  variations, 372
Enzyme, *see also* specific substance
  cluster
    late fetal, 78, 79
    late suckling, 78, 79
    neonatal, 78, 79
    pubertal, 78, 79
  developmental profile, 83–90
    hormonal influences, 91–94
    xenobiotic effects, 94, 95
  differentiation, 78–81
  induction, *see* Induction
  localization, 233–237
  loss, 343, 344
  ontogenesis, 78–81
    classification, 78
    clusters, 78
  organ distribution, 335–337
Enzymology, comparative, 363
Epidermis, *see* Skin
Epinephrine, excretion, 254
Epithelial cell
  alveolar, 123, 134
  bronchial, 122
Epoxidation, 4, 7, 8
Epoxide
  bay-region, carcinogenicity, 47
  formation, 21

  as activation, 38, 39
  hydrolysis rates, 52
  as intermediates, 7
  rearrangement to phenols, 52
Epoxide hydrolase, 14, 39, 40, 351, 359
  activity, 127
  cytochrome P-450 complex, 53
  induction, 52, 116, 117
  localization
    liver, 116, 117
    lung, 121
    skin, 125, 127
  macrophage, 124
  ontogeny, 86, 87
  solubility, 52
  stereospecificity, 52–54
Equilibrium, kinetics, 318
Erythromycin, effect on cholestasis, 265
Escherichia coli, endotoxin, 309
Ester, formation, activation by, 39
Esterase, neurotoxic, 342, 343
Estradiolglucuronide, 264
Estriol, 88
Estrogen, 92, 93, 271
  effect on cholestasis, 265
Estrone, biliary function, 264
Ethacrynic acid, 257
Ethanesulfonic acid, 161
Ethanol, genetic factors, 393
Ether, 168
Ethinyl estradiol, effect on biliary excretion, 265
7-Ethoxycoumarin, 84, 85
7-Ethoxyresorufin, 84, 85
β-Ethyladipic acid, 159
Ethylene formation, 142
Ethylenediaminetetraacetic acid, 159–161
Ethylmorphine, 57, 59, 309, 317
O-Ethyl O-p-nitrophenyl phenylphosphonothioc acid, 352
Ethylnitrosourea, 342
Etodolac, 312
Excision repair enzyme, 35, 36
Excretion, 237–239, *see also* specific organ
  biliary, *see* Biliary excretion
  expiration, 272, 273
  mechanisms, 251–274
  metabolism, 261–263
  milk, 271, 272
  rate, 238

salivary, 270, 271
  plasma level, 270, 271
Expiration, 272, 273
Extraction, ratio, 227, 228
Eye, toxicity target, 326

F

FAD-containing monooxygenase, 4–6
FAD-dependent monooxygenase, 20
FAD monooxygenase, 17
Fasciola hepatica, 317
Fasting, 289, 290
Fatty acid
  autoxidation, 187
  polyunsaturated, 289, 294, 296, 297
Fava bean, 343
Fenfluramine, 332
Fenton-type reaction, 186
Ferredoxin, localization, liver, 117
Ferritin, 187, 222
Fetus
  biastrous formation, 77
  liver, 82
  toxicity target, 326
First-pass effect, 244–246, 384
Flavin, 195
Flavone, 50, 51
  effect on cytochrome P-450, 51
Flavoprotein monooxygenase, 39
Fluroxene, 168
Folic acid, 301
Formamidinesulfenic acid, 12, 22
Free radical, 51
  formation, 8, 10
  reaction, 181–201
    diet, 190, 191
    glutathione, 191–194
    metal in, 190, 191
Freunds' adjuvant, 309, 313
Fructose 1,6-bisphosphate, 56
Furobofen, 312
Furosemide, 106
  excretion, 253
1-(3-Furyl)-4-hydroxypentanone, *see* 4-Ipomeanol

G

D-Galactosamine
  glucuronidation, 63

  as inhibitor, 63
Gallamine, 165, 166
Gastrointestinal tract, toxicity target, 326, 327
Genetic factors
  N-acetyltransferase, 399
  metabolic variation, 393–403
  pharmacogenetics, 402
  sib-sib correlations, 400, 401
  twins, 393
    metabolism, 395
Genetic polymorphism, 396
Gentamicin, 330
Glomerular filtration, 237, 238
  rate, 252
Glomerulus, 252
D-Glucaro-1,4-lactone, 269
D-Glucosamine, 58
Glucose-6-phosphatase, localization, liver, 111
Glucose-6-phosphate dehydrogenase, 56, 343, 344
  localization, liver, 111
Glucosephosphate isomerase, localization, liver, 111
Glucocorticoid, 91, 92
β-Glucuronidase, inhibition, 269
Glucuronidation, 51, 57, 58, 232, 233, 261, 360
  in Gunn rats, 265
  hypoxia, 58
  induction, 261
  nutrition, 58
O-Glucuronidation, 41
  activation by, 39
Glucuronide
  conjugation, in starvation, 290
  enterohepatic circulation, 268, 269
Glutamate dehydrogenase, localization, liver, 111
Glutamic-pyruvic transaminase, localization, liver, 111
γ-Glutamyl cycle, 59
γ-Glutamyl cysteine synthetase, 59
γ-Glutamyltransferase, 89, 91, 95
Glutathione, 54, 59, 60, 188, 190
  concentration, 191–194
    intracellular, 191–193
  conjugation, 43
    effect of diethyl maleate, 263
    of methyl iodide, 263
  detoxication, 40–43
  efflux, 192
  electrophiles, 59

Index

fat oxidation, 191
free radical reaction, 191–194
γ-glutamyl cycle, 58
hepatic cancer, 59
in mitochondria, 193
nonenzymatic reactions, 141–144
ratio to GSSG, 55, 59
reagents, 64
ribonucleotide reductase, 59
synthesis, 59
Glutathione disulfide
efflux, 192
oxidative stress, 192
ratio to GSH, 192
Glutathione peroxidase, 188, 189, 197, 201, 300
cell distribution, 188
ratio to superoxide dismutase, 197
Glutathione reductase, 10
induction, 59
Glutathione synthetase, 59
Glutathione thioether, 261
Glutathione transferase, 241
Glutathione S-transferase, 32, 62, 63, 89, 90, 222, 362
binding, 62
constants, 260
characteristics, 62
DNA modification, 63
effect of glucocorticoids, 92
induction, 260
localization
liver, 117
lung, 121
skin, 125, 127
Glycogen synthase, ontogeny, 79, 81
Glycoprotein, sulfated, 173
Golgi membrane, 255
Griseofulvin, 295
Growth hormone, 93, 271
Guanazole, 167
Guanethidine, 167
Guanidine, unmetabolized, 166, 167
Guanylate cyclase, 143
Gunn rat, 261, 265

H

Haemophilus influenzae, 319
Hair follicle, 125, 127

Halothane, 49, 50, 168, 195
genetic factors, 393
Harmol, 58, 64, 232
Harmol sulfate, excretion, 264
Heart, toxicity target, 326
Heme
degradation, interferon, 315
nitrogen, alkylation, 51
Heme oxygenase, 51, 312, 316
Hemodynamics, 223, 224
Hepatic uptake, 259, 260
Hepatitis, 316, 317
duck, 317
murine, 317
viral, 317
Hepatobiliary excretion, influence of xenobiotics, 263–266
Hepatobiliary transport, 261
Hepatocarcinogen, 117
Hepatocarcinogenesis, 42
Hepatocyte, 107
heterogeneity, 109
Hepatotoxicity, 265
Hepatotoxin, 106
binding, 119
Heptachlor, 356
2,4,5,2′,4′,5′-Hexachlorobiphenyl, 169, 170
Hexachlorocyclohexane, 355
Hexachloroethane, 21
Hexamethonium, 165, 166
excretion, 254
Hexobarbital, 309, 317
Herbicide, 331, 332, *see also* specific substance
Heterocyclic compound, unmetabolized, 162–165
Hippurate, excretion, 253
Histamine, excretion, 254
Homosulfanilamide, 171
Host defense mechanism, 307–321
Hydantoin, formation, 144
Hydralazine, 144
genetic factors, 399
Hydrazine, 9
metabolism, 19, 20
nonenzymatic reaction, 144
oxidation, 5
pyridoxal phosphate and, 19
substituted, activation, 20
Hydrazone, 144
rearrangement, 144
Hydride anion, nucleophilic, 141

Hydrocarbon
  chlorinated, unmetabolized, 169, 170
  halogenated aromatic, 106
  polycyclic, activation, 66
  polycyclic aromatic, activation, 12–14
Hydrochlorothiazide, 163, 164
Hydrocortisone, biliary function, 264
Hydrogen peroxide, 49, 182–184
Hydrolysis
  nonenzymatic, 139, 140, 145
  rate constants, 140
Hydroperoxide, decomposition, 187
Hydroxamic acid
  activation, 60
  as carcinogen, 17
  conjugation, 17
N-Hydroxy-N-acetyl-4-aminoazobenzene, 16
N-Hydroxy-2-acetylaminofluorene, 41, 147, 360
N-Hydroxyamide, 9
N-Hydroxyammonium ion, 9
o-Hydroxyaniline, 16
Hydroxyanisole, butylated, 294
4-Hydroxyantipyrine, 389
  genetic factors, 397
N-Hydroxyarylacetamides, 39
N-Hydroxyarylamine, 17
9-Hydroxybenzo[a]pyrene, 22, 23
4-Hydroxybiphenyl, sulfation, 58
β-Hydroxybutyrate dehydrogenase, localization, liver, 111
9-Hydroxydieldrin, 357
Hydroxyindoleacetic acid, 253
Hydroxylamine, 17
  aromatic, 9
  formation, 9
  oxidation, 5
Hydroxylation, 4
C-Hydroxylation, 3, 4, 8
N-Hydroxylation, 8, 50, 360
  epoxide formation, 38, 39
Hydroxyl radical, 49, 184
3-Hydroxymethylantipyrine, 389
  genetics factors, 397
7-Hydroxymethyl-12-methylbenzanthracene, 47
N-Hydroxy-2-naphthylamine, 39
α-Hydroxynitrosamine, 8
4-Hydroxynonenal, 185
1'-Hydroxysafrole, 39
Hydroxysteroid, sulfation, 60

Hydroxysteroid sulfotransferase, 89
Hydroxytoluene, butylated, 294, 299, see also BHT
5-Hydroxytryptamine, excretion, 254
Hyperoxia, 194
Hypolipidemic compounds, 264
Hypoproteinemia, 291
Hypoxia, 58, 194, 195

I

Imine, prodrug, 146
Iminosulfenic acid, 12
Imipramine, 332
Immunostimulant
  adjuvants, 311, 312
  enzyme changes, 308–311
Indocyanine green, 261
  uptake, 259
Indomethacin, 199, 269, 311
  half-life, plasma, 391
Induction, 47, 343, 344
  bacterial infection, 318
  detoxication enzymes, 364, 379, 380
  in infants and neonates, 96, 97
  for P-450 forms, 44
  by virus, 318
  by xenobiotics, 94, 95
Infection, 316–318, see also specific types
  viral, 318–321
Inflammation, 197–200
  drugs for, 199, 200
Influenza, 318–321
  vaccine, 319, 320
  virus, 317
Influenza A, 319
Influenza B, 319, 320
Inhalation, 244, 245
Insecticide, 329, see also specific substance
  biodegradable, 355
  resistance, 361–363
Interferon, 309, 313–316
  activity, 314
  α-, β, and γ-types, 314
  catalase, 315
  cytochrome P-450, 313–316
  N-demethylation, 315
  heme, 315
  inducers, 313, 315
  species differences, 314

Index

tryptophan 2,3-dioxygenase, 315
Intermediate, transport, 65, 66
Intestine, elimination, 240–244
Intraarterial administration, 244–246
Intramuscular administration, 245
Intraperitoneal administration, 245
Intravenous administration, 244–246
Inulin, 237
Iodipamide, 257
Iodomethanesulfonic acid, 161
4-Ipomeanol, 37, 106, 122, 123, 337, 338
Iron, 298, 299, 300
Isocitrate dehydrogenase, 55
   localization, liver, 111
Isoniazid, 147
   genetic factors, 399
   in twins, 399
Isoprenaline, 157
Isoquinoline, 52
Isosafrole, 50

K

Kelthane, 357, 363
Kepone, biliary function, 264
Keshan disease, 191
Ketamine, 309
Ketone, 144
Kidney, 251
   blood flow, 252
   excretion
      fetal, 254
      in neonate, 254
      xenobiotic, 237–239, 253
   filtration, 252
      glomerular, 237, 238
   toxicity target, 326, 327
   tumors, 344
Kinetics, 213–250
   symbols, 215
Kupffer cell, 107, 312

L

Lactate dehydrogenase, 57
   localization, liver, 111
Lapachol, 195
Latex bead, 309

Lead, toxicity, 298
Leukocyte, polymorphonuclear, 197
Leukotriene, 198
Lidocaine, clearance, 227
Ligandin, 117, 222, 260, *see also* Glutathione
   S-transferase
Linoleic acid, 296
Linolenic hydroperoxide, 186
Lipid
   covalent linking, 21
   dietary, 294–297
      effect on detoxication, 295
      high-fat diet, carcinogenesis, 297
      intake, effect on conjugation, 296
   modification, 37
   peroxidation, 21, 61, 187, 194, 273, 297, 301
      GSH and, 59, 60
      membrane composition, 197
      products, 185
Lipid hydroperoxide, decomposition, 185
Lipid-soluble compounds, excretion, 254
Lipoamide, 193
Lipoprotein, 66
β-Lipoprotein, 218
Lipoxygenase, 198
Liver
   acinus, 107, 109
   bile duct, 109
   bile flow, 110
   biotransformation, 240
   blood flow, 110
   central vein, 109
   efflux rates, 259
   elimination, 239, 240, 242–244
   enzyme distribution, 107–121
   enzyme localization, 111
   excretion, 239
   hepatic artery, 109
   histological models, 107
   necrosis, 326, 344
   portal canal, 109
   portal lobule, 107–110
   portal triad, 106, 107
   portal vein, 109
   regeneration, 269, 270
      glucuronide formation, 269
   sinusoids, 107
   terminal hepatic venule, 109
   toxicity target, 326, 327
   toxin formation, 338, 339

Lung
  elimination, 242–244
  enzyme distribution, 121–124
Lysosome, drug concentration, 255

M

Macrophage, 197, 198
  function, 310
  pulmonary alveolar, 122, 124
Magnesium, deficiency, 301
Malaoxon, 352, 353
Malate dehydrogenase
  cytosolic, 79
  localization, liver, 111
Malathion, toxicity, 352, 353
Maleic anhydride divinyl ether, 309
Malic enzyme, 55
Malondialdehyde, 185, 194
N-Mannich base, decomposition, 146
Mefenamic acid, 199
Membrane
  fluidity, dietary effects, 295
  lipid composition, 295
Menadione, 148
Mengo virus, 317
Mercapturic acid, 16
Metabolite, toxic, see Metabolic activation; specific compound
Metabolic activation, 31–67, see also specific enzymes, substances
  epoxidation, 7, 8
  free radical formation, 10
  C-hydroxylation, 8
  N-oxidation, 8, 9
  S-oxidation, 11, 12
  redox cycling and active oxygen, 10, 11
Metabolism, 261–263, see also specific substance
  environmental factors, 404
  intraindividual variability, 384–390
  nonenzymatic reactions, 138
  polymorphism, 399
  postenzymatic reactions, 138
  sequential first-pass, 234
  xenobiotics, see Xenobiotic metabolism
Metabolite, criteria for detection, 154
Methadone, 262
Methane, production, 20
Methanesulfonic acid, 161, 162

Methemoglobinemia, 2, 18, 327
Methotrexate, 326
  excretion, 253
Methoxyflurane, 168
N-Methyl-4-aminoazobenzene, 38, 39
N-Methylation, 39
Methylazoxymethane, 20
Methylazoxymethanol, 20
2-Methylbenzylhydrazine, 20
3-Methylcholanthrene, 36, 37, 59, 65, 94, 95, 317, 334, 338, 379
  induction, 44, 47, 113, 115, 120
Methyldopa, 157
Methylene dioxybenzene, 50
Methylenedisalicylic acid, 159–161
3-Methylfuran, 122
Methylmercury, excretion, 265
Methylnitrosourea, 297
Methylprednisolone, 200
N-Methylpyridinium, 171
Meticrane, 163
Metyrapone, 50, 52
Microcirculation, 197–200
Microspectrophotometry, 113
Milk
  excretion, 271, 272
  hormones in, 271
  organic anions, 271
  organic cations, 271
  smoking, 272
  species differences, 272
Mineral, in detoxication, 297–302
Mipafox, 343
Mitomycin C, 195
Monoamine oxidase, 20
Monocrotaline, 338, 339
Monooxygenase, 22, 362
  activity, in bees, 358
  dieldrin, 356
  effect of glucocorticoids, 92
  localization, liver, 119–121
  microsomal, 355
  ontogenesis, 83–86
  peroxidative, 49
Morfamquat, 331
Morphine, 51, 88, 253
  excretion, 254
  in milk, 272
Mutagen, 358–361
Mutagenesis, 36, 37, 327
Mycobacterium butyricum, 309, 311

Index

Mycobacterium tuberculosis, 311
Myeloperoxidase, 197

N

NADH, 54
NADPH
  limiting concentration, 56
  regulation, 55–57
NADPH-cytochrome c reductase, 3
  induction, site of, 115
  localization
    liver, 115, 116
    lung, 121, 122
    skin, 125, 127
NADPH/NADP ratio, 56
Nafenopin, 262–264
  as inhibitor, 260
Naphthalene, 122
Naphthoflavone, 50
α-Naphthoisothiocyanate, effect on cholestasis, 265
β-Naphthol, glucuronidation, 63
Naphthoquinone, 197
1,2-Naphthoquinone, 148
1-Naphthylamine, 36
2-Naphthylamine, 16, 39, 64
  bladder tumors, 32, 36
2-Naphthylaminesulfonic acid, 161, 162
Neomycin, 330
Neonate, biastrous formation, 77
Neostigmine, excretion, 254
Nephrotoxicity, 326
Nervous system, toxicity target, 326, 327
Newcastle disease virus, 313, 317
Nicotinic acid, 299–301
Nitrenium intermediates, 16–18
Nitrenium ion, 8, 147
Nitroanisole, 49
4-Nitroanisole, 56, 57, 58, 309, 317
2-Nitrobenzoic acid, 159, 160
Nitro compounds, reduction, 18, 19
S-Nitrocysteine, 143
Nitrofurantoin, 18, 341, 343
  effect on cholestasis, 265
NIH shift, 7, 12
Nitrilotriacetic acid, 159–161
Nitrone, formation, 9
4-Nitrophenylacetic acid, 160
4-Nitroquinoline *N*-oxide, as carcinogen, 18

Nitro reduction, 4, 6
Nitrosamide, 341, 342
Nitrosamine, 341, 342
  activation, 14, 15
  cyclic, 8
N-Nitrosamine, 122
Nitrosoazetidine, as carcinogen, 15
4-Nitrosomorpholine, as carcinogen, 15
S-Nitrosothiol, 143
Nonenzymatic reaction, 138
  with acids, 145
  cyclic derivatives, 146
  with electrophiles, 144
  with nucleophiles, 141–144
  rearrangements, 144
  between xenobiotics, 146–148
Norethandrolone, effect on bile flow, 256
Norethindrone, 147
Norharman, 52
Norpethidine, 341
Nortriptyline, 384
  genetic factors, 393, 396
  half-life, plasma, 391
Novobiocin, effect on cholestasis, 265
Nuclear membrane
  detoxication enzymes, 65
  methylcholanthrene induction, 65
Nucleic acid
  alkylation, 17, 22
  covalent binding, 17
  covalent bonding, 22
Nutrient
  interdependence, 289
  trace, 297–302
    deprivation, 302
Nutrient–nutrient interaction, 289
Nutrition, in detoxication, 287–302

O

Octamethylpyrophosphoramide, 292
Octylamine, 50
Oil, hydrogenated, 294
Ontogenesis, 77–97
  control mechanisms, 82, 83
  developmental profiles, 81, 82
  effect of cell constituents, 81, 82
  of enzymes, 78–81
  hepatic enzymes, 80
  hormonal control, 80–82

membrane constituents, 81
regulation, 80, 82
Oral administration, 244–246
Oral contraceptives, 264
Organic anion
  in milk, 271
  molecular weight
    species variations, 258, 259
    threshold, 258
  renal excretion, 252–254
  transport, 253, 331
Organic cation
  in milk, 271
  molecular weight, 259
Organochlorine insecticides, 355, 356, see also specific substances
Organophosphate insecticides, 351–355, see also specific substance
  resistance, 361, 362
  species differences, 353, 354
Ornithine carbamoyltransferase, ontogeny, 79
Osteomalacia, 326
Ototoxicity, 326
Ouabain, 257
  effect on organic anions, 253
  uptake, 259
Oxidant stress, 10
N-Oxidation, 4, 5, 8, 9
  activity, 123
  effect of glucocorticoids, 92
S-Oxidation, 5, 8, 11, 12
Oxidative stress, 192
  liver, 193
N-Oxide, 8
Oxon, 351, 354, 355
Oxygen
  cytotoxicity, 194, 196
  free radicals, 181–201
  reactive species, 181–201
  singlet, 184–186
    sources, 184
Oxygenation, hyperbasic, effect on SH, 194
Oxy radical, 182

P

Pancuronium, 165, 166
Pantothenic acid, 300
PAPS, see 3′-Phosphoadenosine 5′-phosphosulfate

Paracetamol, 292, 326
Paraquat, 19, 165, 166, 173, 183, 195, 329
  excretion, 254
  toxicity, 331
  pulmonary accumulation, 331, 332
Parenchymal cell, 106
Parkinson's disease, 196
Penicillin, 222
Penicillin G, excretion, 253
Pentachlorophenol, sulfation inhibitor, 63
Pentobarbital, 292, 309
Pentose-phosphate pathway, 55
Peroxidase, 6, 20
Peroxidation, 289
  membrane, 300
Peroxide, glutathione peroxidase, 59
Peroxisomes, 194
Peroxy radical, 186, 187
  formation, 186, 187
Phagocytosis, 312
Phalloidin, effect on bile flow, 256
Pharmacogenetics, 402
Pharmacokinetics, 402, 403
Phenacetin, 384
  binding, 35
  genetic factors, 397
  metabolism, 35
Phenanthrene-9,10-quinone, 148
Phenelzine, genetic factors, 399
Phenformin, 384
Phenobarbital, 45, 52, 59, 255, 260, 262, 265, 269, 294, 379
  hepatotoxicity, 265
  as inducer, 22, 56, 113, 114, 120
Phenobarbitone, 334, 338, 339
Phenol, 171
  acetylation, nonenzymatic reaction, 146
  formation, 13, 14
  sulfation, 60, 61
Phenolphthalein, 258
Phenolphthalein glucuronide, 256
Phenothiazine, effect on cholestasis, 265
Phenotyping, 402
Phenylbutazone, 311, 321, 383
  genetic factors, 393, 396
  half-life, plasma, 388, 391
Phenyl ester
  hydrolysis
    albumin, 140
    water, 140
Phenylhydrazine, 343

Index

Phenyl sulfate, 171
Phenytoin, 309, 310, 318, 326, 383, 384
   genetic factors, 393, 396
   half-life, plasma, 391
Phosgene, 49
Phosphatidylcholine, 295
3'-Phosphoadenosine 5'-phosphosulfate, 54, 55, 58, 59
Phosphoenol pyruvate carboxylase, ontogeny, 79
Phosphofructokinase, 55
6-Phosphogluconate dehydrogenase, 55, 56
Phospholipidosis, 329, 332, 333
Phosphorus, dietary, 289
Phthalazinone, 144
Phthalic acid, 159–161
Physiology, detoxication, 214
Piperonyl butoxide, 50, 358
Pirimiphos methyl, 353–355
Plasma, filtrate, 252
Plasma protein, 217–220
   binding, 253
      blood cell, 221
Plasmodium berghei, 317
Polychlorobiphenyl, chlorinated, 364
Polymorphism, 396
   genetic factors, 399
   race differences, 400
Poly(rI·rC), 309
Polythiazide, 163, 164, 171
Pregnenolone 16α-carbonitrile, 52, 94, 260, 262
   induction, 115
Primaquine, 326, 343
Primidone, half-life, plasma, 391
Probenecid, excretion, 253
Procainamide, genetic factors, 399
Procainamide ethobromide, biliary excretion, 258
Procarbazine, 20
Prodrug, 145, 146
   nonenzymatic reaction, 146
Progesterone, 92, 93, 269, 271
Propoxur, 358
Propranolol
   clearance, 227, 231
   genetic factors, 396
Propylthiouracil, in milk, 272
Prostaglandin, 198
   biosynthesis, 200
   excretion, 253
   metabolism, 295
Prostaglandin endoperoxide synthetase, 340
Prostaglandin synthetase, 6, 49
Protein
   in bile, 261
   binding, 252
      covalent, 2, 21
   dietary, 290–293
   methylation, 20
   modification, 37
Pulmonary alveolar macrophage, 122, 124
Pulmonary fibrosis, 326
Pulmonary toxin, 122
Pyrazole, as inducer, 48
Pyridine, 171
Pyridoxal phosphate, hydrazine and, 19
Pyruvate decarboxylase, 56

Q

Quinacrine, 309
Quinine, excretion, 254
Quinol, 195
Quinone, 49
   concentration, 62
   lipid peroxidation, 61
   sulfation, 60
Quinone reductase, 62

R

Radiation, ionizing, 200
Radical chain reaction, 21
Readsorption, 237, 238
Redox cycling, 189, 195
Reduction
   one-electron, 49, 182
   two-electron, 50
Regioselectivity, 44–52
   microsomal, 50
Renal clearance, 153, 238, 239
Renal excretion, 252–254, *see also* specific organ
   of conjugates, 254
   diuretics, 254
   effect of pH, 254
   organic anions, 252–254
   readsorption, 254

vascular binding, 238
Renal toxicity
  antibiotics, 329–331
  cephaloridine, 329–331
  chemicals, 329–331
Repair function, 341–343
Reproductive system, toxicity target, 326
Respiratory system
  4-ipomeanol toxicity, 337, 338
  paraquat toxicity, 331, 332
  toxicity target, 326
Reticuloendothelial system, 312, 313
Retinol, see Vitamin A
Retinopathy, 326
Ribonucleotide reductase, 59
Rifampicin, effect on cholestasis, 265
Rifoflavin, 299, 300
RNA, modification, 37
Rotenone, 58

S

Saccharin, 163, 171, 173
  excretion, 253
Safrole, 38, 39
Salicylamide, 62
  as inhibitor, 64
Salicylic acid, 222, 224
  excretion, 253
Salivation, 270, 271
Salmonella typhimurium, assay, 359
Schiff's base, 144
Scrotal cancer, 2
Secretin, 256
Secretion, 237
Sebaceous gland, 125, 127
Se-glutathione peroxidase, 59
Selenium
  deficiency, 191, 298
    in chicks, 19
  dietary, 289, 300
Selenocystine, 300
Semiquinone, intermediates, 10
Serine dehydratase, ontogeny, 79
Sesamex, 358
SKF 525-A, hepatotoxicity, 265
Skin
  carcinogens, 125, 127
  enzyme distribution, 124–127
  histology, 125–127

toxicity target 326, 327
Sodium salicylate, genetic factors, 393
Sparteine, 384
Spironolactone, 262
Starvation, 289, 290
Statolon, 309
Stercuronium, 165, 166
Stereoselectivity, 44–52
Sterility, male, 326
Steroid 5α-reductase, ontogeny, 79
Stilbene oxide, 52, 59
  induction, 114, 116
Streptomycin, 326, 330
Streptonigrin, 195
Strychnine, 292
Styrene oxide, 359
Subcutaneous administration, 244–246
Substrate specificity, 223
Succinate dehydrogenase, localization, liver, 111
Sulfamaprine, genetic factors, 399
Sulfamethazine, genetic factors, 399
Sulfamic acid, unmetabolized, 161, 162
Sulfamoylbenzoic acid, 159–161, 171
Sulfanilamide, genetic factors, 399
Sulfanilic acid, 241
Sulfapyridine, in milk, 272
Sulfate
  conjugation, in starvation, 290
  pool, 58
  in vivo concentrations, 58
Sulfate adenyltransferase, 58
Sulfated compounds, 253
Sulfation, 58, 60, 61, 232, 233, 261, 360, see also specific substance
  activation by, 39
  bladder tumor, 64
  carcinogenesis, 64
Sulfene, 12
Sulfide, oxidation, 11
Sulfine, 12
Sulfoacetic acid, 161
Sulfobromophthalein
  excretion, 257
    biliary, 258
  uptake, 259
Sulfonamide, 343
  unmetabolized, 162–165
Sulfonic acid, unmetabolized, 161, 162
Sulfonium analog, 165, 166
N-Sulfonoxy-2-acetylaminofluorene, 40, 41
Sulfotransferase, 32, 60–62

Index

localization, lung, 121
ontogeny, 88, 89
sex differences, 89
Sulfoxide, 12
Sulfoxime, 64
Superoxide, 10, 182
  anion, 49, 182–184
    biocidal activity, 197
    cell concentration, 184
    cytotoxicity, 198
    enzymes producing, 183
    formation, 183
    hyperoxia, 194
    in inflammation, 197–200
    in leukocytes, 197
    microvasculature, 198
    permeability, 198, 199
    sources, 183
Superoxide dismutase, 182, 187, 195, 201
  distribution, 187
  in human, cell concentration, 187
  in inflammation, 199
  ratio to glutathione peroxidase, 197

T

Target organ, see specific type
Tartrazine, 270
Taurocholate, 266
  biliary excretion, 258
  uptake, 259
T cell-mediated response, 310, 311
1,2,4,5-Tetrachlorobenzene, 169, 170
Tetrachlorodibenzo-p-dioxane, 260
2,3,7,8-Tetrachlorodibenzo-p-dioxin, 45, 169, 170, 174
  biliary function, 264
Tetraethylammonium, 165, 166
  excretion, 254
Tetrazolium reductase, activity, 115
Theophylline, 266, 293, 319–321, 383
  clearance, 385
  half-life, plasma, 391
Thioacetamide, 106
  hepatotoxicity, 21
Thioacetamide S-dioxide, toxicity, 22
Thioacetamide S-oxide, hepatotoxicity, 21
Thioacetamide sulfine, hepatotoxicity, 21
Thioamide, oxidation, 11, 12
Thiocarbamate, oxidation, 11

Thiocarbamide, oxidation, 12
Thiol
  oxidation, 11
  toxicity, 12
Thiol S-methyltransferase, 241
Thione, 11
  oxidation, 12
Thiopental, 224
Thiourea, 12
  toxicity, 22
Tilorone, 309
Tissue binding, 221–223
Tissue distribution, of enzymes, 105–127, see also specific tissue, enzyme
Tocainide, 144
α-Tocopherol, 189, see also Vitamin E
Tolbutamide
  genetic factors, 393, 396
  half-life, plasma, 391
Toluene-2-sulfonamide, 171
Toluene-4-sulfonamide, 171
Toxicity
  factors affecting organs, 328
  organ function, 340, 341
  organ predisposition, 327
  organ specificity, 325–345
  organ susceptibility, 325–345
  selective, 349–365
Tracheal mucosa, 121
Transferritin, 187
Triazolo[3,4-a]phthalazine, 144
S,S,S-Tributyl phosphorothioate, 362
Trichloromethyl radical, 21
1,1,1-Trichloropropene oxide, as inhibitor, 359
Trichloropropylene epoxide, 53
Trimethylamine N-oxide, 8, 9
Triphenylacetic acid, 159–161
Trypan blue, 309
Tryptophan 2,3-dioxygenase, 95, 316
  interferon, 315
  ontogeny, 79, 81
Tuberculin, 311
d-Tubocurarine, 165, 166, 255
  biliary excretion, 258
Tubular fluid, 252
Tumor formation, stages, 23
Twins
  genetic factors, metabolism, 393–403
  monozygous, 381, 385
Tyrosine aminotransferase, 81, 95
  effect of cyclic AMP, 94
  of glucagon, 94

U

UDPGA, 54, 57, 58
UDPglucose dehydrogenase, 58
UDPglucosyltransferse, 88
UDPglucuronyltransferase, 32, 40, 52, 58, 60–62, 91, 92
  effect of glucocorticoids, 92
  genetics, 88
  localization
    liver, 118
    lung, 121
    skin, 125
  ontogeny, 79, 85, 87, 88
UDPxylosyltransferase, 88
Ulceration, 326
Umbelliferone, 120, 266
Unmetabolized compounds, 151–175
  acids, 159–165
  attributes, 158–173
  chronic administration, 157
  conjugation, 172
  criteria, 152–158
  defined, 152
  dosage, 153–155
  effect of K_m, 155
  mechanisms, 173, 174
  methodology, analytical, 152, 153
  nonabsorbed, 173
  nonpolar, 169
  polar, 158–168
  retention, 175
  routes of administration, 156, 157
  species differences, 155, 156
  volatile, 168, 169
Urate, excretion, 253
Uridine triphosphate, 55
Urine, formation, 252

V

Variation
  environment, 372, 373
  factors, 370, 371
  interindividual, 369–406
    genetics, 393–403
    magnitude, 405
    sources, 391–406
  intraindividual, 369–406
    sources, 384–390
  metabolism, 384–390
  use of twins, 381, 385
Vinyl chloride, 38
Virus, *see* specific type
Vitamin, in detoxication, 297–302
Vitamin A, 189, 190, 299
Vitamin B_{12}, 301
Vitamin C, *see* Ascorbic acid
Vitamin D, 289
Vitamin E, 189, 190, 289, 299
  as antioxidant, 189
  deficiency, 191
  free radical reactions, 189
  mechanism, 189
Vitamin K, 148, 343

W

Warfarin, 45, 291, 320, 321, 383
  half-life, plasma, 391

X

Xenobiotic, *see also* specific substance
  bacterial metabolism, 268
  binding, 218–221
    intracellular, 260, 261
    organ clearance, 230, 231
    tissue, 221–223
  blood cell, 220
  clearance, 225–237
  enzyme induction, 94, 95
  excretion, *see* Excretion
  in fat, 271
  hemodynamics, 223, 224
  interactions with vascular components, 217–223
  kidney filtration, 253
  metabolism, 214, 225–237
    nutritional effects, 288
    target organ toxicity, 333–340
  nonenzymatic reaction, 146–148
  physicochemical properties, 215–217
  plasma protein, 217–220
  readsorption, 237, 238
  route of administration, 244–246
  substrate specificity, 223
Xenobiotic monooxygenase, *see* Cytochrome *P*-450
Xeroderma pigmentosum, 341

Index

Xylidide anesthetic, 144
X-radiation, 200

Z

Zinc, dietary, 298
Zoxazolamine, 292, 301, 317, 334
 paralysis, 311
Zymosan, 309, 313